BLUETOOTH OPERATION AND USE

Bluetooth Operation and Use

Robert Morrow

McGraw-Hill
New York Chicago San Francisco Lisbon
London Madrid Mexico City Milan New Delhi
San Juan Seoul Singapore Sydney Toronto

Cataloging-in-Publication Data is on file with the Library of Congress

McGraw-Hill 

A Division of The McGraw-Hill Companies

1 2 3 4 5 6 7 8 9 0 DOC/DOC 0 8 7 6 5 4 3 2

ISBN 0-07-138779-X

The sponsoring editor for this book was Judy Bass and the production supervisor was Pamela A. Pelton. It was set in Century Schoolbook by MacAllister Publishing Services, LLC.

Printed and bound by R. R. Donnelley & Sons Company.

This book is printed on recycled, acid-free paper containing a minimum of 50 percent recycled, de-inked fiber.

To my loving wife and daughter for their patience and support

CONTENTS

PREFACE

Bluetooth—what an interesting name! Within that name there is no clue as to what it really is, so I'll tell you up front. It's a short-range wireless technology that replaces a cable, and in doing so, it also bends the world.

I know it's easy to lose heart when you begin to study the Bluetooth specification in detail. The core specification is over a thousand pages long, and that just covers basic operation. Another thousand pages of profile documentation present the various functionalities of this "simple" cable replacement technology. However, despite the fact that many different individuals and committees worked on different parts of the specification, it's surprisingly easy to read and comprehend.

In this book, the operation of Bluetooth is presented with enough detail that it's not really necessary to read the specification beforehand. However, the purpose of this book isn't to serve as a substitute for the specification. Instead, it supplements the specification in two ways. The first is to add insight into topics that are already part of the specification. For example, in Chapter 9, "Bluetooth Security," we show how security is implemented from both a macroscopic and microscopic point of view. The second way this book supplements the specification is by dedicating several chapters and sections within other chapters to material that the specification either covers superficially or not at all. An example of this can also be found in Chapter 9, where various threats to Bluetooth security and their countermeasures are presented.

The lower layers of the Bluetooth protocol stack are covered in great detail in the book. This is where the action is, and an understanding of these layers is essential before any applications or profiles can be created. Conversely, protocol layers above the *logical link control and adaptation protocol* (L2CAP) and the profiles themselves are summarized in the book without repeating implementation details found in the specification. The reasons for this are twofold: first, the length of the book needs to be reasonable, and second, most developers are interested in a small subset of the available profiles, rendering much of this book irrelevant to them should details of all the profiles and higher protocols be included. If there's one thing an author fears, it's irrelevance.

As of this writing, there are two hop-channel sets listed in the Bluetooth specification. The first contains 79 channels and is used for most of the world. The second is a 23-hop set that is used in France. I have complete confidence that the 23-hop sequence will shortly become obsolete, so it is (almost) never mentioned in the book. Of course, all of the details on the shorter sequence are found in the specification.

Over the years, the method of designating a number as hexadecimal has changed several times, but seems to have finally settled on using a 0x prefix; that is, the old 7Fh or 7FH or even $7F_{16}$ has become 0x7F. The latter is certainly a better notation, and it is used throughout the book in spite of the fact that I once thought the x stood for the logical "don't care."

Finally, a word of caution. Bluetooth is still in its infancy, and as a result, the specification is constantly being updated and revised. Those revisions can take place much faster than we can revise this book, and despite our best efforts, some errors or anachronisms may remain. Therefore, if there are conflicts or inconsistencies between the specification and the book, be sure that your designs follow the specification. Also, your comments and opinions would be most welcome.

Robert Morrow
Centerville, Indiana, USA
February 2002
rkmorrow@ieee.org

ACKNOWLEDGMENTS

It has been a wonderful and fulfilling experience to witness the advent of Bluetooth as a viable short-range wireless communication process. A project of this nature requires the coordinated effort of literally thousands of dedicated, professional people. Creating a book of this size is also a daunting effort and would have been impossible had I not received valuable assistance from many individuals to whom I'm humbly indebted.

Les Besser, Jeff Lange, and the other fine people at Besser Associates (www.bessercourse.com) deserve credit for starting this project. They gave me the opportunity to build Bluetooth into several short courses and make presentations around the world. Ted Rappaport, propagation expert and author of a popular textbook on wireless, provided valuable assistance with the art and science of propagation analysis. Ken Wyatt went above and beyond the call by researching antennas and providing links to several radiation pattern analysis packages. June Namgoong created a terrific routine to plot the output of switched diversity antennas to combat Rayleigh fading, and Jim Lehnert provided technical assistance and encouragement spanning many years.

Thanks so much to those who kindly provided material for some of the figures in the book: Eric Meihofer, Ad Kamerman, Wade Gillham, Matthew Shoemake, Allen Hollister, Erik Lundqvist, Carsten Tilm, Peter Stavenick, Moshe Doron, Jacek Wojcik, Ivan Howitt, Vinay Mitter, Paul Kaspian, L.J. Malone, and Max Ammann.

Although too numerous to name individually, I'm most appreciative of the support from the many journalists and editors who bring us the latest news and technical articles in a fast-moving and highly complex wireless industry. My hat's off to the conference organizers and their perseverance in arranging events that require attention to the utmost detail. My editor, Judy Bass, and her assistant, Jessica Hornick, provided the motivation and expertise to bring this project to a successful completion.

Finally, a very special thanks goes to Bill Lae, creator of the brilliant Jocular Science cartoons that appear throughout this book. While studying engineering at Purdue, Bill would often leave one of his gems on the chalkboard for the rest of us to enjoy. More of Bill's fine work can be found at www.jocularscience.com.

CHAPTER 1

Introduction

It could be argued that the age of electronic digital communications began when Samuel F. B. Morse patented the electronic telegraph in 1840. For several decades afterward, Morse code was the most common way to represent letters, numbers, and punctuation, and a wire was the only practical means for carrying information from source to destination. Even now, the wire and similar fiberoptic channels are prevalent in countless applications. However, with today's need for mobile, dynamic communications, wireless transmission is rapidly gaining ground as a ubiquitous means for connecting devices together. Wireless became commercially viable with the advent of broadcast radio, where a powerful and expensive transmitter sends one-way signals to thousands of low-cost receivers. Only recently has two-way wireless become practical and inexpensive enough to break out of the somewhat specialized realm of use mainly by professionals and amateur "ham" radio operators.

Perhaps the best example of widespread two-way communications was manifested by the original cellular telephone network called *Advanced Mobile Phone System* (AMPS), where an extensive matrix of cells communicated over relatively short distances with their mobile and portable phone handsets. As the number of users proliferated, cells became smaller to permit frequency reuse, where two cells share the same channel set, over shorter distances, thus accommodating more phones. Because the radio spectrum has a fixed bandwidth, more users can be accommodated through reduced power and shorter range, enabling others nearby to communicate on the same frequency without causing interference. The concept of frequency reuse has been taken to an even smaller scale with the advent of two-way radio links that are limited to a single building or even to a single room.

One of the most useful applications of short-range two-way digital communications enables computers to talk to each other and to their peripherals such as printers and scanners. This can be as simple as a link between one computer and one or more peripherals or as complex as several devices communicating with each other in the form of a *local area network* (LAN). The LAN can be either wired or wireless and usually has a coverage area of several rooms or floors in a building. The most common *wireless local area network* (WLAN) is the *Institute of Electrical and Electronics Engineers* (IEEE) 802.11b, also called *Wi-Fi*.

Bluetooth carries the WLAN concept to a smaller scale, with its low-power 10 meter range more suited for connecting devices that are located within the same room, or even on a person. This concept, called the *personal area network* (PAN), is intended primarily to replace cables with wireless links. Before we begin examining Bluetooth in detail, it would be helpful to scrutinize the differences between wired and wireless communications so that we can decide where the best applications for each might be.

Differences Between Wired and Wireless Communications

With the advent of cheap and simple two-way digital wireless systems, a choice is now available between wired and wireless, and we might wonder why anyone would choose the former. That's a very good question, and it would be useful to examine the characteristic differences between wired and wireless communications to gain insight into which method is better in a particular application. Although wired systems are commonly thought to use copper conductors for the channel, fiber optics exhibit similar characteristics in several areas, so we will refer to both of these methods as wired.

Advantages of the Wireless Network

When nodes are connected via wire, it's clear that there is very little mobility enabled while wireless breaks the tether and facilitates, within reason, the ability to roam about while continuing to communicate. Similarly, communication can be established using wireless from several different locations without requiring a physical plug into the network. Do you remember the advanced civilizations portrayed in the *Star Wars* movies? Even their wireless communication devices were impressive with their small size and long range, although these seemed to carry only voice conversations. Ironically, the little robot R2-D2 had to locate and plug into a network access point whenever it wanted to communicate with a computer, putting itself in physical danger time after time. Establishing a wireless link would have been far more practical, much safer, and also would have removed a significant source of cinematic excitement.

It is, of course, obvious that a wired network requires installing lots of cable because each node in the network must have physical access to the cable from its particular location. This can cause big headaches when the walls, floors, and ceilings of an existing building must be torn open for cable installation. Many office buildings, especially the old courthouses used by several state and local governments, have historical significance, and altering their structure for cables is often frowned upon. Connecting computers in such a building via a WLAN is often the only reasonable solution. Newer buildings can have LAN wires installed while they are under construction, but the wires themselves can become obsolete in short order, requiring periodic, expensive, and inconvenient cable upgrades. On the other hand, a WLAN can be upgraded simply by installing new hardware and software into the computers.

The cable used for connecting a computer to a peripheral can provide valuable clues about its function. We can look at both ends of the cable and discover the attached devices that are communicating with each other. The connectors on each end are often related to the purpose for which the cable is used, so we can scrutinize the cable and the peripheral attached to it and immediately know the hardware and software required to access that peripheral. Of course, the usual result is that one cable is needed for each peripheral, so all of these cables begin to look like a spaghetti factory (see Figure 1-1). Not so with wireless: the only physical medium is over the air, so connection possibilities are far more versatile without the mess behind the computer.

Figure 1-1 Connecting with wires can be unwieldy. Bluetooth can replace everything except the power cables.

Advantages of the Wired Network

In spite of significant disadvantages, wired networks will probably remain viable for a number of reasons. The wire provides a dedicated, quiet channel for the nodes to access, so reliability is increased. There is little signal attenuation between source and destination on a wire, so nodes can transmit and receive simultaneously. That doesn't mean that several messages can be sent on the wire at the same time (at least not without using special modulation techniques). Instead, the feature enables a transmitting node to determine if another node happens to be sending data simultaneously and a jamming situation (collision) is taking place. Both nodes can then stop sending immediately, reducing the time that the channel wastes in the collision state. Wireless signals exhibit significant attenuation between transmitter and receiver, a characteristic that will be quantified in a later chapter. The result is that when a transmitter is operating, its corresponding receiver is deaf, and collisions cannot usually be discovered until much later. This can cause significant delays and loss of communication efficiency.

High signal attenuation between transmitter and receiver also means that the bit error probability (also called the *bit error rate*, or BER) is much higher in wireless than in wired systems. File transfers over the network must usually take place error free, so additional bits are added to the file for error control. The resulting overhead is usually higher in wireless networks, so their efficiency is lower. Further reductions in BER can be accomplished by reducing data transmission speed in the wireless network to offset the high attenuation. As a rule of thumb, a wired network can send data about 10 times faster than a wireless network of the same vintage. For example, the wired *Universal Serial Bus* (USB) 1.x operates at speeds of about 12 Mb/s, and IEEE 1394 works at 100 to 400 Mb/s. Bluetooth operates at a raw rate of 1 Mb/s, and IEEE 802.11b wireless Wi-Fi works at up to 11 Mb/s. (It's interesting to note that the data rate of a typical fiberoptic channel exceeds its contemporary wired channel also by a factor of 10.)

Security is another issue that is more difficult to address with wireless devices because their transmissions aren't limited to the confines of a cable. Wireless signals are easier to intercept, disrupt, and jam, which requires countermeasures that usually aren't necessary when using a wire.

By now you're probably ready to reverse your position and ask why anyone would use wireless, but take heart. Wireless is clearly here to stay due to its incredible convenience over wired access, and Bluetooth has incorporated several clever techniques to alleviate some of its potential disadvantages compared with communicating over a cable. Because wireless involves using a public resource called the *radio spectrum*, we'll first examine the regulations that Bluetooth and other wireless networks must follow.

Regulation of Unlicensed Bands

Along with many other wireless devices, Bluetooth uses the unlicensed 2.4 GHz frequency band for its operation. Contrary to popular belief, unlicensed doesn't mean unregulated, and indeed most countries strictly regulate the use of unlicensed frequencies. Frequency administration falls under the *Federal Communications Commission* (FCC) in the United States, but other governments often have rules that are quite different from, and sometimes incompatible with, FCC regulations. Naturally, this situation has the potential to cause a regulatory nightmare for Bluetooth in its quest to become a worldwide standard for short-range wireless.

In 1992, the *International Telecommunications Union* (ITU), which is part of the United Nations, formed its *Radiocommunication Sector* (ITU-R) in an attempt to ensure rational, efficient, and economical use of the *radio frequency* (RF) spectrum. Every few years, the ITU conducts a *World Radiocommunications Conference* (WRC), where member nations agree on how the radio spectrum is allocated and used. Furthermore, the Bluetooth *Special Interest Group* (SIG) is working with various governments to bring their regulations in alignment with Bluetooth requirements. (We will discuss the SIG in a later section.)

Bluetooth operates at 2.4 GHz because that is the only practical frequency band that is (mostly) allocated worldwide and requires no license to operate a transmitter. That's good news, of course, but there's a reason that the band is available throughout the world, and that reason is microwave ovens. These ovens became popular long before any other general use of these high frequencies was envisioned, and the microwave oven frequency of 2.45 GHz was chosen because water molecules readily absorb RF energy at this frequency and convert it to heat. These ovens operate at several hundred watts of power, and as we will discover in a later chapter, these can be a significant source of interference to Bluetooth and other wireless users in the 2.4 GHz band.

FCC Part 15 Requirements

The FCC regulates frequency allocation and use in the United States, and its rules often form the basis for regulatory practices in other countries as well, especially in North America. FCC rules are divided into several parts, with Part 15 being devoted to intentional, unintentional, or incidental radiators that are enabled to operate without an individual license. In general, FCC Part 15 devices operating as intentional radiators

- Cannot cause interference to licensed users
- Have no regulatory protection against interference from other users, licensed or unlicensed
- Must have a permanently mounted antenna (preferred) or one that uses a unique connector
- Require government certification prior to marketing

When examining the previous factors, it's obvious that the FCC is most interested in a device's transmission of electromagnetic energy as a potential source of interference to other users. Aside from placing limits on transmit power and modulation techniques within the band of use, the FCC also places limits on how much unintentional radiation can occur outside this band.

The FCC rulebook reads like a legal document (which it is) and can be somewhat difficult to interpret by those without a legal background. However, several web sites contain white papers that can help translate legalese into "engineerese," and the FCC has their own document as well.[1] There are several frequency bands covered by Part 15 regulations, and each band has its associated power limitations and modulation requirements. Because Bluetooth uses the 2.4 GHz band, we will examine the rules for this band in some detail.

Rules for Transmission in the 2.4 GHz Band

The 2.4 GHz band in the United States extends from 2400 to 2483.5 MHz and is one of several *Industrial, Scientific, and Medical* (ISM) bands that are used for a variety of purposes without requiring a license to transmit. For narrowband (nonspread-spectrum) use, transmitter power output is limited to 50 mV per meter measured at a distance of 3 meters, which corresponds to an *effective isotropic radiated power* (EIRP) of about 0.75 mW. Transmit power can increase to 1 W if spread spectrum is used, which can take the form of *direct sequence spread spectrum* (DSSS), *frequency hopping spread spectrum* (FHSS), or a hybrid using a combination of these two methods.

DSSS devices must conform to the following restrictions:

- Transmit output power not more than 1 W or not more than 4 W EIRP using a gain antenna
- Minimum transmitted signal bandwidth of 500 kHz at the –6 dB points
- Peak power density not more than 8 dBm in any 3 kHz bandwidth segment
- Processing gain of at least 10 dB (at least 10 chips per data bit)

We will discover in a subsequent chapter that 802.11b, also called Wi-Fi, uses DSSS and conforms to these rules. On the other hand, Bluetooth uses FHSS, so the FCC requires it to abide by these restrictions:

- Transmit power not more than 1 W or not more than 4 W EIRP using a gain antenna
- At least 75 hopping channels
- Maximum –20 dB bandwidth of 1 MHz
- Hopping channel separation at least the –20 dB hop bandwidth, but not less than 25 kHz
- Hopping rate of at least 2.5 per second
- Cumulative dwell time in each channel not more than 400 ms during each 30 second time period

Some of the requirements placed upon Bluetooth by the SIG are more restrictive than these; such as transmit power cannot exceed 100 mW. This is due in part to the lower maximum power levels allowed in Europe.

Wideband Frequency Hopping (WBFH) Spread Spectrum In August 2000 the FCC supplemented the 2.4 GHz FHSS rules to allow WBFH. This enables FHSS systems to compete favorably with DSSS devices that employ a method called *complementary coded keying* (CCK) to enable much higher data rates (11 Mb/s) than would otherwise be possible if DSSS were to operate in the traditional way (1 to 2 Mb/s). (At this point you may be experiencing a "gobbledygook alert," but more details will emerge when we compare Bluetooth to Wi-Fi in a later chapter.) Anyway, the rules for WBFH are

- At least 15 nonoverlapping hopping channels covering at least 75 MHz
- Transmit power is limited to 125 mW if fewer than 75 hopping channels are used
- Hopping rate of at least 2.5 per second

As a result of these new rules, WBFH devices can employ 25 hopping channels of 3 MHz bandwidth each or 15 hopping channels of 5 MHz bandwidth each. The wider bandwidth in each channel enables a higher data rate for a given modulation technique. Bluetooth Specification 1.1 was released shortly before the FCC approved WBFH; consequently, the Bluetooth data rate of 1 Mb/s assumes that 1 MHz maximum bandwidth per hop channel is enabled.

FCC Product Certification

FCC certification is required before any product can be sold that has the potential to produce RF interference. This includes products containing computer circuitry, broadcast receivers, and Bluetooth devices. Product development can take place without a special license, and certification is actually one of the last steps to be accomplished before the Bluetooth product is brought to the marketplace.

Certification is done on a production-ready prototype. After the prototype is built, it should be tested for FCC compliance, either in-house or via one of the many test facilities available for such purposes. Next, the device is submitted to an authorized FCC testing facility called a *Telecommunications Certification Body* (TCB). The TCB performs the required tests and, if all is well, issues an FCC ID number that must be affixed to each product sold. Several TCBs have been authorized by the FCC, so Part 15 certification consumes only a week or two of time for a properly designed product.

Because Bluetooth devices are supposed to be available for worldwide sale, you may be asking whether the FCC testing process is accepted by other countries in lieu of their own regulatory approval. At this time (early 2002), the answer is . . . maybe. The certification process is in a tremendous state of change, and the eventual goal is to enable one nation's certification to be valid worldwide. A two-phase *Mutual Recognition Agreement* (MRA) has been signed between the United States, the European Union (EU), the

Asian-Pacific countries (APEC), and other countries in the Western Hemisphere (CITEL) in an effort to achieve this goal. Phase 1 enables the countries to accept each other's laboratory test data for checking against their own certification rules, and Phase 2 allows for full acceptance of each other's certification from authorized testing bodies.[2]

The Bluetooth Story

The concept behind Bluetooth had its origins in 1994 when Ericsson began researching the idea of replacing cables connecting accessories to mobile phones and computers with wireless links. As technical details began to emerge, Ericsson quickly realized that the potential market for Bluetooth products was huge, but cooperation throughout the world would be needed for the products to succeed. Therefore, the Bluetooth SIG was formed in 1998, and the first Bluetooth technical specification arrived in 1999.

But why call a wireless system Bluetooth? There's no hint within the name itself that it represents a wireless communication system. Harald Bluetooth was a 10th century Viking monarch who managed to unite Denmark and Norway, and because formalization of the concept of wireless cable replacement began in Scandinavia, it made some sense to identify its Viking origin. Furthermore, Harald's unifying approach to conquest meshed nicely with the goal of uniting computer and peripheral through a specification that would hopefully achieve worldwide acceptance. There's certainly no room for ambiguity in the name: performing an Internet keyword search on Bluetooth results in the return of almost no material unrelated to the associated short range wireless system.

Perhaps the most important reason for naming this specification after King Harald surfaced when a monolith bearing his self-portrait was recently unearthed (see Figure 1-2). Although carved over a thousand years ago, Harald had enough foresight to include a cell phone and a laptop with an antenna in the stone engraving. At least that's what I've been told.

The Bluetooth Special Interest Group (SIG)

Founded by Ericsson, Nokia, IBM, Intel, and Toshiba, the Bluetooth SIG began in February of 1998. Even during its infancy, Bluetooth was clearly

Figure 1-2
Harald Bluetooth's self-portrait. Note the cell phone and laptop, both wireless. King Harald was prescient! (Source: Ericsson)

envisioned as a worldwide communication system as evidenced by Ericsson and Nokia representing Europe, IBM and Intel representing the Americas, and Toshiba representing Asia. The SIG founders were joined in December 1999 by Microsoft, 3Com, Lucent, and Motorola, and these nine entities are now called *SIG Promoters*. At this time there are no plans to add to the Promoter list. Promoters are responsible for upper-level SIG administration, and for providing manpower to run the legal, marketing, and qualification processes.

Some of the major functions of the SIG include

- Petitioning various government agencies to allow Bluetooth to operate in their countries without special requirements or restrictions
- Handling legal issues related to SIG membership, intellectual property, and use of the trademark
- Managing the process that tests devices to insure compliance to the Bluetooth specification
- Managing the interoperability test process to ensure that Bluetooth devices from different manufacturers can communicate with each other
- Managing technical working groups
- Creating and publishing the Bluetooth specification

The highest membership status open to other entities (corporations and other groups) is the Associate Member, which requires signing a legal document and paying an annual membership fee. Associates are allowed to participate in various marketing and technical subgroups and are given access to working group draft documents at version 0.5 and above. Adopters (formerly Early Adopters), on the other hand, join the SIG without paying a fee, although a legal document is still signed, and they are allowed to participate in some regulatory, test, and expert groups. Adopters have access to working group draft documents at version 0.9 and above.

SIG membership is required before the Bluetooth specification can be used in designs, and the member is granted access to the Bluetooth intellectual property and logo without paying royalties. The SIG became a nonprofit corporation in early 2001, and its extensive web site can be accessed at www.bluetooth.org.

Overview of Bluetooth Applications

As the Bluetooth concept began taking shape in the mid-1990s, various usage models were created as possible applications of a short-range digital wireless system with data rates from about 100 to 500 kilobits per second (kb/s). We will examine each of these original models, which were released concurrently with Specification 1.0A. We'll also take a brief look at other possible Bluetooth applications beyond the basic usage models. Additional applications can be found in Chapters 11, "Bluetooth Profiles," and 14, "The Future of Bluetooth."

Three-in-One Phone

The telephone system began as a set of fixed wires connecting phones at fixed locations together through a central office switch. Today this network of phones and wires is called *Plain Old Telephone Service* (POTS). Without the aid of wireless mobility, placing a phone call always meant that a user had to find a phone first, and then call another phone at which the desired person may or may not be physically present. All of us have experienced the character-building frustration of calling several phone numbers in our quest to locate a particular individual.

With the advent of the cellular phone, users could carry their phones with them. This offered two distinct advantages (at least theoretically): A phone is always available to originate a call, and another party can place a call to that phone and have a high probability of reaching the desired person. Of course, significant cellular infrastructure must be present, and use of that infrastructure must be the most convenient and cost-effective way to place a call for this system to predominate. In reality, though, cellular phones can be expensive and yield spotty coverage, so most homes and businesses still use POTS. Even so, a cordless phone connected to POTS can enable roaming throughout the home or office.

The Bluetooth 3-in-1 phone (see Figure 1-3) envisions one phone that can operate with the (non-Bluetooth) cellular system, as a cordless phone through a Bluetooth link to a nearby base station connected to POTS or as a walkie-talkie to another 3-in-1 phone. There would be a seamless transfer

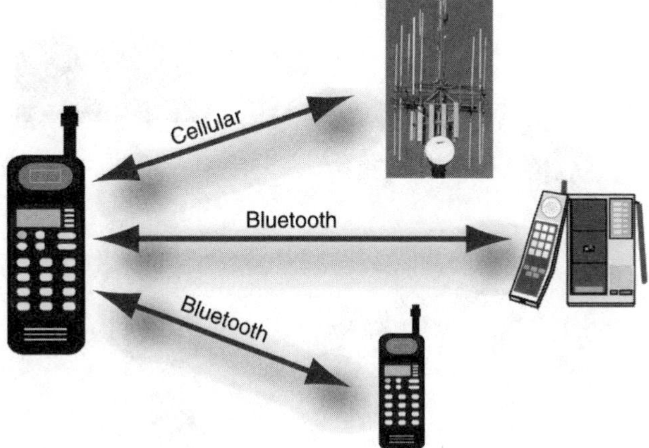

Figure 1-3 The 3-in-1 phone can be used as a cellular phone, cordless phone, or walkie-talkie.

between cellular and cordless phone links as the user moved from place to place. Furthermore, this configuration could enable the assignment of a single personal phone number for all three uses.

At first blush, using Bluetooth as a walkie-talkie, with its 10-meter range, seems a bit strange. I'm reminded of kids a short distance apart shouting into their tin-can phones, "Can you hear me?" and of course the other could, even without the cans. Keep in mind, though, that Bluetooth devices can operate at power levels up to 100 mW, enabling them to communicate at distances greater than 10 meters. Also, when users are separated by walls or other barriers, it may still be more convenient to use wireless devices than to talk through the barrier.

Ultimate Headset

A lightweight headset with a wireless link to its servicing device may prove to be the largest market for Bluetooth devices (see Figure 1-4). Many users have already realized the convenience of connecting a wired headset to the telephone, freeing up both hands for entering data into a keyboard. Others connect corded headsets to their cell phones, some of which are so small that people appear to be walking along enjoying an animated conversation with no one in particular.

As various state and local governments move to restrict cell phone usage while driving, the Bluetooth headset will become an essential device for

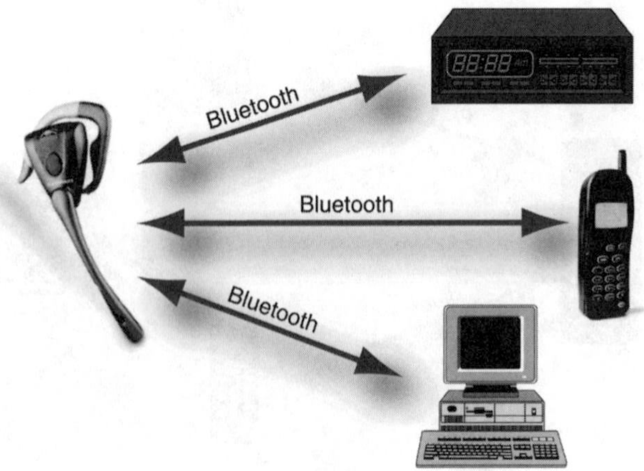

Figure 1-4 The ultimate headset can create an audio link to other Bluetooth-equipped devices. (Source: Plantronics)

making calls from an automobile. The cell phone remains hidden away in a briefcase, and a small, lightweight headset can access the phone via voice commands while both hands are occupied with controlling the vehicle. With hundreds of millions of cellular handsets in use, most of which are eventually carried in a vehicle, it's obvious that if Bluetooth can become the predominant headset-to-handset wireless link, then the sales potential will be vast.

Of course, the Bluetooth ultimate headset won't be limited to connections with one particular cellular phone, but instead could automatically link to any compatible device within range that is capable of audio communication. A user could conceivably don a headset before breakfast and leave it on until retiring for the night. Who knows, maybe a Bluetooth-equipped alarm clock could arouse a person who even wears the headset to bed. Can Bluetooth bioimplants be far behind? Will we begin looking like the Borg in the *Star Trek: Next Generation* series? The thought is intriguing . . .

Internet Bridge

One fairly common method for accessing the Internet while on the go has a laptop computer connected via cable to a cellular phone, which in turn is used to call the *Internet Service Provider* (ISP) dial-up number. The Bluetooth Internet bridge usage model simply replaces the cable with a wireless link, as shown in Figure 1-5. The link must be able to pass connect and disconnect (AT) commands to the phone as well as provide two-way data traffic between the phone and laptop.

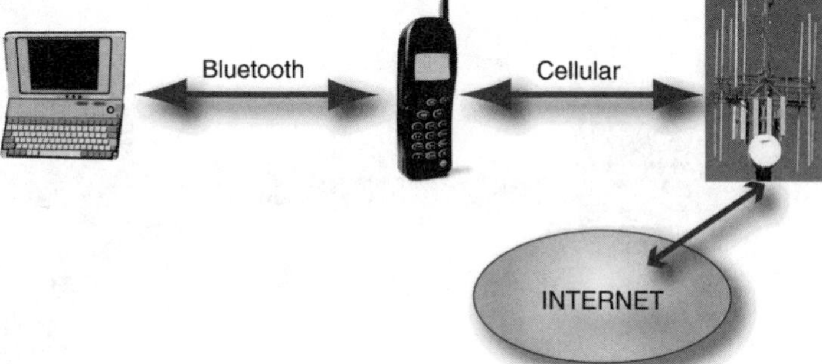

Figure 1-5 An Internet bridge replaces a modem cable with a Bluetooth link through a cellular phone.

Data Access Point

Broadening slightly on the concept of the Internet bridge, a Bluetooth data access point enables a computer to connect to a data service, such as a LAN, via a wireless link (see Figure 1-6). The LAN itself could have any configuration, wired or wireless, and all LAN services would be available to the computer as if it were connected via cable. This method of access would obviate the need to run LAN cables extensively through a building, making it particularly convenient for telemarketers or sales teams to access databases and printers. Can dozens of densely packed users, each with a Bluetooth headset and data access point, operate without significant wireless interference? That question will be answered in Chapter 13, "Coexisting with Other Wireless Systems," when we model and analyze coexistence issues.

Object Push and File Transfer

Bluetooth object push and file transfer usage models build upon similar functions using infrared that have been available in *personal digital assistant* (PDA) devices for several years. After a simple connection process, users are able to transfer information, such as business cards, phone records, or larger files between them. These operations take the form of *push,* where the initiator sends a file, or *pull,* where the initiator retrieves a file (see Figure 1-7). The push and pull operations can be combined into, for example, a business card exchange.

Because Bluetooth is not limited to *line-of-sight* (LOS) operations, the file transfer process enables convenient interactive conferencing. This capabil-

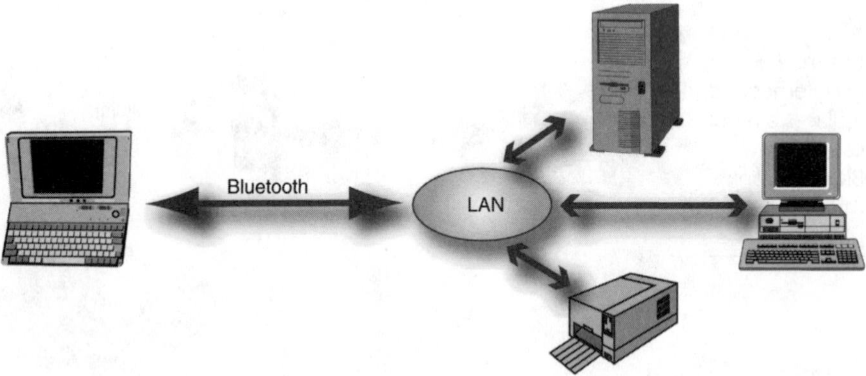

Figure 1-6 LAN access replaces a cable with a Bluetooth link.

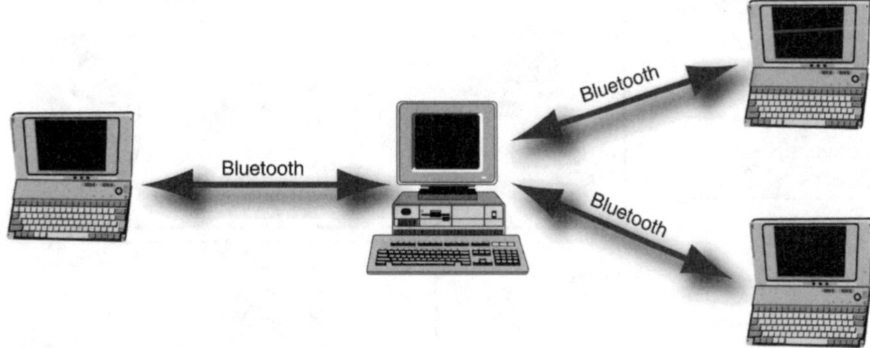

Figure 1-7 Bluetooth can replace a wired or infrared link for transferring files between devices.

ity already exists on many wired LAN implementations, where a file can be opened by several users, and any modification made by one will appear on all screens. For example, a professor could display lecture notes on classroom students' computers, and the professor or student could manipulate graphics on the screen for immediate viewing on all connected computers.

Automatic Synchronization

Many of us have experienced the frustration of trying to manipulate several copies of what is supposed to be the same file on different machines. Does your PDA phone book contain the same numbers as those stored in your cell phone? If all devices that maintain copies of files are Bluetooth equipped, they can, when each is in range, link together and automatically update their files to the latest version.

For example, suppose you want to synchronize files on your office and home computer. While at the office, your PDA can establish a Bluetooth link to the desktop computer and copy new or updated files to its memory. When the PDA is taken home, it can automatically link to the home computer and update its files (see Figure 1-8). Likewise, the process can reverse itself when the PDA is returned to the office.

Other Uses for Bluetooth

The usage models discussed earlier were developed when Bluetooth was still in its infancy. As Bluetooth matures, other applications that are completely new or variations of these usage models will begin to appear. One of

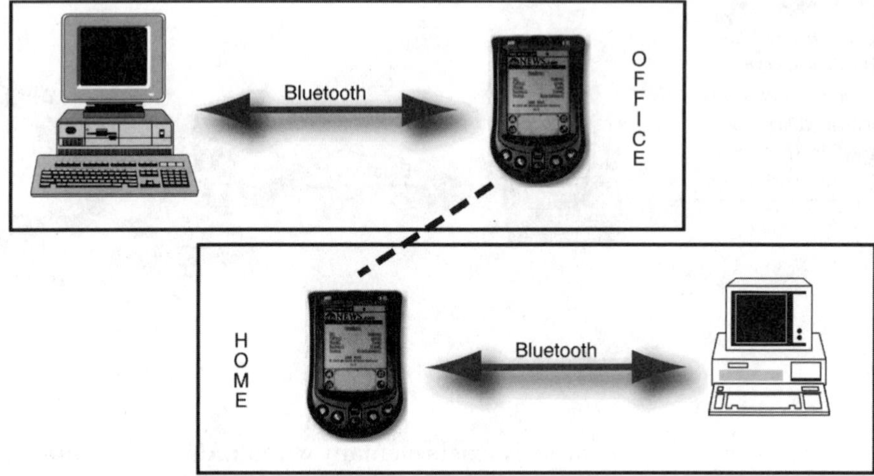

Figure 1-8 Bluetooth can provide convenient file synchronization between several devices.

the broadest applications will surely be realized through variations of the file transfer models. A Bluetooth-equipped PDA could be used for accessing flight information at an airport, ordering off of an electronic menu at a restaurant, or obtaining information about players when attending a sports event. The PDA could provide the means for automated hotel check-in, and it could even serve as an electronic key for unlocking the door when the guest arrives at the room.

As automobiles become more sophisticated and complex, designers have been connecting the various electronic devices in the car with the equivalent of a wired LAN to avoid the expense of weaving harnesses containing hundreds of wires throughout the vehicle. Bluetooth will take this process a step further by eliminating many of the wires altogether. For example, the engine management system could have sensors in the engine compartment communicating with the fuel injection computer without penetrating the firewall with wires. Of course, this application is merely a special case of wireless telemetry, where sensors throughout a building transmit their information to a central location.

Soldiers in the battlefield carry extensive amounts of equipment, much of which needs to be connected. Use of the backpack radio, for example, would clearly be enhanced by operation through a Bluetooth headset. The Bluetooth-enabled PAN can eliminate some of the hazards of connecting the equipment via cables. Military pilots could have the various displays and voice applications sent to their helmet through a Bluetooth link, reducing the number of wires and the risk of injury should the pilot need to eject from the aircraft.

Proximity detection for security purposes lends itself nicely to Bluetooth solutions. Many of us have used a card that provides access to secure areas by holding it over a special pad near the door. These cards typically use *radio frequency identification* (RFID), an ultra-short-range process of interrogation and response. Bluetooth can increase convenience by unlocking the door automatically as the authorized user approaches. In a similar fashion, access to computers within a company can be based upon the proximity of a user carrying the proper Bluetooth-enabled authorization card. When the user walks away, the computer automatically disables access and initiates a screen saver.

Another potentially huge market for Bluetooth is in remote control, where the wireless link can be used in diverse applications ranging from operating lighting and other equipment for stage productions to control of toys. Imagine the excitement of a dozen radio controlled model airplanes involved in a massive dogfight or kids racing miniature Bluetooth-equipped carts from the safety of the sidelines.

Bluetooth Protocols and Profiles

Now that we've examined some of the possibilities of what Bluetooth can do, it's time to begin looking at how Bluetooth can accomplish these goals. Figure 1-9 shows one method of connecting a Bluetooth module to a computer. The module itself is physically attached to the host computer through a wired bus (ironic, isn't it?) such as a USB or *Personal Computer Memory Card International Association* (PCMCIA) card, also known as *PC card*. The host computer runs its usual applications that operate drivers that can communicate with the module through the *host controller interface* (HCI). HCI has provisions for both commands and data to be sent between the module and host. The Bluetooth module connects via RF with another Bluetooth module that in turn is attached to a different host. Data can now be exchanged between the two devices over the wireless link between the modules. The communication process itself must have an orderly structure, and that is accomplished through *protocols*. The type of function that the Bluetooth device is trying to accomplish is determined by its *profile*. Both of these will be covered in much greater detail later, but it would be useful to summarize them now. We begin by examining the rationale behind the standard practice of layered communication system design.

Figure 1-9 An add-on module is one method that Bluetooth can be connected to a computer.

The OSI Model

Take a look again at the Bluetooth-to-host interface in Figure 1-9. A very complex process must occur just to get data from the host application to the radio in the module, across the wireless link, and to the destination host. How should the data be structured? How fast should the data be sent? And (*gasp*) how should the link be established and configured? After thinking about all of these factors together, it's tempting to give up in despair and return to a simpler life of designing vacuum tube oscillators.

In 1977, the *International Standards Organization* (ISO) established a subcommittee to research the need to develop a standardized, layered approach to general computer communications. This work culminated in 1982 with the *Open Systems Interconnection* (OSI) reference model, shown in Figure 1-10. By working through the layers of the model, a designer can gradually create an entire computer communication system without becoming overwhelmed, and (theoretically, at least) a particular layer can be changed without affecting the other layers.[3] The downside to using this method is that redundancy, along with its resulting inefficiencies, can creep into a design. We shall discover this characteristic when we examine the details of Bluetooth packet structure in Chapter 4, "Baseband Packets and Their Exchange."

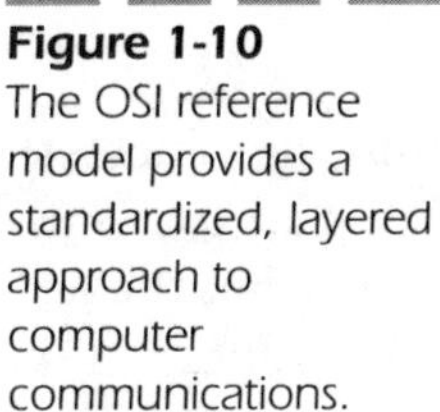

Figure 1-10
The OSI reference model provides a standardized, layered approach to computer communications.

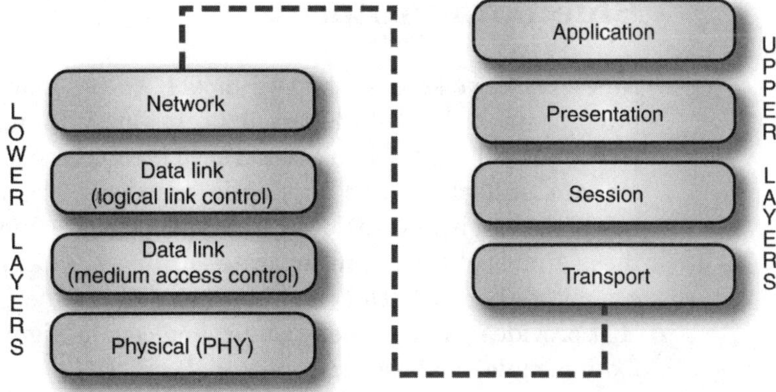

The *physical layer* (PHY) contains the actual physical interface and the rules for its use. In Bluetooth, the PHY is RF and the modulation, and detection processes are listed in the specification. The PHY is made reliable, and link connection and detachment rules are provided by the *data-link layer*. This layer contains *Media Access Control* (MAC), which is a set of rules that determine the structure of basic data packets and how they are sent, and the *Logical Link Control* (LLC), which provides the protocol for link establishment and detachment. The *network layer* provides transparent transfer of data between transport entities on each end of the communication link. A properly implemented network layer relieves the transport layer from requiring any knowledge of the method by which data moves from source to destination. In other words, a Bluetooth device can appear to be a serial cable to the transport layer.

Layers four and above in the OSI model are called *higher layers* (profound, yes?), and their functions start to become less well defined. The *transport layer* includes optimization routines and other *quality of service* (QoS) methods for efficient data exchange, and the *session layer* contains the method for controlling dialog between applications on either end of the link. Finally, the *presentation layer* resolves differences between format and data representation between entities, and the *application layer* provides the means by which applications can access the OSI environment. As we move up the OSI layers, their implementation gradually changes from hardware, to firmware, and finally into software. The Bluetooth protocol stack exhibits the same behavior.

Bluetooth Protocols

Now that we've studied the OSI model, we can move on to the Bluetooth protocol stack shown in Figure 1-11. It's at once apparent that the Bluetooth protocol stack doesn't conform to the OSI model exactly, but the layers are still there and gradually transition from implementation in hardware and firmware (lower layers) to software (higher layers). If each of these groups of layers are separate entities, such as a PC card and laptop computer, then they can communicate with each other through the HCI. HCI provides paths for data, audio, and control signals between the Bluetooth module and host.

The radio completes the physical layer by providing a transmitter and receiver for two-way communication. Data packets are assembled and fed to the radio by the baseband state machine. The link controller provides more complex state operations, such as the standby, connect, and low-power modes. The baseband and link controller functions are combined into one layer in Figure 1-11 to be consistent with their treatment in the Bluetooth Specification 1.1. The link manager provides link control and configuration through a low-level language called the *link manager protocol* (LMP).

The *logical link control and adaptation protocol* (L2CAP) establishes virtual channels between hosts that can keep track of several simultaneous sessions such as multiple file transfers. L2CAP also takes application data and breaks it into Bluetooth-size morsels for transmission, and reverses the process for received data. *Radio frequency communication* (RFCOMM) is the Bluetooth serial port emulator, and its main purpose is to trick an application into thinking that a wired serial port exists instead of an RF link. Finally, the various software programs that are needed for the different Bluetooth usage models enable a familiar application to use Bluetooth. These include *service discovery protocol* (SDP), *object exchange* (OBEX), *telephony control protocol specification* (TCS), and *Wireless Application Protocol* (WAP).

Aside from data communications, Bluetooth has a special provision for real-time, two-way, digitized voice as well. Once these voice packets are created by an application, they bypass most of the data protocol stack and are handled directly by the baseband layer. This prevents unacceptable delay between the time the packets are created and the time they arrive at their destination. Control of the Bluetooth module usually proceeds from the application through HCI to the module, also bypassing the protocol layers used for handling the data communication process itself.

The Bluetooth radio and the baseband/link controller consist of hardware that is typically available as one or two integrated circuits. The firmware-based link manager and one end of the host controller interface,

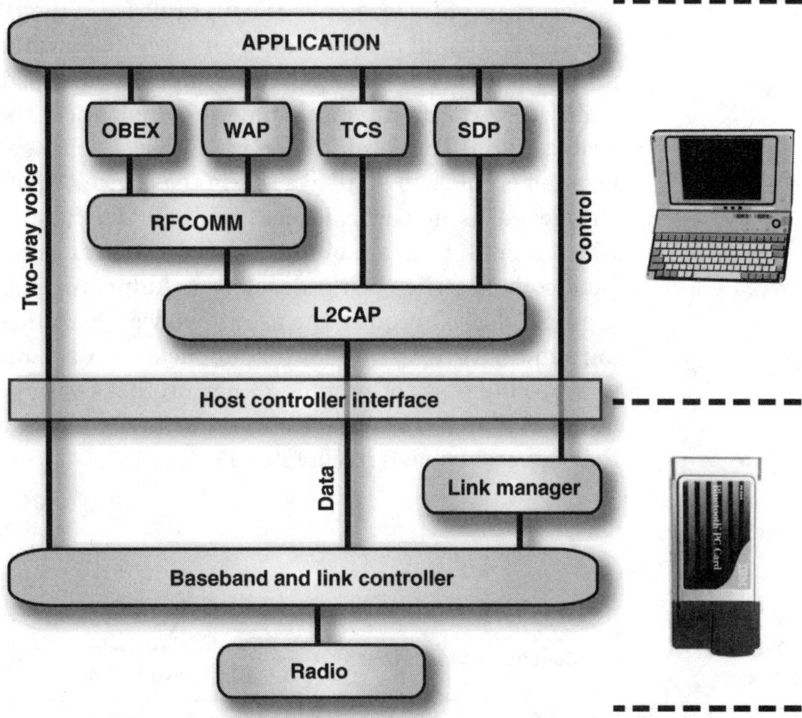

Figure 1-11 The Bluetooth protocol stack

perhaps with a bus driver for connection to the host, complete the Bluetooth module shown in Figure 1-9. The remaining parts of the protocol stack and the host end of HCI can be implemented in software on the host itself.

Bluetooth Profiles

Whereas protocols provide the basic building blocks for Bluetooth operation, the profile is what gives a Bluetooth-equipped device its personality. Do you want the device to be a headset? Use the headset profile. A cordless phone? Use the cordless telephony profile. The purpose of profiles is threefold:

- Lots of options are reduced to those needed for a specific function.
- Procedures for a specific function can be taken from a set of base standards.
- A common user experience is provided across devices from different manufacturers.

Several profiles existed in the first release of the Bluetooth specification, and these original profiles and their interaction with each other are shown in Figure 1-12. For example, if a Bluetooth-equipped device is to have the ability to perform automatic file synchronization, then the Generic Access profile, Serial Port profile, Generic Object Exchange profile, and Synchronization profile will all play a role in the device's capabilities. Profiles can be envisioned as a "vertical slice" through the Bluetooth protocol stack, in which a subset of capabilities in each layer is selected for the particular Bluetooth function being developed. Automatic file synchronization, for example, doesn't require the use of two-way real-time audio, so implementing audio isn't necessary for that application. Each usage model has its own corresponding set of profiles. Other profiles continue to be added as they attain SIG approval. We will take a more detailed look at profiles and how they are constructed in Chapter 11.

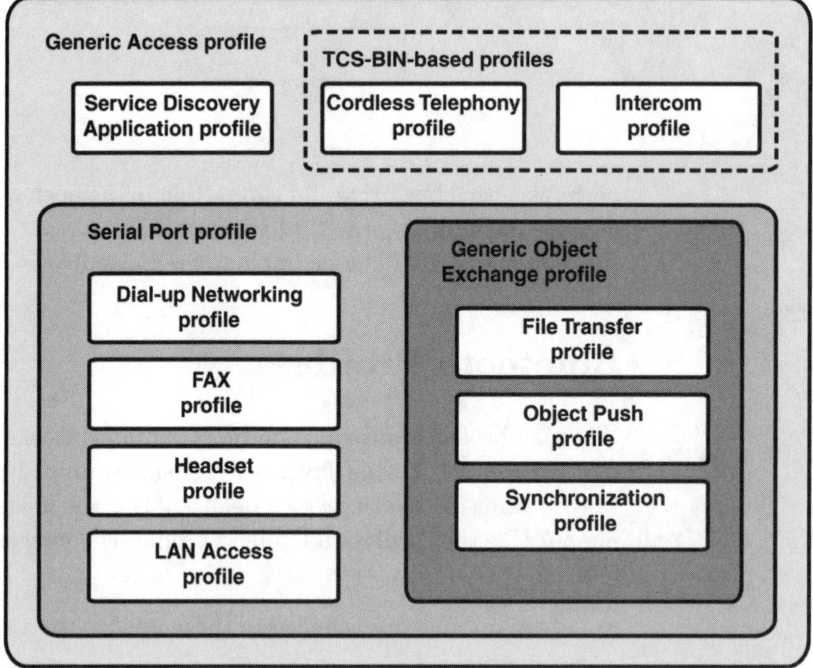

Figure 1-12 Examples of Bluetooth profile interaction

Summary of Bluetooth Operation

Imagine, if you will, a digital communication link between (say) two laptop computers. Let's call one *A* and the other *B*. If data is to be sent only in a one-way direction from *A* to *B*, then we can equip *A* with a transmitter and *B* with a compatible receiver, and the PHY is done. Next, we can program microcontrollers in *A* and *B* to accept a particular data structure, and the MAC is done, and so on up the OSI chain. However, once we remember that the BER using wireless is somewhat high, we may decide that *B* should transmit an acknowledgment to *A* whenever a packet of data is received correctly. That gives *A* a chance to repeat a transmission that *B* had received in error. Clearly, *A* and *B* now require both a transmitter and a receiver. When should *B* send its acknowledgment? If both *A* and *B* use the same frequency for their respective transmissions, then *A*'s transmitter must be off and its receiver on when *B* transmits its acknowledgment. This requires timing coordination between *A* and *B*, but where does this coordination come from? It probably makes sense to give one of the computers some kind of control over the network to prevent chaos due to problems with timing.

Bluetooth controls timing on the network by designating one of the devices a *master* and the other a *slave*. The master is simply the unit that initiates the communication link, and the other participants are slaves. When that link is later broken, the master/slave designations no longer apply. In fact, every Bluetooth device has both master and slave hardware.

The network itself is termed a *piconet*, meaning small network. When there is only one slave, then the link is called *point-to-point*. A master can control up to seven active slaves in a *point-to-multipoint* configuration. Slaves communicate only with the master, never with each other directly. Timing is such that members of the piconet cannot transmit simultaneously, so these devices won't jam each other. Finally, communication across piconets can be realized if a Bluetooth device can be a slave in more than one piconet, or a master in one and a slave in another. Piconets configured in this manner are called *scatternets*. These various arrangements are depicted in Figure 1-13.

The existence of more than one piconet in the same room leads to another significant problem: Can two piconets interfere with each other? Surely they would if everyone operates on the same frequency. Bluetooth

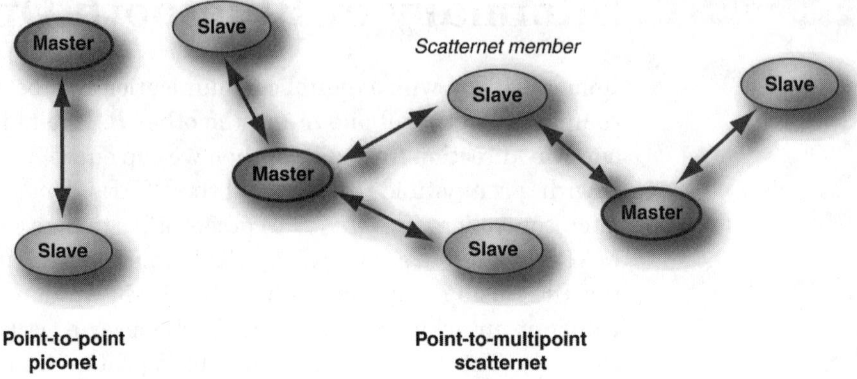

Figure 1-13 Point-to-point, point-to-multipoint, and scatternet topology

prevents this situation by using FHSS, where the 2.4 GHz band is segmented into 79 1-MHz-wide channels, and each piconet, under control of its master, hops from channel to channel in what appears to be a random pattern. In this way cross-piconet jamming occurs only occasionally, and the acknowledgment process discussed earlier provides for error recovery.

Paging and Inquiry

Complexity is increased still further when we realize that FHSS poses significant challenges when a prospective master wants to *page* (initiate a link with) a prospective slave. How does the master know when and on what frequency the slave is listening? One simple solution would be to designate one frequency (channel 1, for example) as a paging channel, so devices meet there to pass hopping parameters from the master to the slave, then they both begin hopping together while communicating. But what if someone puts a sandwich in the microwave oven, and it happens to jam channel 1? Or what if several Bluetooth devices want to initiate a link at the same time? Communications would be paralyzed.

Bluetooth's solution for paging is to have the prospective slave's receiver hop on a certain set of frequencies very slowly (about one hop per second), and the prospective master's transmitter hop on the same set of frequencies very quickly (about 3,200 hops per second). These frequencies are determined by the slave's identity, so different slaves listen for their pages on different frequency sets. The fast-hopping master will then eventually succeed in paging the slow-hopping slave, and communication can begin.

Before a prospective master can page a specific slave, it must determine that slave's identity in the first place so that, among other things, it can calculate which frequencies the slave will be using for its page scan. The prospective master can discover which Bluetooth devices are within range through a process called *inquiry*. When devices respond to an inquiry, they pass to the prospective master the parameters it needs to page them later. The hopping pattern used for an inquiry is the same for all Bluetooth devices.

To summarize, the following steps take place to establish a piconet:

1. The prospective master sends an inquiry to determine who is in range.
2. Devices hearing the inquiry respond with their paging parameters.
3. The prospective master pages a specific device.
4. The paged device responds.
5. Link parameters are exchanged.
6. The communication session begins.

By now you are probably ready to proclaim that Bluetooth is unnecessarily complex, but keep in mind that most of the previous issues have already been implemented by the manufacturers in firmware and software. For a thorough understanding of what Bluetooth can do, it's important to have a background in the basic operation of the piconet, and we will endeavor to do this throughout the book.

Specifications, Standards, and the IEEE

As is the case with many other communication network concepts, Bluetooth began as a specification, which can roughly be defined as a set of features agreed upon by the interested parties. The first Bluetooth specification was version 1.0A, released July 26, 1999, followed by 1.0B on December 1, 1999. Specification 1.1 appeared on February 22, 2001. Each of these documents was over a thousand pages long and covered the entire Bluetooth protocol stack from the radio through the upper layers, including sections on testing and compliance. This level of detail was unusual, as many earlier communication system specifications (such as 802.11) only provided PHY- and MAC-layer information. Because manufacturers were free to implement the higher OSI layers as they wanted, devices from different manufacturers often wouldn't communicate (such as 802.11 again). The Bluetooth SIG realized that this situation could spell disaster for Bluetooth, so its specification, along with associated documents, were carefully constructed to encourage (read: *require*) interoperability regardless of who actually designed and built the device.

In addition to the specification, the Bluetooth SIG also publishes several other documents, all of which are required reading before enough information has been gathered to build a functioning Bluetooth device. These documents discuss in great detail what a device must do to earn the Bluetooth name, but they don't explain how to accomplish these tasks.

Evolution of a Standard

So if Bluetooth is a specification, what is a standard? Generally, a standard is a specification that is adopted by a committee belonging to a (usually) worldwide accepted governing body. Examples of these governing bodies are the *Electronics Industries Association* (EIA), the *International Telegraph and Telephone Consultative Committee* (CCITT), and the IEEE. The IEEE formed the 802 working group several years ago to develop worldwide consensus standards, both wired and wireless, that benefit a so-called network society. To accomplish this goal, the 802 group established formal procedures for submitting a specification for adoption as a standard. For example, wired Ethernet became the IEEE 802.3 standard, and wireless Ethernet is IEEE 802.11. Although these standards focus primarily on the PHY and MAC layers in the OSI model, 802 architecture also includes *Internet Protocol* (IP) at the network layer, *Transmission Control Protocol* (TCP) at the transport layer, and X.400 and X.500 e-mail at the application layer.

The IEEE 802.15 working group has established a charter for the *wireless personal area network* (WPAN) that covers small, short-range, low-power, and low-cost networks that exist within a personal operating space.[4] The activities of this working group include the following:

- 802.15.1 Standardization Task Group for IEEE standardization of the Bluetooth specification
- 802.15.2 Recommended Practice Task Group for coexistence of WPAN and WLAN devices
- 802.15.3 High Rate WPAN Standard Task Group to investigate WPAN solutions with data rates above 20 Mb/s
- 802.15.4 Low Rate WPAN Standard Task Group to investigate 2 kb/s to 200 kb/s WPAN solutions

To convert a specification to an IEEE standard requires formatting the specification to IEEE requirements and matching the protocol stack to the OSI layered model. Finally, comments and comment resolution, followed by balloting, complete the process. In this way Bluetooth became the IEEE 802.15.1 standard.

Health Effects from Exposure to 2.4 GHz RF

As wireless devices become more prevalent, there has been increasing concern over the biological effect of long-term exposure to RF radiation. If microwave ovens also operate at 2.4 GHz, am I being slowly roasted each time my Bluetooth transmitter comes on (see Figure 1-14)? Indeed, one of the possible effects of RF radiation is thermal, where body tissue absorbs energy faster than it can be dissipated. The result is that the tissue heats

Jocular Science
by Bill Lae
© 1987 Lae
Lae
"Look, there's another one. Go get that orange sauce out of the fridge."

Figure 1-14 No, Bluetooth isn't nearly powerful enough to do this. (Source: Bill Lae)

up and can sustain damage. A microwave oven, for example, provides power densities of around 100 mW/cm^2 to cook food.

Guidelines were first issued by the FCC in 1985 to evaluate human exposure to RF energy. These were updated in 1992 and again in 1996 to limit human *maximum permissible exposure* (MPE), measured in mW/cm^2, from an RF source. The formula used to calculate MPE is given by

$$S = \frac{EIRP}{4\pi r^2} \tag{1-1}$$

where power density S is in mW/cm^2, the EIRP is in mW in the direction of the exposed person, and r is that person's distance from the antenna in centimeters.

The FCC, along with regulatory agencies in many other countries, has adopted a two-tiered RF exposure limit. The *occupational (controlled) limit* applies to people exposed to RF as part of their employment and who have been made aware of, and have control over their potential for exposure. The *general public (uncontrolled) limit* applies to the public in nonoccupational situations. For a 2.4 GHz RF source, the occupational MPE limit is 5.0 mW/cm^2, averaged over a six minute period, and general public MPE limit is 1.0 mW/cm^2, averaged over a 30 minute period.[5] These limits are based on the assumption that the exposed person is far enough from the antenna for the RF exposure to be essentially uniform throughout the body (*whole body* exposure).

Would you be able to feel the heat produced by exposure at the MPE limit? Probably not because power densities of about 10 mW/cm^2 are required before measurable tissue heating occurs.[6] In fact, Part 15 transmitter powers are so low that the FCC *Office of Engineering and Technology* (OET) has exempted most Part 15 devices, including Bluetooth, from routine environmental evaluation.[5] These devices still must meet FCC guidelines for compliance, however.

Specific Absorption Rate (SAR)

Although power densities are easy to calculate and apply, exposure limits eventually came to be based upon a quantity called the *Specific Absorption Rate* (SAR), defined as the energy dissipated per unit time per unit mass, with units of *Watts per kilogram* (W/kg). This quantity better reflects the complex nature of near and far field RF and its interaction with human tissue. The FCC occupational SAR limit is 0.4 W/kg, which is one-tenth of the value necessary to raise tissue temperature by one degree Celsius. The limit for the general public, 0.08 W/kg, is one-fifth of the occupational limit.

For partial body exposure, the occupational limit is 8 W/kg, averaged over any 1 gram of tissue. For the general public, the limit is 1.6 W/kg, averaged over any 10 grams of tissue. Partial body SAR limits are higher because normal blood circulation can remove the heat buildup relatively quickly. Many other countries (such as Germany and Australia) also use the previous partial body SAR limits, while others (such as Great Britain) have adopted slightly higher limits.[7]

The FCC has directed that SAR, instead of MPE limits, are to be used for *portable* RF devices, defined as devices (such as those equipped with Bluetooth) where at least part of the user's body would likely be within 20 cm of the antenna. Both partial and whole body exposure SAR limits apply when portable devices are being used.

SAR calculations require knowledge of the RF field distribution within the human body, making these calculations substantially more complex than simply determining the MPE power density at the surface of the skin. Furthermore, SAR takes into account the effects of near field RF (closer than about one-half the wavelength from most types of antennas), which includes electrostatic and induction as well as the usual electromagnetic characteristics of the wave. One method for obtaining SAR information around a person's head is shown in Figure 1-15. A model of the head is built

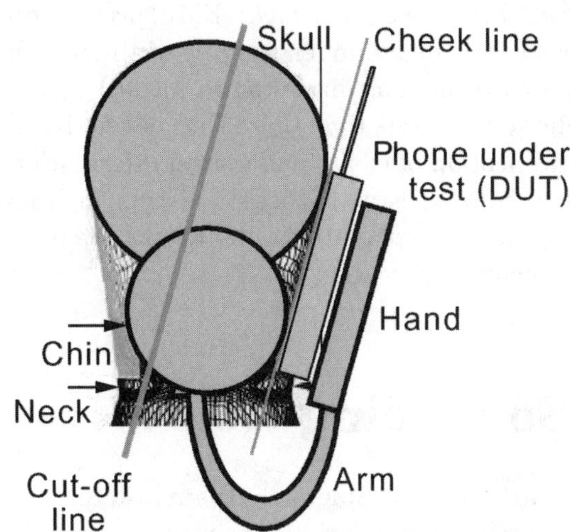

Figure 1-15 Model for determining SAR values within the head of a cell phone user. The head is sectioned along the cut-off line, and a robot-controlled probe measures RF fields inside. (Source: Spectrum Sciences Institute)

out of materials that approximate the various tissue types, and then sectioned for access to the interior. A *device under test* (DUT), such as a cell phone, is placed next to the model, and a robot-controlled probe is moved around inside the head. This probe measures field intensity, from which the various SAR values can be calculated.

Much less understood are the effects of low-power RF radiation on cell structure. This is because it's much more difficult to prove that something doesn't happen than it is to prove that something does happen. A radio wave will be attenuated as it passes through a human body, and the amount of attenuation has been determined for several different RFs.[8] For example, a 2.4 GHz wave will lose about 86 percent of its power density at a depth ranging from 3.5 mm in blood to 13 mm in fatty tissue. At these frequencies, though, the energy in the wave is too low to physically break a cell's molecular bond, so the energy is nonionizing; that is, it cannot strip away electrons. Much higher frequencies (such as X-rays and Gamma radiation, with frequencies higher than about 10^{17} Hz) are required before their energy is sufficient to cause disruption of atomic bonds through ionization.[9] As a result, DNA cannot be modified and genetic alterations, such as cancer, cannot theoretically occur from RF exposure at 2.4 GHz. Some research studies claim to connect RF exposure to physiological problems, such as headaches and sleeplessness and even various forms of cancer, but conclusive evidence linking these events hasn't yet been discovered.[8]

In conclusion, then, there is much work still to be done in researching the health consequences of long-term exposure to RF. The *World Health Organization* (WHO) keeps an extensive database of several hundred RF exposure studies at www-nt.who.int/peh-emf/database.htm. The WHO *International Electromagnetic Field* (EMF) project status report states that, "People with little or no understanding of the health risks of electromagnetic fields view them as an unknown hazard, so they perceive them at a much higher level of risk than those with which they are familiar. Provision and communication of easily understood information on the nature of the health risks from exposure to these fields, placing them in perspective with other risks, and an explanation of how risks are determined, will assist in alleviating people's concerns."[10]

The $5 Pricing Goal

Since the early days of Bluetooth, the marketplace has been full of hype about it becoming a $5 solution for wireless connection. That $5 figure has become so pervasive that we should probably devote some time to putting it into perspective.

Before Bluetooth became a viable alternative, other methods were used for short-range wireless digital communication. One of the most promising solutions was a transmitter module priced at about $15 and a corresponding receiver module priced at about $25, in 1,000-unit quantities. Bits were fed into the transmitter, and they appeared at the receiver. That's it; no sophisticated protocols were included. The devices operated at 1 mW of transmit power in the 900 MHz unlicensed band and could exchange data at up to 50 kb/s on one of eight selectable channels. These modules were used in several wireless products.

In early 2002, a typical Bluetooth chipset cost about $20 in production quantities. This was a complex FHSS transceiver (transmitter and receiver) that could exchange data at rates of 1 Mb/s. The set included PHY and MAC protocol implementations in hardware and firmware, with built-in host controller interface and an extremely small footprint. Some would consider that to be a bargain!

So why is there such widespread opinion that the $5 chipset must be available for Bluetooth to succeed? Perhaps there's an assumption that the cost of a typical $10 cable must be split between the two ends of a wireless link before Bluetooth will become widespread. Of course, this presupposes that there are no additional hardware costs beyond the chipset, that each pair of Bluetooth modules replaces but a single cable, and that wireless adds so little value that customers would rather use a cable if it's a bit cheaper. The first two premises are clearly false, and I think the third one is also. The requirement for additional hardware beyond the chipset will keep the total implementation cost well above $5, at least for the next few years. Certainly, a single Bluetooth module (in a laptop, for example) can replace several cables. Finally, once consumers discover the significant advantage in convenience and performance over the typical cable, they should be willing to pay the higher price.

Summary

By aiming at a market that's already saturated with cheap cables, Bluetooth has assumed a great marketing risk. Can a sophisticated wireless protocol ever replace a communication method already well known for its low cost and reliability? Judging by the excitement and hype surrounding Bluetooth, many thousands of people think so. Once the up-front engineering is finished, even the most complex hardware and software systems are extremely cheap to manufacture in large quantities, and Bluetooth will be no exception. The versatility available when devices are connected without a cable is having a great impact on expanding the traditional usage models beyond their original breadth.

End Notes

1. "Understanding the FCC Regulations for Low-Power, Non-Licensed Transmitters," OET Bulletin No. 63, edited and reprinted February 1996.
2. Cokenias, T. and Judge, B., "FCC and European Certifications: In by 9, Out by 5?" presented at the Wireless Symposium, San Jose, CA, December 2001.
3. Stallings, W., *Data and Computer Communications,* 2nd ed., Macmillan, 1988.
4. Barr, J., "Bluetooth SIG and IEEE 802.15," presented at the Bluetooth Developers Conference, San Jose, CA, December 2000.
5. "Evaluating Compliance with FCC Guidelines for Human Exposure to Radiofrequency Electromagnetic Fields," OET Bulletin No. 65, Edition 97-01, August 1997.
6. TRA Communications White Paper, Warren, MI, 1989.
7. Eger, C., "RF Energy & Health Issues," presented at the Bluetooth Developers Conference, San Jose, CA, December 2000.
8. Scanlon, W., "Health Aspects of Low-Level Exposure to RF Electromagnetic Waves," *RF Design,* July 1999.
9. "Questions and Answers about Biological Effects and Potential Hazards of Radiofrequency Electromagnetic Fields," OET Bulletin No. 56, Fourth Edition, August 1999.
10. The International EMF Project, WHO Status Report, World Health Organization.

CHAPTER 2

Indoor Radio Propagation and Bluetooth Useful Range

One of the first questions that comes to mind whenever wireless is involved is, How far will it go? In wired networks you can be almost sure that adequate signal strength will exist wherever the wire is routed, but wireless can be annoyingly fickle when it comes to reliability.

Low-power indoor wireless has several impediments to successfully designing a link with suitable range. First, the indoor environment is usually quite cluttered, and every obstacle between the transmitter and receiver translates into loss of signal strength. Also, the walls, doors, and furniture can cause severe reflections of the transmitted signal, so multiple copies, with different phase relationships and different time delays, arrive at the receiver. Finally, competition for the 2.4 GHz spectrum can be intense, and the result is often severe interference that further reduces the reliable range.

Propagation analysis is an extremely difficult subject, perhaps because electromagnetic theory itself can be quite intense. During my undergraduate experience, two semesters of e-mag were required, and these performed an admirable task of filtering out those who weren't really interested in becoming electrical engineers. Because the publisher of this book is interested in generating a large quantity of happy readers, our approach here will be to show practical applications with an emphasis on Bluetooth, not involved theory.

In this chapter we will analyze the characteristics of large-scale *path loss* (PL) and how to quickly determine maximum Bluetooth range when accounting for obstructions between the transmitter and receiver. By remembering some simple rules, you can even perform these calculations in your head. Accuracy can be improved further if we surrender the calculations to a computer instead. Next, a look at multipath will tell us what kind of additional compensation is needed to account for signal reflections. Finally, some different antenna designs will be examined for their suitability in a Bluetooth application. The issue of interference will be delayed until Chapter 3, "The Bluetooth Radio," and Chapter 13, "Coexisting with Other Wireless Systems."

Indoor Propagation Mechanisms

When a radio wave strikes a surface that is neither a perfect insulator nor a perfect conductor (in other words, all practical surfaces), some of the wave energy passes through, some is absorbed, and some is reflected. These characteristics give rise to the four paths by which a *radio frequency* (RF) signal

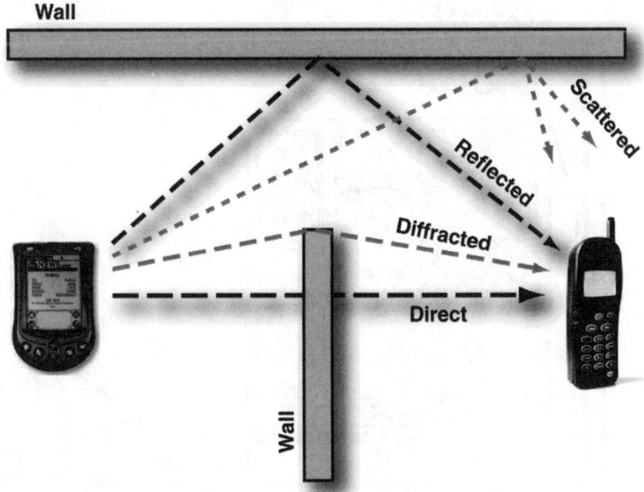

Figure 2-1 Indoor propagation mechanisms can be direct, reflected, diffracted, or scattered.

can travel from transmitter to receiver. The paths are *direct*, *reflected*, *diffracted*, and *scattered*, as shown in Figure 2-1.

The direct path may be *line-of-sight* (LOS) if there are no obstacles between the transmitter and receiver or *obstructed LOS* (OLOS) if the signal must pass through one or more objects en route to the receiver. The other three paths require the signal to travel a longer distance between the transmitter and receiver, so these arrive at the receiver at a later time than the transmitted signal. Often a significant amount of energy is reflected off of various obstructions, especially those constructed of good conductors such as metal. As we shall discover in the section "Reflection, Transmission, and Absorption," the effect of these reflections at the receiver can be significant.

Diffraction occurs when a radio wave encounters a sharp edge, and some of the signal energy is bent around the edge. This effect is most common near doorways and other openings, but is usually less significant than the direct or reflected wave energies. Finally, waves are scattered when they impact a surface that has irregularities that are a significant fraction of the wavelength involved. The scattering effect can often be modeled as a virtual antenna retransmitting the signal at reduced strength. Like their diffracted counterparts, scattered signals are often less significant than the energies that are direct or reflected.

It's extremely easy to become overwhelmed when considering the four propagation methods together in an attempt to perform range calculations for wireless systems. We can simplify our analysis, however, by using the speed-versus-accuracy concept (see Figure 2-2). By considering only the

Figure 2-2 Choosing between speed and accuracy is an important consideration when analyzing indoor signal propagation. (Source: Bill Lae)

The old "speed versus accuracy" dilemma.

direct path, along with some further simplifying assumptions, range estimates can be quickly determined. Next, by retreating from some of the simplifying assumptions, accuracy is improved at the cost of requiring the assistance of a computer. Finally, when multipath is included in the analysis, computation times are increased further, but still greater accuracy is achieved.

Large-Scale Path Loss (PL)

When a receive antenna is located some distance from the transmit antenna, it is logical to conclude that significant signal loss will occur along the path between the two antennas. By considering only the direct propagation mechanism and ignoring the other three, it is relatively easy to calculate the expected power at the receiver given various characteristics at the transmitter, receiver, and along the path. Losses calculated in this manner are called *large scale* because the received signal strengths, both calculated and measured, are considered to be averaged over about 10 wavelengths of distance (about 1.2 meters at Bluetooth frequencies) for each value; thus, they don't take into account the smaller-scale variations from multipath fading.[1]

Link Budget Equation

The classic free-space link budget equation is used extensively in satellite and LOS terrestrial applications, but it can be suitably modified to model an obstructed indoor or outdoor environment as well. This equation is given by

$$P_r = P_t G_t G_r \left(\frac{\lambda}{4\pi d} \right)^2 \tag{2-1}$$

where P_r is the received signal power, P_t is the transmitted power, G_t is the gain of the transmit antenna in the direction of the receive antenna, and G_r is the gain of the receive antenna in the direction of the transmit antenna. Both antenna gains are absolute (not decibels [dB]) in this equation. The carrier wavelength λ is 0.122 meters at 2.45 GHz, and d is the distance in meters between the transmit and receive antennas. Receive power thus conforms to the inverse square law of free-space propagation physics.

Equation 2-1 has an important restriction: The receive antenna must be far enough away from the transmit antenna that the wave is essentially planar so normal propagation physics applies; that is, the receive antenna must be in the *far field*. Although there is some disagreement as to what the minimum far field d should be, the formula is reasonably accurate when $d > \lambda$ for simple half-wavelength antennas.[2] As the antenna design becomes more complex, d must increase for Equation 2-1 to be valid. The formula clearly breaks down for very small values of d because P_r approaches infinity as d approaches zero for any positive (nonzero) P_t.

Using Decibels in the Link Budget Equation

Propagation engineers (and many other engineers for that matter) like to think in terms of dB instead of absolute numbers for three reasons:

- RF propagation exhibits logarithmic characteristics in several respects.
- Large signal strength variations can be expressed within a relatively small range of numbers.

- Mathematical multiplication becomes addition so mental calculations are easier to do.

Remember, though, that dB is a dimensionless quantity, so it is always calculated as a ratio between two values having the same units. In terms of power, dB is calculated by $P_{(dB)} = 10\log(P_2/P_1)$, where P_2 and P_1 are the two powers being compared.

We can now convert Equation 2-1 into its equivalent dB form, giving

$$P_{r(dBm)} = P_{t(dBm)} + G_{t(dB)} + 20\log\left(\frac{\lambda}{4\pi}\right) + 10n\log\left(\frac{1}{d}\right) \qquad (2\text{-}2)$$

The transmit and receive powers are both expressed in dBm, which is defined as power relative to one milliwatt. Because dB and dBm are used so often, we will drop the subscripts in the variables and express powers and gains in dB most of the time. There's also a new factor called the *PL exponent n*, which is included in this equation. We will examine n in detail later.

One question often arises: If the received power is less than the transmitted power, how can the four quantities on the right side of Equation 2-2 be added to P_t and produce a smaller P_r? The answer is that the two logarithm terms in Equation 2-2 are negative, so their values become subtractions. Also, as we'll see later, Bluetooth antenna gains are often slightly negative as well.

Simplified Path Loss (PL) Calculations

Equation 2-2 has several terms, each of which requires careful analysis to ensure that the calculated receiver power is as accurate as possible. Continuing with our speed-versus-accuracy theme, though, we will first simplify the equation by keeping only those terms associated with the *path* between the transmitter and receiver. As such, the antenna gains will be removed, at least for now. Furthermore, we will change terminology slightly and focus on only the PL between the transmitter and receiver, so the actual transmit and receive signal powers can also be ignored. The simplified PL formula becomes

$$PL = 20\log\left(\frac{4\pi}{\lambda}\right) + 10n\log(d) \qquad (2\text{-}3)$$

The first thing to notice about Equation 2-3 is that it's (ahem) simplified, but aside from that the fraction arguments for each log function have been inverted. This is because Equation 2-3 calculates a loss, whereas Equation 2-2 determines a gain. As a result, PL is a positive dB quantity because both terms on the right side are positive at Bluetooth carrier wavelengths and for $d > 1$ meter. Also, note that if both transmit and receive antennas have a gain of 0 dB, then Equation 2-3 is just the ratio of P_t to P_r in dB. This insight will become useful when we make our first attempt at calculating Bluetooth range.

The Effect of Wavelength on Path Loss (PL)

The value of the first term on the right side of Equation 2-3 is determined by the wavelength of the RF carrier, and this term becomes larger as λ decreases. In fact, looking back at Equation 2-1, we notice that P_r is proportional to the *square* of the wavelength. This means that if other factors remain unchanged, doubling the carrier frequency (halving the wavelength) increases PL by 6 dB, which is equivalent to a factor of 4. That seems counterintuitive; isn't the RF power density on the surface of a sphere with radius d from the transmit antenna independent of frequency? That's true, but the reason for this wavelength dependency can be found in an antenna property that remains hidden in Equation 2-3 despite our having removed the antenna gain factors. The physical size of a particular type of antenna is directly proportional to its operating wavelength. When wavelength decreases, the antenna becomes smaller for a fixed gain, and it therefore captures proportionally less of the available RF wavefront at a given distance d from the transmit antenna. Because power is directly proportional to the square of the field strength, an antenna that physically shrinks by half will lose 6 dB of captured signal power.

By substituting $\lambda = 0.122$ meters for a frequency of 2.4 GHz into Equation 2-3, we obtain an even simpler PL formula for Bluetooth frequencies as

$$PL = 40 + 10n \log(d) \qquad (2\text{-}4)$$

At this point you're probably wondering if the formula is now so elementary that it's nearly worthless, but this equation is often reasonably accurate, and with some additional insight we will be able to estimate several PL values and the resulting Bluetooth range quickly and without the use of a calculator.

Another very interesting PL phenomenon can be discovered by setting d to 1 meter. Because $\log(1) = 0$, the PL (assumed to be that of free space) at 1 meter from the transmit antenna is 40 dB. Put another way, only one-ten-thousandth of the transmitted signal power at 2.4 GHz is present at a simple receive antenna located just 1 meter from the transmitter. That's a big loss!

PL Exponent and Typical Values

As mentioned before, the quantity n in Equations 2-2 to 2-4 is called the PL exponent, and this can be adjusted to account for the amount of clutter in the path between transmit and receive antennas. For example, if free-space propagation applies, then $n = 2$, and the three equations relate directly to Equation 2-1. The resulting inverse square law states that signal power drops according to the square of the distance between antennas. It often helps to visualize these changes in a per-octave sense (doubling d) or in a per-decade sense (multiplying d by 10). PL as a function of distance, therefore, increases by 6 dB per octave or 20 dB per decade when free-space propagation applies. These results are summarized in Figure 2-3.

Equation 2-4 can become even more useful when we substitute different values of n to account for various non-LOS situations. If there is clutter between the transmitter and receiver, which is common in both indoor and outdoor environments, then this can be modeled by raising the PL exponent to a value greater than 2. For example, the link budget for urban cellular planning often uses $n = 4$ to account for severe signal attenuation from the numerous high-rise buildings between the cell antenna and phone and to accurately model a direct wave and single reflection from the ground that

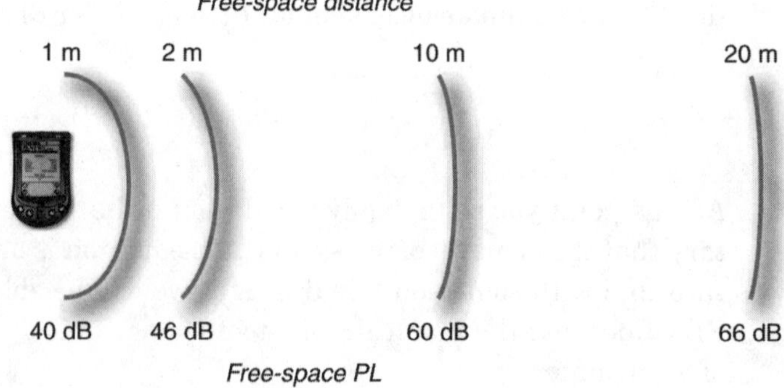

Figure 2-3 Free-space PL values for different distances at a frequency of 2.4 GHz

arrive at the receiver with approximately equal amplitude. Losses over distance in these situations amount to 40 dB per decade.

Surprisingly, there are also occasions where n can be less than 2. This commonly occurs when the transmitter and receiver are located in a long hallway, so multiple reflections off the walls, floor, and ceiling cause the RF wave to sustain itself as if it were in a crude waveguide. The PL in these cases may be as low as 10 dB per decade, dramatically increasing the useful range. I once tested a wireless telemetry link down an office hallway and found the range to be about 100 meters, but from room to room the device would reach only about 10 meters at best.

Selecting the PL Exponent What's the best way to select a good value for n in various indoor environments? One experimental method is to place a transmitter of known output power somewhere in the building and measure the *received signal strength indication* (RSSI), which is the power at the receiver, throughout the area of interest. After compensating for the antenna gains, a curve-fitting technique (such as least squares) can be used to find the best estimate for the PL exponent value. Of course, the RSSI measurements apply to the transmitter located only at that specific position, as it would be for a fixed *wireless local area network* (WLAN) access point or even a fixed Bluetooth device, such as a printer or desktop computer, but most Bluetooth piconets are more dynamic. Consequently, n must be estimated based upon the type of environment in which the piconet is located and by performing many RSSI measurements with the transmitter placed in several different locations. Fortunately, much of that work has already been accomplished, and some of the results are summarized in Table 2-1.[3] These measurements were made on the same building floor at frequencies ranging from 900 to 1,500 MHz; at 2.4 GHz, these values are about the same in some buildings, and in others they tend to

Table 2-1

Typical same-floor indoor PL exponent values

Location	n	σ(dB)	Frequency (MHz)
Retail store	2.2	8.7	914
Office, moveable walls	2.4	9.6	900
Office, fixed walls	3.0	7.0	1,500
Metalworking factory, LOS	1.6	5.8	1,300
Metalworking factory, non-LOS	3.3	6.8	1,300

average about 10 percent higher.[1, 4] The table also includes the standard deviation σ of the set of actual PL measurements against the values found by using Equation 2-4.

There are several interesting insights provided by Table 2-1. A retail store consists of a high ceiling, no interior walls, and shelf units that reach perhaps halfway to the ceiling, leaving large open areas in the upper part of the building for unimpeded propagation to occur. As a result, n is only slightly higher than its free-space value. An office with moveable walls (cubicles) also has open areas just below the ceiling, but n is a bit higher due to increased clutter from more densely packed items. The PL exponent increases further in an office where fixed walls stretch from floor to ceiling, and in this environment RSSI now drops by about 30 dB per decade of distance. The metalworking factory is full of good reflectors, so in LOS conditions the PL exponent drops below that of free space due to the waveguide effect mentioned earlier. These reflectors cause more havoc in non-LOS conditions by effectively blocking much of the signal from the receiver, so n can be quite high as a result.

When the transmitter and receiver are located on different floors of a building, the PL exponent can be even higher for measurements taken at 914 MHz, as shown in Table 2-2.[5] Other researchers who have performed in-building measurements at frequencies close to 2.4 GHz have found average PL exponents throughout a multifloor building to be as low as 3.7 and as high as 7.7, with an average n of 6.0 for all buildings that were measured in Liverpool, England.[4] An RF link often requires transmitter powers in the tens or hundreds of milliwatts to be reliable when the receiver is on a different floor. Keep in mind, though, that these high floor attenuations can be advantageous as well because a low-power Bluetooth piconet on one floor of a building will probably produce very little interference to piconets on other floors and will not be susceptible to interference from these other piconets.

Table 2-2

Average PL exponent values in multifloor buildings at 914 MHz

Number of Floors	n	σ(dB)
Same floor	2.8	12.9
Through one floor	4.2	5.1
Through two floors	5.0	6.5
Through three floors	5.2	6.7

Because the amount of clutter varies, depending upon where an RSSI measurement is being taken, even when the distance d is held constant, the set of measured PL data will exhibit statistical variation around the value calculated by using Equation 2-4. This variation is quantified by σ(dB) in Tables 2-1 and 2-2, which represent the standard deviation of the measured data around the mean given in Equation 2-4 due to *log normal shadowing*. This phenomenon will be discussed later in this chapter.

Calculating Bluetooth Range

At this point we've covered almost enough background to make our first attempt at determining Bluetooth useful range. If we know what PL a Bluetooth receiver (RX) can tolerate and still function, then we can insert this value into Equation 2-4 and solve for the distance d. This requires us to make some assumptions before proceeding:

- The Bluetooth transmitter (TX) power is 0 dBm (1 milliwatt) or +20 dBm (100 milliwatts).
- The RX sensitivity limit is –70 dBm (standard) or –80 dBm (enhanced).
- The Bluetooth transmit and receive antennas each have a gain of 0 dB.

The most common TX power for Bluetooth devices is 0 dBm, as we will discover in the next chapter. The range for this power level is often given as 10 meters, and we will be able to verify this shortly. For longer-range applications, the Bluetooth specification enables TX powers up to +20 dBm. Our range calculations will examine both of these extremes.

An RX must have a sensitivity of at least –70 dBm (for a *bit error rate* [BER] of 10^{-3} or better) to obtain Bluetooth qualification, so the standard receiver just meets that requirement. Actual receivers available from several manufacturers have sensitivities between –80 and –85 dBm, so our enhanced receiver is assumed to outperform the standard receiver by 10 dB.

Finally, the antennas are optimistically assumed to have no gain or loss and a perfect impedance match to their respective transmitter and receiver so no power is lost from transmitter to antenna or from antenna to receiver. This last assumption was implied when Equation 2-2 was simplified into Equations 2-3 and 2-4.

Different values for n can be placed into Equation 2-4 based upon path clutter, and then d can be solved for the highest PL that can be accepted by the transmit and receive specifications given previously. For example, if TX power is +20 dBm and RX sensitivity is –80 dBm, then the highest PL under which the system will still operate is 20 – (–80) = 100 dB. These results are summarized in Table 2-3. Since the Bluetooth link is bidirectional, the transmitter and receiver characteristics given in Table 2-4 must be present at both ends of the link for the listed ranges to be valid.

Rules for Quick Bluetooth Range Calculations Finally, it's time to deliver on the promise to provide some simple rules of thumb for quickly estimating Bluetooth range without resorting to anything other than our quick wits and brilliant intellect. First, look at the entries in Table 2-3 when

Table 2-3

Bluetooth range estimates using the simplified PL model

Type of Clutter	n	TX Power (dBm)	RX Sensitivity (dBm)	Range (m)
None (free space)	2.0	0	−70	31
	2.0	0	−80	100
	2.0	+20	−70	316
	2.0	+20	−80	1,000
Light	2.5	0	−70	16
	2.5	0	−80	40
	2.5	+20	−70	100
	2.5	+20	-80	251
Moderate	3.0	0	−70	10
	3.0	0	-80	22
	3.0	+20	−70	46
	3.0	+20	−80	100
Heavy	4.0	0	−70	6
	4.0	0	−80	10
	4.0	+20	−70	18
	4.0	+20	−80	32

there's no clutter in the transmit-receive path, corresponding to LOS conditions (n = 2.0). By recalling just two rules, we can estimate Bluetooth range for many LOS situations:

- The TX power of 0 dBm and RX sensitivity of –70 dBm yield a base range of about 30 meters.
- The range triples for every additional 10 dB of TX power or RX sensitivity improvement.

For example, suppose TX power is +10 dBm and RX sensitivity is –80 dBm in an LOS environment. The base range of 30 meters triples to 90 meters for the TX power increase and triples again to 270 meters for the RX sensitivity improvement. The estimated range by using Equation 2-4 directly is 316 meters, so our two rules yield answers that are somewhat pessimistic. We will discover later that placing pessimism into our calculations is often a good idea.

For a moderately cluttered path (n = 3.0), the rules change to the following:

- The TX power of 0 dBm and RX sensitivity of –70 dBm yield a base range of about 10 meters.
- The range doubles for every additional 10 dB of TX power or RX sensitivity improvement.

Suppose once again that TX power is +10 dBm and RX sensitivity is –80 dBm, but this time the path is moderately obstructed. The base range of 10 meters doubles to 20 meters for the TX power increase and doubles again to 40 meters for the RX sensitivity improvement. As before, estimates using these rules are slightly pessimistic compared to solving for d by using Equation 2-4.

Finally, for a heavily cluttered path (n = 4.0), the rules become

- The TX power of 0 dBm and RX sensitivity of –70 dBm yield a base range of about 6 meters.
- The range triples for every additional 20 dB of TX power or RX sensitivity improvement.

Again, let TX power be +10 dBm and RX sensitivity be −80 dBm, but this time the path is heavily obstructed. The base range of 6 meters triples to 18 meters when both the TX power increase and the RX sensitivity improvement are considered. Again, estimates using these rules are slightly pessimistic compared to solving for d by using Equation 2-4.

Accuracy of the Simplified Path Loss (PL) Model

By now it's tempting to toss aside the details that we've learned about propagation and just use the simple rules in the previous section for all of our range calculations. Before doing that, we should determine whether or not this process is sufficiently accurate for our needs. There are generally four areas in which errors could creep into Bluetooth range calculations based upon the simplified PL model:

- Clutter variations
- Multipath
- Antenna effects
- Interference

The first three of these will be discussed later in this chapter, and the last will be covered in Chapters 3 and 13. It would be helpful, though, to summarize their effects on the accuracy of simple PL models.

Variations in the amount of clutter will have an effect on actual PL that depends upon where the transmitter and receiver are located within a room or building. The actual range, therefore, could be significantly different from the average predicted value. This is discussed in the section "Log Normal Shadowing."

Because many rooms are highly reflection-prone to RF signals, several copies of a transmitted wave can arrive at the receive antenna at different times and with different phase relationships. As we will soon discover, the resulting multipath can cause signal dropouts (fades) of up to 30 dB. That's equivalent to reducing transmit power by a factor of 1,000! Fortunately, elementary techniques are available to reduce multipath fading to perhaps 10 dB, but its effect on system performance can still be significant.

The antenna in a portable wireless device is often designed to have nearly omnidirectional characteristics, at least in the azimuth plane. Various factors, such as poor matching to the transmitter or receiver, poor efficiency, and RF absorption, often result in a loss of 3 to 5 dB per antenna. Both the transmit and receive antennas together, therefore, can reduce RSSI by perhaps 10 dB. That relates to a 50 percent range reduction in a moderately cluttered path or a 67 percent reduction in free space.

Finally, with the large number of wireless devices competing for access to the 2.4 GHz band, interference is a major cause of reduced performance. The reliability of outdoor cellular systems, for example, has long been interference limited rather than signal-to-noise limited, and indoor systems will soon follow suit.

It's probably obvious by now that using the simplified PL model to find Bluetooth range limits can have significant accuracy problems. We return yet again to the speed-versus-accuracy theme by first attempting to account for clutter variations between the transmitter and receiver. If done correctly, perhaps we can reduce the variance in actual PL measurements compared to what is predicted.

Improving Path Loss (PL) Prediction Accuracy

There are several ways that the simple distance-dependent PL model can be enhanced for greater accuracy, some of which can increase the complexity of the calculations markedly. We will discuss two of these: the multibreakpoint model and the Toledo-Turkmani model.

Multibreakpoint Model Suppose two members of a Bluetooth piconet are both located in the same room. This will probably be a common occurrence, and the PL exponent in such a situation will typically be between 2.2 and 2.5, slightly higher than free space. If one of the piconet members moves to another room, the adjoining wall will cause the PL exponent to increase to (say) 3.0, and if the member proceeds deeper into the building, n could increase yet again to (say) 4.0. This leads to the possibility of adjusting n as distance increases. This is commonly called the *Ericsson multiple breakpoint model*, which had its origins in the behavior of 900 MHz cellular networks.[6]

As an example, let's assume that $n = 2.5$ for d between 1 and 10 meters, 3.0 for d between 10 and 20 meters, and 4.0 for d greater than 20 meters. The total PL can then be found by using a set of formulas given by

$$
\begin{aligned}
PL &= 40 + 25 \log(d), && 1 \le d < 10 \\
PL &= 65 + 30 \log\left(\frac{d}{10}\right), && 10 \le d < 20 \qquad (2\text{-}5) \\
PL &= 74 + 40 \log\left(\frac{d}{20}\right), && d \ge 20
\end{aligned}
$$

The first term in each new formula is the PL at the outer distance boundary from the previous formula, followed by the additional loss from the new value for n. To find the Bluetooth range when the PL is 70 dB, the second formula is used because 70 falls between 65 and 74, and the corresponding range is 14 meters. If the acceptable PL increases to 100 dB, then the third formula is used to find that the range is 89 meters.

Toledo-Turkmani Model The Toledo-Turkmani approach adds terms to the basic PL Equation 2-4 to account for some of the specific layouts encountered in various buildings.[4] The researchers measured the PL at 2.3 GHz in several buildings and developed an equation that fit their measurements more closely than Equation 2-4. Although the model wasn't derived for Bluetooth applications, 2.3 GHz is close enough to Bluetooth carrier frequencies that we will use piconet members in the examples. This new equation is given by

$$PL = 21.6 + 39.1\log(d) + 3.8k_f - 17.8S_w - 8.8C_g - 0.014A_f \qquad (2\text{-}6)$$

The first term, 21.6 dB, is a constant that accounts for the loss at a distance d of 1 meter, and the second term computes additional distance-dependent loss based upon a PL exponent of 3.91. Next, k_f is simply the number of floors separating the transmitter and receiver. The other terms account for the topology of the path itself. The term S_w adjusts for the possibility of RF exiting a window and returning through another window. If both members of the Bluetooth piconet are located on the same windowed side of the building, then $S_w = 1$; if the two members are on windowed sides that are 90 degrees from each other, then $S_w = 0.5$; and if the members are on opposite windowed sides, then $S_w = 0.25$. Piconet members located in adjacent rooms could be modeled with $S_w > 0$ depending upon the nature of any obstructions between them, and the term could be used to compensate for the effects of RF diffraction around corners.

The term C_g is set to 1 when the piconet members are on the first two floors of a building, where signals tend to be stronger. This assumes, of course, that both users are located on the same floor. Finally, A_f is the area of the floor on which the users are located; the larger the area, the stronger the signals tend to be for a given separation d.

As an example, let's put two Bluetooth piconet members 20 meters apart ($d = 20$) in adjoining rooms separated by a corridor ($S_w = 0.5$) of the fifth floor of a building ($C_g = 0$; $k_f = 0$) that has an area of 1,000 square meters ($A_f = 1{,}000$). The Toledo-Turkmani model predicts that the PL will be about 50 dB, which is optimistic compared to results from previous models. Even with its added complexity over Equation 2-4, standard deviations were still 12.4 dB when values in Equation 2-6 were matched against measured PL data.

The multipliers in front of each of these terms can be adjusted for any given building's characteristics based upon actual measurements. However, it's clear that with the additional set of terms and their associated complexity, incorrect assumptions could result in greater errors. Of course, it has now become very difficult to use this equation for PL calculations without the help of a calculator or computer. Therefore, we will return to the simple PL equation given by Equation 2-4 and investigate the nature of its inaccuracy and whether we might use a different approach.

Log Normal Shadowing

Suppose we place a transmitter somewhere in a room and then walk an arc at constant distance *d* away from the transmitter while taking many large-scale PL measurements with a handheld receiver. It's obvious that the PL will vary depending upon the amount of clutter between the two devices: lower when the path is clear and higher when the path is obstructed (see Figure 2-4). A statistical analysis on these measurements will show that if the PL values are expressed in dB, they form a normal (Gaussian) distribution about their mean value. Because converting values to dB requires their logarithm to be taken, this distribution is called *log normal*.

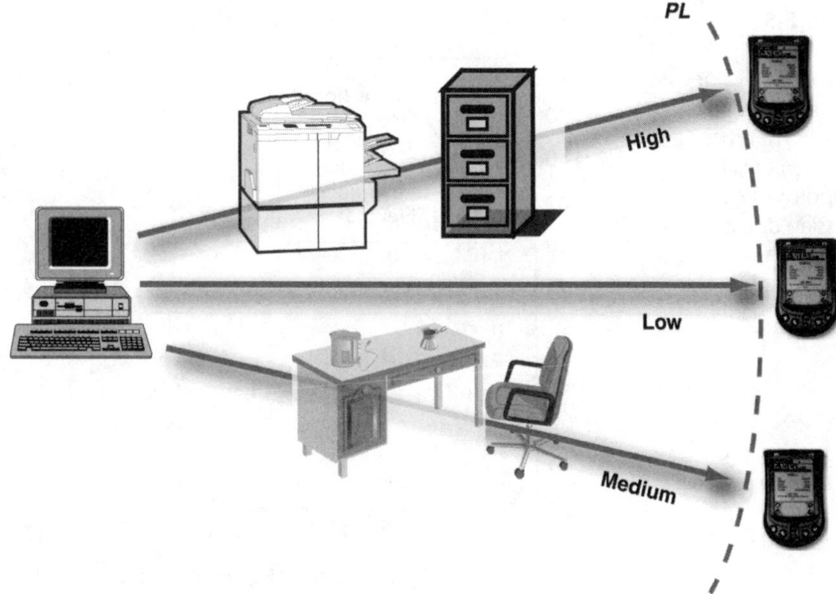

Figure 2-4 The origins of log normal shadowing. Varying clutter means varying PL measurements.

The relative distribution of measured PL values, expressed in dB, can now be plotted as a Gaussian distribution with mean given by Equation 2-4 and standard deviation listed in Table 2-1. For a particular distance *d* and PL exponent *n*, about two-thirds of the actual PL measurements will fall within one standard deviation above and below the value predicted by Equation 2-4. This, of course, assumes that the correct *n* is used in the first place so that Equation 2-4 accurately determines the average PL value. Of the remaining PL measurements, about one-sixth of these will be more than one standard deviation higher than the average value, and about one-sixth will be more than one standard deviation lower.

Figure 2-5 is a plot of actual PL measurements taken throughout an office building at 914 MHz.[5] A PL exponent of 3.54 minimizes the mean square error between the computed average PL, dependent only upon the distance between the transmitter and receiver and the measurements. If a vertical line segment is drawn intersecting the dashed line (average PL), then the distribution of measurements along that line segment is roughly Gaussian with mean value at the intersection of the two lines and standard deviation of approximately 12.8 dB.

Now suppose Bluetooth is being used in an area where the PL exponent is about 2.4, and the standard deviation of actual PL values is about 10 dB (refer to Table 2-1). If TX power is 0 dB and RX sensitivity is –80 dBm, then

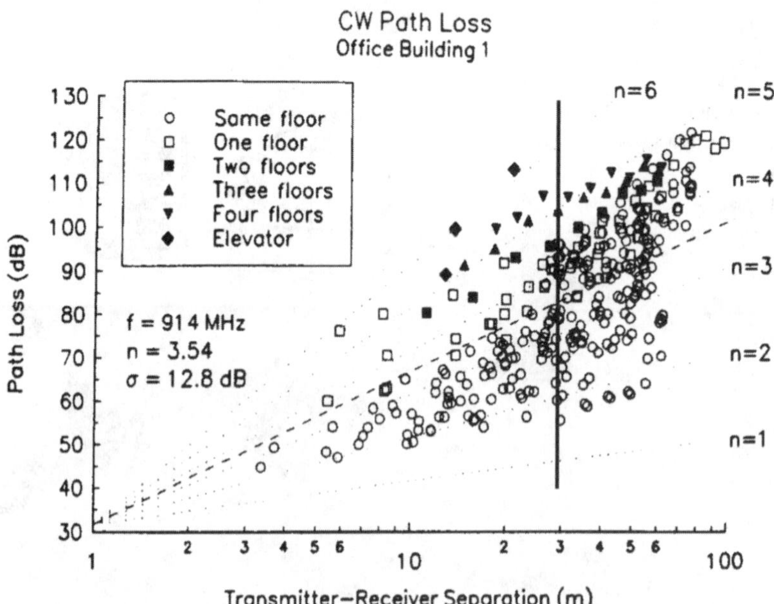

Figure 2-5 Scatterplot of measured PL values with their mean (dashed line). The measurements form an approximate Gaussian distribution along any vertical line segment. (Source: S. Seidel)[5]

Equation 2-4 predicts the average range to be about 46 meters. If the receiver is held 46 meters from the transmitter and a constant-radius arc is walked, then measured large-scale PL will fall between 70 and 90 dB about two-thirds of the time. PL will be less than 70 dB about one-sixth of the time and will exceed 90 dB about one-sixth of the time. Also, about half of the measured PL values will be higher than 80 dB, which raises a significant problem. If the receiver's sensitivity limit is –80 dBm, then the Bluetooth piconet will fail to operate whenever PL exceeds 80 dB. This situation occurs at around half of the possible positions located at the range limit, which is bad news indeed!

Adding Pessimism to the Calculations

It has probably become uncomfortably clear by now that propagation analysis has a significant amount of uncertainty. There is just no practical way to calculate precise PL values in every situation. Also, even if exact values could be determined (by measurement, for example) these would become invalid as soon as an object, such as a person or piece of equipment, moved and thus changed its effect on the path between the transmitter and receiver. Therefore, we have little choice, short of giving into despair, other than to perform various simplifications to the PL problem and accept the inevitable statistical variations. In doing so, we can attempt to compensate for the resulting inaccuracies by biasing the calculations either optimistically or pessimistically.

Given a choice between optimistic or pessimistic bias, it makes far more sense to proceed pessimistically. The results we obtain can then be considered to model, if not a minimum performance, at least a level of performance that the system will exceed most of the time. Conversely, the danger of adding optimism is that the actual system may perform worse than that calculated, and that can cause lots of implementation headaches.

Pessimism can be inserted into Equation 2-4 in two ways:

- A higher PL exponent can be inserted.
- A lower PL value at the range limit can be assumed.

These methods can be employed separately or together to yield range results that are more likely to be met or even exceeded in actual situations. Using a higher PL exponent increases the probability that the Bluetooth link will still operate when the path has a greater-than-average number of obstructions. Assuming a lower PL value at the range limit can compensate for such factors as antenna orientation and losses and inefficiencies from coupling the transmitter and receiver to their respective antennas. Care must be taken, of course, when adding pessimism because it's easy to go overboard. I once worked with an engineer who put so many pessimistic assumptions into the propagation model that the calculated Bluetooth range limit was only 1 meter!

Continuing with our previous example, if we use a PL exponent of 2.5 instead of 2.4, the calculated range limit is reduced from 46 to 40 meters. Additional pessimism could be included by limiting the PL to 70 dB instead of 80 dB, further reducing the range to 16 meters. At this range, signal strength will be adequate most of the time because the receiver is now significantly closer to the transmitter. Also, a 16-meter range is sufficient for many uses for which Bluetooth was conceived, and even this slightly pessimistic calculation exceeds the 10-meter range popularized in the Bluetooth literature.

Even with this new approach we can still perform many PL and range calculations mentally. By working with pessimistic assumptions, though, we run a further risk of overengineering the system. For example, placing an unwarranted 10 dB increase in the PL value may require an additional 10 dB of transmitter power before the equations are satisfied, increasing cost, power supply consumption, and the potential for interference. Rather than simply using a pessimistic outlook, therefore, we may choose instead to turn to the computer to calculate PL values with more accuracy. The next step on the speed-versus-accuracy quest is the use of ray tracing methods to model the actual path between the transmitter and receiver.

Primary Ray Tracing for Improved Accuracy

Up to now we have lumped all clutter between the transmitter and receiver into a quantity called the PL exponent. By selecting n carefully and using Equation 2-4, we can quickly find a corresponding PL for a given distance or a corresponding distance for a given PL. PL standard deviations can exceed 10 dB using this method, translating into a useful range in many situations that is well below what we expected. To counter this, pessimism can be added to reduce the calculated useful range for a given set of circumstances.

It is possible to improve accuracy substantially by sacrificing the ability to perform these calculations in our head and turning instead to the computer. If the significant obstacles to RF propagation are placed into a *computer-aided design* (CAD) drawing of a building floor plan, then each path of interest can be traced on the drawing and a corresponding PL can be calculated based upon the actual set of obstructions between the transmitter and receiver. This process is called *primary ray tracing*.

Direct Modeling of Partition Losses

Focusing once again on the direct wave, the path between the transmitter and receiver is characterized by both distance and the type and number of obstructions through which the wave must pass. In a typical environment in which Bluetooth devices are used, signals may pass through people, computers, and furniture within a room, and through walls, doors, and/or windows if the piconet users are located in different rooms. It's generally convenient to calculate PL from partitions that tend to remain fixed over long periods of time and less so from objects that are easily moved such as people, tables, and chairs.

The concept behind finding total PL values using primary ray tracing employs the following rules:[7]

- Signal strength drops suddenly as it passes through a partition, and the amount of drop depends upon the type of partition.
- Free-space PL occurs between partitions.
- Losses between floors exhibit special behavior and are modeled separately.

These rules can be shown mathematically as

$$PL = 20 \log\left(\frac{4\pi d}{\lambda}\right) + \sum_i [(P_i)(AF_i)] + FAF \qquad (2\text{-}7)$$

For Bluetooth frequencies, we can once again substitute $\lambda = 0.122$ meters into Equation 2-7 and obtain

$$PL = 40 + 20 \log(d) + \sum_i [(P_i)(AF_i)] + FAF \qquad (2\text{-}8)$$

This formula looks more intimidating than it actually is, and it will certainly make more sense after the terms are defined. The first term on the right side of Equation 2-8 represents the free-space PL at 1 meter, and the second term includes the additional free-space PL at a distance d beyond 1 meter. These values are the same as those given by Equation 2-4 for $n = 2$.

The third term represents additional losses from the partitions through which the signal must pass. A partition of type i is given an attenuation factor AF_i, and there are P_i partitions of type i in the path. For example, suppose there are two walls, each with an attenuation factor of 3 dB, and one door, with an attenuation factor 2 dB, between the transmitter and receiver. The total partition attenuation is then 2 walls times 3 dB each plus 1 wall times 2 dB, or 8 dB total. The last term in Equation 2-7, called the *floor attenuation factor* (FAF), represents an additional attenuation if a signal must pass through one or more floor/ceiling partitions en route to the receiver.

Table 2-4 shows some typical attenuation factors for various partitions found in a typical home or office environment.[6] Although most of these measurements were taken at 1.3 GHz, their values are much more dependent upon construction materials than upon carrier frequency.

Of course, the PLs in Table 2-4 are average values measured in many different buildings, and these could be inaccurate for a specific situation. This can be remedied by directly measuring partition losses using a continuous wave (CW) transmitter and a receiver with RSSI capability. First, the transmitter is operated at a known distance from the receiver, and the RSSI is noted. Next, the partition of interest is placed between the transmitter and receiver, which are kept at the same distance from each other as before, and the new (weaker) RSSI is recorded. Finally, the partition attenuation factor is the difference between the two RSSI values in dB.

Table 2-4

Typical partition attenuation factors

Partition	Loss (dB)
Fixed walls	3.0
Moveable walls	1.4
Doors	2.0
Metal partitions	5.0
Windows	2.0
Exterior walls	10.0
Basement walls	20.0

Signal attenuation through the floors/ceilings in a building doesn't follow the rules given by the third term in Equation 2-8. Instead, losses from the floor/ceiling partitions taper off as the signal passes through multiple stories of a building, as shown in Table 2-5.[7] Results in this table were measured in a specific multistory building. The loss through one floor is quite high at 13 dB, which is to be expected from a partition constructed from thick reinforced material. Furthermore, the standard deviation of PLs measured through the single floor is also somewhat high at 7.0 dB. Adding a second floor increases the FAF by only 6 dB, and the standard deviation is significantly reduced. Finally, the FAF through four or more floors becomes constant at about 27 dB, with a very small standard deviation.

Why does the FAF exhibit such behavior? Researchers aren't completely sure, but they think that perhaps the signals make their way from floor to floor through elevator shafts or heating, ventilation, and air conditioning (HVAC) ductwork. Or perhaps the signal passes through an external window, reflects off of a nearby high-rise building, and returns through other windows. Either method would circumvent the assumption that each partition adds a separate, independent contribution to total signal attenuation.

It's important to realize, however, that a constant FAF does *not* mean that a signal passing through several floors isn't attenuated more as the number of floors increases. Take another look at Equation 2-8. The FAF is one of four terms that compose the entire attenuation picture. A transmitter and receiver that are located several floors apart will be separated by a significant distance, so the second term in Equation 2-8 becomes larger as the total number of floors increases.

Finally, let's look at the examples portrayed in Figure 2-6. Both the transmitter and receiver are on the same floor, so the FAF is zero. They are separated by a distance of 10 meters, but one path has one wall of 3 dB

Table 2-5

Typical FAFs

Number of Floors	Loss (dB)	σ(dB)
1	13	7.0
2	19	2.8
3	24	1.7
4 or more	27	1.5

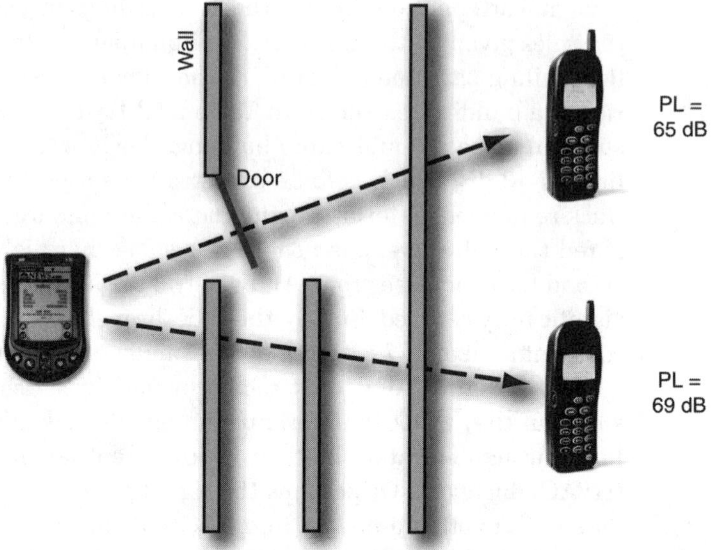

Figure 2-6 Example of PL calculations using primary ray tracing

attenuation, and one door with 2 dB attenuation. Plugging these values into Equation 2-8, we find the total PL to be

$$PL = 40 + 20 \log(10) + 1 \times 3 + 1 \times 2 = 65 \text{ dB} \tag{2-9}$$

If the receiver is moved slightly, then the signal must pass through three walls and the PL increases to 69 dB.

One of the difficulties of using Equation 2-8 for finding PL values is that the terms in the equation depend upon the topology of the building and placement of the transmitter and receiver. We are no longer working with numbers based upon just a distance and PL exponent, but instead we must now seek the help of a computer to make several hundred calculations for each situation being analyzed. First, a CAD drawing of one floor of an office building is entered into a site-planning software package. This floor plan shows numerous fixed walls and doors. Next, a transmitter is placed at a particular position in the building. The various partition attenuation factors are entered, and the software generates contours that represent constant PL values.

Figure 2-7(a) represents PL contours using Equation 2-4 as a strict distance-dependent-only model. The transmitter location is indicated by a dot in the center of the room. The contours are concentric circles and show

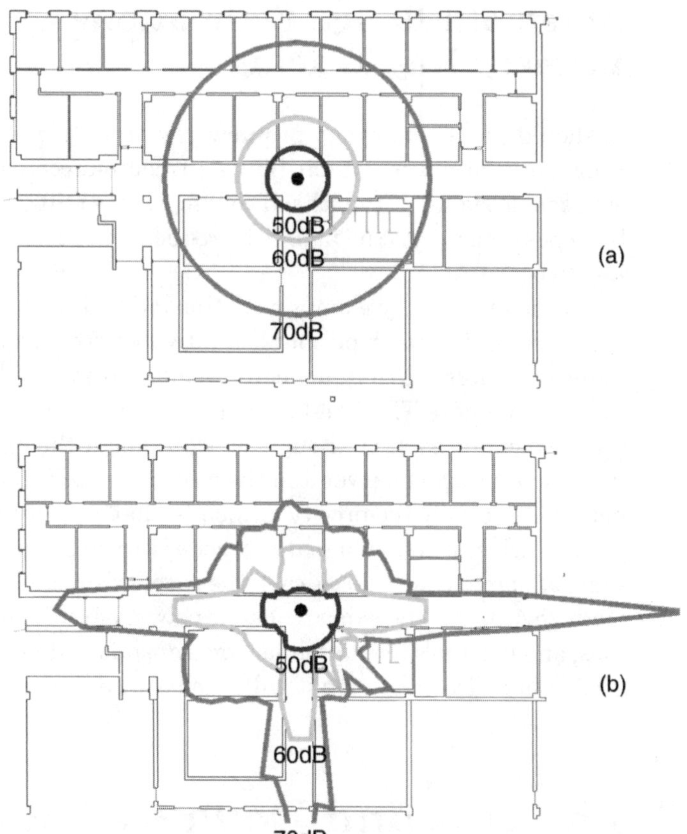

Figure 2-7
PL contours generated by computer analysis of a building floor plan, with the transmitter located at the central dot. The distance-dependent model in Equation 2-4 produces concentric circle contours (a), and primary ray tracing given by Equation 2-8 shows more realistic irregular contours (b). (Source: Wireless Valley Communications, Inc.)

the same PL values in hallway LOS situations as in obstructed paths within the rooms. This is one reason why the PL standard deviations from using Equation 2-4 are around 10 dB. The most obvious feature about Figure 2-7(b) is that the contours are no longer concentric circles because Equation 2-8 is now being used. Instead, signals are weaker in the rooms and stronger in the hallways, as we would expect. Directly modeling the partition losses requires a large number of calculations, but one of the benefits is an improvement in PL accuracy to a standard deviation of 3 to 5 dB.

The accuracy of primary ray tracing can be improved by using measured partition attenuation values in place of the estimates in Table 2-4. Also, some software packages enable actual PL measurements within the building to be used in an optimization program that adjusts the primary ray tracing values to improve accuracy further.

Bluetooth Range Estimations Using Primary Ray Tracing

It should be obvious that primary ray tracing can easily accommodate range calculations by establishing a PL contour corresponding to the weakest useable signal at the RX. Of course, it may still be wise to incorporate a bit of pessimism into the figures by reducing the tolerable PL by (say) 10 dB just to be safe.

Although accuracy is improved, this method works best when one of the nodes in the Bluetooth piconet is relatively fixed, such as in a desktop computer or printer. PL contours can be plotted from this location to determine the region where TX signal strength is adequate. Bluetooth is a two-way communication system, though, so what about the reverse link, where the fixed device is the receiver and the mobile device is the transmitter? Fortunately, there is a reciprocity principle that usually applies to the direct path, which states that signal characteristics in one direction are nearly the same as signal characteristics in the reverse direction. That is, if both Bluetooth devices use the same transmit power, have the same receive sensitivities, and use the same antennas for transmit and receive, and if the link is good in one direction, then it will also be good in the other direction.

The Ups and Downs of Multipath

Until now we have included only the direct path (primary ray) in our PL calculations, but we implied earlier, somewhat ominously, that significant signal energy can be included in the reflected paths as well. This opens up a whole new dimension to determining accurate PL figures because we know that reflected signals can have a phase difference with respect to the transmitted signal. As a result, the signal strengths can't just be added together to find the total power at the receiver. (If they could, then the direct path would represent a minimum received signal power, but no such luck.) Instead, the significant signals must be added together as vectors to find the total received signal strength.

These vector additions can change markedly when the receiver is moved over distances that are only a fraction of the RF carrier wavelength. The resulting fluctuations in received signal power can occur quite rapidly at Bluetooth frequencies, even when the receiver moves at a walking speed, because the carrier wavelength is a relatively short 0.122 meters. As such, these fades are called *small scale* to differentiate them from the large-scale PL caused by signal attenuation from obstructions.

We begin our study of multipath by quantifying the phase relationship between the various reflected signals at the receive antenna. Next, the two major characteristics of the multipath channel, Doppler spread and delay spread, are examined. These characteristics are applied to the Bluetooth channel to determine its fading properties. Finally, a statistical analysis of the multipath-prone channel is given, which can help us calculate the probability of outage for the Bluetooth piconet.

Characterizing the multipath channel can include mathematics that will bring sweat to the brow of even the most robust individual (see Figure 2-8), but fear not, we will once again focus on an intuitive grasp of what's actually occurring.

Jocular
Science
by Bill Lae
© 1987 Lae
QED
"Finally, skipping a few minor algebraic steps, it is intuitively obvious that we arrive at A = B. ...... check this at your leisure."

Figure 2-8 Multipath analysis can be filled with intimidating mathematics. (Source: Bill Lae)

Reflection, Transmission, and Absorption

When a radio wave strikes the surface of a typical obstruction, part of the wave's energy passes through the obstruction and continues on its way. This forms the direct wave studied earlier. For example, Table 2-4 shows that half of the signal's power continues through a typical plaster wall for a loss of 3 dB. What happens to the other half? Some of the power is absorbed by the wall and some is reflected, either specularly (that is, like a mirror) or through scattering. To simplify the analysis (where have we heard that before?), we will assume that all reflections are specular, so that the angle of incidence equals the angle of reflection, and that the direct path angle is unaltered as it passes through the obstruction (see Figure 2-9).

Within a particular room, it's obvious that multiple reflections can occur. Suppose a transmitter and receiver are located in a simple room with four walls, as shown in Figure 2-10. Various paths exist between transmitter and receiver, including the direct path (LOS in this example), several first-order reflections (one reflection), second-order reflections (two reflections), and so on. Perhaps one of the first questions that comes to mind at this point is, How many reflections must be considered before the reflected path signal strength is too small to matter? That's an important question because the answer determines how many reflections the computer software must track. Fortunately, research has shown that only the first reflection has significant power under most situations.[8]

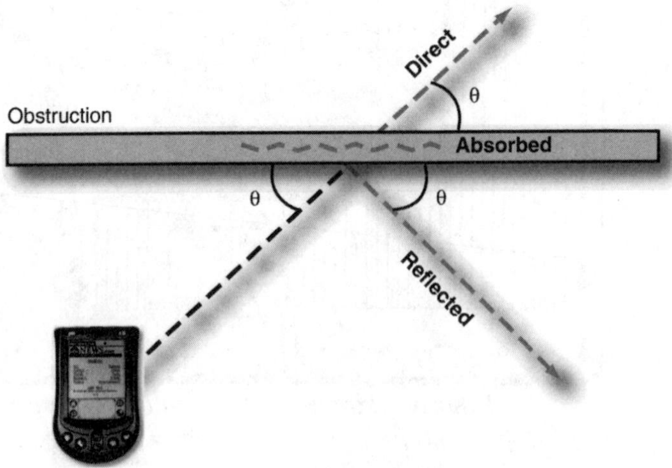

Figure 2-9 Radio waves striking most indoor obstructions are partially directed, partially absorbed, and partially reflected. Diffraction and scattering are ignored.

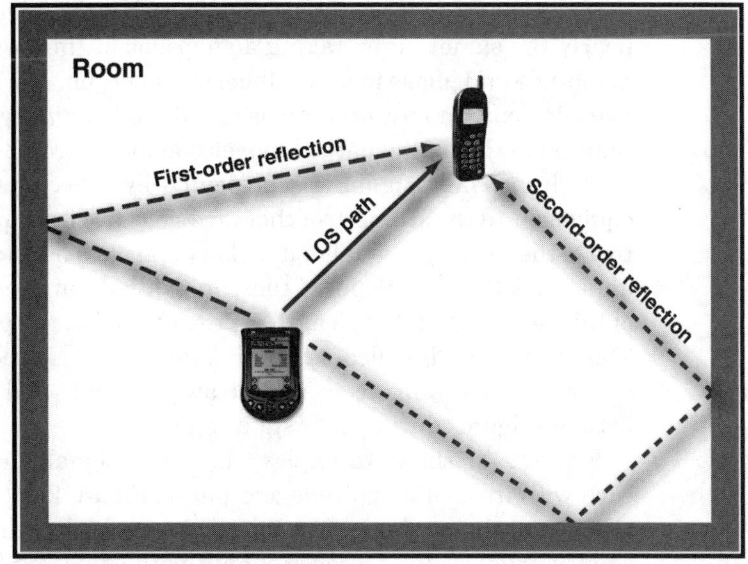

Figure 2-10 A transmitter and receiver located in a simple room can have an LOS path, first-order paths, second-order paths, and so on between them.

Continuing with our plaster wall example, suppose its attenuation factor is 3 dB as before, so half of the incident signal power (the part not making its way through the wall) is either reflected or absorbed. If half of this remaining power is absorbed by the wall, then the first-order reflection will be 6 dB below the total incident power; thus, it will contain one-fourth of the power in the incident wave. The reflected wave travels farther than the LOS path between transmitter and receiver, which results in additional PL. A second-order reflection will contain less than one-sixteenth of the power in the incident wave from the first of the two reflections, and the wave travels farther while becoming even weaker. Note, though, that if the partition is highly reflective (remember the metalworking factory?), then most of the incident power is indeed reflected and higher-order reflections must be modeled to obtain accurate RSSI values.

Phasor Diagram for Multiple Signal Arrivals

When multiple copies of the same transmitted signal arrive at the receiver from multipath reflections, each of these copies will have its own amplitude, frequency, and phase associated with it. If we wanted to diagram all three of these values, then a three-dimensional plot would be needed. Fortunately, however, the frequency associated with each incoming signal is the same (or nearly the same), so by taking a snapshot in time we need plot only the amplitude and phase in a two-dimensional graph. The abscissa is called the *real* (Re) axis, and the ordinate is called the *imaginary* (Im) axis in order to conform to the mathematical convention for portraying phase.

Each incoming signal is represented by a vector whose magnitude is equivalent to the strength of the wave's electric field and whose angle relative to the Re axis is equal to its relative phase. The total field at the receive antenna is the vector sum of these individual components, and the square of this resultant vector's length is proportional to the power at the receiver. This type of plot is called a *phasor diagram* and is used extensively among communications engineers to portray amplitude and phase relationships between signals during an instant in time.

Figure 2-11 shows examples where two signals arrive at the receiver, each with its own magnitude and phase. Figure 2-11(a) diagrams a situation where the total power at the receiver is higher than the power in any single arrival, so the presence of multipath helps RSSI. Figure 2-11(b) presents a case where the total power is less than the power in the strongest arrival, and here multipath is detrimental to RSSI. Notice that if Path 1 and Path 2 are equal in amplitude and exactly opposite in phase, then they cancel each other completely.

Multipath Effect on Received Signal Strength

If the effect of multipath sometimes helps and sometimes hurts RSSI, will the two effects essentially average out? If so, can we ignore the effects of multipath? The answer to the first question is, unfortunately, no. Furthermore, even if the two effects did average out over time or space, we still couldn't ignore multipath. If the average RSSI provides adequate system performance, then a stronger signal doesn't really do much good. A weaker signal, though, could drop RSSI below the minimum performance level, and the Bluetooth piconet will stop communicating. Even worse, multipath often hurts RSSI far more than it helps, which we will now discover.

Suppose a transmitter is placed near a floor that's highly reflective to RF, and a moveable receiver is placed at distance d away. As shown in Figure 2-12, signals can proceed along the LOS path from transmitter to receiver, and they can also reflect from the floor. Both arrive at the receive antenna, and their phases add vectorially, as shown in Figure 2-11. The position of the receiver can be changed horizontally such that any desired phase relationship between signals on LOS and reflected paths can be achieved.

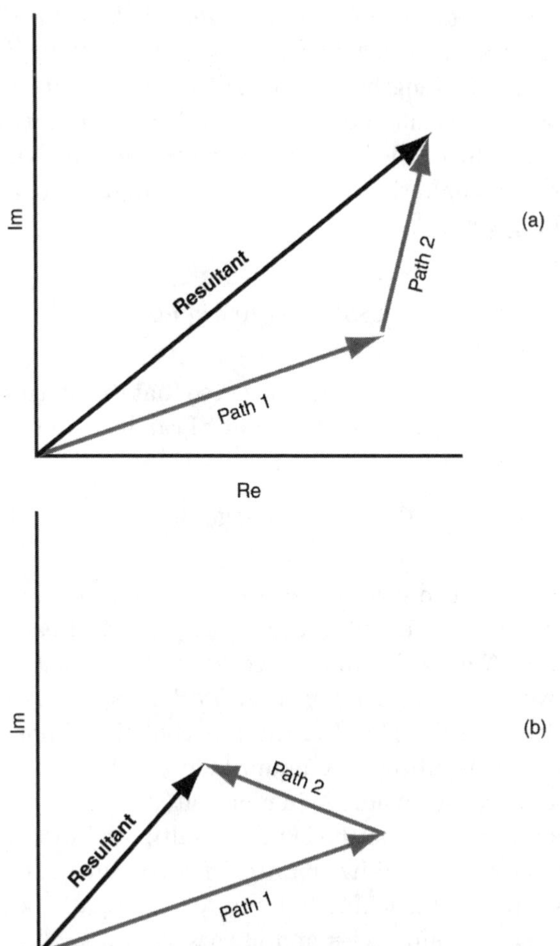

Figure 2-11 These phasor diagrams show that two signals arriving at a receiver can add constructively (a) or destructively (b). Signal amplitude is proportional to vector length, and signal phase is the angle of the vector relative to the Re axis.

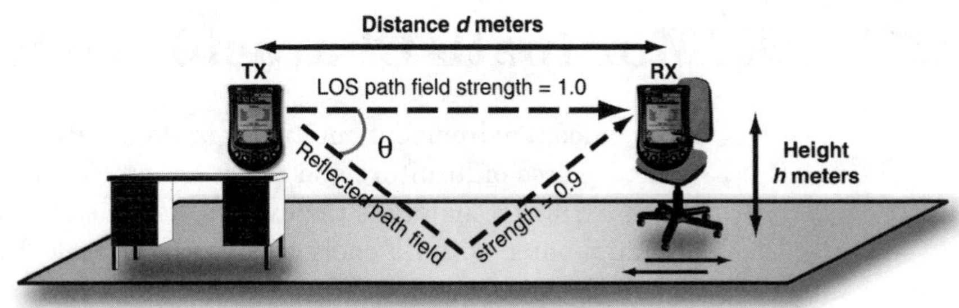

Figure 2-12 A multipath example in which one reflected signal combines at the receiver with the LOS signal. The receiver can be moved such that the two waves combine with any possible phase relationship.

To gain some insight into the extent that the RSSI changes when multipath is present, we arbitrarily set the LOS path field strength to 1.0 (0 dB) and the field strength of the reflected signal to 0.9. These strengths are proportional to the magnitude of the electromagnetic wave's E-field. First, let's calculate the RSSI from both signals together when they are perfectly phase aligned. In this case, the field strengths add, and the best-case RSSI in dB is given by

$$RSSI_{max} = 20 \log(1.0 + 0.9) = 5 \text{ dB} \quad (2\text{-}10)$$

Now suppose the receiver is moved so that the phases of the two signals are totally opposed. The worst-case RSSI can be found as

$$RSSI_{min} = 20 \log(1.0 - 0.9) = -20 \text{ dB} \quad (2\text{-}11)$$

We mentioned before that even if multipath can increase RSSI, it may not be very helpful if the Bluetooth piconet already has adequate signal strength. Now we've discovered with this simple example that RSSI is improved by a relatively modest 5 dB at best. However, we can be in big trouble when the phase of the reflected signal opposes the LOS signal's phase. The resulting maximum fade is 20 dB in our example, which is enough to cause an outage in many situations. We are rapidly approaching the uncomfortable conclusion that multipath hurts more than it helps.

Of course, the real-live multipath environment is far more complex than this simplified model, with many reflected paths arriving at the receiver with distinct amplitudes and phases. As we shall soon see, though, measured fading values are surprisingly close to those in this simple example.

Multipath Characteristics

In an indoor environment, multipath is almost always present because the various pieces of furniture and other equipment in a building, along with the construction materials themselves, are prone to reflections when they are illuminated with RF energy. Unlike large-scale fading, which averages signal strength over distances of about 10λ, small-scale fading can occur when a receiver is moved only a fraction of a wavelength. Furthermore, the movement of transmitter, receiver, and/or objects affecting the communication path cause changes in the signals arriving at the receiver, both in amplitude and frequency.

The effects of multipath can be placed into two broad categories. *Doppler spread* characterizes *frequency shifts* among the various signal components due to motion somewhere within the physical communication channel. *Delay spread* characterizes the *time delays* between the arrivals of each significant signal at the receive antenna. Each of these two characterizations can be examined in both time and frequency domains, and several similarities can be found between them by doing this.

Doppler spread is best analyzed from a narrowband point of view, where each path combines vectorally at the receive antenna into a composite result. To measure fading from Doppler spread, a simple RF oscillator can be tuned to the carrier frequency and placed somewhere in the building. RSSI is recorded as a receiver is moved about, similarly to large-scale PL measurements but with no signal strength averaging.

Delay spread, on the other hand, is best grasped as a wideband process, where each path is examined separately and the time delays between the arrival of each is scrutinized. This can be measured by sending a short pulse of RF energy and measuring the various arrival times and signal strengths of each at the receiver.

Doppler Spread

Suppose the chair with the receiver in Figure 2-12 is placed next to the transmitter and then rolled away with a constant velocity v. Also, let's assume for now that there's no reflected signal yet. The frequency of the CW signal along the LOS path will shift lower by the amount

$$f_D = \frac{v}{\lambda} \tag{2-12}$$

where f_D is the Doppler shift, v is the velocity of the receiver away from the transmitter, and λ is the carrier wavelength. If the transmit frequency is 2.45 GHz and the receiver is moving away from the transmitter at 1 meter/second (a moderate walking speed), then the receiver will see the transmit frequency along the LOS path shifted down by about 8 Hz.

But what is the Doppler shift on the reflected path? That depends upon where the receiver is relative to the transmitter. Suppose the transmit and receive antennas are each an identical h meters above the floor, and we neglect the length of the antennas themselves. If the LOS path is increasing with velocity ν_{LOS}, then the reflected path is growing with velocity

$$v_{ref} = v_{LOS}\cos\theta = v_{LOS}\frac{d}{\sqrt{(2h)^2 + d^2}} \qquad (2\text{-}13)$$

where d is the length of the LOS path. When the transmitter and receiver are close together, the Doppler shift on the reflected path is nearly zero because ν_{ref} is small. When d is 1 meter and ν_{LOS} is 1 meter/second, then ν_{ref} is 0.24 meter/second, producing a (instantaneous) Doppler shift of about 2 Hz compared to 8 Hz along the LOS path. As the LOS path becomes longer, ν_{ref} approaches ν_{LOS} and the two Doppler shifts become nearly identical.

It may be tempting to ignore such small Doppler shifts altogether. After all, what effect can 8 and 2 Hz Doppler shifts have on a 2.45 GHz carrier? Plenty, as it turns out. The Doppler shifts affect only the *rate* of signal fading, but the *depth* of fading is determined by the relative strengths and phase relationship in the LOS and reflected signal paths. Figure 2-13 shows a plot of the change in RSSI as a function of distance as the receiver moves away from the transmitter. Because our reflected signal has a field strength of about 0.9 that of the LOS signal, every time the phases of the two arrivals cancel, there's about a 20 dB drop in RSSI, and every time the two phases reinforce each other, there's about a 5 dB increase in RSSI, compared to the strength of the LOS signal alone. The small-scale RSSI fluctuates rapidly at

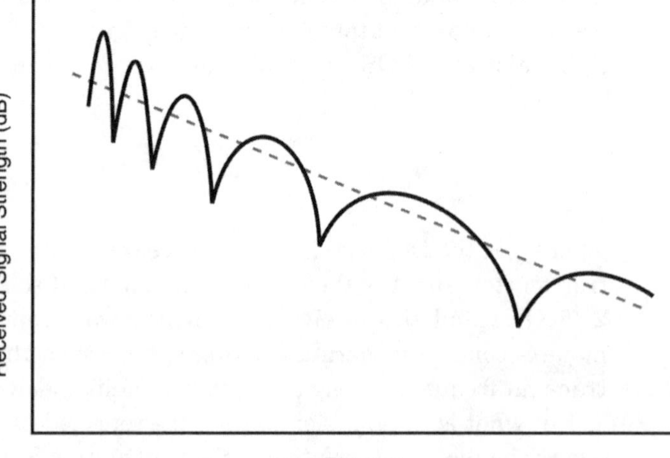

Figure 2-13 As the receiver in Figure 2-12 moves away from the transmitter, the rate of fading gradually reduces as the Doppler shifts of the two paths become nearly identical.

first, and then more slowly as the two Doppler shifts become closer to each other.

What happens to the phase relationship between the LOS and reflected waves as d becomes large? Intuitively, it seems that the two signals will gradually align in phase and reinforce each other, but that, unfortunately, is not the case. Instead, grazing reflections with the ground cause a 180 degree phase shift in the reflected wave, so the two actually combine destructively for large d. The result is a 40 dB per decade loss of signal strength for the composite waves compared to a 20 dB per decade loss over distance for the LOS wave alone.[8] Once again, multipath makes things worse than we had hoped. Fortunately, though, this phenomenon is far more common in outdoor cellular systems, with their larger distances, than in short-range indoor wireless.

In actual situations it's not necessary to know the Doppler shift in each of the many paths; instead, Doppler spread, defined as twice the largest Doppler shift, is used. A signal that fades over time does so because of Doppler spread, and this in turn implies that motion occurs somewhere. The movement can be the transmitter, the receiver, or any nearby object that can affect RSSI. Clearly, Doppler spread will exist to some extent in almost every Bluetooth piconet link. Figure 2-14(a) and (b) show the fading situation and equivalent Doppler spread when a transmitter and receiver move relative to each other. These motions and the resulting random multipath arrivals cause fades that can approach 30 dB at times.

If the transmitter and receiver are fixed relative to each other, then fading can still occur from the movement of nearby objects. This situation is shown in Figure 2-14(c) and (d). Now there is significant power in the carrier from the direct path, which doesn't have any Doppler shift. Consequently, the fading, although still occurring at about the same rate, isn't as deep and reliability is improved.

Outage Probability One of the first questions that comes to mind when scrutinizing Figure 2-14 is, How often will the fades be so severe that the incoming signal is too weak for the receiver to detect? Fortunately, the answer is rather easy to calculate provided the following assumptions hold:

- All frequencies within the bandwidth of the transmitted signal fade together.
- All incoming paths are subject to random Doppler shifts.

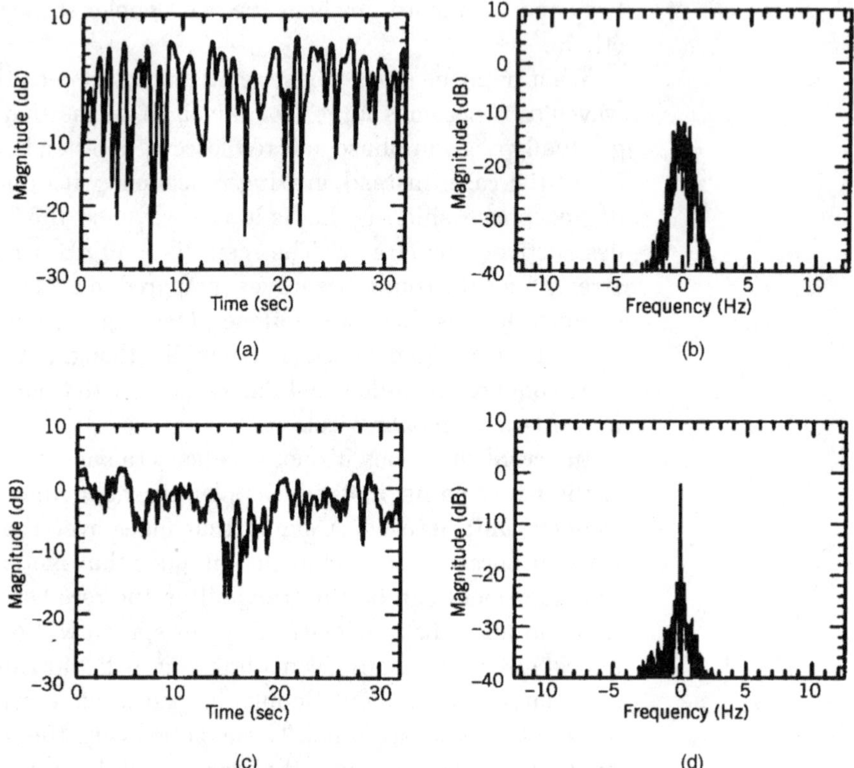

Figure 2-14 Doppler spread is portrayed in the time domain (a) and (c), and in the frequency domain (b) and (d). Fading can be more severe if the transmitter and receiver are moving relative to each other (a) and (b) than if their positions are fixed (c) and (d). (Source: S. Howard)[9]

The first assumption will be addressed in more detail in the section "Delay Spread," and the second assumption is satisfied by the situation depicted in Figure 2-14(a) and (b), where the transmitter and receiver are randomly moving relative to one another and perhaps other nearby objects are randomly moving as well. If these two assumptions hold, then the channel exhibits what is called *Rayleigh fading*.

For a Rayleigh-faded channel, the probability Pr(outage) that the signal power has faded below a certain threshold level P_{th} can be expressed in terms of the receiver's average signal power level P_{ave} by the formula[6]

$$\Pr(\text{outage}) = 1 - \exp\left(-\frac{P_{th}}{P_{ave}}\right) \tag{2-14}$$

If P_{th} and P_{ave} are expressed in dB units, then they must be converted back to watts before their ratio is taken. For example, suppose the average received power is –60 dBm and the receiver will lose the signal if it falls below –70 dBm. The actual ratio of the two powers is 0.1. According to Equation 2-14, the probability of outage in a Rayleigh-faded channel is about 0.1, so the channel will be unusable about 10 percent of the time.

The situation depicted in Figure 2-14(c) and (d) is a bit different. Because the transmitter and receiver are fixed relative to each other, the direct (LOS or OLOS) path between them has no random Doppler shift, and the direct signal is strong enough to dominate the Doppler-shifted multipath reflections from nearby moving objects. The channel now exhibits *Rician fading*, and as a result, the fades, although they still occur about as often, aren't as deep as they were in the Rayleigh case.

Fading characteristics in a Rician channel clearly depend upon the ratio of the power in the dominant path compared to the power in the Doppler-shifted paths. If the ratio is high, then fades are relatively shallow, and if the ratio is low, then fades are relatively deep. If the dominant path disappears entirely, then channel fading is Rayleigh. Consequently, Pr(outage) in a Rician channel will be less than that in a Rayleigh channel given the same average signal power at the receiver, so Equation 2-14 can be used to find an upper bound on this probability.

Coherence Time When a data symbol is sent from the transmitter to the receiver in a digital communication system, it would be nice if the signal representing that data symbol arrived at the receiver with minimum distortion. Sure, there will be amplitude changes and time delays along the way, but as long as these affect the entire transmitted signal equally, then the receiver has a good chance of detecting the data correctly. One way that distortion can creep into the transmitted signal is if significant amplitude fluctuations (fades) occur during the time a single data symbol is being sent. Therefore, it would be helpful if we could calculate the time during which RSSI remains essentially invariant, which is called the *coherence time*, and check to see if the symbol duration is less than the coherence time.

The coherence time is related to the reciprocal of the maximum Doppler shift, but there's a problem: The channel is continuously changing when fading is present, so a small amount of fading will occur even during the coherence time. The amount of tolerable fading will dictate the formula that is used for calculating the coherence time. One that has gained recent acceptance is given by[6]

$$Tc = \frac{0.4}{f_D} \tag{2-15}$$

where T_C is the coherence time and f_D is the maximum Doppler shift. In words, the coherence time is about 40 percent of the reciprocal of the maximum Doppler shift.

Now we can relate coherence time to the data symbol rate to determine whether the channel fades significantly during transmission of a data symbol. There are two possibilities:

- **Slow fading** The coherence time is greater than the symbol period, so the baseband signal (the symbols representing the data bits) varies faster than the channel does. This means that each symbol has nearly constant amplitude while it is being received, and distortion from fading is negligible.
- **Fast fading** The coherence time is less than the symbol period, so the channel varies faster than the baseband signal does. This means that each symbol has a high probability of experiencing a fade while it is being received, and distortion from these fades may be significant.

To reduce distortion from Doppler fading, therefore, a slow-faded channel is preferred. This can be achieved by transmitting a relatively fast symbol rate for a given set of fading conditions. Keep in mind, though, that even with a slow-faded channel the possibility exists that deep fades will occur from time to time, resulting in outage. This outage could affect several consecutive channel symbols when fading is slow.

Delay Spread

The second major way that multipath can affect communication performance can be characterized by the channel's delay spread. This is a measure of the spread of time it takes for reflected signals of significant amplitude to arrive at the receiver. Unlike Doppler spread, delay spread does not require any motion and is strictly a function of arrival times. Of course, if there is motion, then the delay spread characteristics will change with time.

Delay spread can be measured, at least theoretically, by transmitting a very brief pulse and plotting the arrival times of all incoming copies of the pulse at the receiver. Figure 2-15 shows an example of this in an indoor situation.[6] The first arrival is set to a time of 0 nanoseconds (ns) and an amplitude of 0 dB. Additional pulses that reflect off the various obstructions then arrive later with their own amplitudes, eventually tapering off into the noise floor or threshold level of the receiver.

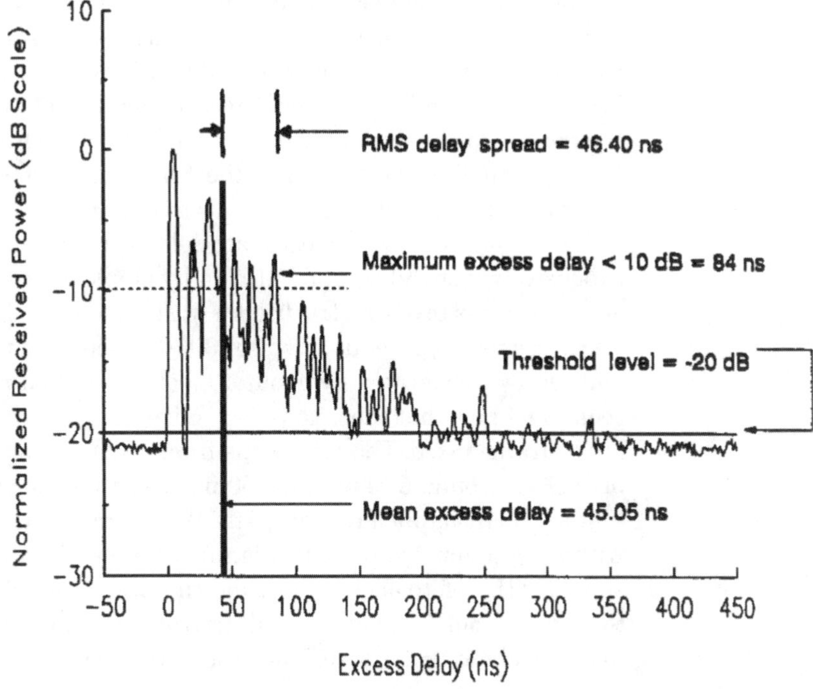

Figure 2-15 Delay spread is the time required for the various multipath components to arrive at the receiver. (Source: T. Rappaport)[6]

Characterizing delay spread is a bit more difficult to do than Doppler spread. For instance, when do we stop counting incoming pulses? Because their amplitudes taper off with time, it seems that eventually the pulses will be too weak to affect system performance. Also, Figure 2-15 is just an example of delay spread for one specific situation, and many of these plots must be generated from different locations and some sort of statistical processing must be accomplished on the results.

Delay spread can be quantified in one of two ways: *maximum excess delay* and *rms delay spread*. Maximum excess delay is the time between the first arrival and the last arrival that occurs above a particular threshold, usually 10 dB below the strongest. The quantity is easy to determine by simply looking at the excess delay plot. In Figure 2-15, for example, there are a total of six arrivals within 10 dB of the power in the strongest arrival, and the last occurs 84 ns after the first. Therefore, the maximum excess delay is 84 ns. This quantity has limited use, though, because it provides no indication of the number or relative amplitudes of the various arrivals within the 10 dB window. Consequently, two channels with the same excess delays can behave very differently.

The rms delay spread can alleviate some of these problems at the expense of a more complex calculation method. This is done by finding the second central moment about the mean excess delay. This quantity adjusts for the number, relative amplitudes, and times of arrival of the various multipath components. The calculation process can be found in Reference 6, but because I promised to limit the level of mathematical panic, it won't be repeated here. The rms delay spread for the arrivals in Figure 2-15 is 46.4 ns.[6]

Believe it or not, we can estimate delay spreads without the complex and expensive wideband equipment usually employed for such measurements. With apologies to our metric friends, just remember that light travels about 1 foot per ns in space. In a typical 30 $\times$ 30 foot room with the transmitter and receiver separated by about 10 feet and placed near the center of the room, we know that the first pulse arrives at the receiver about 10 ns after it was transmitted. The reflections from walls, ceiling, and floor will probably arrive around 5 to perhaps 50 ns after that, assuming only one reflection for each significant multipath. We assume that first-order reflections within the room are the only significant reflections present. Excess delay spread is therefore about 50 ns, and rms delay spread, although impossible to find without knowing actual arrival times and amplitudes, will most likely be about 35 ns or so. Now wasn't that easy?

Coherence Bandwidth Now that we have a grasp of delay spread fundamentals, it's time to look at the effect delay spread has on a digital communication system. Figure 2-16 depicts a transmitter and receiver in which the signal takes two paths: one is LOS and the other bounces off of a highly reflective surface. Let's disregard the phase relationship between the two signals for now, because that was covered in the section on Doppler spread, and instead concentrate on *timing*. Suppose the transmitter sends a 0

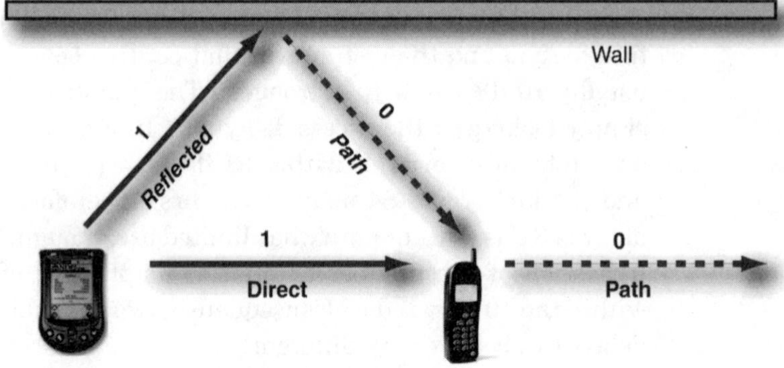

Figure 2-16 Delay spread can cause intersymbol interference (ISI) at the receiver, increasing the probability of bit error.

followed soon afterward by a 1. Furthermore, suppose that the symbol duration is the same as the time difference between signal arrivals on the LOS and reflected paths; that is, if the reflected signal arrives at the receiver 50 ns after the LOS signal, then the symbol duration is also 50 ns, corresponding to a data rate of 1/(50 ns), or 20 megabits per second (Mb/s).

First, the 0 appears at the receiver along the LOS path. Next, the 1 along the LOS path and the 0 from the reflected path appear *together* at the receiver, as shown in Figure 2-16. Finally, the 1 from the reflected path appears. Do you see the problem when the 1 first appears at the receiver? The receiver is trying to detect a new 1 from the LOS path and the previous 0 from the reflected path at the same time! Which symbol will win? If the 0 is detected over the 1, then a bit error occurs. This phenomenon is called *intersymbol interference* (ISI). ISI leads to *irreducible BERs* because increasing transmitter power won't improve the situation; the ratio between LOS and reflected signal powers remains unchanged.

We suspect that ISI is somehow related to the delay spread, and this relationship is determined by the *coherence bandwidth* of the channel. If significant ISI occurs, then the transmitted signal is distorted at the receiver, increasing the probability of bit error. The coherence bandwidth is a measure of the range of frequencies over which the channel response is *flat*, meaning that the distortion over the channel that results in ISI is negligible. The formula can be expressed as

$$B_c \approx \frac{0.1}{T_{rms}} \tag{2-16}$$

where B_C is the coherence bandwidth and T_{rms} is the rms delay spread. The coherence bandwidth is essentially the widest bandwidth that a transmitted signal can have with negligible ISI. The multiplier 0.1 can actually range between 0.02 and 0.2, depending upon the amount of correlation desired between the various parts of the transmitted signal bandwidth. Clearly, larger rooms (with larger rms delay spreads) have smaller coherence bandwidths.

The concept of coherence bandwidth can sometimes be more easily grasped by returning to the associated rms delay spread. In these terms, Equation 2-16 states that if the transmitted symbol duration is greater than 10 times the rms delay spread, then ISI will be small. In other words, the first 10 percent of the duration of a data symbol will be significantly affected by the previous symbol. The remaining 90 percent will be received in the clear because all of the significant reflections now contain the same data symbol.

Like its coherence time counterpart, coherence bandwidth can also be divided into two major categories:

- **Flat fading** The bandwidth of the transmitted signal is less than the channel coherence bandwidth. Equivalently, the symbol period is greater than 10 times the rms delay spread. The transmitted waveform is not significantly altered by multipath-induced ISI.
- **Frequency selective fading** The bandwidth of the transmitted signal is greater than the channel coherence bandwidth. Equivalently, the symbol period is less than 10 times the rms delay spread. The transmitted waveform may be significantly altered by multipath-induced ISI.

To reduce distortion from ISI, a flat-faded channel is preferred. This is accomplished for a given set of parameters by reducing the channel symbol rate. Incidentally, reducing the channel symbol rate doesn't necessarily require reducing the data rate, as we will discover in the next chapter.

Doppler Spread Versus Delay Spread

We can summarize the relationship between Doppler spread and delay spread in the following way:[6]

- Doppler spread leads to *frequency dispersion* and *time-selective fading*. Fading can be either *slow* or *fast*.
- Delay spread leads to *time dispersion* and *frequency-selective fading*. Fading can be either *flat* or *frequency selective*.

These processes are independent from each other, but both are caused by multiple signal paths.

Bluetooth Channel Fading Analysis

Finally, let's examine the Bluetooth channel for Doppler spread, delay spread, and outage probability.

Doppler Spread If a Bluetooth piconet exists in an indoor environment where objects can move at speeds up to a fast walk (1.3 meters/second or so), then maximum Doppler shifts will be about 10 Hz. According to Equation 2-15, the coherence time is about 40 ms. The Bluetooth baseband symbol rate is 1 μs, which is much smaller than the coherence time. Therefore, the Bluetooth channel is slow faded. In fact, it's *really* slow faded because about 40,000 data symbols can be sent during the coherence time.

Delay Spread To find the coherence bandwidth, let's assume that the rms delay spread in a rather large room is 50 ns. Equation 2-16 gives a corresponding coherence bandwidth of about 2 MHz. The maximum bandwidth of a Bluetooth signal during each hop is 1 MHz, so the Bluetooth channel is flat faded. Note that if the Bluetooth symbol rate is more than doubled (which could happen if higher data rates are adopted in a future specification), then the Bluetooth channel begins to experience frequency selective fading and a process called *equalization*, discussed later in this chapter, may become necessary to combat ISI.

Outage Probability Because the Bluetooth channel is flat faded, Equation 2-14 can be used to find the outage probability Pr(outage) in a Rayleigh faded channel if we know the average received power P_{ave} and the threshold signal power P_{th} below which the receiver will fail to operate with an acceptably low BER. The average receive power can be found using Equation 2-2 or one of its simplified variants, and the threshold signal power is provided by the manufacturer of the receiver being used.

To return to an earlier example, if the average received power is –60 dBm and if the RX will lose the signal if it falls below –70 dBm, then the probability of signal loss in a Rayleigh-faded channel is 0.1. If the piconet members move closer together so that their average received power increases to –50 dBm, then the probability of outage is reduced by an order of magnitude to 0.01. Also, if both members of the piconet are a fixed distance from each other with a strong LOS or OLOS path, then the channel will probably exhibit fading statistics that are Rician instead of Rayleigh, so the actual outage probability will be lower than that calculated using Equation 2-14.

Large-Scale Path Loss (PL) Using Multiple Ray Tracing

At this point we return for a moment to the problem of determining the large-scale PL in a particular indoor environment. It's clear by now that multipath plays a significant role in finding even the large-scale PL values because it is always present and affects every measurement taken despite our attempts to average them out. In fact, any adjustments that are made to the free-space PL relationship given by Equation 2-1 based upon actual measurements can't help but include the effects of multipath.

Since computing power has increased markedly over the last few years, software packages have been developed that can find large-scale PL values by considering not only the direct path, but several reflected paths as well for a particular building or room layout. This is done by looking at the reflected signals (rays) that pass from transmitter to receiver, with the number of rays limited by computational time and desired accuracy.

For higher frequencies, such as those used in Bluetooth, though, it's extremely difficult to track precise phase angles from these reflections. With a wavelength of only 0.112 meters, slight inaccuracies in the placement of building partitions within the CAD drawing will translate to gross inaccuracies in the vector combinations of the various signal paths at the receiver. Fortunately, research has shown that simply adding the powers in the multipath components is sufficient for large-scale PL calculations that are fairly close to those actually measured within the building.[10] Of course, even this technique cannot account for objects that move from time to time unless the simulation is run again.

Multipath Mitigation

There's good news and bad news in the results obtained from an analysis of multipath on the Bluetooth channel. The good news is that the channel is slow faded and flat faded, so there isn't a significant distortion problem caused by the channel on the transmitted signal. The bad news is that fade depth from Doppler spread can be as high as 30 dB, and it's unrealistic to simply increase transmit power by a factor of 1,000 to compensate. Therefore, we must come up with a better way.

Multipath mitigation, which is fancy terminology for "fixing it," can fall into several major categories. Among these are including a fade margin into PL calculations, diversity combining, equalization, and error control.

Fade Margin

The easiest way to compensate for multipath fading effects is to just live with them. This is done by building a *fade margin* into the PL equations when estimating maximum communication range. As we discovered in the section on Doppler effects, fading can occasionally be as high as 30 dB, so by subtracting that amount from the maximum acceptable PL, the receiver will still have (barely) adequate signal strength during these periods of deep fades to maintain the link.

Including a high fade margin is often used in outdoor communication systems with high-power transmitters, such as paging systems, because there's room to include relatively high fade margins in the model. For example, if a 100 W (+50 dBm) paging transmitter links to receivers that have a sensitivity of –90 dBm, then the maximum acceptable PL, including a 30 dB fade margin, is still 110 dB. For low-power indoor wireless, though, there isn't enough signal strength overhead to accommodate fade margins that high. Suppose our typical TX power is 0 dBm and the associated receiver sensitivity is –70 dBm. Maximum PL with a 30 dB fade margin is only 40 dB, and we know from previous analysis that a 2.4 GHz wave in free space has a PL of 40 dB over a distance of just 1 meter.

Although we can't include fade margins high enough for the deep fades, it certainly makes sense to use 10 dB or so of wiggle room in the PL calculations to account for shallow fades and other inefficiencies, as we've discussed earlier.

Diversity

One of the most versatile ways to compensate for multipath is through *diversity combining*, which is a process that adds redundancy somewhere in the communication system such that the effect of multipath is reduced. Diversity can be in space, time, angle, frequency, or path.

Space diversity exploits the fact that deep fades usually occur over a relatively small physical space, often over a distance of less than one-half wavelength in any direction.[11] Therefore, if a receiver employs multiple antennas placed far enough apart, then the probability is very high that at least some of these antennas will capture strong signals. The receiver could then select the antenna with the strongest signal (*selection diversity*) or simply check the antennas in sequence until one is found where the signal is strong enough to yield the desired performance (*scanning diversity*). BERs can be reduced further by using signals at all of the antennas. The signals are first phase aligned and then their contributions are combined after each has its gain set according to its signal strength (*maximal ratio combining*), or all antenna elements are given equal gain (*equal gain combining*) to reduce signal processing requirements. Multiple antennas can also be used at the transmitter, but very sophisticated signal processing is required at the receiver to realize a significant performance improvement from this arrangement.

Time diversity is based on the assumption that fades last a relatively short time. If a set of data is repeated over time, then the probability is high that at least one of the data sets will be successfully received. Rather than blindly transmitting data over and over, a process called *Automatic Repeat Request* (ARQ) is often used, where only the transmissions received in error are repeated.

Angle diversity uses a directional antenna to enhance the capability of a receiver to hear better in the direction of the strongest signal and to reject signals arriving from other directions. In this way, the detrimental phase cancellation effect of multipath is reduced. An added benefit is that interfering transmissions from other sources can also be reduced at a receiver using angle diversity. Incorporating angle diversity into a portable device using Bluetooth is a challenge, though, because antenna directivity must compensate for the device's change in orientation during use.

Frequency diversity works by transmitting the same information using several different carrier frequencies, each with independent fading. If a deep fade exists on one of the frequencies, the probability is low that a fade exists on the other frequencies too. A benefit of the Bluetooth *frequency hopping spread spectrum* (FHSS) ARQ process is that transmissions that aren't received correctly by a piconet member are retransmitted on a different hop frequency and at a later time. Does Bluetooth have a wide enough hopping frequency range to gain the benefit of frequency diversity? We'll soon find out.

Path diversity uses a process called *direct sequence spread spectrum* to enable a receiver to actually discern between different copies of the same transmitted signal separated by only a few nanoseconds in time. As such, the receiver is no longer held hostage to the possibility of multipath phase cancellation and the resulting deep fades. Instead, the receiver can pick out the strongest signal and decode it, or even combine several incoming copies into a composite from which the transmitted data bits can be extracted with high reliability. The latter process is used in a device called a *RAKE receiver*, which is implemented in *Code Division Multiple Access* (CDMA) cellular phones.

As signal-processing techniques improve along with associated cost and power reductions, this processing power will certainly find its way into Bluetooth devices. As such, we can expect to see more sophisticated diversity techniques being employed to combat interference and multipath.

Diversity Combining Using Two Antennas One of the most powerful methods for combating multipath is with a simple space diversity technique that uses two antennas at the receiver. Inside, a circuit monitors the rela-

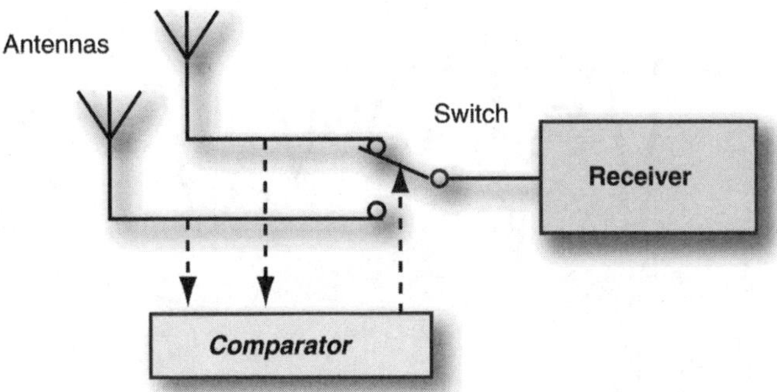

Figure 2-17 Block diagram of a receiver using selection combining with two antennas

tive signal strengths on each antenna and switches the one with the higher signal level into the receiver front end (see Figure 2-17). This rendition of selection combining is common in devices such as wireless microphones and some 802.11b Wi-Fi WLAN implementations.

Considerable performance improvement is possible with two receive antennas separated far enough in space that they experience independent fading processes. If path clutter is evenly distributed between the transmitter and receiver, then this separation can be as small as one-half wavelength.[11] Because a deep fade lasts a short time compared to the time between fades, there is only a small probability that both antennas will be deeply faded at the same time. Figure 2-18 shows typical independent Doppler fades at each of the two antennas, along with a trace of the fading that occurs at a receiver using selection combining. The average received signal strength is normalized to 0 dB. Notice that although each antenna occasionally experiences fades in excess of 20 dB, the selection combining process limits fades to about 5 dB.

At Bluetooth frequencies, two antennas must be separated by a minimum of 6 cm for them to collect independently faded signals. Unfortunately, even that small spacing may not be practical in many Bluetooth applications. However, as long as the two antennas are far enough apart that they're not deeply faded together, then receiver bit error performance will be improved over using a single antenna. When two antennas are this close together, they can influence each other to the point where they become somewhat directive, though, so it's important to measure the relative sensitivity of the antenna system from several different directions to ensure adequate coverage.

Bluetooth FHSS as an Antimultipath Technique The Bluetooth frequency hopping channel set ranges from 2,402 to 2,480 MHz in 1 MHz steps, so a total bandwidth of about 79 MHz is eventually covered during

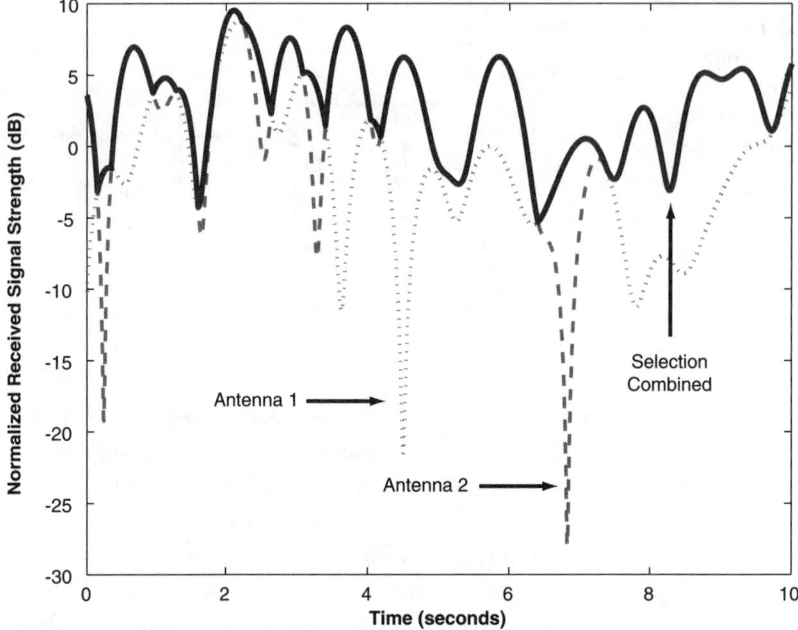

Figure 2-18 Selection combining, even with only two receive antennas, reduces fading significantly.

a long hopping sequence. The transmitted bandwidth on any one hop frequency is about 1 MHz. Because Bluetooth uses FHSS, it would be interesting to discover if frequency hopping can be employed successfully as an antimultipath technique similar to frequency diversity. If one set of data is sent on a frequency that's in a deep fade, will a hop to a new frequency change the phase relationships of the received signal paths enough that RSSI improves significantly? Or do all the Bluetooth channels typically fade together?

Before embarking on a detailed quantitative analysis using reflected rays, recall that the coherence bandwidth for a fairly large room at Bluetooth frequencies was about 2 MHz, which is much less than 79 MHz. That tells us that portions of the 79 MHz band that are more than 2 MHz apart fade somewhat independently in this environment. So far, so good. Recall, however, that the coherence bandwidth increases for smaller rooms, and the coefficient in Equation 2-16 could be as high as 0.2 if we want to apply more generous correlation criteria. So let's throw caution to the wind and consider the coherence bandwidth of a rather small room with an rms delay spread of only 10 ns. The coherence bandwidth, using the broad criteria, is 0.2 / 10 ns = 20 MHz, so we're still fairly encouraged that channels separated by more than 20 MHz may still fade independently. Of course, delay spreads that are even smaller would produce still wider coherence bandwidths, but the room size under such conditions would be rather small, and the piconet participants would be so close together that fades deep enough to shut down the link will occur very rarely.

Now that we've discovered that, even under very pessimistic conditions, the entire Bluetooth channel set probably won't fade at the same time, it's still important to determine whether hopping out of a channel affected by a deep fade will put us into another channel that has useable signal strength. A comprehensive analysis would be quite difficult (not to mention rather boring), so let's set up an example using a realistic situation instead.

Consider a point-to-point Bluetooth piconet with the users separated by a direct path of 3 meters and a single reflected path of 6 meters total length (see Figure 2-19). For simplicity, we'll assume that the paths produce equal field strengths at the receiver, so if they arrive at the receiver with 180 degrees of phase difference, then they cancel each other completely, a *really* deep fade. Now let's assume that we are no longer deeply faded when the two RF fields combine to produce a fade of 10 dB or less compared to the power in one of the paths alone. Trigonometry tells us that this situation occurs if the phase difference between the two waves changes by about 18 degrees or more on either side of 180—that is, if the phase difference is outside of the range 162 to 198 degrees. Now comes the fundamental question: How much must the hop carrier frequency change before these phase criteria are met?

The easiest way to approach this is to note that the reflected path is about 25λ longer than the direct path at Bluetooth frequencies, so if the hop frequency increases such that each of these 25 wavelengths is 18/25 degree shorter, then the phase change criteria will be met, just barely, over 25 wavelengths. This corresponds to a frequency increase of about 5 MHz;

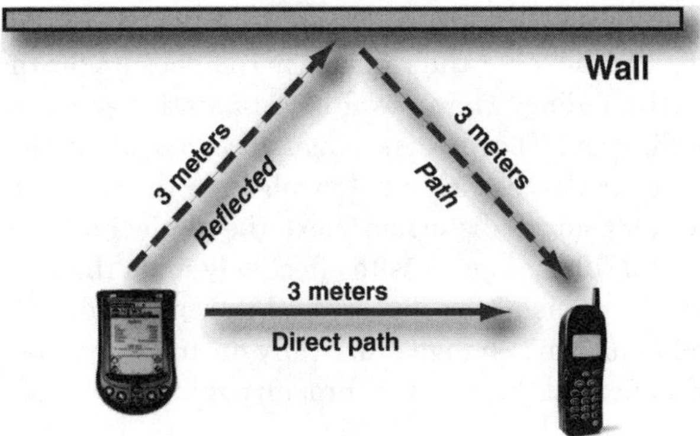

Figure 2-19 Two piconet members separated by a 3-meter direct path and a 6-meter reflected path

likewise, a 5 MHz decrease in hop frequency will also just meet the criteria for escaping the deep fade. Because members of the Bluetooth piconet hop randomly within the 79 channel set, the probability of escaping the deep fade during the next hop is approximately [(79 – 10)/79], or 75 percent in this particular situation. Of course, longer or shorter path-length differences will increase or decrease the probability of escape, respectively.

Although the previous analysis is crudely simplified, it does tell us that the Bluetooth piconet members can, under realistic situations, hop away from a deep fade. Of course, there's also a chance that the members will hop back into the fade, so deep fades within the Bluetooth channel set will reduce the rate at which data can be transferred from one piconet member to another unless another antimultipath technique such as selection combining is also used.

Incidentally, the antenna diversity method of selection combining is made more complicated by frequency hopping. We know that the probability is high that each hop will be independently faded, so antenna selection must be reaccomplished at the beginning of each hop for best performance. However, because the dwell time in each hop frequency is far less than the channel coherence time, the same antenna can be used for the entire time spent on each hop frequency.

Equalization

As we've discovered earlier, both fast-faded and frequency-selective-faded channels distort the transmitted signal en route to the receiver, so it would certainly be nice if the multipath situation always gave us slow, flat-faded channels instead. We also discovered that transmitting higher data rates in accordance with some future Bluetooth specification may indeed transform the flat-faded channel into one that is frequency-selective despite our wishes to the contrary.

Is it possible for the receiver to compensate for the effects of frequency-selective fading? The answer is yes, and this is done through the process of equalization. The process usually begins when the transmitter sends a known sequence of channel symbols to the receiver, where they are analyzed for channel distortion. Next, the receiver adjusts the characteristics of a digital filtering process to effectively undo the distortion in the channel. Now the transmitter can proceed to send its data, and provided channel conditions don't change drastically during this time, bit error probabilities are reduced at the receiver through equalization. The training sequence can be repeated periodically, if necessary, for the receiver to readjust its equalizer to fast-changing channel characteristics.

Furthermore, the transmitter itself could predistort its signal prior to transmission, and this distortion is (hopefully) canceled out by the distortion introduced by the channel. This alternate equalization process is more complex than receive-only equalization, though, because the receiver of the data must periodically and efficiently send its channel distortion assessment to the transmitter and, of course, the channel must remain relatively stable between receiver assessment periods.

Error Control

Error control is a process that compensates for channel imperfections by providing several ways for the communication link to improve communication reliability. In general, error control attaches redundancy to the transmitted bits that can be used to improve the BER at the receiver. Somewhat surprisingly, this improvement can be made arbitrarily high as long as arbitrarily long delays can be accepted. Error control can be implemented in three general ways, all of which will be examined in greater detail in the next chapter:

- **Error detection codes** Additional bits that are sent as part of a data string that the receiver uses to check the accuracy of the incoming message. The receiver performs an efficient mathematical test out of which a pass-fail verdict is reached.
- **Error correction codes** Additional bits that are sent as part of a data string that the receiver uses to actually correct errors that may appear in the incoming message. These codes are often substantially more complex than those used for error detection, but their capability to correct bad data makes them more powerful as well.
- **ARQ** Used in conjunction with error detection codes to enable a receiver to request a bad set of data to be retransmitted. If a set of incoming data is good, then the receiving node returns an *acknowledgment* (ACK) to the transmitting node; otherwise, the receiving node returns a *negative acknowledgment* (NAK). If an ACK is returned, the transmitting node continues with the next part of the message; otherwise, the last part of the message is sent again.

Bluetooth usually uses all three of these error control methods when piconet members exchange data. First, the error correction code is applied to an incoming data set to correct as many errors as possible, then the data set is checked by the error detection code for any remaining errors. Finally, an ACK is returned if all is well. As we will discover in the next chapter, each message and each ACK or NAK is transmitted on a different hop frequency. This ingenious technique enables the Bluetooth piconet to combine advantages of the three error control methods with both frequency and time diversity to combat multipath and other imperfections in the wireless communication process.

Antennas for Bluetooth

No propagation study would be complete until the effects of the transmit and receive antennas are included. Up to this point we've assumed that the transmit antenna effectively radiates all of the power presented to it, and the receive antenna couples all of its power to the receiver's input stage for many of our PL calculations. Obviously, these idealized criteria cannot be met, so it's important to discover where antenna inefficiencies originate and how they can be minimized.

The purpose of the antenna is to radiate signal power from the transmitter into space and gather signal power from space and send it to the receiver. The antenna has several characteristics, such as gain and directivity, that model its operation, and fortunately, these characteristics are similar whether the antenna is used for transmitting or receiving. Because the same antenna is usually switched between the transmitter and receiver, we will concentrate on antenna performance during signal transmission.

We start by examining the reasons behind power coupling losses between the transmitter and antenna and between the antenna and receiver. Next, the antenna's radiation pattern and antenna placement issues are discussed, followed by some examples of antennas designed for the 2.4 GHz frequency band.

Power Transfer and Impedance Matching

When a signal of incident power P_i is sent to an antenna, some of the power (P_t) is coupled into free space and some (P_r) is reflected back to the transmitter, as shown in Figure 2-20. At the receiver, the antenna sends power P'_i to the receiver's input stage, of which P'_t is actually coupled to the circuit and P'_r is reflected back to the antenna. The ratio P_i/P_t at the transmitter, or P'_i/P'_t at the receiver, is called the *mismatch loss*, and (neglecting heating losses, which are usually small) lower mismatch losses mean that more power is radiated into space at the transmitter or coupled to the input stage at the receiver. If the mismatch loss is 3 dB at the transmitter, for example, then half of the incident power is reflected back to the transmitter and the other half is radiated. If both transmitter and receiver have 3 dB mismatch losses in their antennas, then we can assume that there's 6 dB signal loss at the receive end of the communication link from these mismatches.

For maximum radiated power (mismatch loss of 0 dB), the antenna must be *impedance matched* to the device connected to it. Most RF circuits are designed for a 50 ohm impedance, so for best performance the antennas should also have a 50 ohm feed point impedance. The study of antenna impedance can become involved mathematically due to the presence of phase shifts between the voltage and current values. If an antenna design fails to present an input impedance of 50 ohms, capacitors and/or inductors can be added to the terminals such that the transmitter sees a 50 ohm impedance and thus transfers maximum power to the antenna for radiation. These components are called the *matching network*.

Radiation Pattern and Antenna Placement

An ideal omnidirectional antenna will radiate its signal equally well in any direction, and this antenna is called an *isotropic source*. A physical antenna never performs like this, but instead will always exhibit *directivity*. Thus, in some directions the radiated signal power is less than that of the isotropic source, and sometimes (but not always) in other directions, the physical antenna's radiated signal power exceeds that of the isotropic source. Antenna power measured relative to an isotropic source is called the *effective isotropic radiated power* (EIRP) with units given in dBi. The *gain* of an

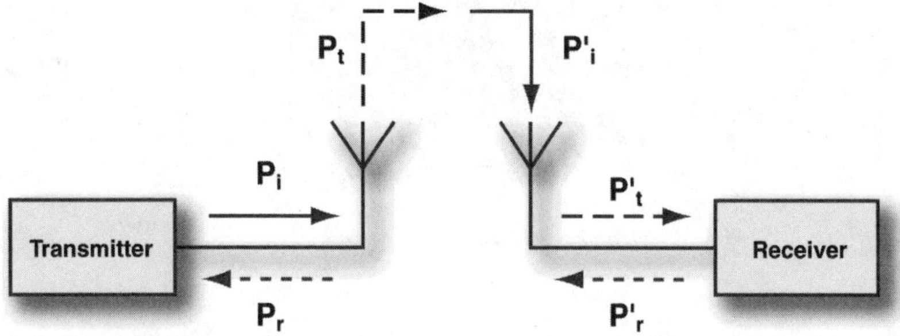

Figure 2-20 Power sent to the antenna is either transmitted or reflected back to the source.

antenna is the maximum power that the antenna radiates relative to that from an isotropic source. For example, if an antenna has a gain of 10 dBi, then in the direction of maximum signal strength the power radiated is 10 dB above that of the isotropic source. It's important to realize, however, that antennas don't violate the laws of physics, and their gain is simply a concentration of power in a particular direction. Consequently, there are always other directions in which the gain antenna will actually perform worse than the isotropic source.

The radiation pattern of any antenna is three-dimensional in nature (see Figure 2-21), but is usually expressed as a set of two polar plots: one for azimuth and one for elevation. Sometimes these two plots are combined into a single set of axes. The plots will often be normalized to a maximum

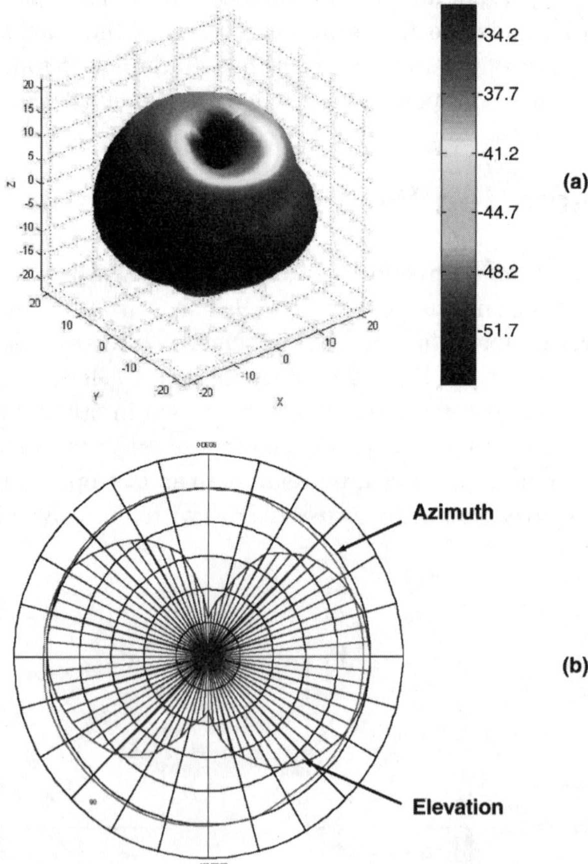

Figure 2-21 Three-dimensional radiation pattern (a) and combined azimuth and elevation plots (b) of a vertically mounted half-wave dipole antenna. (Source: GigaAnt)

radiation power of 0 dB so the power transmitted in other directions can be easily compared to this maximum. The dipole has a maximum gain of 2.14 dBi, and that occurs in the azimuth plane at 0 degrees of elevation for a vertically oriented antenna. Signal strength is lower at other elevation angles, and becomes very low at elevation angles approaching +90 or –90 degrees. Incidentally, sometimes antenna gains are expressed relative to the dipole (dBd) rather than relative to an isotropic source (dBi). To convert dBd to dBi, just add 2.14 dB.

Because all physical antennas have directivity to some degree, signal strength along a path will be affected, sometimes greatly, by the orientation of the wireless device containing that antenna. Fixed devices, such as desktop computers, can perform well with antennas that produce good signals over 360 degrees of azimuth and in the upper 90 degrees of elevation because most links will be in those directions. Portable wireless devices, especially those that operate while they're being held, have much more difficulty maintaining a reliable wireless link for all possible orientations, especially near the range limit.

When objects, such as people, are close to Bluetooth antennas, they can significantly reduce performance in two major ways: RF absorption and antenna impedance changes. Although we've already studied partition attenuations in some detail, partitions located very close to the transmitter can reduce signal power over a relatively large arc, adversely affecting a wide coverage area. Furthermore, antenna impedance changes caused by the nearby objects will almost always increase the mismatch loss, further decreasing the power radiated by the antenna. The Bluetooth headset is an example of such a device. Radiated power levels are significantly reduced in azimuth directions that require the signal to pass through the user's head.

Bluetooth Antenna Implementation

There has been a trend started by the cell phone manufacturers to conceal the presence of the antenna. Remember when cell phones had long, fixed antennas? These often got in the way and were easily broken. Next, phones were built with pull-out antennas that supposedly could be left in the down position if the cell signal was adequate. That didn't happen often, so the extended antenna was still subject to breakage. Phones were then built with very small exposed antennas that were rugged and fairly unobtrusive. Finally, phones began appearing in which the antennas were completely concealed.

Many of the lessons learned in the design of cellular phone antennas can be applied to Bluetooth as well. The selection criteria consists of the following:

- Should the antenna be internal or external?
- What physical size is acceptable?
- What radiation pattern should it have?
- What are the placement restrictions?

Bluetooth antennas generally fall into three design categories: the half-wave antenna, the *planar inverted F antenna* (PIFA), and the patch antenna. The first one is usually quite obvious as a rigid or flexible plastic rod protruding from the device's case. The second and third antenna types are built on the *printed circuit board* (PCB) and are typically concealed from view. A printed circuit style of antenna can even be placed upon the surface of the Bluetooth radio *integrated circuit* (IC).

One of the significant drawbacks of antennas that are placed near or on a circuit board and over the board's ground plane is that their radiation pattern is usually confined to the hemisphere above the board, and very little signal is transmitted or received through the board itself. If a nearby ground plane isn't needed for proper antenna operation (the half-wave dipole antenna is an example), then the antenna can be mounted on a section of the PCB where the ground-side conductive surface has been removed. Now the antenna can efficiently radiate through the board itself.

Half-Wave Antenna Construction of the half-wave antenna consists of a wire or conductive strip that is (surprise) one-half wavelength long. As a half-wave dipole, the antenna is fed at its center, with one side connected to the transmitter output (or receiver input) and the other to ground. At this center point the antenna's impedance is close to 50 ohms. Alternatively, the antenna can be fed at or near one end, but then a suitable matching network is needed.

An example of a Bluetooth half-wave antenna, along with its radiation pattern, is shown in Figure 2-22. This antenna is about 5 cm tall, which is shorter than a half wave at 2.45 GHz in free space (6.1 cm), but the use of special materials for the antenna's construction, along with a phenomenon called *end effect*, can make an antenna appear longer to RF than it really is.

The antenna has good omnidirectional characteristics in the azimuth plane at about 0 dBi. There is about a 2 dB loss compared to an ideal dipole, probably due to its 75 percent efficiency rating. The elevation plane shows

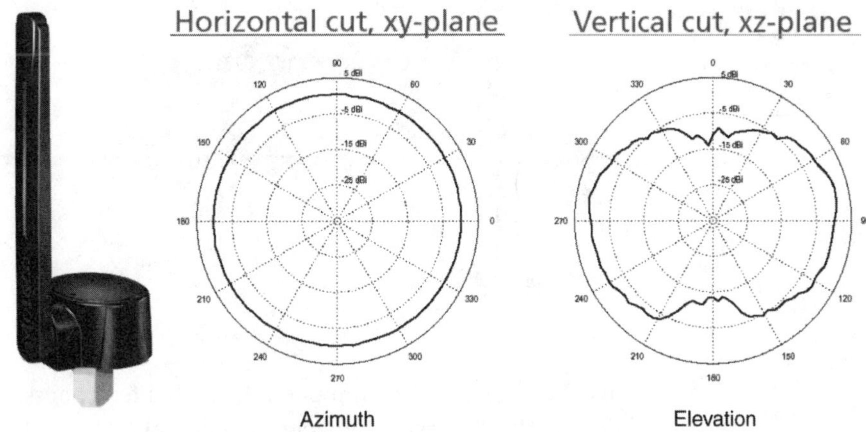

Figure 2-22 A half-wave antenna at 2.4 GHz, along with its azimuth and elevation radiation patterns (Source: GigaAnt)

the typical drop in signal level at angles approaching +90 and –90 degrees, but in this case antenna inefficiencies work in our favor and the actual antenna performs better than an ideal dipole at these high elevation angles. Other renditions of the half-wave antenna can be mounted completely inside the Bluetooth device, but some efficiency is usually sacrificed in such a configuration due to the proximity of other components.

PIFA The planar inverted F antenna is a simple metal structure that has the shape of the letter *F* that appears to be inverted when the antenna stands vertically. Figure 2-23 shows the antenna in its horizontal position as viewed from the side. For 2.4 GHz, typical dimensions are 25 mm long, 4 mm wide, and 1 mm high, but the actual dimensions depend upon the type of material in the PCB and the proximity of other components.[12] The PCB upon which this antenna is attached usually has the traditional two-layer RF construction, where the lower layer is a ground plane (ground side) and the upper layer contains most of the circuit traces (component side). The feed point of the PIFA is attached to the upper layer, and the ground contact passes through the board and connects to the ground plane.

Patch Antenna The patch antenna consists of a pattern etched or placed onto the component side of a PCB and is separated from the ground plane side by the board itself. The etching process enables almost any two-dimensional shape to be constructed, but most patch antennas are simple squares, rectangles, or strips. In fact, the PIFA pattern can be etched into the board as a patch antenna. A square patch antenna at 2.4 GHz might

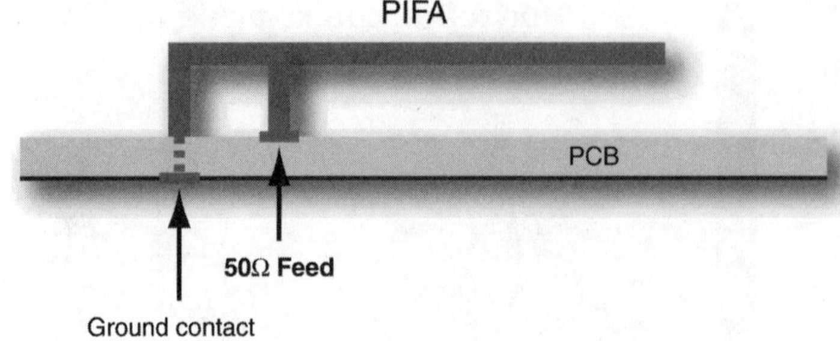

Figure 2-23 Typical construction of the PIFA

have dimensions of 37 mm per side, with a feed point near one of the edges. Like the PIFA, these dimensions depend heavily on the type of material out of which the PCB is constructed and the effect on the antenna from nearby components.[13]

One of the shortcomings of the patch antenna is its relatively narrow bandwidth before the mismatch losses become significant, so even if the antenna is optimized at 2.45 GHz, mismatch losses may approach 3 dB at the band edges. Furthermore, even away from the influence of other components on the circuit board, patch antennas can be quite directional in the azimuth plane as well as the expected directivity in the elevation plane from the board itself (see Figure 2-24).

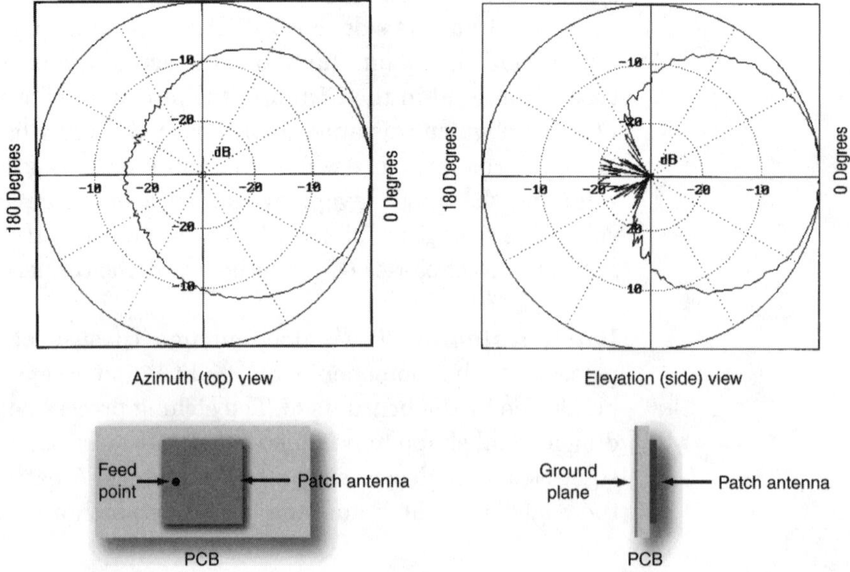

Figure 2-24 Radiation patterns from a horizontal square patch antenna at 2.45 GHz (Source: M. Amman)[13]

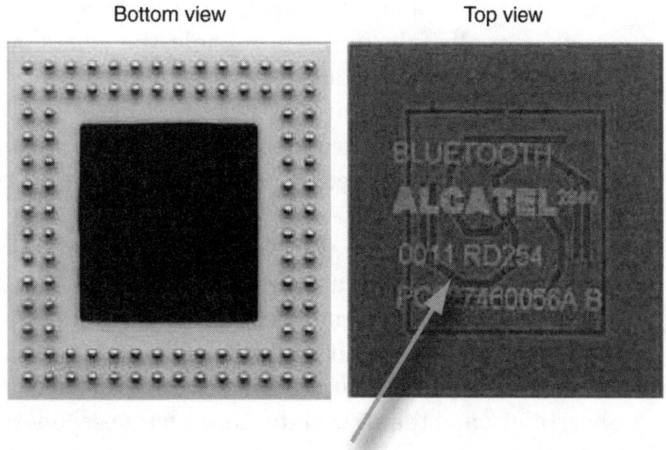

Figure 2-25 Bluetooth radio and controller IC can also include the antenna itself. (Source: Alcatel)

On-chip Antenna Several manufacturers offer complete Bluetooth radio transceivers as a single IC, and others even include some of the digital control circuitry with the radio. If the physical size of the IC is large enough, the antenna can also be part of the IC package (see Figure 2-25). By selecting this style of IC, the Bluetooth systems engineer doesn't need to design an antenna, feed mechanism, and matching network. Drawbacks include the need for mounting the entire chip in a location suitable for the antenna and the requirement that the chip size remain large enough to incorporate a reasonably efficient antenna. Furthermore, the inner workings of the IC must be carefully shielded to prevent interaction with the antenna.

Summary

Bluetooth is meant to be a dynamic, ad-hoc network, so it's going to be used under vastly varying conditions with both fixed and portable piconet members. Formal propagation modeling to determine the range of a Bluetooth device is most applicable for fixed devices such as a printer, desktop computer, or LAN access point. The analysis will help determine where a device should be placed within a room or building for best access by its intended users and what specifications the transmitter, receiver, and antenna should have.

For portable Bluetooth devices, the models are less useful in a formal sense. After all, if the link is unreliable, then just move closer. On the other hand, the quick range calculations facilitated by the simplified PL Equation 2-4 can prevent unrealistic expectations. Bluetooth users shouldn't expect, for example, to use their 1 mW headset to communicate throughout an office building; indeed, reaching the next room could be a hit-or-miss proposition.

The type of antenna also plays a major role in determining the useful range. Many concealed antennas are mounted on a PCB, and these often have very limited performance in directions that require them to transmit or receive through the board itself.

It's also important to realize that propagation analysis is filled with uncertainty, and that translates into a wireless link that won't always meet user expectations. Fortunately, even the nontechnical among us have already discovered that the cellular telephone has its limitations in coverage and reliability, so no one should be surprised when Bluetooth occasionally fails as well.

End Notes

1. Kim, S., et al. "Pulse Propagation Characteristics at 2.4 GHz Inside Buildings," *IEEE Transactions on Vehicular Technology*, August 1996.
2. Bansal, R. "The Far-Field: How Far is Far?" *Applied Microwave & Wireless,* November 1999.
3. Anderson, J., et al. "Propagation Measurements and Models for Wireless Communication Channels," *IEEE Communications Magazine*, November 1994.
4. de Toledo, A. and Turkmani, M. "Estimating Coverage of Radio Transmission into and within Buildings at 900, 1800, and 2300 MHz," *IEEE Personal Communications Magazine*, April 1998.
5. Seidel, S. and Rappaport, T. "914 MHz Path Loss Prediction Models for Indoor Wireless Communications in Multifloored Buildings," *IEEE Transactions on Antennas and Propagation*, February 1992.
6. Rappaport, T., *Wireless Communications: Principles & Practice,* 1st ed., Upper Saddle River, NJ: Prentice Hall, 1996.
7. Seidel, S. and Rappaport, T. "Site-Specific Propagation Prediction for Wireless In-Building Personal Communication System Design," *IEEE Transactions on Vehicular Technology*, November 1994.

8. Pahlavan, K. and Levesque, A. *Wireless Information Networks*, New York: John Wiley & Sons, 1995.

9. Howard, S. and Pahlavan, K. "Doppler Spread Measurements of Indoor Radio Channel," IEE *Electronics Letters*, 18th January 1990.

10. Valenzuela, R., et al. "Estimating Local Mean Signal Strength of Indoor Multipath Propagation," *IEEE Transactions on Vehicular Technology*, February 1997.

11. Jakes, W. "A Comparison of Specific Space Diversity Techniques for Reduction of Fast Fading in UHF Mobile Radio Systems," *IEEE Transactions on Vehicular Technology*, November 1971.

12. Kurz, H. P. "The Bluetooth Radio and Antennas," *SIGnal Newsletter*, February 2000.

13. Amman, M. "Design of Rectangular Microstrip Patch Antennas for the 2.4 GHz Band," *Applied Microwave & Wireless,* November/December 1997.

CHAPTER 3

The Bluetooth Radio

The radio is the device in a Bluetooth system that enables communication to occur without wires. As shown in Figure 3-1, the radio is also at the bottom end of the Bluetooth protocol stack. Indeed, all of the blocks in the protocol stack that are above the radio are implemented in one way or another in wired networks as well, but the radio itself is unique to wireless.

Despite the importance of the radio to wireless communication, the Bluetooth specification is surprisingly sparse when discussing this part of the system. The specification dictates *what* the radio must do, but is silent on *how* those requirements are to be implemented. Many of these requirements are actually quite difficult to accomplish, so manufacturers have developed a substantial amount of *intellectual property* (IP) associated with the design of an inexpensive Bluetooth radio on a single chip. Many of these designs are carefully guarded secrets. The datasheet itself is often available only under a *nondisclosure agreement* (NDA) and even then only to agencies that are actively participating in Bluetooth development work. Some of these fabrication processes and architecture issues will be discussed in Chapter 12, "Module Fabrication, Integration, Test, and Qualification."

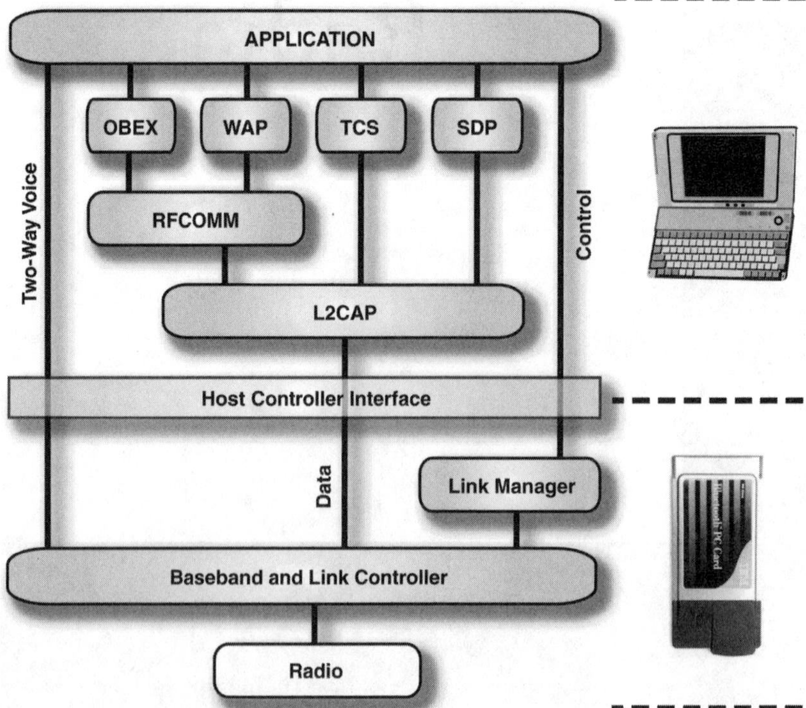

Figure 3-1 The radio is at the bottom end of the Bluetooth protocol stack.

Because the purpose of the radio wave is to convey information from one point to another, we examine some of the ways in which this information can be modulated onto the carrier and the relative performance of each. The advantages of *Gaussian-filtered frequency-shift keying* (GFSK) used in Bluetooth are presented, followed by a look at how faster data rates can be achieved. Next, we develop some of the groundwork in determining the best possible performance of a wireless communication system, measured by the maximum data rate given an arbitrarily small bit error probability.

In Chapter 2, "Indoor Radio Propagation and Bluetooth Useful Range," we mentioned that interference will play a greater role in determining the performance of wireless devices in the 2.4 GHz band. We will point out in this chapter how interference can affect the *bit error rate* (BER) at a receiver. Bluetooth devices attempt to avoid partial-band interference though *frequency hopping spread spectrum* (FHSS), and we show how this is done and give some simple formulas for calculating the probability that interference will occur from other users. Finally, the Bluetooth transmitter and receiver specifications are presented.

Placing Information onto the Radio RF Carrier

In digital communication systems the data (consisting of binary digits, or *bits*) is represented in some way as electronic signals that are sent from transmitter to receiver. Perhaps the simplest representation is to let the binary 1 be a high voltage and a 0 be a low voltage. These voltages can be sent one by one down a wire from transmitter to receiver. A sequence of raw data represented in this way is called the *baseband signal*.

More effort is needed, though, for wireless communication to take place. The baseband signal must alter an RF carrier wave at the transmitter through a process called *modulation*. At the receiver, the baseband data is extracted from the carrier through *demodulation*. Modulation and demodulation require additional circuitry, but are needed for the following reasons:

- The frequencies in the baseband signal are not high enough to propagate over reasonable distances.
- Antennas would be excessively long at baseband frequencies.
- If everyone transmitted their information at baseband, then there would be massive interference and pandemonium.

In this section we will examine the features of some of the modulation techniques used for digital communication so that the reasons behind the method chosen for Bluetooth will become apparent.

Amplitude-, Frequency-, and Phase-Shift Keying

An RF carrier is represented mathematically as a sinusoid having amplitude, frequency, and phase. Any one of these three quantities or a combination of them can be controlled by the baseband data signal. Because a data bit can be either a binary 1 or 0, a set of two different signals composes the simplest possible modulation scheme. One of these is called $s_1(t)$, corresponding to a 1, and the other is called $s_0(t)$, corresponding to a 0. The receiver attempts to determine which of these was actually sent by processing a copy of the transmitted signal that often becomes extremely weak and noisy during its journey to the receiver. Occasionally, the noise will be such that the receiver will mistake $s_0(t)$ for $s_1(t)$ or vice versa, and a bit error occurs. The bit error probability, also called the BER, is an extremely important measure of the performance of a digital communication system. We will now examine the three most common simple modulation schemes and compare their respective BER values.

Amplitude-Shift Keying (ASK) If the transmitted signal is amplitude-shift keyed, then it can be expressed mathematically as

$$s_1(t) = a_1 \cos[2\pi f_c\, t + \phi] \tag{3-1}$$

$$s_0(t) = a_0 \cos[2\pi f_c\, t + \phi] \tag{3-2}$$

where *s(t)* is the modulated carrier, a_1 is the amplitude corresponding to a 1, and a_0 is the amplitude corresponding to a 0, f_c is the carrier frequency, and ϕ is the (fixed) phase relative to some reference. If $a_1 = 1$ and $a_0 = 0$, then the modulation process is called *on-off keying* (OOK) because the carrier is simply turned on for a 1 and off for a 0. This is by far the most common form of ASK.

Frequency-Shift Keying (FSK) If the data bits change the frequency of the carrier instead of its amplitude, we have

$$s_1(t) = A\cos[2\pi(f_c + f_d)t + \phi] \tag{3-3}$$

$$s_0(t) = A\cos[2\pi(f_c - f_d)t + \phi] \tag{3-4}$$

where the amplitude is now fixed, and the frequency varies with the bit being sent. The quantity f_d is called the *frequency deviation*, which determines the amount of shift up or down from the nominal carrier frequency f_c that occurs when a 1 or 0 is sent, respectively. For example, a carrier frequency of 1 MHz might be shifted up slightly to 1.01 MHz to represent a 1 and down slightly to 0.99 MHz to represent a 0. In this case, $f_d = 0.01$ MHz. This process is called *binary frequency-shift keying* (BFSK), or sometimes 2FSK.

Phase-Shift Keying (PSK) Finally, the baseband data signal may instead modulate the phase of the RF carrier, represented mathematically as

$$s_1(t) = A\cos[2\pi f_c t + \phi_1(t)] \tag{3-5}$$

$$s_0(t) = A\cos[2\pi f_c t + \phi_0(t)] \tag{3-6}$$

For example, the phase could be $\phi_o = 0$ degrees when the data bit is a 0 and $\phi_1 = 180$ degrees when the data bit is a 1. This is BPSK or 2PSK.

All three of these modulation schemes use only a single data bit to make each change to a transmitted data symbol. Therefore, the *symbol rate*, which is the rate at which new transmitted signals are sent, is equal to the *bit rate*, which is the rate at which new data bits are created. Examples of data with their corresponding baseband signal, along with waveforms produced by all three modulation techniques, are given in Figure 3-2. We shall later explore the possibility of using multiple data bits to produce a single change to the transmitted signal, in which case there are multiple data bits per channel symbol.

Complexity and Performance Each of the previous three modulation techniques has its place in wireless communication. OOK is extremely simple in concept and execution, and inexpensive devices, such as garage door openers, use this modulation method. FSK is slightly more complex and is employed by Bluetooth in a modified form. Finally, PSK is even more complex because the receiver must extract phase information. Variants of PSK are used for 802.11 *wireless local area network* (WLAN) systems and in many digital cellular networks. The PSK phase tracking process in the receiver is called *coherent detection*. Detection without phase tracking, used in the majority of OOK and FSK receivers, is called *noncoherent detection*.

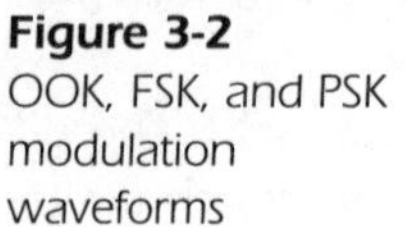

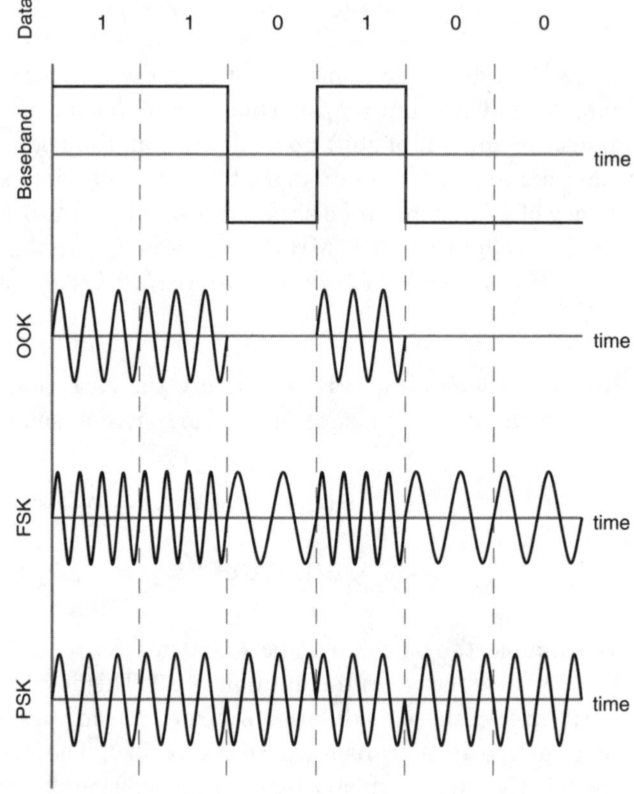

Figure 3-2 OOK, FSK, and PSK modulation waveforms

(FSK can also be detected coherently, as we will discuss in the next section. For that matter, OOK can also be detected coherently, but that's almost never done.)

So why would anyone want to use modulation schemes more complex than OOK? The reason is that FSK performs better than OOK, and PSK performs better than FSK, under most conditions. To quantify performance, though, requires some care because the modulation processes are different enough that it's easy to fall into the trap of comparing apples to oranges. To avoid this problem, we assume that the following characteristics are common to all three modulation schemes:

- The ever-present noise is white (meaning equal energy exists across the entire frequency spectrum), the distribution of its instantaneous values are Gaussian with zero mean, and the noise adds to the received signal. This type of noise is called *additive white Gaussian noise* (AWGN).

- *Signal-to-noise ratios* (SNRs) are expressed as the ratio between the energy in a data bit, E_b , relative to the AWGN power in the receiver per Hz of bandwidth, called N_0.
- In each case an optimum receiver is assumed; that is, a receiver having circuitry that can detect the data in AWGN with the minimum possible bit error probability.

Now we can compare the BERs of the modulation schemes against the SNR. The results are given in Table 3-1, which shows the SNR required to obtain a BER of 10^{-3} and 10^{-6}. In general, moving from OOK to FSK or from FSK to PSK has the benefit of about a 3 dB reduction in the required peak transmit power for a given BER. As an example, if an OOK transmitter requires 10 mW of power to achieve a 10^{-3} BER at the receiver, then switching to FSK can achieve the same BER with a transmit power of only 5 mW, and using PSK reduces the transmit power requirement further to about 2 mW. It's important to realize, however, that these results are somewhat sanitary and don't take into account fading, non-AWGN noise and interference, and nonideal transmitters and receivers.

Now here's a trick question. Suppose our transmitter is battery powered. Which scheme provides a longer battery life, OOK with 10 mW transmit power or FSK with 5 mW transmit power? FSK, right? No, not really. The reason is that an FSK transmitter is on all of the time, regardless of whether a 1 or 0 is being sent. For the same BER, the OOK transmitter is transmitting at twice the power, but only when a 1 is being sent; otherwise, the transmitter is off. Therefore, the average power is the same for both FSK and OOK (provided, of course, that there's an equal distribution of binary 1s and 0s). Still, the OOK transmitter must be designed for twice the peak power, so it is generally less efficient and more costly than an equivalent FSK transmitter.

FSK has another advantage over OOK that is not evident from Table 3-1. Consider a receiver attempting to detect an OOK signal in the presence of interference. The receiver's detector is programmed to output a binary 1

Table 3-1

SNR versus BER in AWGN

Modulation	SNR for BER of 10^{-3}	SNR for BER of 10^{-6}
OOK	14 dB	17 dB
FSK	11 dB	14 dB
PSK	7 dB	11 dB

when the incoming RF carrier is present and a 0 when the carrier is absent. Due to interference, the OOK detector may decide that a carrier is always present and respond by outputting a string of 1s with a correspondingly high BER. An FSK detector, on the other hand, attempts to decide which of the two signals given by Equations 3-3 and 3-4 is stronger. Therefore, as long as the interfering transmission affects both $s_0(t)$ and $s_1(t)$ about equally, then it may have very little effect on the FSK detector's performance. For similar reasons, the FSK detector performs well when the incoming signal is faded, as long as both $s_0(t)$ and $s_1(t)$ are faded about equally and are still strong enough to be detected.

Like FSK, PSK also has the advantages over OOK described in the last paragraph. Furthermore, phase tracking at the receiver gives PSK a performance advantage over FSK of about 3 dB. However, this phase-tracking requirement is difficult to implement in FHSS because the transmitter must be off while hopping to a new carrier frequency. As a result, the receiver must reacquire the incoming signal's phase at the beginning of each hop, and the resulting delays can reduce communication throughput. There is a type of PSK, called *differential phase-shift keying* (DPSK), that doesn't require precise phase tracking at the receiver, but its BER performance is slightly worse than that of coherent PSK. Therefore, FSK is the modulation process most often used for commercial FHSS radios.

To summarize, then, FSK is an excellent choice for use in Bluetooth for the following reasons:

- FSK is superior to OOK in BER performance for a given peak transmit power.
- FSK is superior to OOK in interference rejection.
- FSK is superior to OOK in a Rayleigh-faded environment.
- FSK is much less complex than PSK at about a 3 to 4 dB SNR penalty.

FSK Modulation Index and Bandwidth

As mentioned before, the quantity f_d is the amount of frequency deviation away from the carrier to represent a data bit: up in frequency for a 1 and down in frequency for a 0. This quantity is different from the data rate, so how should f_d be selected? It probably makes intuitive sense that a larger f_d gives the receiver a better chance of deciding whether $s_0(t)$ or $s_1(t)$ was transmitted (which is good), but it is also obvious that a larger f_d equates to a greater transmitted bandwidth for a given data rate (which is bad). What if the data rate increases? Must f_d increase as well for the communication system to perform properly?

One way to help quantify the relationship between the frequency deviation f_d and the data rate R (or the bit duration $T = 1/R$) is through the *modulation index k*, which is defined as

$$k = \frac{2f_d}{R} = 2f_dT \tag{3-7}$$

This assumes, of course, that there is one data bit per transmitted symbol. Many digital communication systems that use FSK (including Bluetooth) define a value, or range of values, for the modulation index. The data rate R is usually fixed, so Equation 3-7 can be used to find the deviation f_d given the other two quantities. In general, inexpensive, noncoherent FSK transmitters use a modulation index of about 1 (frequency deviation equal to half the data rate) for reasonable receiver performance. If the more-sophisticated coherent (phase-tracked) FSK detection is used, then the modulation index can be as small as 0.5, which is the minimum possible for the signals $s_0(t)$ and $s_1(t)$ to remain *orthogonal*, or completely separated, at the receiver's detector. For this reason the aforementioned process is called *minimum shift keying* (MSK).

The bandwidth of an FSK signal is theoretically infinite, but the energy density in the transmitted waveform becomes insignificant for frequencies far removed from the carrier frequency. For an unfiltered baseband signal used in an FSK modulator, the bandwidth B_T of the transmission is approximated by *Carson's rule*, which says that

$$B_T \approx 2(f_d + R) \tag{3-8}$$

so the transmitted bandwidth is approximately twice the sum of the deviation and the data rate.

Now let's apply Carson's rule to the Bluetooth data rate of 1 *megabit per second* (Mb/s), assuming a modulation index of 1 for noncoherent detection. According to Equation 3-7, the corresponding deviation is 500 kHz; substituting this value into Equation 3-8 yields a bandwidth B_T of 3 MHz.

However, the FCC restricts transmitted bandwidth to 1 MHz per hop channel. Oops.*

Reducing FSK Bandwidth with Gaussian Baseband Filtering

One of the goals in the design of many digital communication systems is to cram the fastest data rate possible into a given bandwidth or, equivalently, to use the minimum possible bandwidth for a given data rate. When the Bluetooth specification was written, FCC rules limited the maximum bandwidth in each FHSS hop channel to 1 MHz, so a modulation scheme had to be chosen to operate within this constraint while transmitting data as fast and as far as possible with cheap components. Now that's asking a lot!

Jocular Science
by Bill Lae
welve megawatt fibulator can cause aliasing... This may
e cross-mapped by sorting all hypermorphisms and solv-
ng the np iteratively... Just use a quad oversample and a
Carbotchy Nyquist Filter, I did and you should see the
© 1987 Lae
Lae
Technobabble

Figure 3-3 Networking at a Bluetooth conference (Source: Bill Lae)

*FCC rules require power levels in any 100 kHz segment outside the 1 MHz bandwidth limitation to be at least 20 dB below the 100 kHz wide maximum power level in the main lobe. Carson's rule determines the approximate width of the main lobe out to the first zero crossing in the frequency domain, so it will overestimate FCC bandwidth values if the second and subsequent lobes are weaker than –20 dB compared to the main lobe. The bottom line from all of this technobabble (see Figure 3-3). A 1 Mb/s data rate using ordinary FSK won't fit into 1 MHz of bandwidth.

The three simple modulation schemes of OOK, FSK, and PSK have a characteristic that's detrimental to bandwidth control: They all transition abruptly between $s_0(t)$ and $s_1(t)$ when the baseband data signal changes from a 0 to 1 and likewise from $s_1(t)$ to $s_0(t)$ when a transition from 1 to 0 occurs. Fourier transform theory states that fast transitions in time correspond to wide spans in frequency, so the corresponding data rate would need to be rather slow to conform to FCC bandwidth rules. Data speeds could increase if we could somehow reduce the abruptness of the modulated signal's transitions in response to data changes. Various methods can be used to perform this task, but we will concentrate on the Gaussian-filtered frequency-shift keying technique used for Bluetooth modulation.

GFSK first passes the baseband signal though a low-pass filter that has a Gaussian response curve in the frequency domain (see Figure 3-4[c]). The Gaussian filter's response $H(f)$ as a function of frequency f can be expressed mathematically as

$$H(f) = \exp\left(\frac{-1.4f^2}{B^2}\right) \tag{3-9}$$

where B is the –3 dB bandwidth of the filter. The bandwidth $B = 500$ kHz for the Bluetooth Gaussian filter. This filter smoothes the sudden transitions in the baseband signal so that the FSK-modulated RF carrier also smoothly transitions between a frequency of $f_c + f_d$ and $f_c - f_d$, thus reducing bandwidth. Figure 3-4(a) and (b) show representative examples of the unfiltered and filtered baseband signal.

Although transmitted bandwidth is reduced by this filtering process, the price for this reduction is the introduction of intersymbol interference by the modulation process itself. Refer again to Figure 3-4. After filtering, the shape at the baseband of a particular bit is influenced by its neighboring bits. For instance, a binary 1 with 1s on either side of it has maximum amplitude (corresponding to maximum deviation of the modulated carrier) for the duration of the symbol period; however, a 1 surrounded by 0s has time periods on either side of its midpoint where the amplitude is less than maximum. ISI increases as B in the Gaussian filter decreases for a given symbol rate and will increase with an increasing symbol rate for a constant B. Because the waveform in Figure 3-4 depends upon both filter bandwidth B and symbol duration T, it is customary to express the Gaussian filter's characteristics as a BT product. As a general rule, ISI is fairly small if the BT product is greater than about 0.5.[1] The BT product for the Bluetooth Gaussian filter is 0.5, so the modulation is called 0.5 GFSK.

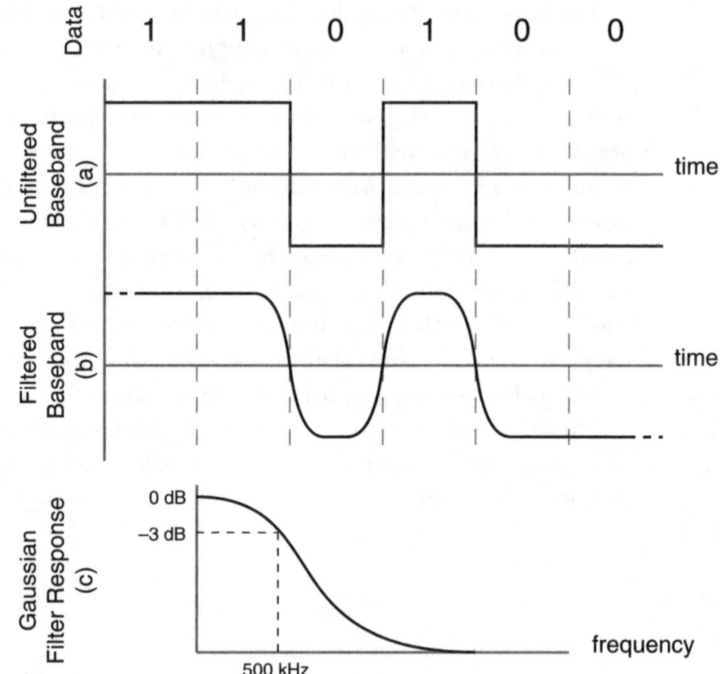

Figure 3-4 An unfiltered baseband signal (a) is passed through a Gaussian baseband filter (c) producing a filtered baseband signal (b). Although the Gaussian filter response is shown in the frequency domain, its time domain impulse response is also Gaussian.

Calculating the bandwidth occupied by a GFSK signal is a bit involved, but we can at least gain some insight into the improvements offered by GFSK by scrutinizing Table 3-2.[2] This table compares the bandwidths of MSK and 0.5 GFSK as a fraction of the data rate R. For a 1 Mb/s data rate, 90 percent of the transmitted signal's power is within only 780 kHz of the carrier frequency for MSK, but this is reduced further to 690 kHz for 0.5 GFSK. The GFSK advantage grows for higher power percentages, with 99.9 percent of the signal's power located within about half the bandwidth in the GFSK signal compared to MSK. Also, remember that MSK requires a coherent detector in the receiver for proper performance, making it less suitable for FHSS applications, whereas GFSK can be detected noncoherently.

Table 3-2

MSK versus GFSK relative bandwidths as a fraction of data rate R2

Modulation	90%	99%	99.9%
MSK	0.78	1.20	2.76
0.5 GFSK	0.69	1.04	1.33

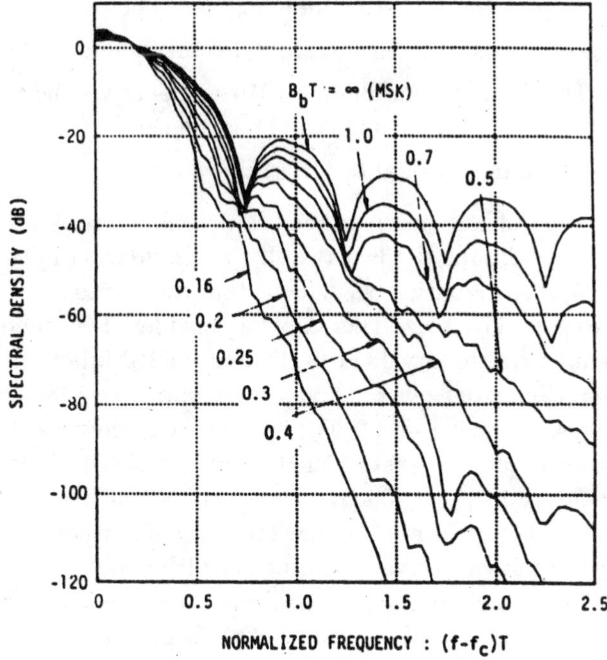

Figure 3-5 MSK versus GFSK spectral density plots (Source: K. Murota[2])

Figure 3-5 shows the results in Table 3-2 in greater detail, along with the normalized spectral densities of GFSK signals with *BT* products (labeled B_bT in the figure) less than 0.5, which have correspondingly higher ISI. Note that the second lobe of the MSK signal is only 20 dB below the peak of the main lobe, but the second lobe of the 0.5 GFSK signal is about 30 dB below the main lobe. Furthermore, the bandwidth of the main lobe itself is less for GFSK than MSK.

Bluetooth Modulation Requirements

Now, finally, it's time to present, with great fanfare, the details of the Bluetooth modulation process. To receive Bluetooth qualification, the radio must have the following modulation characteristics:

- GFSK with $BT = 0.5$.
- Symbol rate of 1 Ms/s, corresponding to a data rate of 1 Mb/s.
- Modulation index k between 0.28 and 0.35.
- Binary 1 has a positive f_d and 0 has a negative f_d.

- Symbol timing better than ±20 parts per million (ppm).
- Zero crossing error not greater than one-eighth of a symbol period.
- The f_d corresponding to a 1010 sequence will be at least 80 percent of f_d corresponding to a 00001111 sequence.
- Minimum f_d equal to 115 kHz.

Most of the previous modulation characteristics are self explanatory, except perhaps the last two. If the modulation index is between 0.28 and 0.35, then Equation 3-7 tells us that the corresponding f_d must be between 140 and 175 kHz. This can be considered a steady state f_d for a long sequence of binary 1s or 0s. Due to the ISI inherent in GFSK, it's possible that f_d may not reach its maximum value for a binary sequence that alternates between 1 and 0. The specification requires that f_d under these conditions reach at least 115 kHz, which is about 80 percent of the minimum steady-state f_d of 140 kHz. Reduced bandwidth at the expense of ISI is manifested by the relatively slow transitions in frequency when a 1 is followed by a 0 and vice versa. The Bluetooth ISI penalty for 0.5 GFSK is only about 0.2 dB, which means that an FSK transmission with no ISI can be 0.2 dB less powerful for the same BER performance.

As we discovered earlier, the Bluetooth radio is required to have a sensitivity of at least –70 dBm for a BER of 10^{-3} or better, and that the data rate is 1 Mb/s in a bandwidth of 1 MHz. Given the bandwidth and signal-level constraints at the receiver, does Bluetooth have a reasonably fast data rate? Or can a more clever modulation or coding scheme be used that will significantly increase the data rate within a 1 MHz hop bandwidth? To find the answers we first turn to a means by which we can find, in theory at least, the fastest possible data rate given a set of bandwidth and SNR constraints.

Channel Capacity Limits

We devoted much of Chapter 2 to answering the following question: How far will the radio transmit? In this section we will discuss the next logical question: How fast can the radio send and receive data? This is a very important issue, not the least because there are several misconceptions associated with finding the answers. One of the most prevalent is the erroneous idea that faster data rates require increased signal bandwidth. We will discover that the maximum possible data rate depends upon both signal bandwidth and SNR.

The second misconception is that, for a given data rate, there is an associated minimum BER below which we cannot go. This is also false: It is actually possible, under the right conditions, to achieve a BER that is *arbitrarily small*. This concept sounds counterintuitive and smacks of trying to invent the perpetual motion machine. However, if a communication system is sending data at a rate below the *channel capacity*, then it's possible to achieve BERs that are as small as desired.

Shannon Limits as a Function of Bandwidth and SNR

In his epic paper, "The Mathematical Theory of Communication," first published in 1948, Claude Shannon laid the foundation for the study of information theory.[3] In the paper Shannon defined the concept of information and quantified its characteristics mathematically. Once these concepts were understood, it became possible to develop a formula for finding the fastest rate at which information may be sent over a communication channel. This surprisingly simple formula is given by*

$$C = W \log_2\left(1 + \frac{S}{N}\right) \tag{3-10}$$

where C is the channel capacity in bits/second, W is the receiver's bandwidth in Hz (ideally, equal to the transmitted signal's bandwidth), and S/N is the SNR as an actual ratio (not dB). SNR is related to E_b/N_0 by

$$\frac{E_b}{N_0} = \frac{S}{N}\left(\frac{W}{R}\right) = \frac{S}{N}(WT) \tag{3-11}$$

The channel capacity, then, is a function of both signal bandwidth and SNR. Signal bandwidth actually appears in two places in the formula: once as the quantity W and again as part of the total noise power because N is equal to the noise power N_0 per Hz multiplied by the received bandwidth W. As an example, suppose the SNR is 20 dB (signal power 100 times greater

*To find the logarithm base 2 on your calculator, use the relationship $\log_2(n) = \log_{10}(n)/\log_{10}(2) \approx 3.32 \log_{10}(n)$.

than noise power) and the signal's bandwidth is 50 kHz. Channel capacity in this case is about 333 kb/s.

Although Equation 3-10 is simple, its implications are actually quite subtle. If data is sent across a channel at a rate less than channel capacity, then a method of sending the data exists such that the BER at the receiver can be made arbitrarily small. In other words, a modulation and error control process can be employed to reduce bit error probability as much as desired. On the other hand, if the data rate across the channel exceeds capacity, then bit errors will inevitably occur at the receiver and must be tolerated.

There is, of course, no hint in Equation 3-10 of how to actually achieve arbitrarily low bit error probabilities when data is sent at a rate less than channel capacity, and much research effort has been devoted to finding modulation and error control coding methods to make the BER as small as possible. It has been noted, though, that lower BERs are obtained at the expense of greater delays in processing the incoming data at the receiver. So we don't get a free lunch after all.

Communication systems, either wired or wireless, can be roughly divided into two categories: *power limited* and *band limited*. The method of constructing signals to represent the data is different for each of these two categories, which we now show.

Power-Limited Channels

Low-power transmitters, such as Bluetooth and others that often operate from a battery, can produce low SNR values at the receiver. These systems can still achieve fairly high data rates by taking advantage of the bandwidth part of Equation 3-10. Even wireless systems with rather high transmit powers can still fall into the power-limited category due to the high path loss across the communication channel. These devices usually require somewhat simple modulation schemes to keep the BER low in low SNR situations; thus, the number of bits per channel symbol is low (typically 1 or 2 bits per symbol).

Good modulation techniques for power-limited channels include OOK, 2FSK, and 2PSK, each having 1 bit per channel symbol. Both FSK and PSK can easily be extended to include 2 bits per channel symbol for improved data rates in a power-limited situation. For 4FSK the four channel symbols are

$$s_{10}(t) = A\cos[2\pi(f_c + 3f_d)t + \phi] \tag{3-12}$$

$$s_{11}(t) = A\cos[2\pi(f_c + f_d)t + \phi] \quad (3\text{-}13)$$

$$s_{01}(t) = A\cos[2\pi(f_c - f_d)t + \phi] \quad (3\text{-}14)$$

$$s_{00}(t) = A\cos[2\pi(f_c + 3f_d)t + \phi] \quad (3\text{-}15)$$

Each symbol in the alphabet is spaced $2f_d$ in frequency away from the adjacent symbols, just like 2FSK, so the data rate for 4FSK is doubled by increasing the bandwidth and using the same symbol rate as 2FSK. The subscripts on $s(t)$ denote the 2 bits corresponding to that particular channel symbol. Also notice that the subscripts progress from bottom to top in such a way that only 1 bit changes for each adjacent symbol. This type of binary counting (00, 01, 11, 10) is called the *Gray code*, and this is used to attempt to prevent multiple bit errors from occurring in response to a symbol error. If noise corrupts one of the symbols, the probability is very high that an adjacent symbol will be mistakenly selected by the receiver, and an adjacent symbol error will result in only a single bit error. Similarly, FSK can be extended further to 8FSK, 16FSK, and so on, by increasing bandwidth through adding more members to the symbol alphabet.

PSK can also be extended to an alphabet with 2 bits per symbol with the following representations:

$$s_{10}(t) = A\cos[2\pi f_c t + \phi_{10}(t)] \quad (3\text{-}16)$$

$$s_{11}(t) = A\cos[2\pi f_c t + \phi_{11}(t)] \quad (3\text{-}17)$$

$$s_{01}(t) = A\cos[2\pi f_c t + \phi_{01}(t)] \quad (3\text{-}18)$$

$$s_{00}(t) = A\cos[2\pi f_c t + \phi_{00}(t)] \quad (3\text{-}19)$$

For these symbols the four phases $\phi_{00}(t)$, $\phi_{01}(t)$, $\phi_1(t)$, and $\phi_{10}(t)$ are separated 90° in phase, for example, 45°, 135°, 225°, and 315°, respectively. Once again the progression uses Gray coding to reduce the chance that a symbol error will cause a double bit error. This modulation process is quite common and is called *quadrature phase-shift keying* (QPSK). For identical signal amplitudes and symbol rates, QPSK produces bit errors twice as often as BPSK. However, because the QPSK transmitted bit rate R is double that of BPSK, the BERs for the two processes are actually identical. Just think: Switching from BPSK to QPSK gives twice the data rate using the same transmit power without affecting the BER. That comes perilously close to obtaining something for nothing, don't you think?

Like its FSK counterpart, PSK can also be extended to even more bits per symbol by increasing the size of the symbol alphabet. This is done at the expense of smaller phase differences between adjacent symbols. Therefore, for the same BER performance, transmit power must be increased to obtain a higher SNR. The bandwidth remains the same regardless of how many members are in the PSK alphabet. Therefore, this form of PSK may be better suited to band-limited channels, which we discuss next.

Band-Limited Channels

Band-limited communication channels are those in which the signal bandwidth is severely restricted, but SNR values can be quite high. Perhaps the best example is the ordinary telephone line, where the bandwidth is only about 3 kHz, but SNR values can exceed 30 dB. Because of the narrow bandwidth, channel symbols must change relatively slowly so they can reach the receiver without unacceptable distortion; therefore, to achieve these relatively high data rates, multiple bits must be encoded on each channel symbol.

One of the most common ways to encode lots of data bits per channel symbol is to combine ASK and PSK into a process known as *quadature amplitude modulation* (QAM). An alphabet containing a large number of symbols is created, each of which has a unique amplitude-phase pair. If the number of symbols in the alphabet is M, then $\log_2 M$ bits can be sent per symbol. QAM is one way that so-called M-ary keying can be accomplished. For example, if there are 256 possible channel symbols, each with a unique amplitude and phase, then 8 bits can be sent per symbol. Transmitted bandwidth is low because symbols change much more slowly than the data rate, but SNR must be high for the receiver to accurately detect which symbol was sent out of the many possibilities. A form of QAM called *trellis-coded modulation* (TCM) is used for 56 kb/s modem communication on a telephone line.

As we discovered in Chapter 2, the SNR in a wireless link depends strongly on the distance between the transmitter and receiver. Low-cost wireless devices typically experience a high SNR when the transmitter and receiver are only a few meters apart, but in these situations they can support a higher data rate than when they are farther apart. Some wireless protocols, most notably 802.11b Wi-Fi, vary their data rates in response to different SNR values.

Capacity of the Bluetooth Channel

Finally, let's use Shannon's formula to determine the capacity of the Bluetooth channel. This will help us determine whether the 1 Mb/s data rate that Bluetooth uses is close to capacity or whether the channel is capable of much faster rates.

To find the Bluetooth channel capacity, we'll first calculate the noise power in the 1 MHz Bluetooth channel bandwidth. Although the noise floor in a receiver is determined by its construction, we'd like to find the lowest possible noise power in 1 MHz so that Equation 3-10 will give the highest possible data rate. This minimum noise power N is given by the formula

$$N = kTB \tag{3-20}$$

where k is Boltzmann's constant (1.37×10^{-23} Joules/°K), T is the temperature in °K, and B is the bandwidth of interest. For a room temperature of 292 °K and a bandwidth of 1 MHz, the noise power is 4.0×10^{-15} watts, or -114 dBm. Signal power at the receiver is determined by Equation 2-1, but let's for now use the value of -70 dBm, which is often the RSSI at a range of 10 meters with 0 dBm transmit power. The SNR in this case is 44 dB, or 2.5×10^4, and the associated channel capacity is 14.6 Mb/s. This capacity is significantly higher than the 1 Mb/s that Bluetooth offers. In fact, a channel capacity of 1 Mb/s in 1 MHz of bandwidth requires an SNR of only 1 (0 dB)! Figure 3-6 plots channel capacity as a function of SNR for a 1 MHz wide bandwidth, and it's obvious that Bluetooth's data rate is significantly below channel capacity at any reasonable SNR value.

So why is the Bluetooth data rate so slow? Achieving data rates close to channel capacity requires complex modulation and coding processes that aren't part of the Bluetooth specification in order to keep costs low. Indeed, perhaps the most common complaint about Bluetooth is its limited data rate. As we'll see in Chapter 14, "The Future of Bluetooth," the Radio2 working group in the Bluetooth SIG is researching methods for increasing data rates.

The capacity values that we've calculated up to this point are based upon the assumption that AWGN is the only impediment to a perfect channel. In many wireless systems, interference can be a greater impediment to performance than noise. We now take a look at interference and how the Bluetooth piconet can limit its effect.

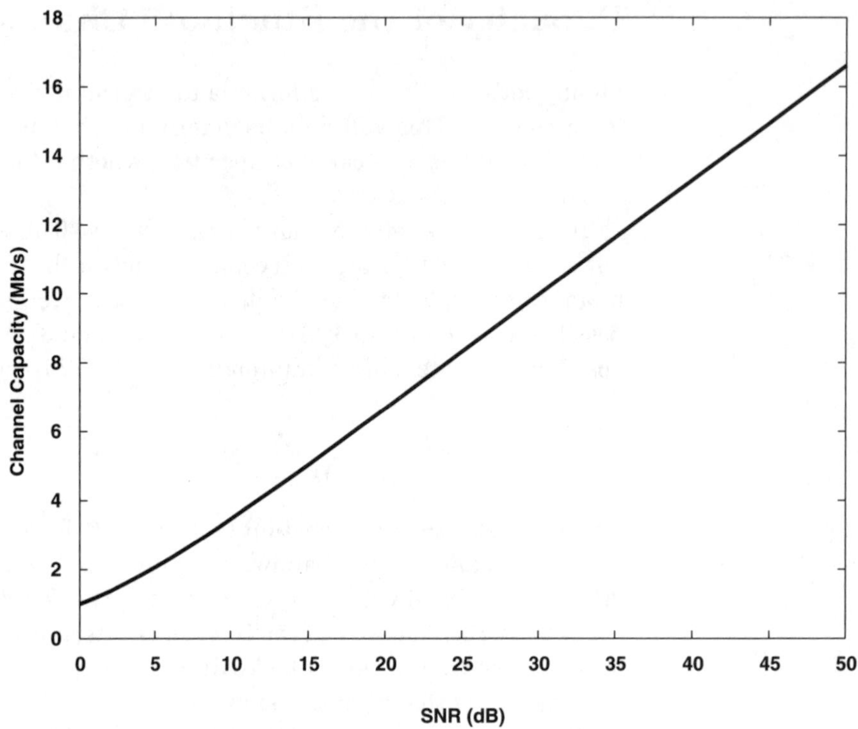

Figure 3-6 Capacity versus SNR in a 1 MHz wide bandwidth channel

The Nature of Interference

As more and more wireless devices appear in the 2.4 GHz band, interference is expected to play a major role in determining system performance. As an example, the cellular telephone network is so saturated with users in many urban sites that performance is limited from *carrier to interference* (C/I) ratio more so than from SNR. Interference is different from noise in a number of ways. Interference usually varies in intensity and frequency, and it isn't caused by thermal effects. It can be especially disruptive if it has relatively high power compared to thermal noise. Finally, interference can even be caused by deliberate jamming, although in most cases interference is accidental. As we discovered earlier, FCC rules require Part 15 devices to cope with interference on their own because they have no regulatory protection against it.

A Bluetooth piconet can experience interference from several different sources that operate in the 2.4 GHz band, such as

- Microwave ovens
- Other Bluetooth piconets
- 802.11b/g networks
- HomeRF networks
- Cordless telephones
- Custom devices
- Licensed users such as amateur radio operators

Interference can appear in different ways, with characteristics in both time and frequency. In time, interference can be either continuous or intermittent, and in frequency, it can be either narrowband or broadband. We examine specifics from these sources in Chapter 13, "Coexisting with Other Wireless Systems," but it's clear that Bluetooth devices need to incorporate some means of operating reliably in an interference-prone environment. This is accomplished through FHSS.

Frequency Hop Spread Spectrum (FHSS)

Suppose you're designing a communication system to operate in an environment that has a high potential for interference. What should the carrier frequency be? One possibility is to select a frequency and simply hope that the interference isn't severe enough to significantly affect data throughput, but that approach will probably fail often enough to be unsatisfactory. A better idea would be to create several carrier frequencies, or channels, and select the one that has an interference level below some threshold prior to the start of each communication session. The channel selection process could even be automatic. This method is used by many cordless phones in the United States to avoid interference from neighbors' phones. It works well because there is usually a limit to the number of cordless phones within a particular area, and enough channels are designated to almost guarantee that a clear one will exist at any particular time within a home.

For a Bluetooth piconet, though, it's likely that interference will change from moment to moment in the 2.4 GHz band, and many uses will try to coexist in this band within range of each other. Choosing a single channel and then remaining there for an entire session runs the risk of experiencing such a sudden, catastrophic increase in interference that the Bluetooth participants may not even be able to coordinate a channel change, and communication will then be lost. To avoid this situation, Bluetooth uses FHSS as an interference avoidance technique.

A block diagram of a Bluetooth FHSS communication system is shown in Figure 3-7. Binary baseband data is GFSK modulated and transmitted using a carrier determined by the frequency synthesizer. Instead of producing only a single carrier frequency, the synthesizer is controlled by a hop code generator that causes it to change carrier frequency at a nominal rate of 1,600 hops per second. One Bluetooth data packet is sent per hop.

This hop pattern itself appears to be random but is actually created by a pseudorandom algorithm in the hop code generator. The generator is duplicated at the receiver and will create the same hopping pattern that the transmitter uses. While communicating, then, the transmitter and receiver hop together from channel to channel. Furthermore, the two devices have agreed upon the hop sequence ahead of time, so even if some hop channels contain catastrophic interference, the piconet will survive because all members will soon hop together out of that channel.

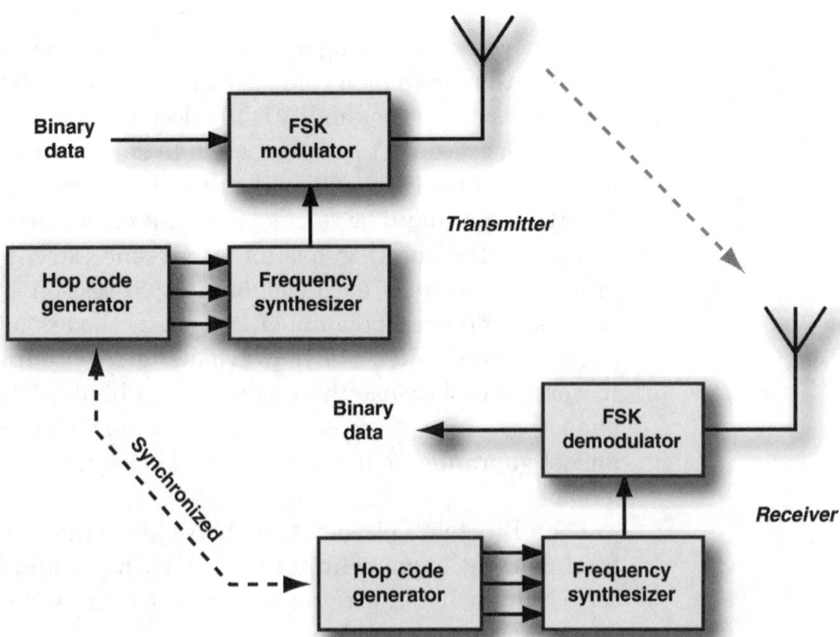

Figure 3-7 A block diagram of an FHSS communication system

Synchronization Between Communicating Devices

For two devices to communicate using FHSS, they must be properly synchronized so that they hop together from channel to channel. This means that the devices must

- Use the same channel set.
- Use the same hopping sequence within that channel set.
- Be time synchronized within the hopping sequence.
- Ensure that one transmits while the other receives, and vice versa.

All of these synchronization items are determined by the piconet master. The master passes the FHSS synchronization parameters to a slave during the page process, which we cover in detail in Chapter 5, "Establishing Contact Through Inquiry and Paging."

Hop Channel Set and Period An FHSS radio is programmed to operate on a certain set of frequencies, which is called the *channel set*. For Bluetooth, the channel set consists of the carrier frequencies

$$f_c = 2{,}402 + k \text{ MHz};\ k = 0,1,\ldots,78 \qquad (3\text{-}21)$$

Thus, there are 79 possible frequencies in the channel set, each spaced 1 MHz apart and covering 2,402 to 2,480 MHz. The piconet hops pseudorandomly within these channels as shown in Figure 3-8. Each channel is 1 MHz wide, and the Bluetooth GFSK data transmission occupies this bandwidth.

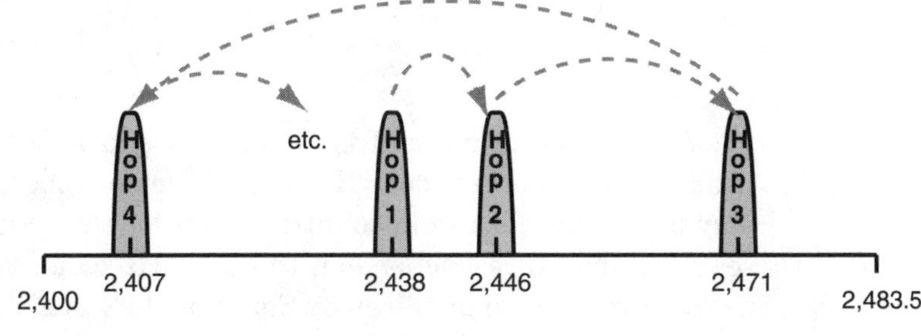

Figure 3-8 Radios in a Bluetooth piconet hop from frequency to frequency in a pseudorandom pattern. Depicted bandwidths are not to scale.

The sequence itself is determined by a pseudorandom hop generator that repeats its sequence after a certain number of hops. The period must be at least equal to the number of hop channels, but can also be much longer. The Bluetooth hop period during normal piconet communications is 2^{27} hops long, so at a rate of 1,600 hops per second the pattern will repeat after about 23.3 hours. More details on the hop generator and how the sequences are determined will be given in Chapter 4, "Baseband Packets and Their Exchange."

The period is long to prevent accidental hop synchronization between two different piconets within range of each other. This can be viewed in the following way. Suppose two piconets happen to hop into the same channel at a particular time. With a long pseudorandom sequence that's different for both piconets, the probability is only about 1 in 79 that the next hop will also be into the same channel. Therefore, both piconet hop sequences, because they are long, can be treated as completely random.

FHSS Cross Interference

Although we cover coexistence issues in Chapter 13, it would be helpful to develop a brief analysis of the nature of interference between multiple Bluetooth piconets. Because devices in a piconet transmit one packet per hop, we will assume that if two or more piconets within range of each other transmit on the same channel, then all affected packets are lost. Also, even though actual piconets are not time synchronized with each other, for now we will assume that all of them hop at the same time instant, but each uses a sequence that is randomly determined. Thus, the probability of packet error P_E from interference produced by another Bluetooth piconet becomes the probability that two or more transmissions occur within one channel. This is given by the formula

$$P_E = 1 - \left(1 - \frac{1}{M}\right)^{K-1} \tag{3-22}$$

where M is the number of hop channels and K is the number of users transmitting simultaneously. For most Bluetooth operations $M = 79$, but M may be smaller for use in countries that don't permit the full 79-channel sequence due to restrictions on the 2.4 GHz band. Figure 3-9 shows the packet error probability given by Equation 3-22 as a function of the number

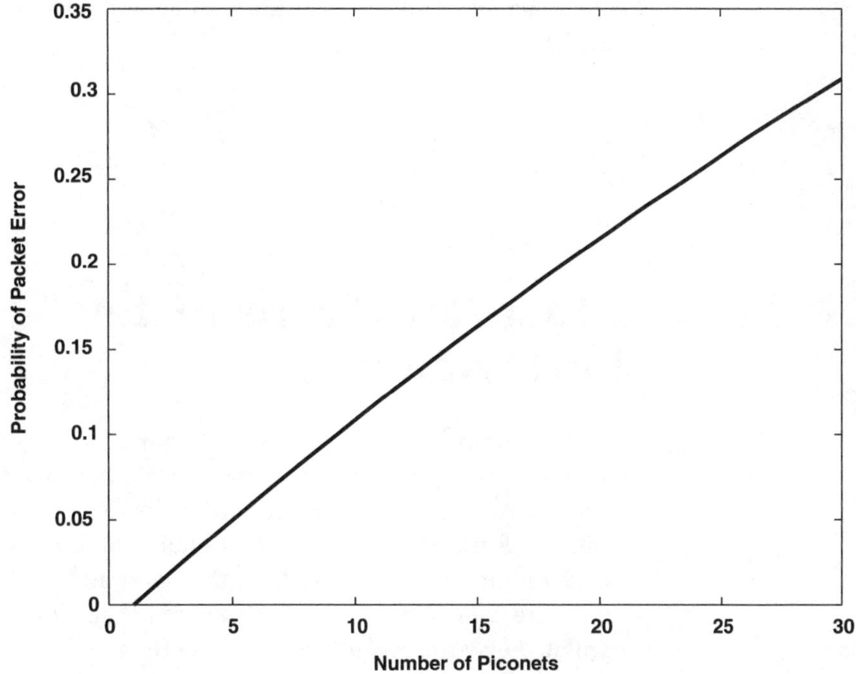

Figure 3-9 Packet error probability as a function of total piconets for the 79-channel Bluetooth hop set

of piconets, all of which can interfere with each other, for the Bluetooth 79-channel hop set. A more detailed analysis of Bluetooth piconet mutual interference will be given in Chapter 13, where we will discover that the error probabilities given in Figure 3-9 are optimistic.

As with many formulas involving probabilities, Equation 3-22 looks strange with all of the "one minus" parts to it. Its interpretation, though, is quite straightforward. The quantity $1/M$ is the probability that two piconets will hop into the same channel at any given time, so $1 - 1/M$ is the probability that the two will hop into different channels. Now let's focus on one of the piconets, which we call the *desired piconet*. (I know, that's a terrible term, but it has become the de facto standard in this kind of analysis.) The quantity $1 - 1/M$ taken to the $K - 1$ power is the chance that none of the other piconets will hop into the desired piconet's channel, so that becomes the probability of packet success. Finally, one minus this quantity is the probability of packet error. Whew.

If there are lots of hop channels compared to the number of piconets within range of each other, then we can assume that the probability is very low that three or more piconets will hop into any particular channel. This will probably be the case in most Bluetooth situations, so Equation 3-22 simplifies to

$$P_E \approx \frac{K-1}{M} \tag{3-23}$$

Now isn't that one easier?

Bluetooth Transmitter Performance

The Bluetooth transmitter exists, of course, as an intentional radiator of electromagnetic signals within the 2.4 GHz ISM band. Transmitter specifications are often given as quantities in both the time and frequency domains. Furthermore, any intentional radiator also generates unintentional radiation in other parts of the band and outside the band as well. These are called *spurious emissions*, and their amplitudes are stringently regulated by government agencies. The Bluetooth specification devotes only a few pages to these values, which are selected to meet government regulations without, hopefully, demanding high-cost solutions.

Transmitter Power Classes

Bluetooth transmitters fall into three basic classes that are determined by their maximum power output. The class 1 transmitter is the most powerful at 100 mW (20 dBm) maximum power, the class 2 transmitter has a maximum power of 2.5 mW (4 dBm), and the class 3 transmitter produces 1 mW (0 dBm). Class 1 transmitters must have a power control feature to reduce power to a level adequate for communication in order to prevent excessive interference to other users in the band. The other two classes don't require power control, but it can be employed as an optional feature.

Class 2 and Class 3 Devices The class 2 Bluetooth device is enabled to have a maximum power output of 4 dBm, with a nominal power output of 0 dBm and minimum power (at the maximum power setting, if power control is implemented) of –6 dBm. Maximum power output for the class 3 Bluetooth transmitter is 0 dBm. Power control is optional, but will probably not find its way into many class 2 and class 3 Bluetooth devices.

Class 1 Devices Maximum transmit power output for class 1 Bluetooth devices is up to 20 dBm, and power control is required down to 4 dBm, or lower if desired. The lower power limit is suggested to be –30 dBm (1 μW), but is not mandatory for levels below 4 dBm. (According to the analysis in Chapter 2, a transmit power of –30 dBm will result in a nominal range of about 1 meter.)

Power control is implemented using a feedback mechanism between the master and a slave in the piconet. In describing its operation, let's assume that the master's transmitter power is being adjusted by a slave. For power control to be possible, the master's transmitter must have the capability to change its power level automatically, and the slave's receiver must have a calibrated RSSI. Furthermore, there must be a means for the slave to direct the master to adjust its power, either up or down. This is done using the *Link Manager Protocol* (LMP) packets, as described in Chapter 7, "Managing the Piconet."

A power-controlled transmitter must have the ability to adjust its output level in steps that range in size between 2 and 8 dB. The adjustment range should be between 4 dBm (or, optionally, lower) and the maximum power level (up to 20 dBm). As an example, suppose a transmitter's maximum power output is 20 dBm. The power control requirement could be met by implementing a step size of 8 dB, in which case there would be only three power levels: 20, 12, and 4 dBm. If a step size of 2 dB is used instead, then nine power levels would be needed.

A receiver participating in the power control process attempts to place the incoming signal's power level within the *Golden Receive Power Range*. (No, I didn't make up that name; it's actually in the Bluetooth specification.) The lower range is selected to be a specific value between −56 dBm and 6 dB above the actual sensitivity of the receiver. For example, if a receiver has a sensitivity of –80 dBm, then the lower threshold of the range is between −74 and −56 dBm. The upper threshold is 20 dB above the lower threshold, to an accuracy of ±6 dB. See Figure 3-10 for a graphical display of the Golden Receive Power Range. As you can see, this power range has a rather wide span and loose accuracy constraints to keep implementation costs down.

Continuing with our example, suppose the slave's Golden Receive Power Range is between −60 and −40 dBm and the master's transmit power step size is 8 dB. Now assume that the master sends a packet to the slave, and the slave's RSSI shows −65 dBm. The slave will then return an LMP packet to the master, asking it to raise its transmit power by one step. Assuming no changes occur in the propagation path, the slave receiver's RSSI will now read −53 dBm upon arrival of the next packet from the master, which is

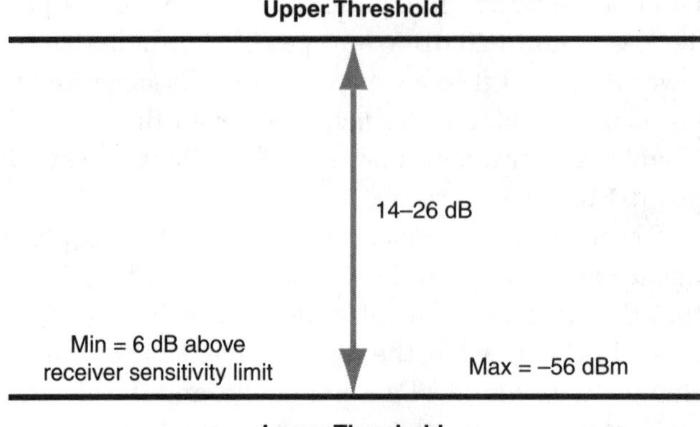

Figure 3-10 The Golden Receive Power Range

within the desired range. Of course, power control can also compensate for changes in path loss as the master and slave change positions relative to each other.

What if a class 1 Bluetooth device is communicating with a device without RSSI? Power control cannot be accomplished in this case. When the class 1 device discovers the lack of RSSI capability at the other end of the link, then it must reduce its transmitter power to class 2 or 3 levels. This discovery is made during the early stages of link setup through the "LMP supported features" packet, as described in Chapter 5. A device without RSSI probably has a class 2 or 3 transmitter, so if the class 1 device can hear the signal, then higher power levels are probably not needed for reliable communication.

During the inquiry and page processes, which take place before link setup, high transmit power may overload the receiver of nearby Bluetooth devices. Therefore, it may be prudent to reduce power sometime during the inquiry and (perhaps) paging process to improve the chance that nearby receivers will successfully decode the incoming messages.

Consequences of Higher Transmit Power

While designing a short-range wireless system, it's tempting to build a transmitter with the highest power possible. After all, won't that improve the chance that we'll achieve adequate range? Perhaps, but there are several consequences to using higher power. In no particular order, these are

- In moderate clutter, increasing range by a factor of 4 requires increasing transmit power by a factor of about 100.

- Both transmitters on the link must have higher power before range can increase (assuming identical receiver performance).
- Interference potential to other users increases rapidly with increased transmit power.
- Components in the transmit circuit become more complex and expensive.
- Power control is required for levels higher than 4 dBm.
- Battery life can be significantly shortened.
- RF power levels higher than a few milliwatts can severely interfere with the operation of nearby digital and analog circuits.

For these reasons, most Bluetooth devices will probably be class 2 or class 3. Class 1 units will most likely be installed in client devices such as laptop and desktop computers, and in fixed servers such as printers and access ports.

Transmitter Spurious Emissions and Drift

As mentioned earlier, an intentional radiator, such as a Bluetooth transmitter, also produces unintentional radiation at various frequencies, and these spurious emissions are tightly restricted by various governments to prevent interference to other users. Spurious emissions can occur either within the 2.4 GHz band or outside of it. Within adjacent channels, power levels in any 100 kHz band segment must be −20 dBc (decibels below carrier level), compared to the 100 kHz band segment within the transmitter's channel that contains the highest power level when the signal is modulated with pseudorandom data. For example, if f_c = 2,410 MHz, then the transmitter is intentionally radiating between 2,409.5 and 2,410.5 MHz. Suppose the highest-power 100 kHz band segment within that channel contains 5 dBm of power measured at the antenna connector. Then power levels in any 100 kHz segment within the two adjacent channels (2,408.5 to 2,409.5 MHz and 2,410.5 to 2,411.5 MHz) are required to be below −15 dBm.

Adjacent channel spurious emissions must total less than −20 dBm within a channel that is 2 MHz away from the tested transmitter's channel and less than −40 dBm for channels that are 3 MHz or more away. These power levels are the total within the entire 1 MHz bandwidth of the channel. An exception is made for up to three channels that are 3 MHz or more away, and these power levels can total up to −20 dBm in each channel. In all cases, FCC or other applicable government regulations must also be met.

For out-of-band spurious emissions, measurements are made using a 100 kHz bandwidth throughout a wide range of frequencies, and the resulting power levels cannot exceed those in Table 3-3. Note that the last two entries are additional restrictions placed upon some of the frequencies covered in the second entry of the table. The *Personal Communications Service* (PCS) operates in the 1.8 GHz band, and the *Mobile Satellite Service* (MSS) uses the 5.15 GHz band. Bluetooth devices will probably be placed into phones that employ either of these services, so spurious emissions from the Bluetooth transmitter must be extremely low to avoid interference.

Notice that both in-band and out-of-band spurious emissions are absolute (dBm) and not relative (dBc). This factor places an additional burden on class 1 transmitters because their relative spurious emissions must be 20 dB (100 times) less than spurious emissions from a class 3 transmitter.

When a transmitter first turns on following a hop to a new frequency, it must be accurate to within ±75 kHz of the actual f_c. After startup, the transmitter will drift in frequency while it's sending a Bluetooth packet. The maximum drift rate during the transmission of any packet can be as high as 400 Hz/μs, but the total drift cannot exceed the limits given in Table 3-4. (For definitions of packet lengths, see Chapter 4.)

Table 3-3

Out-of-band spurious emission limits for the Bluetooth transmitter

Frequency Band	Operating Mode	Idle Mode
30 MHz–1 GHz	−36 dBm	−57 dBm
1 GHz–12.75 GHz	−30 dBm	−47 dBm
1.8 GHz–1.9 GHz	−47 dBm	−47 dBm
5.15 GHz–5.3 GHz	−47 dBm	−47 dBm

Table 3-4

Frequency drift limits during packet transmission

Type of Packet	Frequency Drift
One time slot	±25 kHz
Three time slots	±40 kHz
Five time slots	±40 kHz

Bluetooth Receiver Performance

The circuitry in the receiver contains the greatest amount of IP within the Bluetooth hardware set provided by most manufacturers. Even from the early days of wireless communication, the receiver was far more difficult to design than the transmitter. The demands on the receiver, especially in an interference-prone environment, are significant. The receiver must amplify and decode an extremely weak signal containing a few picowatts of power and reject hundreds, if not thousands, of unwanted signals, some of which exceed the desired signal's power by several orders of magnitude.

One of the best performance measures of a digital communication receiver is its BER. BER versus input SNRs are developed mathematically as part of early university-level digital communications courses, and many manufacturers of Bluetooth receivers include BER for a given input signal power level in their specification sheets. The goal of any digital receiver design is to get the BER as low as possible.

The Bluetooth specification uses -70 dBm as the reference sensitivity level. The actual sensitivity level for a particular receiver is the weakest signal at which a BER of 0.1 percent, or 10^{-3}, is achieved. Several performance criteria are to be met without exceeding the 10^{-3} raw BER out of the receiver's detector circuitry. Of course, a BER that averages 1 in every 1,000 received bits is unacceptable for file transfers and most other types of data exchange between two Bluetooth users. Remember, though, that Shannon's theorem states that a communication system operating at a bit rate below capacity can be coded in some way so that the effective BER is as low as desired. We shall see in Chapter 4 that error control coding can be implemented to reduce the effective BER to negligible levels.

The procedure for testing a receiver for BER will be discussed in detail in Chapter 12. However, the concept is straightforward: A known bit sequence is sent to the receiver as a Bluetooth-modulated signal at some calibrated power level, and the raw data is checked at the receiver's output for a match. The estimated BER is the ratio of bad bits to the total number of detected bits.

Receiver performance requirements are listed in the Bluetooth specification in terms that deviate somewhat from traditional receiver specifications. Where the Bluetooth specification lists sensitivity and interference levels that are required for a certain BER performance, traditional specifications have focused on circuit performance instead. These include items such as receiver noise figure, 1 dB compression, and third-order intermodulation intercept points, and others with equally impressive names. Many of these criteria have their roots in analog communication, and their effect upon a receiver's BER is sometimes not very obvious. Fortunately, some of the chipset manufacturers are beginning to list their receiver performance figures in terms of those given in the Bluetooth specification, so an easy comparison can be made.

Sensitivity

For Bluetooth certification, the receiver must at least meet the *reference sensitivity level* of −70 dBm for a raw BER of 10^{-3}. This sensitivity level must be met using *any* transmitted signal that adheres to the Bluetooth specification, not just a signal from the same manufacturer's transmitter. This sensitivity level is actually quite poor compared to state-of-the-art receivers, which can have sensitivities of −100 dBm or better. Bluetooth, though, is intended for inexpensive, short-range links, so the demands upon receiver performance is comparatively low. Bluetooth chipset manufacturers are producing receivers with typical sensitivities of −80 to −85 dBm.

Because a receiver must be able to detect weak signals, it's obvious that the total amplification level within the receiver is quite high. As a result, there's also a limit to the highest signal level that a receiver can demodulate. Antenna signal levels that are too high will drive some of the receiver's amplifiers into saturation. Fortunately, FSK demodulators can often perform quite well under these conditions, but eventually, the signal at the detector could become so distorted that the resulting BER is excessive. For Bluetooth certification, the receiver must be able to detect signal levels as high as −20 dBm without exceeding a BER of 10^{-3}.

Interference Immunity

As interference in the 2.4 GHz band becomes more prevalent, the capability of a Bluetooth receiver to selectively detect the desired signal and reject the others will become more critical to the proper operation of the piconet. Interference performance generally falls into three categories: C/I ratio, blocking, and image rejection. For each of these interference scenarios, the Bluetooth specification requires that the interfering signal be either an unmodulated carrier, also called *continuous wave* (CW), or a *reference Bluetooth-modulated signal*. This reference interfering signal is defined by the following:

- GFSK with a modulation index of 0.32 ±1 percent and BT product of 0.5 ±1 percent.
- Bit rate of 1 Mb/s ±1 ppm.
- Modulating data is a pseudorandom binary sequence called PRBS15 (discussed in Chapter 12).
- Frequency accuracy better than ±1 ppm.

A reference *desired* Bluetooth signal has characteristics that are the same as those previously discussed except the modulating data is a different pseudorandom binary sequence called PRBS9. These sequences (PRBS9 and PRBS15) are used to simulate the random data patterns that are sent during a typical piconet session.

C/I Ratio The C/I ratio is the ratio of desired (C) to interfering (I) signal power levels that the receiver must be able to tolerate without exceeding a BER of 10^{-3}. C/I values depend upon the proximity in frequency of the interfering signal's carrier to the desired carrier. For a channelized radio system, such as Bluetooth, these values are given for co-channel (same channel) and adjacent channel interfering signals that are Bluetooth modulated. A summary of these values is given in Table 3-5.

Co-channel and the first two adjacent channel C/I values are measured with the desired signal level at −60 dBm. The last C/I value is measured with the desired signal at −67 dBm to keep the combined signal powers well below the maximum specified level of −20 dBm.

The required C/I value drops as the carrier frequency of the interfering signal moves farther away from the desired signal. In other words, the farther the separation in frequency becomes between desired and interfering carriers, the higher the allowable interfering power level before it causes the desired BER to reach 0.1 percent. For example, an interfering Bluetooth

Table 3-5

C/I performance for a Bluetooth receiver

Requirement	Value
Co-channel C/I	11 dB
Adjacent (1 MHz) channel C/I	0 dB
Adjacent (2 MHz) channel C/I	−30 dB
Adjacent (≥3 MHz) channel C/I	−40 dB

transmission 1 MHz away from the desired signal can have the same power level as the desired signal before the desired BER reaches 10^{-3} (C/I = 0 dB), but when the interferer is 2 MHz away, then its power level can be 1,000 times greater (C/I = −30 dB) before the BER is exceeded. Incidentally, receiver designers have discovered that the adjacent channel C/I values are often more difficult to achieve than the co-channel C/I.

Receiver Blocking Performance Although Table 3-5 implies a higher allowable interfering signal power as its carrier frequency moves farther away from the desired carrier frequency, common sense tells us that eventually the interfering signal will be powerful enough to cause a general detrimental effect on desired receiver performance regardless of the interfering signal's frequency. The front end in most receivers contains a wideband amplifier that will saturate for high signal levels within its passband, and amplifier or filter stages beyond the front end may saturate as well. When this happens, the receiver becomes overwhelmed and can't properly amplify and detect a weak desired signal. This effect is called *blocking*.

The out-of-band blocking signals are measured with the desired signal at −67 dBm, which is 3 dB above the −70 dBm reference signal. Unlike the C/I measurements, the blocking signal is assumed to be CW. The allowable blocking signal power within each associated band is given in Table 3-6. These represent the out-of-band power levels under which the Bluetooth receiver's BER is not to exceed 0.1 percent. Note that for the range of 2.0 to 3.0 GHz (minus the 2.4 GHz ISM band), the CW blocking signal's power level is equivalent to a C/I of −40 dB, which is the same as that required for Bluetooth-modulated interfering signals that are 3 MHz or more away from the desired carrier frequency within the 2.4 GHz ISM band.

Many receivers, especially those controlled digitally (as Bluetooth receivers are), can be more sensitive to blocking signals at certain out-of-band frequencies because of various harmonic effects from component selection and their orientation on the circuit board, and from crystal frequency

Table 3-6

Out-of-band blocking power levels for a Bluetooth receiver demodulating a −67 dBm desired signal

Interfering Carrier Frequency	Power Level
30–2,000 MHz	−10 dBm
2,000–2,399 MHz	−27 dBm
2,498–3,000 MHz	−27 dBm
3,000–12,750 MHz	−10 dBm

harmonics in both the analog and digital parts of the circuit. As a result, the Bluetooth specification allows exceptions to the blocking levels in Table 3-6 for 24 frequencies at integer multiples of 1 MHz. At 19 of these frequencies, a −50 dBm blocking signal may be sufficient to bring the BER to 10^{-3}, and at the remaining 5 frequencies, the receiver is allowed to be blocked at any power level.

Image Rejection As we will discover in Chapter 12, many receivers are designed to convert the RF to one or more *intermediate frequencies* (IFs), where most of the amplification and filtering takes place. This type of receiver is called the *superheterodyne*. (Doesn't that sound like a term right out of a 1950s sci-fi movie?) The IF is fixed at a value that is usually significantly lower than the carrier frequency, so inexpensive high-performance filters and amplifiers can be used in the IF section of the receiver.

The conversion from the RF frequency f_{RF} to the IF frequency f_{IF} is performed by the *mixer*, as shown in Figure 3-11. (The frequency f_{RF} is identical to the carrier frequency f_c that we've used up to this point, but we'll follow receiver design convention and use the former terminology here. Also, I realize that terms such as RF frequency and f_{IF} are redundant, but once again we'll adhere to convention.) This conversion can be modeled mathematically as a multiplication between f_{RF} and the *local oscillator* (LO) frequency, f_{LO}. When two sinusoidal signals are multiplied together, the result is two new frequencies equal to $f_{RF} + f_{LO}$ and $f_{RF} - f_{LO}$. The higher (sum) component is filtered out, and the lower (difference) component becomes the IF; thus, $f_{IF} = f_{RF} - f_{LO}$. This implies, of course, that $f_{RF} > f_{LO}$, in which case the receiver uses *low injection*. Because f_{IF} is fixed, the receiver is tuned to different RF frequencies by changing the LO frequency. All of the components of Figure 3-11 are contained in the FSK demodulator depicted in Figure 3-7.

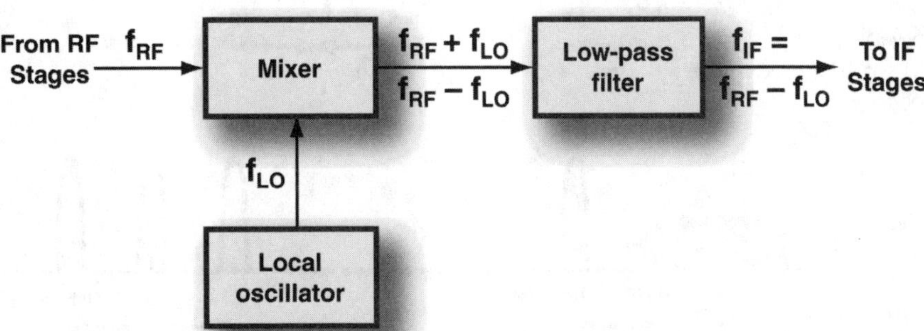

Figure 3-11 Conversion from RF to IF in a superheterodyne receiver

Unfortunately, there also exists yet another frequency, called the *image frequency* f_{Image}, such that $f_{IF} = f_{LO} - f_{Image}$. This means that another transmitter using f_{Image} as its carrier frequency could severely interfere with the desired signal with carrier frequency f_{RF}. After a bit of thought, we can conclude that $f_{Image} = f_{RF} - 2f_{LO}$. To make matters even more complex, it's also possible for a receiver to have $f_{IF} = f_{LO} - f_{RF}$ (*high injection*) and in this case $f_{Image} = f_{RF} + 2f_{LO}$. Both of these situations are depicted in Figure 3-12.

Are you still with me after all of this? Really? Wait until we get to third-order intermodulation distortion in the next section . . . .

Anyway, if a Bluetooth receiver has a desired signal power level of −67 dBm, then an interfering signal at the image frequency is required to be at or above −58 dBm before the BER is driven above 10^{-3}, corresponding to a C/I_{Image} of −9 dB. If the image frequency isn't an integer multiple of 1 MHz, which occurs if f_{IF} isn't an integer multiple of 1 MHz, then for test purposes f_{Image} is selected to be the closest 1 MHz to the actual image frequency. Finally, (did you think this would ever end?) if the interfering signal's frequency is ±1 MHz away from f_{Image}, then its signal level must be above −47 dBm before the BER exceeds 10^{-3}. This corresponds to a $C/I_{Image\pm1\,MHz}$ of −20 dB.

Notice that the image C/I performance figures sometimes conflict with the adjacent channel C/I requirements given in Table 3-5. In these cases, the more relaxed specification applies.

Intermodulation Distortion (IMD) All amplifiers have some nonlinearity associated with them, and, of course, the amplifiers used in Bluetooth

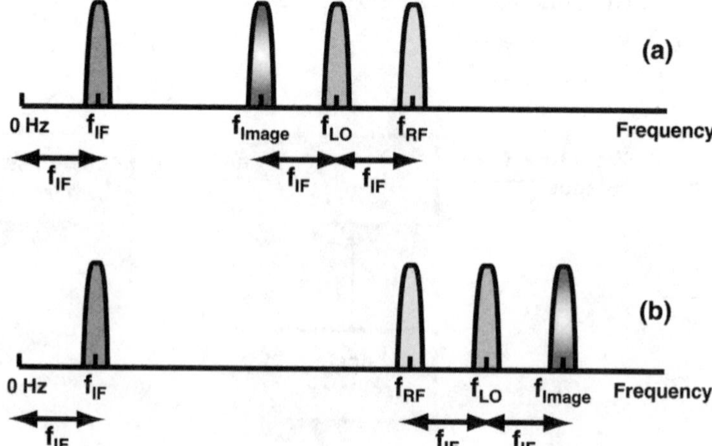

Figure 3-12 The location of image frequencies for high-injection (a) and low-injection (b) superheterodyne receivers

receivers are no exception. An amplifier with nonlinearities will produce new frequencies at its output that weren't present at its input. If one of the new frequencies happened to be the same as the desired signal's carrier, then interference could occur.

Often the most troublesome new frequency, from an interference standpoint, is generated as a third-order product from two other frequencies, which we call f_1 and f_2. A third-order product f_3 has the relationship

$$f_3 = 2f_1 - f_2 \tag{3-24}$$

The burning question is this: Can two interfering signals at f_1 and f_2 combine to form a third-order product at the desired signal's carrier frequency f_c? The answer is yes, indeed, and this can be done in many different ways. For example, suppose the desired receiver hops to f_c = 2,410 MHz. Then interfering signals at f_1 = 2,413 MHz and f_2 = 2,416 MHz can produce a third-order product at 2,410 MHz. In other words, if three Bluetooth piconets are within range of each other and if they hop to the previous frequencies, then the desired piconet may experience interference from the other two. This type of interference is called third-order *intermodulation distortion* (IMD). You could argue that the probability of the previous event happening is exceedingly small, but note that third-order IMD could also occur under many different hop frequency combinations, one or another of which could happen quite often.

The Bluetooth specification requires that third-order IMD be tested under the following conditions:

- The desired signal f_c has a power level of −64 dBm.
- The interfering signal f_1 is CW with a power of −39 dBm.
- The interfering signal f_2 is reference Bluetooth modulated at a power of −39 dBm.

The desired receiver's BER cannot exceed 10^{-3} when the relationship given in Equation 3-24 holds and the difference between f_1 and f_2 is selected to be any one of the following: 3, 4, or 5 MHz.

Spurious Responses and Emissions As we will discover in Chapter 12, Bluetooth receivers (and most other digital receivers, for that matter) contain oscillators, timers, and other circuitry that can produce their own RF emissions. The result is that a receiver may attempt to detect its own generated signals, and it may transmit unintended signals as well. The former are *spurious response frequencies*, and the latter are *spurious emissions*.

Table 3-7

Receiver spurious emission limits

Frequency Band	Limit
30 MHz–1 GHz	−57 dBm
1 GHz–12.75 GHz	−47 dBm

A receiver will probably fail to meet the C/I requirements in Table 3-5 and in the section "Image Rejection" on its spurious response frequencies. Up to five of these are allowed at frequencies at least 2 MHz away from the frequency to which the receiver is tuned. A relaxed C/I value of −17 dB is permitted for these frequencies.

Spurious emissions are listed in Table 3-7 for two different band segments. The maximum power requirement must not be exceeded for any 100 kHz band segment within the ranges listed.

Summary

The Bluetooth radio modulates data onto a carrier using FSK as a compromise between cheaper but less capable OOK and better-performing but more expensive PSK. The Bluetooth baseband signal is Gaussian filtered for a higher data rate within a 1 MHz bandwidth, but at the expense of a slight increase in intersymbol interference. FHSS is used to reduce the detrimental effect of interference. A total of 79 channels, each 1 MHz wide, are employed in a pseudorandom hop pattern.

According to Shannon's theorem, channel capacity is based upon both occupied bandwidth and SNR. Most short-range wireless systems, including Bluetooth, have capacity limitations based upon the relatively low transmit power, so bandwidth must be fairly high for fast data rates. The capacity of the 1 MHz wide Bluetooth channel at a 44 dB SNR is almost 15 Mb/s, but obtaining data rates close to this figure requires more complex modulation and coding schemes than those currently used in the Bluetooth radio.

Bluetooth transmitters can have power levels up to 20 dBm, but an output power greater than 4 dBm requires power control via a feedback mechanism between devices at each end of a link. Higher transmit power also increases cost and may interfere with other circuitry in close proximity to the transmitter.

The Bluetooth receiver's specifications are generally based upon a raw BER of 0.1 percent. Sensitivity and interference limits are specified at levels that are supposed to be achievable with inexpensive circuitry. However, together these specifications are stringent enough that several new approaches to radio design have been developed by the various manufacturers who provide Bluetooth radio ICs.

End Notes

1. Rappaport, T. *Wireless Communications: Principles & Practice,* 1st ed., Upper Saddle River, NJ: Prentice Hall, 1996.
2. Murota, K. and Hirade, K. "GMSK Modulation for Digital Mobile Radio Telephony," *IEEE Transactions on Communications*, July 1981.
3. Shannon, C. "The Mathematical Theory of Communication," *Bell System Technical Journal*, July and October, 1948.

CHAPTER 4

Baseband Packets and Their Exchange

Bluetooth sends data in a packet format, where a digital message is broken into several smaller packets and sent one by one to the destination. Unlike the *Internet Protocol* (IP), where packets are routed from node to node between source and destination, Bluetooth baseband packets are sent directly from their source (master or slave) to their destination (slave or master). Each packet is transmitted on a new hop frequency. Baseband packets are constructed to take advantage of the *time division duplexing* (TDD) process that provides for an orderly exchange of data between master and slave. The master and one of the slaves in the piconet take turns transmitting, so two or more transmitters are never on at the same time within a piconet.

Where data integrity is of primary importance, *packet switching* is used. This means that each incoming packet is checked for data bit errors by the destination node, and if errors are present, the destination asks the source to repeat the packet. Throughput is thus influenced greatly by the channel's *bit error rate* (BER). On the other hand, if low latency (small delay) is required, *circuit switching* is employed by the piconet. Real-time two-way voice communication falls under this category. Digitized voice packets are assigned specific transmission time slots, and repeat transmissions aren't allowed. Throughput is therefore unaffected by channel integrity; instead, voice packets sent over a channel with a high BER will produce distortion when they are converted back to analog audio.

The *baseband protocol* consists of the basic piconet functions such as

- Assembling packets from higher protocol layers and sending them to the radio
- Receiving bits from the radio and assembling packets for processing by higher protocol layers
- General piconet timing
- Frequency hop selection
- Channel control processing
- Error control
- Data whitening
- Basic security operations

These functions are performed by the *link controller*, which sits above the radio in the Bluetooth protocol stack (see Figure 4-1). The piconet is controlled by the application through the *link manager* (see Chapter 7, "Managing the Piconet"). Data is passed to the link controller through the *logical link control and adaptation protocol* (L2CAP, see Chapter 8, "Transferring

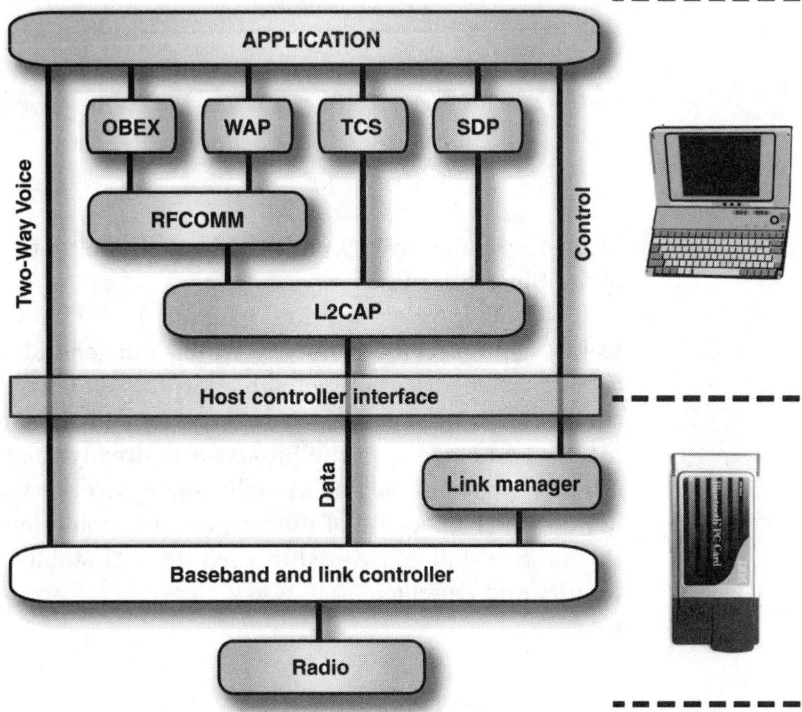

Figure 4-1 The baseband and link controller processes are above the radio in the Bluetooth protocol stack.

Data and Audio Information"), and real-time two-way voice is usually sent directly to the link controller from the application to reduce latency.

Bluetooth baseband packets follow the general packet structure in many other applications. A data packet contains an *access code*, *header*, and *payload*, as shown in Figure 4-2. The access code (72 bits nominal) is used for initial synchronization of the receiver circuitry, but it also contains other information, such as the piconet identity or recipient's address, depending upon the context of the application. The header (54 bits) includes the destination address, the type of payload to follow, and some error control information. Finally, the payload is a variable-length field that contains the message. The quaintly named "Little Endian" transmission order is used here, which means that the *least significant bit* (LSB) is always sent first.

It's important to keep in mind that only part of each baseband packet contains the actual message that the applications are attempting to exchange. Other parts of the packet (such as the access code and header), although necessary for the proper operation of the piconet, are overhead and don't contribute to the data throughput of the piconet. For this reason

Figure 4-2 General Bluetooth baseband packet structure

alone, the actual throughput will always be below the 1 Mb/s raw data rate of the Bluetooth radio.

In this chapter we will first examine how TDD operates, followed by a look at how data bit errors are controlled in general Bluetooth communications. Next we discuss Bluetooth addresses and how the packets are structured and sent from one device to another. At this point we assume that the master and slave(s) in the piconet are already communicating with each other. As you can imagine, establishing a wireless link in the first place is quite involved because of the frequency hopping nature of Bluetooth, but this process will be covered in Chapter 5, "Establishing Contact Through Inquiry and Paging."

Time Division Duplexing (TDD)

One of the most important criteria in a two-way communication system is determining how and when the radio units at each node can exchange information. One possibility is to equip both nodes with a transmitter and receiver and operate them simultaneously on different frequencies. This process, called *frequency division duplexing* (FDD), is a form of full-duplex transmission and is used in the traditional cell phone system. Full duplex enables users on each end of the conversation to interrupt each other, which is considered by some to be an advantage. FDD can be somewhat expensive because the transmitter and receiver must operate independently, requiring two frequency synthesizers, and a device called a *duplexer* must be used to combine an outgoing transmitted signal and an incoming receive signal with a single antenna.

A cheaper and physically smaller system can be built by using a process called *half duplex*, which means that one node transmits while the other node receives and vice versa. Half duplex is cheaper because only one frequency synthesizer is needed per radio and it is smaller because an antenna switch replaces the bulky duplexer. TDD assigns disjoint slots of time to each transmitter in a two-way communication system such that the users take turns sending data to each other. Although similar to the way walkie-talkies work through their push-to-talk buttons, TDD uses electronic *transmit/receive* (T/R) switching and thus can transition much faster between transmit and receive. This is the method used for communication within the Bluetooth piconet. Advantages of TDD include the following:

- Only one frequency synthesizer is required.
- A cheaper antenna switch replaces the duplexer.
- Fast TDD switching can masquerade as FDD.

Single-Slave Operation

We'll begin by examining the simplest possible Bluetooth piconet, consisting of one master communicating with a single slave in a point-to-point configuration, as shown in Figure 1-13. Time is divided into slots that are nominally 625 μs in length and numbered with consecutive integers. The master transmits to the slave in even-numbered time slots, and the slave transmits to the master in odd-numbered time slots. Each transmission takes place at a new hopping frequency, and a complete packet of data is sent in each time slot. This process is depicted in Figure 4-3.

By using this particular rendition of TDD, both the master and slave have equal access to the channel, and each is ready to receive when the other transmits. Communication thus proceeds in an orderly manner.

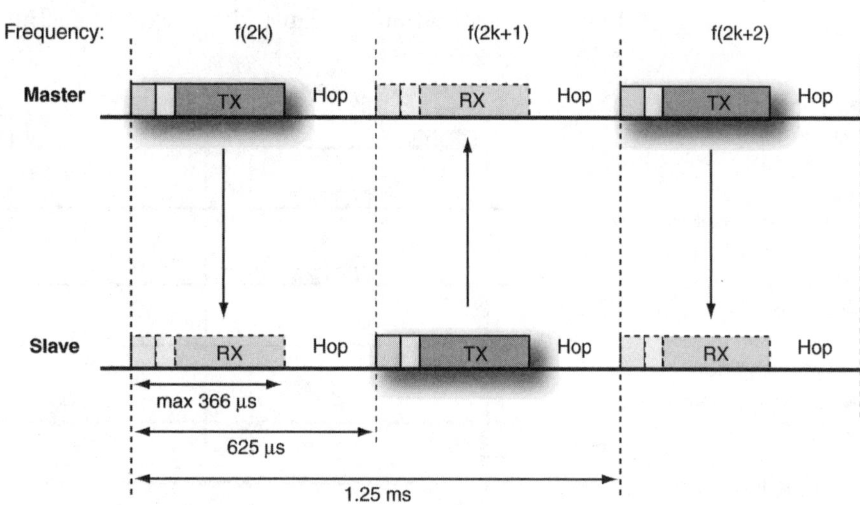

Figure 4-3 Bluetooth TDD communication process in a point-to-point configuration. The master transmits (TX) in an even time slot, starting at time (2k), where k is some integer index, and the slave receives (RX) in that slot. The slave transmits and the master receives in the odd-numbered slots.

Switching between transmit and receive at any one Bluetooth node is so fast that it appears to its human user to be communicating full duplex.

Each packet is allowed up to 366 μs for its transmission, equating to a maximum one-slot packet length of 366 bits. The additional 259 μs is used by the radio to change to the next frequency in the hop sequence. During the time nominally needed to perform a hop, no communication occurs in the piconet. The result is, of course, reduced throughput. Because the timing circuitry in each Bluetooth device is inexpensive, drift and jitter may occur between master and slave clocks even within relatively short periods of time. To accommodate this, a receiving node must allow for a window of 10 μs on either side of the expected packet arrival time for actual arrival to occur.

Multislave Operation

When the piconet consists of a master and two or more slaves, point-to-multipoint communication is taking place, and the total throughput is multiplexed between the piconet members. Once again, TDD is employed (albeit in a slightly more complex form) to prevent piconet members from jamming each other from simultaneous transmissions.

As shown in Figure 4-4, the master transmits on even-numbered time slots as before, but a slave can transmit only when it is specifically addressed by the master in the previous time slot. If, for example, the master sends a packet to slave 1, slave 2 will keep its receiver on long enough to decode the packet access code (identifying the piconet) and header (identifying the destination). Because the packet is destined for slave 1, slave 2

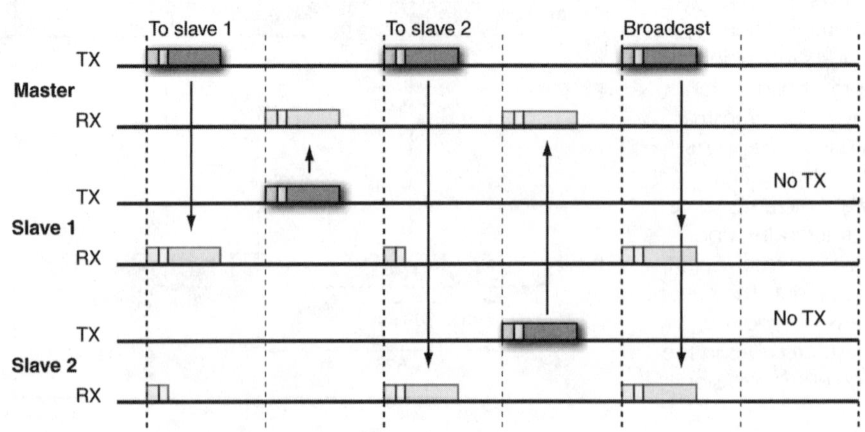

Figure 4-4 A multislave operation. Slaves communicate only with the master, and a slave can transmit only when addressed by the master in the previous time slot. Slaves not addressed can turn their receivers off after decoding the header.

turns its receiver off after decoding the header and waits for the beginning of the next even-numbered time slot. In short-range wireless devices, the receiver is often the most power-hungry device, even more so than the transmitter. The multislave TDD method is designed for power efficiency by requiring slave receivers to be active only when necessary. Broadcast packets are received by all slaves, but none can respond in the next time slot.

It's also apparent from Figure 4-4 that slaves communicate only with the master in the piconet. If two slaves want to exchange data, they can either go through the master or form their own independent piconet.

Multislot Packets

One of the obvious problems with a packet that's limited to 366 bits is the relatively high percentage of its total length that must be used for the access code and header. Both of these fields together require 126 bits, or 34 percent of the total packet length. We're left with the uncomfortable fact that only 240 payload bits can be accommodated within a 625 μs time slot, corresponding to an actual bidirectional data rate of only 384 kb/s, equivalent to 192 kb/s per node in a point-to-point configuration. As we will discover later, this data rate is reduced further when error control is employed, and the even lower rate must be divided between all members of the piconet.

For significant throughput enhancement, Bluetooth devices can transmit multislot packets of either three or five time slots in duration. Figure 4-5 shows examples of one-, three-, and five-slot packets and their timing. Every

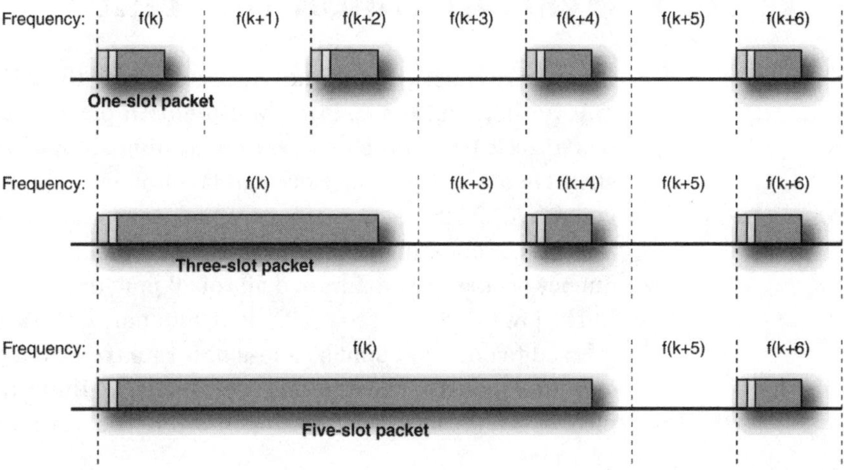

Figure 4-5
Single- and multislot baseband packets. All packets are sent on a single hop frequency.

baseband packet, regardless of its length, is transmitted on a single hop frequency. These channels (frequencies) are depicted as $f(k + n)$, where $k + n$ represents a particular time slot index, which is even for master-to-slave transmission and odd for slave-to-master transmission. When transmission of a multislot packet is completed, the hop sequence resumes at the channel it would have used had no multislot transmission occurred, so these longer packets have minimal disruption on piconet operation. Either the master or slave can transmit multislot packets if the other can support their reception.

Now the total number of bits in a single baseband packet can be much higher. If a five-slot packet is sent, four of the slots can be filled completely with bits, and about 366 μs of the last slot can be occupied, making a total length of $4 \times 625 + 366 = 2{,}866$ bits. (This number can be extended to 2,871 bits to accommodate the longest allowable packet.) The access code and header still require 126 bits, but this overhead now requires only about 4 percent of the total packet length. Throughput is improved considerably.

So why doesn't Bluetooth always use five-slot packets? The reason is that there's a high price to pay when a long packet is received with errors. In this event the entire packet must be resent, which requires another five slots (and a sixth slot for the recipient to acknowledge reception). Furthermore, long packets are more likely to be corrupted than short packets. Therefore, if the communication channel is error prone, throughput may be higher when one- or three-slot packets are used instead. We will quantify this proposition in the section "ACL Throughput Comparisons—Imperfect Channel."

System Clock for Timing Control

In order to meet the timing requirements of the TDD process, along with many other timing needs of the Bluetooth piconet, every Bluetooth device has a clock from which the various timing signals are derived. Figure 4-6 shows one way that this clock can be implemented. A 28-bit counter is triggered by 3.2 kHz square wave, which corresponds to an increment every 312.5 μs or one-half of a time slot. This reference oscillator must have a frequency accuracy (maximum drift) of 20 ppm and a maximum jitter of 10 μs. In the low-power modes (sniff, hold, and park), the accuracy requirement is relaxed to only 250 ppm so a less accurate (read: cheaper and lower power) oscillator can be used to trigger the clock. Important timing values for piconet operation can be tapped from the counter at various flip-flop locations.

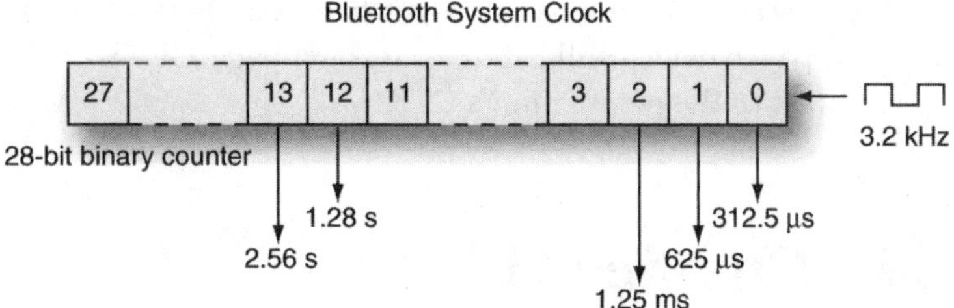

Figure 4-6 The Bluetooth clock from which timing signals are generated

The type of Bluetooth clock used for various piconet functions is different, depending upon the type of operation being performed by the device. All devices have a native clock, called CLKN, that runs whenever the Bluetooth device is powered up, and it is never adjusted; that is, the individual flip-flops are never preloaded with a particular value as part of piconet operation. CLKN turns over (overflows) about once a day.

Piconet timing is under control of the master, so the slaves must have a way to synchronize their clocks to that of the master. The piconet timing clock is called CLK, and for the master, CLKN = CLK. Slaves, however, must derive CLK from their own CLKN, and this is done by adding an offset to their CLKN to form CLK. For example, if the piconet CLK happens to be 0x5E421A3 and a slave's CLKN is 0x33B206A, the offset is CLK − CLKN = 0x2A90139. (I knew that old hexadecimal calculator would come in handy one of these days.) Remember, though, that the slave's CLKN is never adjusted. Instead, a new 28-bit register is loaded such that its value is equal to the slave's CLKN + offset = CLK.

Adding an offset to CLKN to get CLK cannot account for any offset that may be between the independent 3.2 kHz triggering instants of the master's and slaves' respective CLKN circuits that use the implementation shown in Figure 4-6. Therefore, a fine-tuning mechanism must also be employed to align the master's CLKN and slaves' CLK transitions. A slave accomplishes this by noting when each master's transmission begins and adjusting its timing signal transitions accordingly. For example, the 3.2 kHz triggering signal for CLK could in turn be derived from an auxiliary counter triggered by a faster oscillator. The slave could then add an offset to the auxiliary counter for timing resolution equal to the auxiliary counter's LSB transitions. The goal is for each slave in the piconet to have on hand a clock called CLK that has the same value as the master's CLKN, and these clocks all increment together every 312.5 μs.

A third clock, called CLKE, is an estimate of a paged device's CLKN by a paging device. As we will discover in Chapter 5, this estimate is important because it enables the pager to guess the time and frequency at which the paged device will listen for its next page. CLKE is derived by adding an offset to the paging unit's CLKN.

Physical Links

Two different physical links can be established between Bluetooth devices: *asynchronous connectionless* (ACL) and *synchronous connection-oriented* (SCO) links. The ACL link is used for data communication, and the SCO link is used for real-time two-way voice. In either case, transmission rules are established so that piconet members almost never jam each other by transmitting at the same time.

Synchronous Connection-Oriented (SCO) Link

If low latency is more important than data integrity, a SCO (pronounced "sko") link is established between the master and slave. Latency is the time between the creation of a new packet at the transmitting node and its successful reception at the destination node. The SCO link is a circuit-switched, point-to-point link between a master and single slave. Latency is guaranteed to be a small, fixed value through two methods:

- Packets are scheduled for transmission in specific time slots.
- Packets are never retransmitted.

A circuit-switched environment is required for real-time two-way voice communications where latencies in excess of a few tens of milliseconds can significantly impede the ability to communicate. Fortunately, voice reproduction from a digitized bit stream can tolerate a fairly high percentage of bit errors, so under most conditions the lack of packet retransmissions shouldn't be a significant detriment to performance.

SCO packets are exchanged in pairs, first from master to slave and then from slave to master, in consecutive time slots. The slave can transmit a digitized voice packet in its reserved slot even if the master doesn't transmit in the previous slot, but cannot if the master transmits a packet to a different slave in that slot.

Asynchronous Connectionless (ACL) Link

The ACL link is used where data integrity is more important than latency. Packet switching is used on the ACL link, where a packet received with uncorrectable bit errors is usually retransmitted until it's error free. The average number of retransmissions increases with increasing channel BER, so latency is variable and can occasionally be quite long.

A slave can transmit an ACL packet to its master only if it was specifically addressed by the master in the previous master-to-slave time slot. If a slave fails to decode a packet in a master-to-slave slot, it cannot transmit in the next slot. A master can also send broadcast packets that are messages intended for more than one slave. Active slaves cannot transmit in the slave-to-master slot following a broadcast packet.

The ACL link is also used for transmitting isochronous data, which is data that has timing issues that are less critical than real-time two-way voice. An example would be transferring streaming audio, such as an MP3 file, that's being played during reception. In this situation a buffer is filled with MP3 data before it begins playing, which enables some packet retransmissions to occur without interrupting the sound. However, if a certain packet requires too many retransmissions, it can be flushed, enabling the transmitting node to move on to the next packet in the file.

Error Control Coding

An understanding of error control is essential before Bluetooth baseband packet throughput figures can be calculated. Remember, for Bluetooth certification the receiver is required to have a raw BER of 10^{-3} or less at a signal level of -70 dBm. This BER is much too high for reliable file transfers, so a means must be used to reduce the BER by several orders of magnitude. This is accomplished through *error control coding*. We will look first at error control in general and then examine details of the Bluetooth implementation when packet structures are presented.

An error control code can consist of an *error detection code* for determining whether or not errors occur within a block of data, and/or an *error correction code* that can both find and correct a limited number of errors in the block of data. Both codes work by adding bits to the data to form redundant patterns. The sending node runs an algorithm to generate the redundant bits associated with the error control process, and these bits are transmitted along with the data. The receiving node performs another algorithm on the received bits to correct any errors and then to check the packet for any remaining uncorrected errors. If additional errors remain that cannot be corrected, then the receiving node requests a retransmission of the same information from the sending node.

Bluetooth has the capability to use both of these methods to reduce errors. Either a *binary repetition code* or a *shortened Hamming code* enables the receiver to correct, within limitations, the errors that appear in a data packet, and an *Automatic Repeat Request* (ARQ) process enables the recipient to request a retransmission of packets containing uncorrectable errors. A *cyclic redundancy check* (CRC) field at the end of data packets is used by the recipient to check a packet for bit errors as part of the ARQ process.

The foundations for error control coding are based on an area of mathematics called *finite fields*, discovered by a French mathematician named Evariste Galois (1811–1832). He was only 20 years old when he was killed in a duel. On the evening before he died he outlined his mathematical accomplishments and gave the information to a friend. The material was finally published in 1846, thus laying the groundwork for error control codes and cryptography. Who said mathematicians lead a dull life?

Error Detection

The CRC consists of a group of j bits (typically 8 or 16 bits) appended to the end of a *frame*, or block, of k bits of data. This group of bits is sometimes called the *frame check sequence* (FCS). The FCS bits are calculated such that the resulting $k + j$ bits are exactly divisible by some predetermined number. The receiver then divides the block of received data by the same number and checks the remainder. If it is zero, then the block is considered to be error free. If not, then the receiver asks for a retransmission of the data block.

One of the first questions that probably comes to mind is this: Do circumstances exist under which the CRC will check good, but the data block actually contains errors? This situation could be extremely serious because it could result in erroneous data being passed on to the Bluetooth host. To find the answer, we must first describe mathematically how the CRC is calculated.[1]

Mathematical Description of the CRC Process A string of bits can be represented as a *polynomial*, also called a *vector*, in terms of a dummy variable X such that each exponent corresponds to the position of each 1 in the string. For example, the binary string 11001 can be represented as $1X^4 + 1X^3 + 0X^2 + 0X^1 + 1X^0 = X^4 + X^3 + 1$. By doing this, we can perform special mathematical operations by using familiar polynomial algebra. Multiplication follows the usual rule of adding the exponents, such as $X^4 \times X^2 = X^6$. Addition is done with modulo-2, which is the same as the *exclusive-OR* (XOR) function: $0 + 0 = 0$, $1 + 0 = 1$, $0 + 1 = 1$, and $1 + 1 = 0$. The first three of these are the same as ordinary addition, but in the fourth the 1s add together to produce 0 with no carry bit.

Armed with these rules, we can now multiply two polynomials together, such as $(X^4 + X^2 + 1) \times (X^2 + 1)$ to get $X^6 + X^4 + X^4 + X^2 + X^2 + 1$. This reduces to $X^6 + 1$ because modulo-2 algebra cancels the duplicate terms. Notice that the disappearance of the two X^4 and two X^2 terms via modulo-2 arithmetic means that the result is different from actually multiplying the equivalent binary strings, or their decimal representations, together, but this special finite-field algebra is used to build the FCS. This algebra has the significant advantage of being lightning fast when performed by a microprocessor.

If the message in polynomial form is represented as $M(X)$ and the bit pattern $G(X)$ is the polynomial divisor (the *generator polynomial*), then the associated quotient $Q(X)$ and remainder $R(X)$ have the relationship

$$\frac{X^j M(X)}{G(X)} = Q(X) + \frac{R(X)}{G(X)} \tag{4-1}$$

The quantity X^j is a left-shift operator by j bits to make room at the end of the message for the FCS $R(X)$. If $R(X)$ has j bits, then $G(X)$ must have $j + 1$ bits. Now the sending unit transmits the sequence

$$T(X) = X^j M(X) + R(X) \tag{4-2}$$

As an example, suppose $M(X) = X^9 + X^7 + X^3 + X^2 + 1$ (1010001101) and $G(X) = X^5 + X^4 + X^2 + 1$ (110101). After performing the operation in Equation 4-1, we obtain $R(X) = X^3 + X^2 + X$ (1110), and the resulting $T(X) = X^{14} + X^{12} + X^7 + X^5 + X^3 + X^2 + X$, equivalent to 101000110101110. Note that the FCS is actually 01110 to make it the required 5 bits long.

After a bit of thought, it's easy to see (don't you just hate that phrase?) that if the sequence $T(X)$ is received error free, then

$$\frac{T(X)}{G(X)} = 0 \tag{4-3}$$

and if there are errors, then the result should not be zero. Therefore, the receiver performs the division in Equation 4-3 and asks for a retransmission if the result isn't zero.

Implementing CRC with a Linear Feedback Shift Register (LFSR) The CRC process can be easily implemented by creating a shift register bank with XOR feedback that emulates $G(X)$. A shift register using this topology is called a *linear feedback shift register* (LFSR). The data string is entered into the end of the LFSR starting with the MSB, and the LFSR is clocked once for each data bit entry. After k clock cycles, the last data bit will have been entered into the LFSR. Next the register bank is clocked j times with binary 0 at the input. Now the content of the LFSR flip-flops is the FCS. An LFSR implementation of $G(X) = X^5 + X^4 + X^2 + 1$ is shown in Figure 4-7. We'll leave it as an exercise for the reader (that's you) to clock the data string 1010001101 into this LFSR and obtain 01110 as the FCS.

LFSR techniques are used in a number of places in the Bluetooth link controller. Among these are

- Producing the CRC FCS
- Creating error correction codewords
- Generating random numbers
- Performing key generating operations for Bluetooth security

Although these various functions could be accomplished with different LFSR hardware implementations, they are usually done with software algorithms for increased efficiency. The shift-and-add (actually, shift-and-

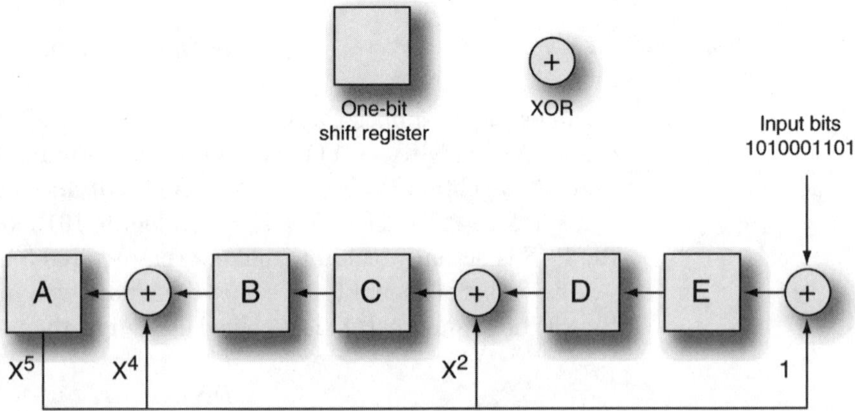

Figure 4-7 An LFSR implementation of an FCS encoder for a CRC using $G(X) = X^5 + X^4 + X^2 + 1$ (Source: W. Stallings[1])

XOR) routines that comprise the heart of LFSR operations are easy and fast in a microcontroller and require only a few lines of assembly code.

Bluetooth CRC Error Detection Capability Returning to our original question, we can now conclude that if $T(X)$ can be corrupted by a certain pattern of errors such that division by $G(X)$ produces zero anyway, then the CRC process fails and bad data could be accepted by the Bluetooth host. It turns out that all of the following bit patterns are *not* divisible by $G(X)$ and thus will be detected as errors in the data:[2]

- All single-bit errors
- All double-bit errors as long as $G(X)$ has a factor with at least three terms
- Any odd number of errors as long as $G(X)$ contains the factor $X + 1$
- Any burst error that has a length less than the length of the FCS
- Most burst errors of longer length

A burst is defined as a binary sequence beginning and ending with 1 and having, at worst, randomized bits in between. If the received data block contains a burst error longer than the FCS length, then it's possible that the block will be divisible by $G(X)$ with no remainder, and thus, the burst error will go undetected. For a burst error of length r, where r is greater than the length of the FCS, the probability that the CRC will fail is approximately equal to 2^{-r}. As an example, suppose the FCS is 16 bits long. All burst errors of 15 bits or less will be detected by the CRC with certainty, but if the burst is 20 bits long, then the CRC will erroneously indicate that the data is good with a probability of only 2^{-20}, or about one in a million. This probability may seem high, but remember that the chance of 20 bits in a row being randomized from a noise burst is usually low to begin with. Also, many data files contain their own error-checking mechanism independent of that included in Bluetooth, so the end result is an exceedingly tiny chance that bad data will be accepted as good.

The baseband packet header consists of 10 bits protected by a CRC in the form of an additional 8-bit FCS. The *header error check* (HEC) generator polynomial is

$$G(X) = X^8 + X^7 + X^5 + X^2 + X + 1 \tag{4-4}$$

$$= (X + 1)(X^7 + X^4 + X^3 + X^2 + 1)$$

so $G(X)$ contains the factor $X + 1$. Therefore, we can conclude that the HEC will detect all errors that are odd in number, all double-bit errors, burst errors of 7 bits or less, and most other error patterns. Having 10 bits of data protected by an additional 8-bit FCS may at first appear to be overkill, but it's very important to the proper operation of the piconet that no undetected errors occur in the header.

A baseband packet payload can have as many as 2,745 bits, which includes a 16-bit FCS. The generator polynomial used to create the FCS is called CRC-CCITT, defined as a CRC using the generator polynomial given by

$$G(X) = X^{16} + X^{12} + X^5 + 1 \tag{4-5}$$

$$= (X + 1)(X^{15} + X^{14} + X^{13} + X^{12} + X^4 + X^3 + X^2 + X + 1)$$

As with the HEC, $G(X)$ contains the factor $X + 1$, so this CRC will detect all odd errors and all double-bit errors. Burst errors of 15 bits or less and most other error patterns will also be discovered. Notice, however, that longer payloads have a correspondingly higher probability that the CRC will fail to detect a certain pattern of errors, but in any case the chance of this happening is extremely small—so small, in fact, that we will assume for later analysis that the CRC will never fail to detect errors in the payload.

ARQ The TDD nature of communication between the master and slave is ideally suited for the implementation of a simple ARQ process for repeating the transmission of bad packets. Here's how it works:

1. The sending node transmits a data packet.
2. The receiving node checks packet integrity through the FCS.
3. The receiving node returns an *acknowledgment* (ACK) packet in the next time slot if the packet is good; otherwise, it returns a *negative acknowledgment* (NAK) packet.
4. The sending node transmits the next data packet in the following time slot if an ACK is returned; otherwise, it repeats the same packet in the following time slot.

The ACK/NAK information is contained in the baseband packet header. Note that each packet is transmitted on a different hop frequency, so if one of the frequencies is unusable because of interference, then the resulting NAK, the packet retransmission, and the ACK probably won't use this particular frequency again.

Now here's a question. (No, don't panic, we don't ask that many, and there aren't even any problems at the end of each chapter.) Will the ARQ process guarantee that a file will eventually be transferred completely error free? The answer is, somewhat surprisingly, yes, provided the following criteria are met:

- The FCS will never fail to discover a packet containing errors.
- The ACK/NAK process always operates correctly.

We've already made the reasonable assumption that the FCS will almost always work, and that additional error control beyond Bluetooth is usually present as a backup. For the ACK/NAK process to be reliable, we must assume that a NAK is never mistaken for an ACK (the HEC will prevent this). Furthermore, we must also assume that the sending node never fails to receive an incoming ACK. If an incoming ACK is missed, then the sending node will retransmit the previous packet, but the receiving node will think that it's getting a new packet. A corrupted file will then result.

So what happens when a hop frequency has interference and the sender misses an ACK? The baseband packet header also contains an SEQN bit that is toggled (changed from binary 1 to 0 or from 0 to 1) for each new packet transmitted by the sending node. The receiving node can now check for duplicate packets by examining their SEQN bits and discarding any that arrive due to a lost ACK. The entire ARQ sequence of events is shown in Figure 4-8. We can now conclude that the probability is extremely high that a data file will eventually be transferred from the sender to receiver error free.

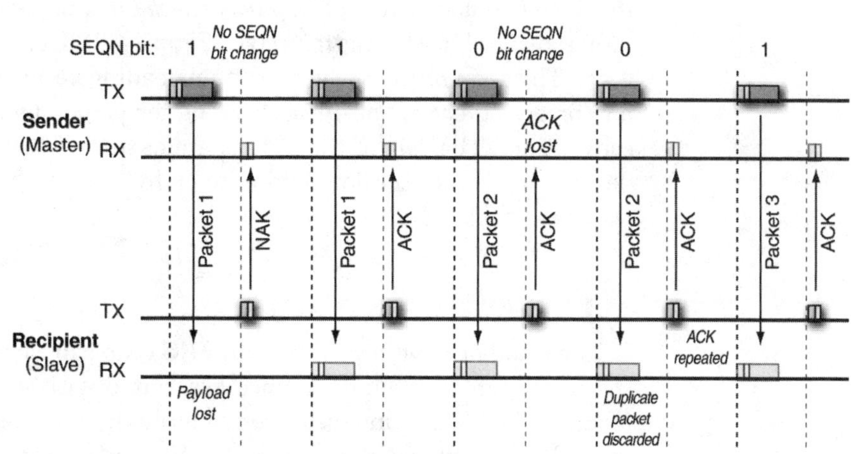

Figure 4-8 The Bluetooth baseband ARQ process

Here's a tip to save a bit of energy. The sender will repeat a packet transmission if a NAK is returned by the recipient or if the recipient doesn't respond at all. Therefore, if the recipient has nothing to say except NAK, then it could save power by remaining silent in that time slot instead. This technique is efficient when the sender is the master and the recipient is a slave. If the recipient is the master, then it should return a NAK for missed packets to authorize that slave to retransmit the packet in the following time slot.

Error Correction

Error detection is essential for a wireless communication link to operate with an effective BER low enough that file transfers can be accomplished essentially error free. Unfortunately, the only way to recover from a bad packet so identified is though its retransmission using ARQ, and this can increase delay and decrease throughput markedly when the raw BER across the communication channel is somewhat high.

An alternative to simply having a go/no-go verdict (CRC) on a packet of data is to include an error correction code with the packet's data bits. This code is generated by an algorithm at the sending node that creates additional bits that the receiving node can use in another algorithm to correct up to a certain number of errors per block of data. Because the recipient performs error correction without any further assistance from the transmitting node, the process is called *forward error correction* (FEC).

Bluetooth uses a type of FEC called a *systematic block code*, where a block of data of length k bits (usually smaller than the length of the packet itself) has an additional $n - k$ *parity check bits* appended to it such that the total codeword block length is n (see Figure 4-9). This code is called an (n, k) code. (The term *parity check* is a bit misleading because these bits are used for correcting errors, not just checking for errors. Once again we'll follow convention.) The *rate* of the code is a measure of the fraction of the block devoted to actual user data and is given by

$$r = \frac{k}{n} \tag{4-6}$$

The number of bit errors that an FEC code can correct is determined by the minimum Hamming distance associated with its set of codewords of length n. The Hamming distance is simply the number of bits between two equal length strings of binary data that differ. For example, the strings

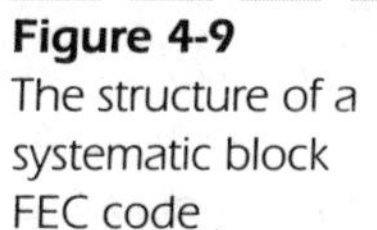

Figure 4-9 The structure of a systematic block FEC code

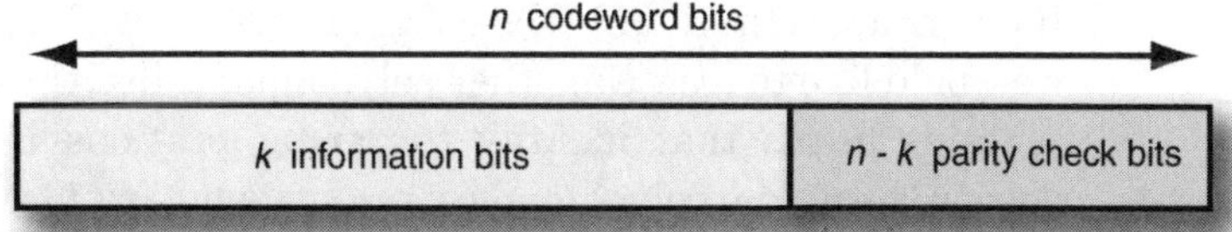

11010 and 10100 have a Hamming distance of three because the middle three bits are different between the two strings. The Hamming distance between arbitrary codewords in an error correction code may vary, depending upon which codewords are selected. The minimum Hamming distance is the smallest distance between any two codewords in the set. If this distance is called d, then the maximum number of errors t that the code can correct is given by

$$t = \left\lfloor \frac{d-1}{2} \right\rfloor \tag{4-7}$$

where the fancy brackets mean "integer part of." As an example, if the minimum Hamming distance of a particular code is three, then it can correct one error per encoded block of data.

Bluetooth uses three types of block FEC codes in its packets. The first is a modification of a (63,30) *Bose-Chadhuri-Hocquenghem* (BCH) code into a (64,30) code that is part of the access code and will be discussed in the next section. The (3,1) binary repetition code always protects the packet header and can be optionally used for SCO packet payloads. Finally, the (15,10) shortened Hamming code can be optionally used to protect SCO or ACL packet payloads.

Mathematical Description of the FEC Process You may think that the mathematics behind error correction codes is similar to error detection codes, and you would be right. A generator polynomial $G(X)$ is used on the message vector $M(X)$ to create a remainder $R(X)$, as shown in Equation 4-1. The parity check bits that comprise $R(X)$ are appended to the end of the message, just like the FCS bits are for CRC. The resulting codeword vector $T(X)$ is transmitted. Because FEC is more complex than CRC, the FEC generator $G(X)$ is usually longer than that used for the FCS.

Decoding the FEC codeword is also more involved because we must somehow identify where the errors are within the received binary string. Suppose the received binary string is written as a polynomial vector $V(X)$.

If no errors exist, then $V(X) = T(X)$. However, if $V(X)$ contains errors, then we would like to create another polynomial called the *error vector* $E(X)$ that contains terms that identify the error locations in $V(X)$. Remember, a nonzero term in a polynomial represents a binary 1, so if we simply perform a term-by-term XOR operation between $V(X)$ and $E(X)$, the result will be $T(X)$. Why is that, you ask? A binary 1 in $E(X)$ will cause the bit in the same corresponding position in $V(X)$ to invert, correcting the error at that location. Neat, yes? The bad news is that finding $E(X)$ requires finite-field matrix algebra, which is beyond the scope of this discussion. However, you can find a practical discussion of this subject in *Digital Communications: Fundamentals and Applications* by B. Sklar[3] and a more in-depth analysis in *Theory and Practice of Error Control Codes* by R. Blahut.[4]

Binary Repetition Code The binary repetition code is one of the simplest FEC codes. The transmitting node simply repeats each data bit n times, where n is odd. At the receiver a majority vote is taken to determine the actual bit that was sent. The (3,1) binary repetition code used by Bluetooth has two codewords. A 1 is encoded as 111 and a 0 is encoded as 000. The minimum Hamming distance (in fact, the only Hamming distance) for this code is three, so according to Equation 4-7, the code can correct a single-bit error within each block.

Notice that if a double- or triple-bit error occurs in a codeword block, then the decoder will fail and output a data bit error. This situation is rare if the bit errors are independent and the overall BER is low, but a burst of errors caused by high noise or interference could cause consecutive errors to occur within a single codeword. Therefore, if data integrity is critical, then an error detection scheme, such as a CRC, is still required to check for uncorrected errors after the FEC process is completed.

The binary repetition code is not particularly efficient because the highest rate possible with this code is only 1/3. Because the blocks are quite short, it runs the risk of failing even when a short burst error occurs. For example, two consecutive bit errors have a two out of three chance of occurring within one 3-bit block, and three consecutive bit errors will always cause one block to be decoded in error. The advantage of this code is that the encoding and decoding algorithms are extremely simple and fast, so there is very little delay between the time the signal arrives at the receiver and the original information can be extracted. It is also fairly powerful with its capability to correct one bit error for every 3 bits received. Both of these features are important during baseband packet reception because the payload immediately follows the header, so it's important to reliably decode and then interpret the header with minimal delay.

Shortened Hamming Code An (n, k) Hamming code has a structure given by[4]

$$(n, k) = (2^m - 1, 2^m - 1 - m) \tag{4-8}$$

For example, if $m = 5$, then the codeword length n is 31 bits, and the corresponding data block length k is 26 bits with an additional 5 bits for error correction. This code can correct one bit error or detect two bit errors anywhere within the 31-bit codeword.

Now put your thinking cap on. No, not that one. Get the big, heavy one. Suppose the first 16 bits of the 26-bit data block are always 0. What will a codeword look like for this code? Well, because the systematic codeword consists of the data itself plus 5 extra bits for error correction, it will be 31 bits long, but the first 16 bits will always be 0. Now suppose we simply drop these 16 bits rather than waste time transmitting them. By dropping these extraneous bits, we've created a (15,10) shortened Hamming code, which is one of the FEC codes available for use on baseband data packet payloads. The generator polynomial for this code is

$$\begin{aligned} G(X) &= X^5 + X^4 + X^2 + 1 \\ &= (X + 1)(X^4 + X + 1) \end{aligned} \tag{4-9}$$

which is used in Equation 4-1 to generate the correct 5 parity check bits for any 10-bit message vector. This shortened code can actually be more powerful than the original (31,26) code because the one correctable error or two detectable errors can occur over a shorter 15-bit field. In other words, a higher BER can exist before this new code becomes overwhelmed.

Figure 4-10 gives examples of how this code operates. The transmitting node uses its error correction encoder on 10 data bits to create a 15-bit codeword that is sent over the air. The receiver feeds the 15 bits into its error correction decoder. Provided there is at most one error in the incoming codeword, then the decoder will output the correct data sequence. If two errors exist, then the receiver's decoder will set an error alarm to inform the Bluetooth controller that the data block is bad, but the location of the errors is unknown and, hence, can't be corrected. A NAK (or no response at all) is returned and the sender retransmits the packet. Finally, if three or more errors occur in the 15-bit block, then the decoder may output the incorrect data block without necessarily triggering the error alarm. All is not lost, however, because the CRC is still available to check data integrity after the

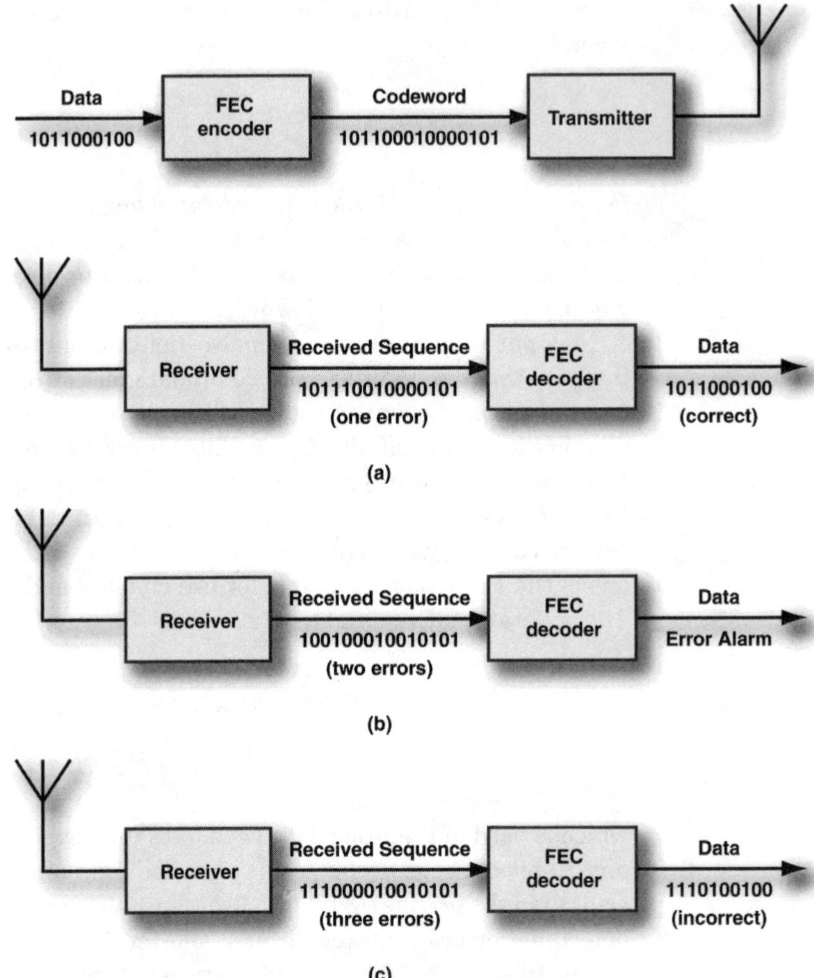

Figure 4-10 Operation of a (15,10) systematic block FEC code in which one error (a), two errors (b), and three or more errors (c) occur within a block of 15 codeword bits. The code can correct one error and detect two errors. (The illustrated bit patterns do not necessarily belong to the Bluetooth [15,10] shortened Hamming code.)

FEC decoder submits its output for all blocks within a single baseband packet. Once again, a NAK (or stony silence) is returned, and the sender retransmits the packet.

Let's summarize the error control process as it applies to a baseband packet payload. The transmitting node first appends a CRC to a set of data, and then divides it into blocks of k bits and runs the FEC encoding algorithm on each block. (If the last block is shorter than k bits, then the encoder appends binary 0s to the string until its length is k bits.) Next, the sequence of codewords, each of length n, is concatenated and transmitted as a packet payload. The receiving node separates the payload from the rest of the packet, divides the sequence into blocks of n bits each, and uses its FEC decoder on each block to obtain k bits of data. These are then concatenated to form the (hopefully) original data set. Finally, the CRC is checked to assess data integrity and a retransmission is requested if necessary. Isn't that easy? Well, almost easy; there's a data whitening process as well, but we'll cover that in the section "Data Whitening" in this chapter.

Bluetooth Addresses and Names

Bluetooth baseband packet structure will make a lot more sense if we first examine how Bluetooth addressing is accomplished. The important addresses are the *Bluetooth device address* (BD_ADDR), the *active member address* (AM_ADDR), the *parked member address* (PM_ADDR), and the *access request address* (AR_ADDR). These addresses are all binary numbers usually represented in hexadecimal form, but a Bluetooth unit can be given a plaintext name as well for better interface to us humans.

Bluetooth Device Address (BD_ADDR)

BD_ADDR (pronounced "bee-dee-adder") is a 48-bit address that's unique to each Bluetooth device. Its format is given in Figure 4-11, which follows the IEEE 802 standard. The BD_ADDR is divided into three fields, which are

- *Lower address part* (LAP) containing 24 bits
- *Upper address part* (UAP) containing 8 bits
- *Nonsignificant address part* (NAP) containing 16 bits

The NAP is nonsignificant only in that it isn't used to determine things such as the Bluetooth hop channel set or various access codes, but it is still

Figure 4-11 Fields of the BD_ADDR

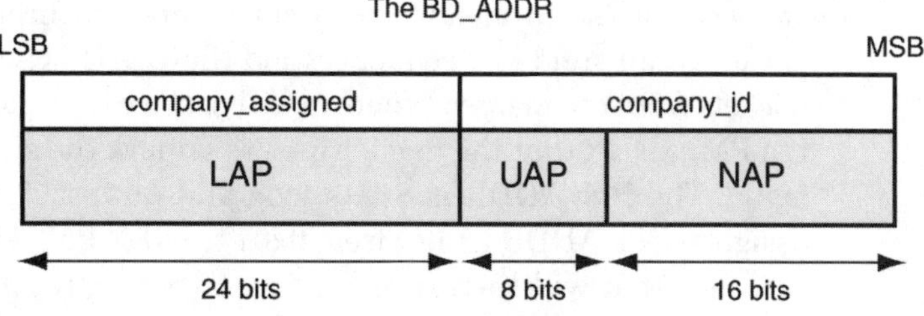

part of the BD_ADDR and helps make each address unique. The NAP is also used for Bluetooth security. The other two fields are used for items such as identifying the piconet, paging particular Bluetooth devices, and generating the frequency hop channel set. Some of the LAP fields are reserved and therefore cannot be assigned to a specific Bluetooth device.

The UAP and NAP together form a 24-bit entity called the *company_id* that is assigned by the IEEE 802 group as an *organizationally unique identifier* (OUI). The 24-bit LAP, which is company_assigned, is appended to the company_id to form the BD_ADDR, so a single company_id can support over 16 million Bluetooth devices. In fact, the 48-bit Bluetooth address space is big enough that every person on earth could own over 50,000 Bluetooth devices, each with a unique BD_ADDR. In spite of the vast addressing space available, address conflicts can still occur across Bluetooth piconets, as we shall soon discover.

Active Member Address (AM_ADDR)

A master needs the capability to address each slave in a piconet individually, but it would be a waste of transmit time to use the 48-bit BD_ADDR for this purpose. Because there are at most seven active slaves in a piconet, it's possible to assign each slave a unique address within only a 3-bit address space. The AM_ADDR is a 3-bit address assigned by the master to slaves as they enter the piconet. Up to seven active slaves can be assigned AM_ADDR values from 001 to 111. The AM_ADDR value 000 is reserved for broadcast packets from the master to multiple slaves. The AM_ADDR is the first field in the baseband packet header.

Parked Member Address (PM_ADDR)

Although the piconet is limited to seven active slaves (slaves actively communicating with the master), an unlimited number of slaves can be *parked members* of the piconet. Parked slaves are synchronized to the master's packet timing and hop sequence, and they periodically listen for broadcast packets from the master. When an active slave is parked, the master assigns it a PM_ADDR that the master uses to unpark the slave and make it active again. The PM_ADDR is 8 bits long and up to 255 parked slaves can be assigned PM_ADDR values from 0x01 to 0xFF. PM_ADDR 0x00 is assigned to slaves that will then respond only to their BD_ADDR for unpark commands from the master. Chapter 6, "Advanced Piconet Operation," provides more information on the low-power modes sniff, hold, and park.

Access Request Address (AR_ADDR)

One of the most powerful features in Bluetooth piconet organization is the capability of parked slaves to request an unpark command from the master, equivalent to a peripheral requesting service by interrupting the *central processing unit* (CPU). This is done through the AR_ADDR. This address is assigned by the master when parking a slave and is used by the slave to determine its *access window*, which is a special half slot of time in which the slave can send an unpark request to the master. Details of the access request procedure are covered in Chapter 6.

Bluetooth Device Name

The *Bluetooth device name* is a user-friendly (finally!) name that can be assigned to a particular Bluetooth device. This name consists of a maximum of 248 bytes of data and is assigned according to the *Unicode Standard Transformation Format Eight* (UTF-8) standard. The UTF-8 dictates how *American Standard Code for Information Interchange* (ASCII) characters are to be encoded into an 8-bit format. Typical Bluetooth names might be "Jim's cell phone" or "Printer, Room 5-D."

Bluetooth Baseband Packet Format

As we mentioned earlier, there are three parts to a baseband packet: the access code, header, and payload (refer to Figure 4-2). Not all packets contain all three parts, however. Instead, packets can be constructed in one of three ways:

- Access code only (68 bits)
- Access code and header (126 bits)
- Access code, header, and payload (up to 2,745 bits)

Because the payload can be of variable length, a baseband packet containing a payload also has a variable length with its maximum determined by whether one, three, or five time slots are used for its transmission. As each new part of the packet is introduced, we'll discuss how these different packet types are used.

Access Code

In almost every wireless packet communication system, the packet itself begins with a special pattern of bits that is known by the receiving node. These bits provide synchronization for the RF carrier phase (if necessary) and help the receiving node find both the bit and word (byte) boundaries. In this way, when the actual information begins arriving, the receiver's timing circuitry is calibrated and ready to detect subsequent bits having values that aren't necessarily known in advance.

All baseband packets begin with an *access code* that provides bit and word synchronization. (Of course, carrier phase synchronization is not necessary because noncoherent bit detection is used.) In general, the access code

- Can be used by a slave to resynchronize its CLK to the piconet's CLK (master's CLKN)
- Provides bit and word synchronization
- Sets DC compensation for the receiver's detector so BER is minimized
- Includes basic piconet identification information

The access code consists of three parts: the *preamble*, the *sync word*, and the *trailer*, as shown in Figure 4-12. The preamble and the first bit in the sync word combine to form a 5-bit sequence of alternating 1s and 0s that provide bit synchronization and give the receiver's detector a chance to set

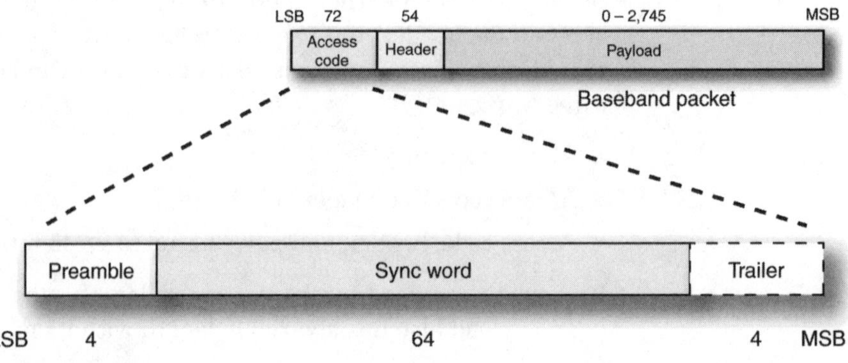

Figure 4-12 The access code is made up of a preamble, sync word, and (if more bits follow) a trailer.

its decision threshold midway between the one and zero voltage levels for the lowest possible BER. Likewise, the trailer, which is present only if a header follows the access code, couples with the last bit in the sync word for another 5-bit alternating binary sequence to prepare the receiver's detector for correctly decoding the packet header.

Successfully detecting the access code is of critical importance to piconet operation because if the access code is missed or decoded improperly, then the rest of the packet will also be missed. In order to improve the probability that the access code is detected, it has the following features:

- The complete access code is always known by the recipient. Therefore, the receiver can duplicate the expected access code's bit pattern and place it into a correlator. Incoming data is then shifted into the correlator and checked for a match after each shift.
- The minimum Hamming distance between two different sync words is 14. As such, if the receiver's correlator detects a match of 57 bits or more within an incoming sync word, then it can be nearly certain that the expected access code has arrived.

Figure 4-13 shows how a correlator works conceptually. To simplify the mathematics, we will represent a binary 1 as +1 (no surprises there) and a binary 0 as −1. We will also use a binary sequence shorter than 64 bits for clarity. An incoming bit sequence is lined up against the expected sequence, the terms are multiplied bit by bit, and the product terms are added together. Next, the result is tested against a threshold. If the threshold is exceeded, then a match is assumed to have occurred.

If a perfect match occurs over all 64 bits of an incoming sync word, then the correlator will output a value of 64. Up to seven of these bits could be received in error, and the correlator will still output a value of 57 or higher. On the other hand, if an undesired sync word is placed into the correlator, which we could model as a string of random +1 and −1 values, then the average correlator output will be 0, well below the threshold value for a match. The same low average output value will occur when the desired sync word isn't aligned with the expected sequence in the correlator, so the correlator's output includes precise bit alignment timing information as well as detection of the desired sync word itself.

For a correlator to work well, an incoming desired bit sequence shouldn't exceed the threshold at shifts other than correct alignment. In other words, the desired sequence needs to have low *autocorrelation* values at shifts other than precise alignment. Also, it's important that undesired sequences don't exceed the threshold at any shift value, so undesired sequences should have low *crosscorrelation* with the desired sequence.

Figure 4-13 A correlator operation when the incoming sequence is correct and properly aligned (a), correct but not aligned (b), or incorrect (c)

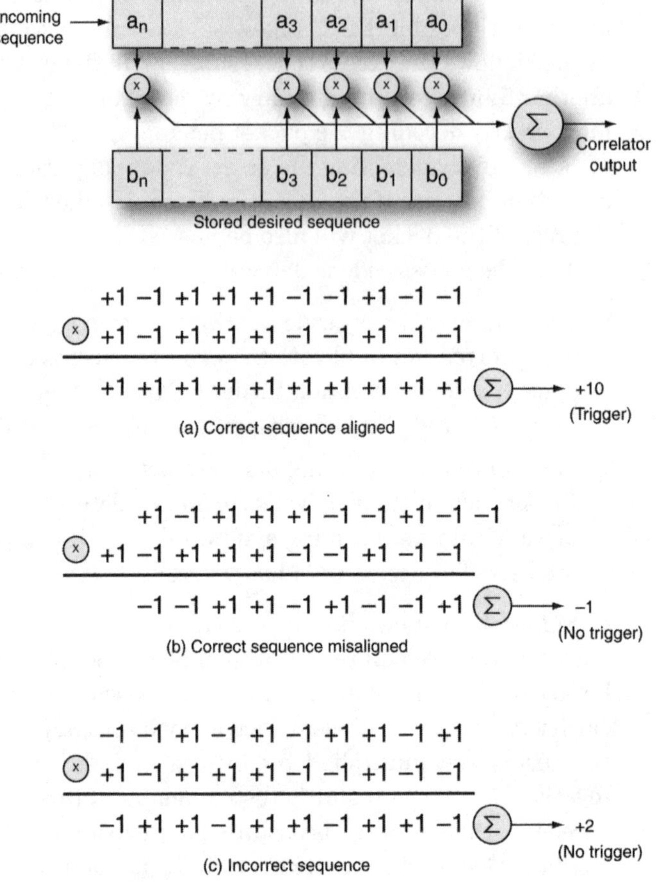

Access Code Types The access code is derived from the 24-bit LAP of a device's BD_ADDR. But which device's BD_ADDR? That depends upon how the access code is used. There are four contexts in which different access codes are required:

- ***Channel access code*** **(CAC)** Every packet sent from master to slave and from slave to master in the piconet begins with the CAC. This access code is derived from the LAP of the master's BD_ADDR. Piconet members use the CAC to check for the correct LAP before accepting the rest of the packet. Because the LAP is not necessarily unique between two Bluetooth devices, it's quite possible that separate piconets within range of each other could have the same CAC. There's a risk, then, that a piconet member could accept a packet from the wrong piconet. As we'll discover in the next section, the baseband packet header has additional safeguards to prevent this from happening.

- ***Device access code* (DAC)** The DAC is used by the master for paging a specific Bluetooth device for entry into its piconet. The master knows the paged device's BD_ADDR via an inquiry process and can assemble the correct DAC from that device's LAP. The paging and inquiry processes are covered in Chapter 5.
- ***General inquiry access code* (GIAC)** The general inquiry process enables a prospective master to put out a call inviting every Bluetooth device within range to reply. Devices answer with, among other things, their BD_ADDR which the prospective master can use to build a DAC for a subsequent page. The inquiry is accomplished by transmitting a GIAC, which is an access code built from the reserved 24-bit LAP of 0x9E8B33.
- ***Dedicated inquiry access code* (DIAC)** The DIAC enables a prospective master to narrow its inquiry down to devices having a specific capability. For example, if a printer is being sought, it makes little sense to send a GIAC and sift through all the responses looking for a printer. Devices having a particular capability will respond to the associated DIAC. These 63 reserved addresses are 0x9E8B00 to 0x9E8B3F, except for the one belonging to the GIAC.

An *inquiry access code* (IAC) can refer to either the GIAC or DIAC. Furthermore, an *ID packet* can refer to either a DAC or IAC—that is, a DAC, GIAC, or DIAC, transmitted by itself. (Isn't this fun?) An ID packet is always 68 bits long. On the other hand, the CAC is almost always followed by at least a header, in which case its length is 72 bits due to the appended 4-bit trailer. The DAC, GIAC, and DIAC can be used alone as a short 68-bit packet (ID packet) or as a 72-bit string (with attached trailer) along with a header and payload to form a *frequency hop synchronization* (FHS) packet. The FHS packet will be covered in the section "FHS Packets."

Sync Word Derivation Building the sync word requires a sequence of steps, all of which involve the familiar shift-and-add routines that can be implemented with LFSR techniques. Figure 4-14 gives a pictorial representation of this method. The steps are as follows:

1. *Obtain the desired LAP*. The 24-bit address is retrieved that corresponds to the type of access code (CAC, DAC, GIAC, or DIAC) desired. This LAP is arranged so that the LSB is sent first, so it appears on the left side of Figure 4-14.

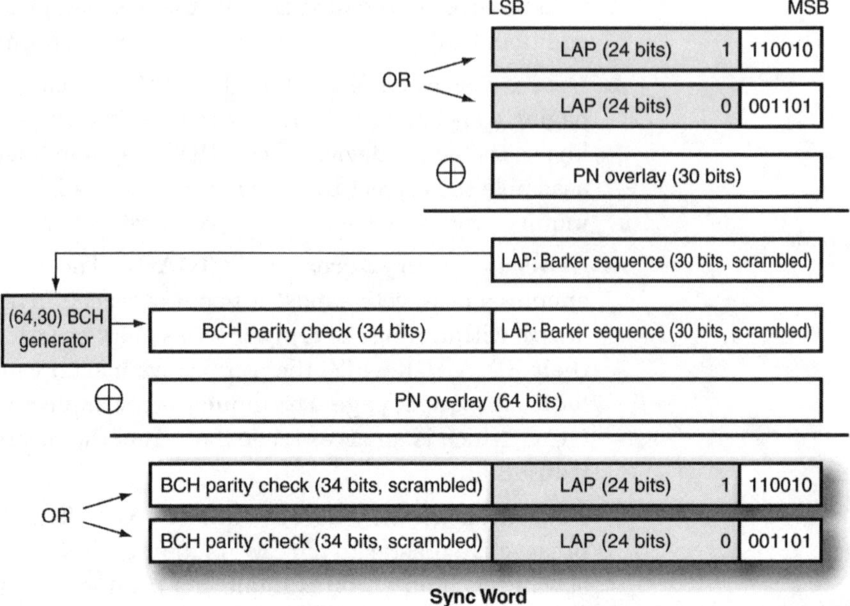

Figure 4-14 Generating the sync word

2. *Append the partial Barker sequence*. A 6-bit sequence is appended to the MSB (right) end of the LAP such that the last 7 bits form a Barker sequence. The two possible sequences are just binary complements of each other. Barker sequences have very desirable autocorrelation properties. If two identical 7-bit Barker sequences are exactly aligned with each other in a correlator, then the output will be +7, but if one of the sequences is offset by one or more bits, then the correlator will output either +1 or −1, depending upon the offset value. Therefore, including a Barker sequence onto the LAP will improve the sync word's autocorrelation properties.
3. *XOR the resulting sequence with part of the pseudonoise (PN) overlay*. When the access code is finished, part of it will be encoded with a pseudorandom sequence, also called PN. This is a sequence of binary 1s and 0s that appear random, but are actually generated by an LFSR using the polynomial $G(X) = X^6 + X^4 + X^3 + X + 1$. The PN sequence is always the same for all sync word constructions, but it has the effect of scrambling the 30-bit LAP:Barker sequence into what appears to be a random number.
4. *Generate the codeword*. The scrambled 30-bit sequence is now sent to a (64,30) modified BCH encoder, out of which a 64-bit codeword emerges.

This code is systematic, so the scrambled sequence appears as the most significant 30 bits of the 64-bit codeword.

5. *XOR the codeword with the entire PN overlay*. After the BCH codeword is XORed with the PN overlay sequence, the original 30-bit unscrambled LAP:Barker sequence reappears with 34 additional (scrambled) codeword bits appended to it.

The resulting 64-bit sync word thus consists of three parts. These are, from LSB to MSB, 34 bits of a scrambled BCH codeword, 24 bits of the selected LAP, and (along with the MSB of the LAP) a 7-bit Barker sequence. The sync word has good autocorrelation and crosscorrelation properties, helped in part by the Barker sequence and scrambled BCH codeword.

Although reversing the sync word generation process to extract the error-corrected LAP would be straightforward, remember that the receiver doesn't actually need to decode the sync word because its pattern is already known. Instead, it detects the presence of the sync word with its correlator. Both transmitting and receiving nodes create identical sync words from the same LAP. The transmitting node transmits an access code using this sync word. The receiving node is notified of its arrival when the correlator output exceeds a predetermined threshold value.

Using the Access Code in Short Hopping Sequences During the inquiry and paging processes, a prospective master tries to find (inquire) or connect with (page) other Bluetooth devices within range. As we will discover in Chapter 5, the time for a successful inquiry or page to take place can be reduced significantly if the usual 79-channel frequency set is shortened to only 32 channels. However, we learned in Chapter 1, "Introduction," that, until recently, the FCC required at least 75 hopping channels for a FHSS communication system to be legal in the 2.4 GHz ISM band. Even with wideband frequency hopping, 32 channels that are each 1 MHz wide won't meet the FCC requirement to cover at least 75 MHz of the band. So how does Bluetooth get away with using 32-channel sequences for paging and inquiry?

The answer is that the access code, which is used alone for paging and inquiry transmissions, does meet the FCC rules for a hybrid spread spectrum system. A hybrid system combines both direct sequence and frequency hopping techniques. (Direct sequence, used in 802.11b Wi-Fi, will be covered in more detail in Chapter 13, "Coexisting with Other Wireless Systems.") FCC rules require hybrid spread spectrum to have a *processing gain* of at least 17 dB from the combined direct sequence and frequency hopping techniques. For direct sequence, processing gain is the ratio of the number of transmitted bits to the number of actual information bits, expressed in dB. For frequency hopping, the processing gain is the number of hop channels expressed in dB.

The access code transmitted by itself has 68 bits, of which 24 bits are the LAP of the associated BD_ADDR. The remaining bits are part of a PN sequence that adds 10 $\log_{10}(68/24) = 4.5$ dB of processing gain. Likewise, the 32 hopping channels increase processing gain by another 15 dB for a total of 19.5 dB. (It could be argued that the preamble and trailer should be considered information rather than part of the PN sequence, but even so the FCC requirements for the hybrid spread spectrum processing gain are still met.) Bluetooth is, therefore, a hybrid spread spectrum system during inquiries and pages and an FHSS system during normal piconet operation.

Header

The header, if it exists, always immediately follows the access code. The header contains real information—that is, information that isn't necessarily known ahead of time by the recipient. The actual information field is only 10 bits long, but its correct reception is extremely important to proper piconet operation.

Description of Fields The baseband packet header consists of 10 bits divided up into five fields, along with an 8-bit HEC. These 18 bits are then encoded with the (3,1) binary repetition code, making a total of 54 bits that are actually transmitted.

The various fields in the header are shown in Figure 4-15. These fields are

- **AM_ADDR** This 3-bit address distinguishes between the active slaves in the piconet. This address is temporary and is assigned to a slave during the page process (see Chapter 5). A packet sent from the master will have the destination slave's AM_ADDR listed in the header. Other slaves in the piconet, upon decoding a header with AM_ADDR different from theirs, can turn off their receiver until the next even-numbered time slot in which the master will send a new packet (refer to Figure 4-4). The AM_ADDR of 000 isn't assigned to any slave; instead, this address is used for broadcast packets from the master to multiple slaves. (The FHS packet may also have AM_ADDR = 000, but it is not a broadcast packet.) Slaves that either leave the piconet or are parked by the master relinquish their AM_ADDR.

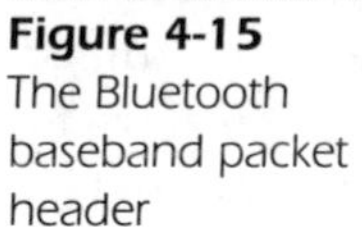

Figure 4-15 The Bluetooth baseband packet header

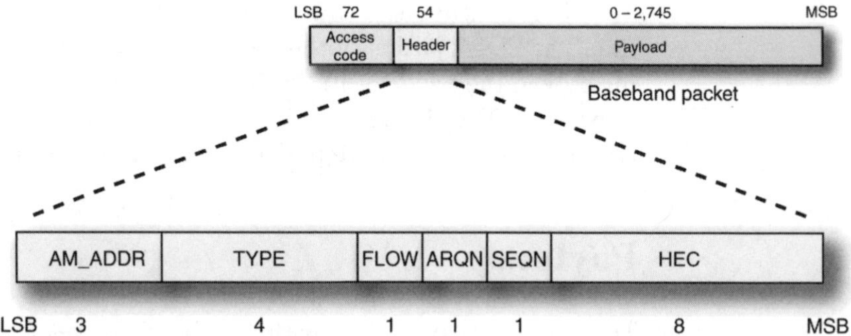

- **TYPE** The 4-bit TYPE of packet field enables up to 16 different packet types to be named. The packet types are different for data and real-time two-way voice applications. The number of slots occupied by the packet is also included in the TYPE field, so a nonaddressed slave knows when to begin listening for the next master transmission. Details of the packet types will be covered in the next section.
- **FLOW** This bit controls the flow of ACL packets. For example, if the receiver buffer fills with incoming packets, then flow is stopped (FLOW = 0) in the return header. The other node can send only ID, POLL, NULL, or SCO packets until flow is resumed (FLOW = 1).
- **ARQN** The ARQN bit tells the source that a packet was successfully received (ARQN = 1) or had errors (ARQN = 0) and is used in the ARQ process.
- **SEQN** This is a sequential numbering bit to prevent duplicate packets from being accepted by the receiving node due to a failed ACK.
- **HEC** The 8-bit HEC sequence is created with the generator polynomial given by Equation 4-4. To provide additional protection from inadvertently accepting a packet from another piconet using the same access code, the HEC LFSR is preloaded with the master's BD_ADDR UAP during normal piconet operation. The HEC value is then calculated in the usual manner. The UAP is part of a device's company_id, so the probability is greatly reduced (but not completely eliminated) that a packet from another piconet with the same CAC will be mistakenly accepted.

NULL and POLL Packets Two types of baseband packets can be assembled that consist of only an access code and a header, with no payload. These are the NULL and POLL packets, both of which are a fixed 126 bits long.

The NULL packet is used to convey ARQ or FLOW information across the link. The POLL packet, although similar in structure to the NULL (differing only in the header's TYPE field), is used by a master to ping a slave for a response. The slave is required to respond with any packet; this can be a NULL packet if the slave has no information to convey to the master.

Payload

The third part of a Bluetooth baseband packet consists of the payload, which is where the user information is located. The payload structure differs, depending upon whether the packet is an FHS, ACL, or SCO type. We'll discuss the FHS packet in this section, followed by ACL and SCO packets in the next section.

Synchronization Within FHSS Devices We learned in the last chapter that for two devices to communicate using FHSS, they must be properly synchronized so that they hop together from channel to channel. This means that the devices must

- Use the same channel set.
- Use the same hopping sequence within that channel set.
- Be time synchronized within the hopping sequence.

When devices are time synchronized and hopping together, they are said to be using the same hop sequence and phase. The FHS packet contains the information needed to meet the previous criteria. Before discussing the FHS payload fields, we'll first look at how Bluetooth devices generate their hop sequences.

While a member of a piconet, either as a master or as a slave, a Bluetooth device's hop generator (see Figure 4-16) has as one of its inputs the least significant 28 bits of the master's BD_ADDR, which determine the hop sequence being used by the device. The hop phase is determined by the most significant 27 bits of CLK, created by combining the unit's CLKN and an offset. The LSB of the CLK bits used for hop phase changes every 625 μs, resulting in a hop rate of 1,600 per second. Of course, because CLK = CLKN for the master, its offset is zero. The hop generator's output is the value sent to the frequency synthesizer of the unit's radio. The goal, then, is for every participant in a piconet to have the same input values to their hop generators at each instant in time so that they hop together from frequency to frequency. In other words, they are using the same hop sequence and

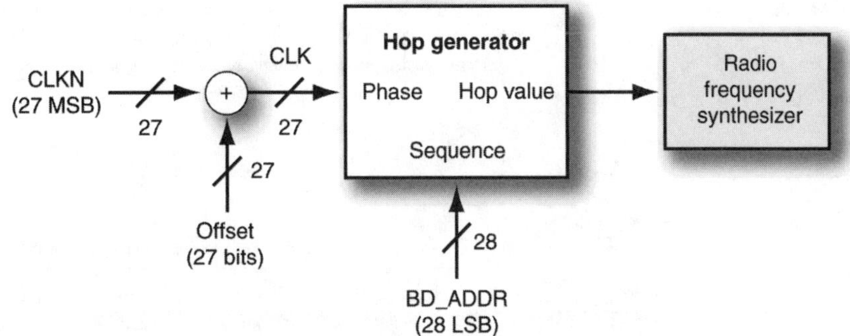

Figure 4-16 For a piconet member, the Bluetooth hop generator takes as its inputs the most significant 27 bits of CLK and the least significant 28 bits of the master's BD_ADDR and outputs a hop frequency value to the radio's synthesizer.

phase. (We'll look at some of the other hop generator input values in the next chapter.)

FHS Packet The FHS packet is the means by which the master sends its frequency hopping information to a slave during the page procedure. The FHS packet is also sent from devices that are responding to an inquiry. This packet consists of 144 bits of user data and a 16-bit FCS that are all protected by the rate 2/3 FEC for a total payload length of 240 bits. The access code, header, and payload combine to make 366 bits, completely filling a single time slot.

The data fields of the FHS payload are shown in Figure 4-17 and are described in the following list:

- **Parity bits** These are the same as the first 34 bits in the sync word of the unit that sends the FHS packet.
- **LAP** This is the LAP of the BD_ADDR of the device that sends the FHS packet. The sender's sync word can be easily reconstructed by using these first two FHS fields.
- ***Scan repetition* (SR)** This indicates the time interval between two successive page scan windows, described in Chapter 5.
- ***Scan period* (SP)** This is the time period in which the mandatory page scan mode will be used by the device after it responds to an inquiry, described in Chapter 5.
- **NAP** This is the NAP of the BD_ADDR of the device sending the FHS packet.
- **Class of device** The class field identifies the major function of the device sending the FHS packet. Classes include services such as networking, rendering, capturing, object transfer, audio, telephony, and

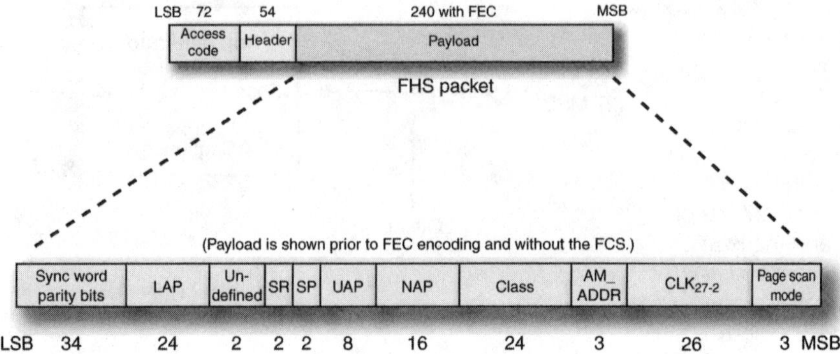

Figure 4-17 Payload fields in the FHS packet

information. These values are contained in *Bluetooth Assigned Numbers* by the Bluetooth SIG.[5]

- **AM_ADDR** This is the active member address assigned by the master to the slave during the page procedure. The field is set to 000 when the FHS packet is associated with a response to an inquiry.
- **CLK_{27-2}** These are the most significant 26 bits of the CLKN of the device sending the FHS packet. Because the two LSBs aren't included, the resolution of this field is two time slots. CLK_{1-0} can be assumed to be 00 when the master sends its FHS packet to a slave during page because this transmission occurs at the start of an even-numbered time slot.
- **Page scan mode** This informs the recipient of the type of page scanning that the sender of the FHS packet uses at times other than when the mandatory page scan mode is required. These will be discussed further in Chapter 5.

Notice that the FHS packet contains the sender's entire BD_ADDR and enough bits of the sender's CLKN to enable a recipient to program its hop generator to follow the sender's hopping from channel to channel. Although we discuss Bluetooth security in detail in Chapter 9, "Bluetooth Security," it's clear at this point that because the FHS packet is sent over the air, Bluetooth FHSS doesn't really enhance security of the transmitted data. Any receiver with the capability to hop and possessing the algorithms given in the Bluetooth specification can hop along with the piconet members and intercept their transmissions. Furthermore, this interception can occur without the knowledge of the piconet users.

Packets on the Physical Links

The two types of physical links that Bluetooth supports, ACL and SCO, have different latency, ARQ, and packet structure requirements. Indeed, these two links are treated separately by the piconet participants. We will first look at ACL packets, followed by a comparison of throughput using the various ACL options available. Next, we'll look at the SCO packet format and how several of these links can be multiplexed.

ACL for Data

User data other than real-time two-way voice are usually transferred using ACL packets. This includes commands and responses associated with controlling the piconet. Seven ACL packet types have been defined, six of which have a CRC code and thus can employ ARQ for reliable communication.

The payload part of an ACL packet usually has three fields in it. These are the payload header (which is different from the baseband header that we've already discussed), user data, and FCS. The FCS is often called the CRC code in the Bluetooth specification. Figure 4-18 shows this structure

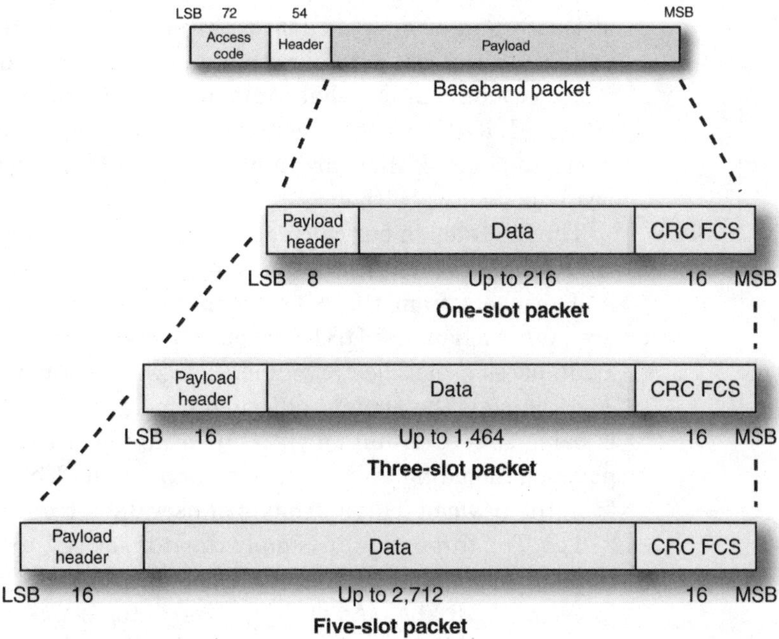

Figure 4-18 Payload structure for ACL packets using one, three, and five time slots and no payload error correction coding. The maximum payload lengths are slightly different in three- and five-slot packets when FEC is used. The baseband header and payload header are separate fields.

for packets without any payload error correction coding. The content of the payload header will be covered in Chapters 7 and 8.

ACL packets have a three-character identifier consisting of two letters followed by a number. The first character is always a *D*, which means *data*. The second character can be either an *H* for *high speed* or *M* for *medium speed*, and the last character can be a 1, 2, or 3, which identifies the number of time slots used by the packet. The packet is high speed if it has no FEC, and it is medium speed if the (15,10) shortened Hamming code is applied to the payload. As an example, DM3 is a data, medium speed, and three-slot ACL packet. Notice that ACL packet payloads never use the (3,1) binary repetition error correction code; that code only protects the baseband packet header.

A special ACL packet, called the *AUX1*, is used when transferring raw data between two Bluetooth devices. The payload of this single-slot packet has an 8-bit header but no FCS. No ARQ is possible, and link quality cannot be guaranteed with these packets. The AUX1 packet is rarely used to transfer data, but can be handy when testing the BER across a channel. Known bit patterns can be placed into AUX1 packets and the received data compared to the expected data over a period of time.

ACL Throughput Comparisons—Error-Free Channel The throughput of any of the ACL packets can be calculated in a straightforward way, at least in an error-free channel. First, remember that the Bluetooth raw data rate is 1 bit per microsecond, so any time duration expressed in microseconds contains that same number of bits. We assume for the calculations that the piconet has one master communicating with one slave (point-to-point). If there are multiple slaves, then the total throughput must be divided between them.

The first step in our calculations requires identifying what we'll call the *cycle time* T_{cycle}, defined as the time between the start of two consecutive ACL packets from the same node. For example, if a master and slave are both exchanging DH1 packets, then each transmits one packet every 1,250 μs, so that's the cycle time. Next we need to determine the *data time* T_{data}, which is the amount of time during a cycle in which actual user data is being sent. Continuing with our example, a DH1 packet has a 240-bit payload, including an 8-bit header and 16-bit FCS. Subtracting these bits from the payload leaves 216 bits of user data, transmitted over a period of 216 μs. The throughput is simply the ratio of T_{data} to T_{cycle} in Mb/s. For DH1 packet exchange, then, each user has a throughput of 216/1,250 = 0.1728 Mb/s, or 172.8 kb/s. (As we'll discover later, some of the higher protocols, such as L2CAP, have their own overhead, so the user throughput values that we're calculating here may still be optimistic.)

The previous example assumed that the master and slave both use the same type of packet for their data exchange in what is called a *symmetric channel*. There are many situations, such as file transfer, in which one node will transmit data as three- or five-slot packets, but the other node will return only a single-slot NULL or perhaps POLL (if the node is the master) packet containing the ACK/NAK information in the header, or perhaps a DH1 or DM1 packet if a small amount of data is to be returned as well. This *asymmetric channel* has a *forward direction* for the long packets and a *reverse direction* for the single-slot return packets. After a little reflection it's easy to see (there we go, using that phrase again) that the forward throughput on the asymmetric channel is higher than it would be if the reverse channel also contained multislot packets for symmetric communication.

Now take a look at Table 4-1. This table lists the different ACL packet types with their supported user data bits, the T_{data} and T_{cycle} for symmetric and asymmetric channels, and the associated maximum throughputs. Asymmetric data is assumed to contain single-slot packets in the return channel that have the same FEC as the packets in the forward channel; for example, if the forward channel uses DM3 packets, then the reverse channel contains DM1 packets.

According to Table 4-1, the lowest maximum throughput comes with using DM1 packets, where a very large percentage of time is dedicated to overhead. As a result, maximum throughput is only about 10 percent of the 1 Mb/s raw transmission rate in each direction. On the other extreme, an asymmetric channel using DH5 packets results in a maximum forward throughput of more than 700 kb/s, or about 13 times faster than a 56 kb/s telephone modem.

To paraphrase a question we asked earlier, why would any Bluetooth node use anything other than DH5 packets, either on an asymmetric or symmetric channel, depending upon the needs? We will address this question next.

ACL Throughput Comparisons—Imperfect Channel After finishing the arithmetic required to calculate the numbers in Table 4-1, we realize with a sinking heart that perhaps most of the numbers there are of academic interest only. Why would a Bluetooth node settle for only 387.2 kb/s by using DM3 packets on an asymmetric channel when merely changing to DH5 packets nearly doubles the maximum throughput? The answer, of course, is that a practical channel has a BER greater than zero, so we should try to devise a way to account for poor channel conditions in the throughput calculations.

Table 4-1

Maximum throughput of ACL symmetric and asymmetric channels with BER = 0

	Symmetric Channel			Forward Asymmetric Channel			Reverse Asymmetric Channel		
Type	**T_{data} μs**	**T_{cycle} μs**	**Throughput kb/s**	**T_{data} μs**	**T_{cycle} μs**	**Throughput kb/s**	**T_{data} μs**	**T_{cycle} μs**	**Throughput kb/s**
DM1	136	1,250	108.8	–	–	–	–	–	–
DH1	216	1,250	172.8	–	–	–	–	–	–
DM3	968	3,750	258.1	968	2,500	387.2	136	2,500	54.4
DH3	1,464	3,750	390.4	1,464	2,500	585.6	216	2,500	86.4
DM5	1,792	6,250	286.7	1,792	3,750	477.8	136	3,750	36.3
DH5	2,712	6,250	433.9	2,712	3,750	723.2	216	3,750	57.6

Once this realization hits us, we begin to experience the familiar symptoms of mathematical panic that arise anytime we try to perform an analysis where probability is involved. At least I did. It was almost enough to make me try physics instead (see Figure 4-19). Once we recover and begin to collect our thoughts, it becomes apparent that we should begin the analysis by finding the probability of packet success given various BER values likely to be encountered on the Bluetooth channel. Table 4-1 contained the implicit assumption that the BER was zero, resulting in a packet success probability of one. Even so, the throughput values in this table will be useful in the upcoming analysis.

Let's define a few terms before tackling the actual mathematics:

- The channel BER is represented by bit error probability p.
- The packet success probability is called Q_E.
- There are L total payload bits in a packet, including payload header and FCS.
- If FEC is used, then the block length is n bits, within which t errors can be corrected.

Jocular Science
by Bill Lae
EE=(mc²)²
© 1987 Lae
"Breaking new ground again, eh Phillips?"

Figure 4-19 Which is more difficult, physics or engineering? (Source: Bill Lae)

Keep in mind that the FEC protects the entire payload, including the payload header and FCS. We will also assume for ease of analysis that the access code and header, with their more robust error control, will always be received correctly and that bit errors in the payload are independent of each other. (Chapter 13 contains some results when interference, instead of independent bit errors, destroys Bluetooth baseband packets.)

The probability of packet success Q_E can now be determined using binomial analysis techniques. If no FEC is used, then a packet of length L is received successfully if and only if (sometimes called *iff* in the vernacular of formal logic) all bits are received correctly. This probability is

$$Q_E = (1 - p)^L \tag{4-10}$$

If an error correction code is used that can correct t errors within a block of n bits, then a packet is successful if and only if every block has t or fewer errors in it. The probability that a single block will be successful, which we call Q_n, is given by

$$Q_n = \sum_{i=0}^{t} \binom{n}{i} p^i (1 - p)^{n-i} \tag{4-11}$$

Equation 4-11 simply determines the chance that there are t or fewer errors within a block of n bits of data given a BER of p.

Because there are L/n total blocks of data (where L/n is assumed to be an integer), then the probability that the entire payload is received successfully is equal to the probability that none of the blocks has more than t errors. This is given by

$$Q_E = (Q_n)^{L/n} \tag{4-12}$$

By carefully scrutinizing Equation 4-10, we can conclude that it is just a special case of Equations 4-11 and 4-12 with $t = 0$ and $n = 1$.

Now the actual user data throughput is found by multiplying the result in Equation 4-12 by the throughput figures in a perfect channel given by Table 4-1. Notice that as L/n becomes larger, which it will for multislot packets, Q_E becomes smaller and performance is diminished. It would be interesting, then, to see whether there are cases where user data throughput will actually be higher when shorter packets are used. Furthermore, the error correction code overhead in a DM packet reduces the amount of user data within that packet, but its Q_E is higher than that of a similar-length DH packet when the BER is greater than zero. Is there a BER above which data throughput will be improved by using FEC?

To find out, let's first put some numbers specific to Bluetooth into the equations for Q_E. We have $t = 1$ and $n = 15$ for DM packets, and $t = 0$ and $n = 1$ for DH packets. The total number of payload bits L in a one-, three-, and five-slot packet is given in Table 4-2, along with values for t, n, and L/n. The reason L is sometimes different for DM and DH packets with the same slot duration is from the requirements that the total number of bits (including FEC-encoded payload header and FCS) in a DM payload be divisible by 15 so there's an integral number of (15,10) FEC codewords, and the number of bits in a DH packet be divisible by 8 for an integral number of bytes.

For DM packets, we can now simplify Equation 4-11 to

$$Q_n = (1 - p)^{15} + 15p(1 - p)^{14} \tag{4-13}$$

and then insert this result and the values for L/n from Table 4-2 into Equation 4-12 to obtain the Q_E. Throughput for a given BER is obtained by multiplying Q_E by the appropriate DM throughput values in Table 4-1. For DH packets, we insert the $L/n = L$ values from Table 4-2 into Equation 4-10 and multiply the resulting Q_E by the DH throughput values in Table 4-1.

Figure 4-20 shows a plot of the different DM and DH packet throughput figures for various BER values. (Note to reviewers: It's true—a technical book *can* have a good plot.) For multislot packets, we've plotted the forward channel throughput only. The different packet types exhibit various degrees of robustness as channel conditions deteriorate. As long as bit errors are

Table 4-2

Payload length and error correction parameters for ACL packets with CRC

Type	L	t	n	L/n
DM1	240	1	15	16
DH1	240	0	1	240
DM3	1,500	1	15	100
DH3	1,496	0	1	1,496
DM5	2,745	1	15	183
DH5	2,744	0	1	2,744

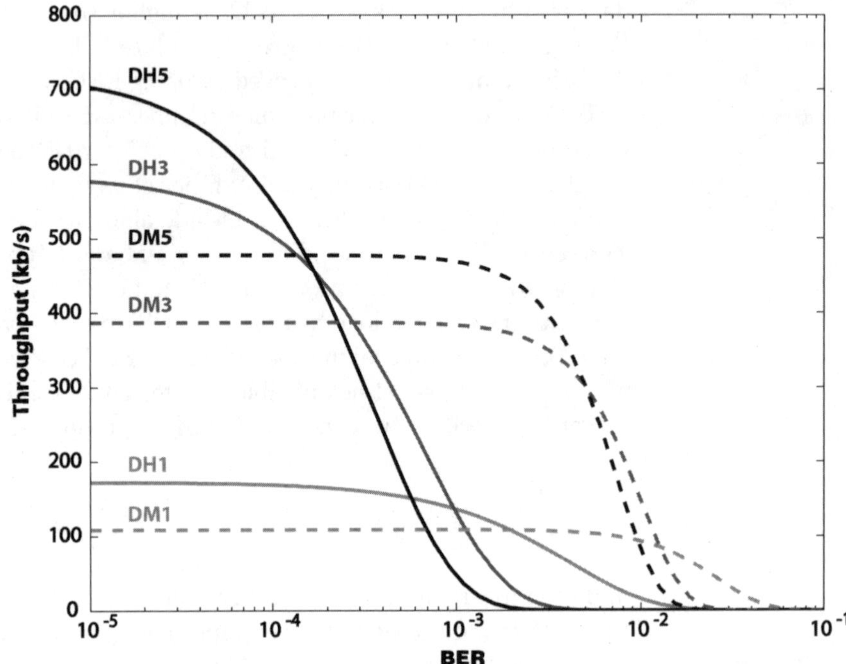

Figure 4-20 Throughput for various ACL packet types when the BER is greater than zero

independent, the following rules could be used to obtain reasonable throughput:

- Use DH5 packets for BER values less than 10^{-4}.
- Use DM5 packets when the BER is between 10^{-4} and 10^{-2}.
- Use DM1 packets when the BER exceeds 10^{-2}.

Although there is a BER interval in which DM3 packets outperform DM5, that BER interval is small, and the performance gain is fairly low. In a nutshell, then, the rule for ACL packet selection is simple: Use the longest packet possible, either with or without error correction, until the BER gets really bad; then switch to the shortest packet with error correction to keep the link alive. Of course, a BER worse than 0.01 will probably affect the reliability of access code and header reception as well, so actual throughput under these abysmal conditions will most likely be lower than that shown in Figure 4-20. Also notice that the DM5 packet performs best when the BER is 10^{-3}, which is the standard used for measuring receiver sensitivity for Bluetooth certification.

SCO for Real-Rime Two-Way Voice

User data consisting of digitized real-time two-way voice are transferred using SCO packets. Three SCO packet types have been defined, along with a fourth type that combines SCO and ACL data within the same payload field. An SCO link is symmetric between a master and a single slave, so slots are reserved to support a 64 kb/s digitized voice stream in each direction. Bluetooth sends digitized voice in a manner similar to that used in DECT. (If the Boolean variable `you_are_European = true`, then DECT means *Digital European Cordless Telephone*, but if `you_are_European = false`, then DECT means *Digital Enhanced Cordless Telephone*.) Each baseband SCO packet contains a few dozen bits of the voice stream, and because the raw Bluetooth data rate is 1 Mb/s, the relatively slow 64 kb/s digitized voice stream can be sent in periodic short bursts, leaving plenty of time to listen for a return SCO packet from the other end of the link. Because Bluetooth devices exchange single-slot packet pairs at the rate of 800 per second, a SCO link sounds like it's full duplex to the user.

The payload part of a SCO packet has only one field in it, consisting of 240 bits of digitized voice data, with or without error correction. These packets don't contain a payload header, nor is there an FCS. As a result, SCO packets are never retransmitted, and no ARQ exists. This technique reduces latency to a few milliseconds at the expense of having to live with the inevitable bit errors. As the BER increases, then, the SCO link will begin to suffer from distorted audio.

Like their ACL counterparts, SCO packets also have a three-character identifier consisting of two letters followed by a number. Unfortunately, some of the letters and numbers are the same as ACL packets, but their meaning is completely different. The first two SCO packet characters are always *HV*, which means *high-quality voice*. The last character can be a 1, 2, or 3, which identifies the type of error correction in the payload. The payload of a HV1 packet has the (3,1) binary repetition code, the HV2 payload contains the (15,10) shortened Hamming code, and the HV3 payload has no error correction capability. The easiest way to keep these straight is to realize that the digit is the numerator in the FEC code rate; that is, HV1 uses rate 1/3, HV2 uses rate 2/3, and HV3 uses rate 3/3 (that is, no) FEC. Is that clear? Never mind . . . .

The structures of the three types of SCO packets are shown in Table 4-3. The analog voice stream is sampled at 64 kb/s and the data is sent to the link controller for encoding into HV packets. These all have identical payload sizes of 240 bits (30 bytes), but the amount of digitized voice contained in each one differs due to FEC overhead.

Table 4-3

Structure of SCO HV packets

Type	Bits of Digitized Voice	FEC	Payload Size in Bits	Packets per Second for 64 kb/s
HV1	80	(3,1)	240	800
HV2	160	(15,10)	240	400
HV3	240	none	240	267

A special SCO packet called *data / voice* (DV) is used when it's necessary to include some data other than digitized voice in one of the SCO time slots. For example, some link control commands and responses can be piggy-backed in this way if desired. The DV packet contains in its payload 80 bits of digitized voice with no FEC, along with an 8-bit header, up to 72 bits of ACL data, and a 16-bit FCS that is calculated for the ACL data alone. Only the ACL part of the packet is subject to ARQ, so if a retransmission is required, then the old data is attached to a subsequent DV packet with new voice information or placed within a pure ACL packet in a non-SCO time slot.

It's obvious from Table 4-3 that a single HV1 SCO link will fully load the Bluetooth piconet, with packet timing depicted in Figure 4-3 way back at the beginning of this chapter. The link can be managed via commands sent in the payload of a DV or DM1 packet, either of which replaces the HV1 packet scheduled for a particular time slot. Short commands as part of a DV packet won't interrupt the voice stream, but longer commands that require a DM1 packet will interrupt the sound for 80 bits (1.25 ms). The receiving node can easily interpolate the voice waveform during this time, and the user shouldn't even notice as long as the interruption occurs only occasionally. More information on link management can be found in Chapter 7.

For a SCO link with HV2 packets, two consecutive time slots out of every four are used for SCO packet exchange. If HV3 packets are used instead, then two consecutive time slots out of every six are used. Therefore, up to two bidirectional voice links can be supported with HV2 packets, and up to three can exist when HV3 packets are used, before the Bluetooth piconet is fully saturated. These multiple voice links can exist between a master and single slave or between a master and multiple slaves. Of course, the piconet master must be at one end of each of these voice channels.

For scatternet operation, a single device can support one HV3 link in each of two piconets. Because different piconets aren't time synchronized, a user in one piconet needs to be away for at least three consecutive time slots to be able to exchange two SCO packets in a different piconet. Therefore, scatternet operation isn't possible for devices exchanging HV1 or HV2 packets.

Packet TYPE Codes

The TYPE code is a 4-bit field in the header that tells the recipient what kind of packet follows. Packets fall into three general categories: those on ACL links, those on SCO links, and those that can be used on either link. These are listed in Table 4-4 with their TYPE codes given from MSB to LSB, but remember that the codes are transmitted from LSB to MSB.

Table 4-4

Packet TYPE codes

TYPE Code	Slot Occupancy	ACL Link	SCO Link
0000	1	NULL	NULL
0001	1	POLL	POLL
0010	1	FHS	FHS
0011	1	DM1	DM1
0100	1	Undefined	DH1
0101	1	HV1	Undefined
0110	1	HV2	Undefined
0111	1	HV3	Undefined
1000	1	DV	Undefined
1001	1	Undefined	AUX1
1010	3	Undefined	DM3
1011	3	Undefined	DH3
1100	3	Undefined	Undefined
1101	3	Undefined	Undefined
1110	5	Undefined	DM5
1111	5	Undefined	DH5

Packet Broadcasting

As we've already learned, packets having an AM_ADDR = 000 (except FHS) are broadcast packets that can include information for multiple slaves. For example, the master in a wireless telemetry network may broadcast a request for all slaves to take a certain measurement simultaneously.

Because a broadcast message isn't addressed to one particular slave, no active slave can respond in the following slave-to-master slot, so broadcast packets have no built-in ARQ to ensure that the message was received by all slaves. Aside from the usual error control, *quality of service* (QoS) (that's another way of saying "reliability") is improved by sending a broadcast message a number of times to improve the chance of reception. A message that would ordinarily require sending M baseband packets to each slave in the piconet could instead be broadcast by transmitting each of the M packets N_{BC} times, for a total of MN_{BC} broadcast packets. If an ACK is required, then the master must poll each slave in turn for a response.

If a separate poll is required for each slave to acknowledge successful reception of a broadcast packet, won't it be more efficient simply to send duplicate messages to each slave separately and automatically receive a packet-by-packet ACK through the ARQ process? That's true for shorter messages to fewer slaves, but broadcasting and separate ACK polling may be more efficient for longer messages to many slaves.

Calculating the threshold message length beyond which it's more efficient to broadcast is straightforward as long as (ahem) the channel BER is zero. Let's first (re)define a few terms:

- M is the number of packets in the broadcast message.
- N_{BC} is the number of times each broadcast packet is sent.
- S is the number of slaves to receive the broadcast message.
- T_{BC} is the number of time slots needed to complete the message transaction using broadcast packets.
- T_S is the number of time slots needed to complete the message transaction by sending separate copies to each slave.

If a broadcast message is sent and then slaves are individually polled for reception, the number of time slots to complete the process, assuming all slaves receive the message successfully, is

$$T_{BC} = 2MN_{BC} + 2S \qquad (4\text{-}14)$$

and if the message is sent separately to each slave using conventional packets, then the time for completion, again assuming all slaves receive the message successfully, is

$$T_S = 2MS \tag{4-15}$$

Equation 4-15 essentially assumes that the communication channel's BER is low enough that no packet retransmissions are needed. To be fair, we should assume the same for Equation 4-14 as well by setting $N_{BC} = 1$; thus, Equation 4-14 reduces to

$$T_{BC} = 2(M + S) \tag{4-16}$$

Now it's a simple matter to discover under what conditions $T_{BC} < T_S$; this occurs when $M + S < MS$. Therefore, it's more efficient to broadcast a message and poll each slave separately for ACK when the sum of the number of message segments and the number of slaves receiving the message is less than their product, at least when the BER is low. If we make the reasonable assumption that a broadcast message is directed to at least two slaves, then any message contained in three or more separate packets will be sent more efficiently using packet broadcasting and subsequent polling rather than sending the message separately to each slave, at least when the BER is low.

If the BER is not particularly low, then T_{BC} may be large because the message must be repeated individually or sent as another broadcast message to those slaves returning a negative reply to their subsequent poll. In these situations it may be more efficient to simply send the same message to each slave separately.

Data Whitening

Before any baseband packet is actually transmitted, it is scrambled through a procedure known as *whitening*. This is done by feeding the packet header and payload into an LFSR with generator function $G(X) = X^7 + X^4 + 1$. This LFSR makes the packet appear to be composed of random data, so the whitening process

- Removes redundant patterns from the actual data
- Removes any DC bias due to a predominance of either binary 1s or 0s in the original payload

Whitening is not done for security reasons because the whitening algorithm is publicly disclosed in the Bluetooth specification. Instead, the main motivation for whitening is to enhance the performance of the FSK detector in the Bluetooth radio receiver. Many inexpensive detectors set their binary 1/0 decision threshold to a value based upon the assumption that the data contains approximately the same number of 1s and 0s. This assumption must hold for relatively short data strings as well as over the packet as a whole for the BER to be minimized. By removing redundant bit patterns and DC bias, the whitening process accomplishes both of these criteria, making the receiver's detector blissfully happy.

Figure 4-21 shows how error control and whitening fits into the transmission and reception sequence of baseband headers and payloads. Even though CRC and FEC may be optional for payloads, whitening is always performed on the entire packet. It's obvious that Bluetooth never sends plaintext over the air; all packets must be dewhitened before they can be interpreted.

Logical Channels

The physical links available over Bluetooth can be used in a number of different ways to become *logical channels*. These are channels that use the

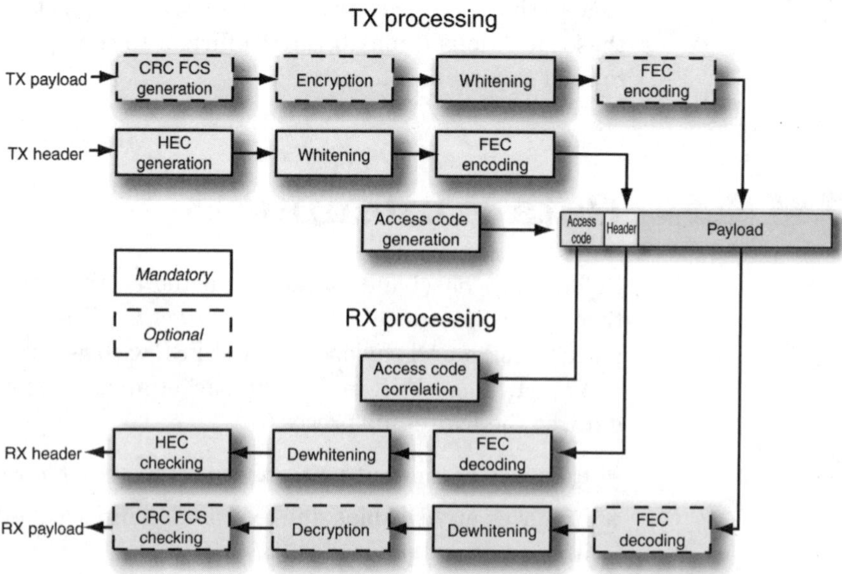

Figure 4-21 The bitstream processing for the Bluetooth baseband packet header and payload

various physical attributes of Bluetooth baseband packets to best meet the differing requirements for the type of data that is being exchanged.

The five different logical channels can be described in the following way:

- ***Link controller*** **(LC)** This channel carries the low-level link control information such as ARQ, flow control, sequencing, and payload type. This channel is mapped to the packet header.
- ***Link manager*** **(LM)** This channel contains control information that is exchanged between the link managers of communicating Bluetooth devices (see Chapter 7). This channel is mapped to DM1 packet payloads.
- ***User asynchronous*** **(UA)** This channel carries user data where integrity is of higher priority than latency. An example is file transfer. These packets always include an FCS with optional FEC. Each packet is retransmitted until it is received error free within certain link timeout parameters. This channel is mapped to ACL packet payloads.
- ***User isochronous*** **(UI)** This channel carries user data where integrity and low latency are both about equally important. An example is *streaming* (one-way) real-time audio. These packets always include an FCS with optional FEC. Each packet received in error is retransmitted until a *flush timeout* occurs (see Chapter 8), after which the next packet is sent. This channel is mapped to ACL packet payloads.
- ***User synchronous*** **(US)** This channel carries data where low latency is of higher priority than integrity. An example is real-time two-way voice communications. The packets have optional FEC, but never include an FCS field, so no retransmissions take place. This channel is mapped to SCO packet payloads.

Summary

The Bluetooth link controller is a state machine that handles the low-level tasks associated with the basic communication link. Packets are divided into an access code, header, and payload. Data is exchanged through TDD at a nominal switching rate of 1,600 packets per second, although overhead can be reduced by using three- or five-slot packets that have a much larger percentage devoted to user payload.

The master always transmits in even-numbered time slots determined by its own native clock value. A slave addressed in that time slot can reply in the next (odd-numbered) time slot. Slaves never communicate directly with each other.

The ACL link promises high data integrity and is primarily used for file transfers, while the SCO link promises low latency and is primarily used for real-time two-way voice interaction. On either link, error correction is optional, whereas error detection is available only for ACL traffic. Packets that check bad are retransmitted through ARQ procedures.

In the next chapter we will examine how the various Bluetooth devices find each other and set up a communication link in the first place.

End Notes

1. Stallings, W. *Data and Computer Communications,* 2nd ed., New York, NY: Macmillan, 1988.
2. Peterson, W. and Brown, T. "Cyclic Codes for Error Detection," *Proceedings of the IRE,* January 1961.
3. Sklar, B. *Digital Communications: Fundamentals and Applications*, Englewood Cliffs, NJ: Prentice Hall, 1988.
4. Blahut, R. *Theory and Practice of Error Control Codes,* Reading, MA: Addison-Wesley, 1983.
5. *Bluetooth Assigned Numbers,* Version 1.1, published by the Bluetooth SIG, February 22, 2001.

CHAPTER 5

Establishing Contact Through Inquiry and Paging

Up to now we've assumed that the Bluetooth piconet already exists, and the master and slaves are hopping together and communicating. No doubt you've been wondering how the Bluetooth devices get to that point in the first place. That happens through the processes of *inquiry* and *page*. Both of these occur before the piconet has been officially established, so the devices involved aren't really master and slave(s) yet. Instead, we'll call them *prospective master* (p-master) and *prospective slave* (p-slave).

Because Bluetooth is a dynamic, ad-hoc network, devices are highly mobile and require some kind of automated or semi-automated method of finding and connecting with each other. This will usually consist of a two-step process. The first step is for a p-master to discover which other Bluetooth devices are in range (inquiry), and the second step is for the p-master to initiate connection with a particular device that responded to the inquiry (page).

What makes one device a master and another device a slave? Every Bluetooth device contains the ability to be either a master or a slave, so there's nothing inherent in the hardware that dictates who the master is. Instead, the master is defined as the device that initiates the establishment of the piconet, and slaves are the devices that enter the piconet at the request of a p-master. In other words, the master had initiated the connection via a page, and the slave had answered the page.

When first contemplating how the inquiry and page process can take place we quickly run into a dilemma. We already know that the p-master must send its *frequency hop synchronization* (FHS) packet to a p-slave so the latter can use the same hop sequence and phase used by the master, but on what frequency should the p-slave listen to receive the FHS packet? We discussed in Chapter 1, "Introduction," some solutions, one of which was to establish a separate channel for pages and inquiries. The problem, of course, is that interference on this particular channel could destroy the capability for Bluetooth devices to connect to each other. Furthermore, several devices attempting to perform pages or inquiries simultaneously could effectively jam each other.

A better solution, of course, is to establish a special hop sequence, or set of sequences, to use for paging and inquiries. This would make it much more difficult for one interfering signal to prevent any piconet from being established. For inquiries, the p-master may know nothing about nearby devices, so a single, common hop sequence (actually, one sequence for sending an inquiry and another for responding to the inquiry) is used by all devices for initial device discovery. A p-slave responding to an inquiry sends its FHS packet, within which is its *Bluetooth device address* (BD_ADDR). Now the p-master can create a new hopping sequence based upon this BD_ADDR for transmitting a subsequent page for establishing a piconet with that p-slave.

We learned in Chapter 3, "The Bluetooth Radio," that, for two FHSS devices to communicate, they must use not only the same hop channel set and sequence, but they must also be coordinated in time (phase) so that when one device transmits, the other is listening on the same channel. How can the correct phase be determined during inquiry and page? We'll leave that as an open question for now. After all, we need to include *some* information in this chapter beyond the introduction.

In this chapter we'll first examine the various states associated with Bluetooth piconet activity, followed by the methods for selecting the hop sequence and phases for inquiries and pages. Next, we examine the page process in detail because a clear understanding of paging is required before the inquiry process can be properly presented. Therefore, we assume that the paged unit's BD_ADDR is already known by the p-master when it begins the page. This BD_ADDR is retrieved by the pager during inquiry, which we will discuss near the end of this chapter.

When designing any type of communication system, wireless or wired, it's very important to insure that no device can enter an infinite loop while waiting for another device to respond to some request (see Figure 5-1). During

Jocular Science
by Bill Lae
MoM!
© 1987 Lae
"Here son, let me show you how to break out of this loop."

Figure 5-1 It's important that timeouts are included in Bluetooth to avoid infinite loops. (Source: Bill Lae)

the page and inquiry process there are several places where the p-master or p-slave is waiting for a response from the other, so timers must be used to initiate a backup plan if no response arrives. After the link is established, either the master or one or more slaves could disappear without warning so another timer needs to be established to recover from such an event. A Bluetooth timer is usually an 8-bit or 16-bit register containing an unsigned integer value that represents a number of 625 μs time slots. We'll look at the timers associated with page and inquiry near the end of each of their respective sections.

Bluetooth Piconet Activity and States

The Bluetooth link controller is a state machine, which is a fancy way of saying that it operates like a computer and can usually do only one thing at a time. Of course, the Bluetooth device's radio is supposed to be inexpensive, and as a result, it can't receive and transmit simultaneously. Be sure to keep these limitations in mind as we describe the page and inquiry processes. A device can't, for example, listen for its page and participate in a piconet at the same time. Instead, it must multiplex between these activities in some manner such that (hopefully) nothing is missed.

State Diagram: Point-to-Point Link Activity

Suppose a p-master wants to search the vicinity for available Bluetooth devices and then perhaps connect to one of these in a point-to-point piconet. The sequence of steps involved in such an endeavor can be drawn in the form of a *state diagram*. This diagram contains a set of *states* associated with each step of establishing a piconet and a set of *state transitions* that depict the allowable movement from state to state (see Figure 5-2). A device can be in only one state at a time, but it can (where enabled) move between states, sometimes quite rapidly.

When a Bluetooth device is first powered up, it enters the STANDBY state, where its hardware and software are initialized. From this state, a p-master can enter either the PAGE state or the INQUIRY state, depending upon whether or not the paged unit's BD_ADDR is known. The INQUIRY state enables the p-master to discover the BD_ADDR of all

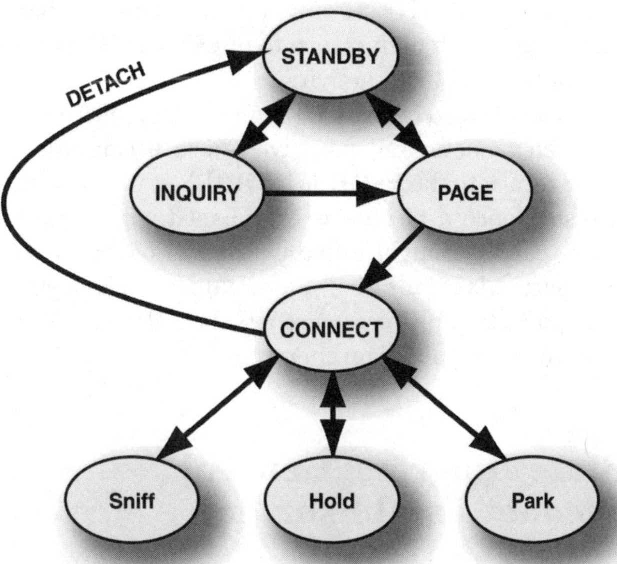

Figure 5-2 A state diagram for establishing a point-to-point piconet

respondents, and from there it can either return to the STANDBY state or enter the PAGE state to establish a point-to-point piconet with a p-slave. If the page is successful, then the p-master becomes the master and the p-slave becomes the slave of the new piconet, and the devices begin to exchange network setup parameters (see Chapter 7, "Managing the Piconet") and user information (see Chapter 8, "Transferring Data and Audio Information"). During this time the two devices are in the CONNECT state.

A p-slave can periodically move out of the STANDBY state to listen for inquiries and pages. If paged successfully, it transitions to the CONNECT state along with the new master. While in the CONNECT state the slave may arrange with its master to enter one of three low-power modes called *sniff, hold,* and *park* (see Chapter 6, "Advanced Piconet Operation"). The sniff mode enables the slave to check for a master transmission less often than in every even-numbered time slot. The hold mode is a time during which the slave can temporarily exit the piconet without disconnecting from the master. After the hold time expires the slave once again performs normal piconet duties. A slave that is in the park mode relinquishes its AM_ADDR and activates its receiver only to periodically resynchronize its CLK with that of the master. The master must unpark the slave with a new AM_ADDR before normal communication can resume.

From the CONNECT state either master or slave can initiate a disconnect, from which both return to the STANDBY state if they aren't involved in another Bluetooth activity.

We mentioned before that Figure 5-2 applies only to a point-to-point link between a master and one slave. This is because there's no way for the master to return to the PAGE or INQUIRY states from the CONNECT state. As a result, once a piconet is established with a single slave, no additional slaves can be brought into the piconet. Some early Bluetooth chipsets have this particular limitation. Later chipsets are more versatile, enabling not only point-to-multipoint piconets, but also scatternet operations. We will examine this state diagram next.

State Diagram: General Piconet Establishment

The most versatile Bluetooth devices can move readily between the different states belonging to either a p-master or a p-slave, depending upon what the goals of the end user may be. As an example, suppose a master of a piconet with a single slave wants to bring a new slave into the piconet. Clearly, the master requires the capability to move from the CONNECT state back into the PAGE or INQUIRY states. It's also clear that sending a page and listening for a page are separate states because the former belongs to a p-master and the latter belongs to a p-slave.

Figure 5-3 shows a state diagram for a general Bluetooth device that's capable of point-to-point, point-to-multipoint, and scatternet operation. To keep the diagram from being any more busy than it is we've removed the low-power modes and just show the various routes from the STANDBY to the CONNECT state and back.

As before, devices first power up into the STANDY state and from there can enter one of four states: PAGE, PAGE SCAN, INQUIRY, or INQUIRY SCAN. These are defined as follows:

- **PAGE** Used by a p-master to establish a piconet with a particular p-slave whose BD_ADDR is known
- **PAGE SCAN** Used by a p-slave to listen for its page
- **INQUIRY** Used by a p-master to discover the BD_ADDR and other information of devices in range
- **INQUIRY SCAN** Used by a p-slave to listen for an inquiry

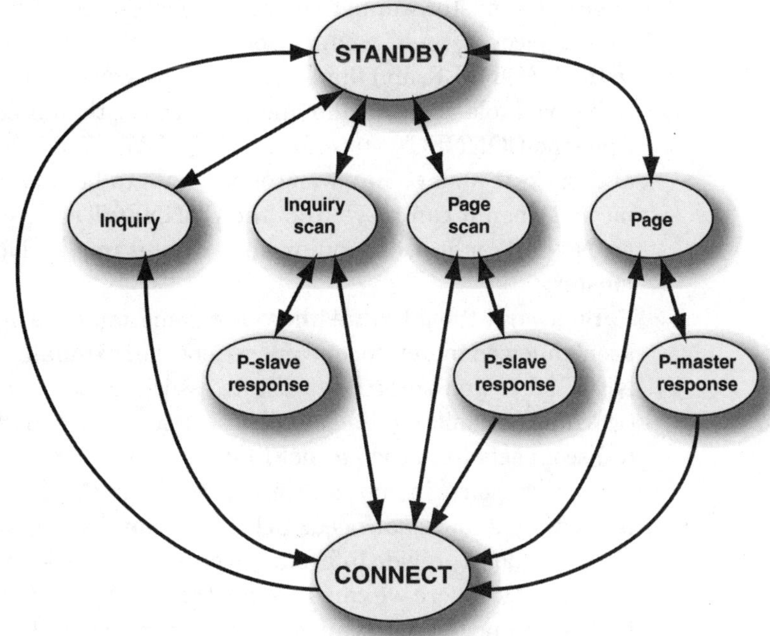

Figure 5-3
A general state diagram for connecting Bluetooth devices with each other

The PAGE, PAGE SCAN, and INQUIRY SCAN states all have responses associated with them, which are

- **P-slave response after PAGE SCAN** If a p-slave hears its page while in the PAGE SCAN state, then it responds with its own *device access code* (DAC).
- **P-master response after PAGE** When the p-master hears the p-slave's DAC response, then it responds in turn with its own FHS packet, giving the p-slave enough information to hop with the master during normal piconet operation.
- **P-slave response after INQUIRY SCAN** If a p-slave hears an inquiry while in the INQUIRY SCAN state, then it responds with its own FHS packet. The inquiring device doesn't acknowledge this response.

Finally, if all is well, both p-master and p-slave enter the CONNECT state and begin normal piconet operation.

Aside from showing more detail, the state diagram in Figure 5-3 differs from the diagram in Figure 5-2 in that devices can move from the CONNECT state back into either of the four states associated with inquiries and pages. If a master wants to bring another device with a known BD_ADDR into its piconet, then it moves from the CONNECT to the PAGE state, then to the P-MASTER, and finally back to the CONNECT state. If a slave wants to be available as a slave in another piconet, then it can periodically move from the CONNECT state to the INQUIRY SCAN or PAGE SCAN states and back. If a slave in one piconet wants to become a master in another piconet, then it can move from the CONNECT state into either the PAGE or INQUIRY states, depending upon what is required for it to become a new master.

Be aware, though, that there are some multiple state transitions that aren't allowed under normal operations. For example, a device in the CONNECT state won't proceed to a state associated with an inquiry or page and (warning: archaic legal term follows) thence to the STANDBY state unless it disconnects from the piconet first.

At this point in our adventure through the intricacies of page and inquiry, we know *what* needs to be done, but not yet *how* to do it. The state diagrams don't provide information on which frequencies should be used for this activity. Before we can dive into these details, though, it would be helpful to examine the hop sequence selection processes for pages and inquiries.

Hop Sequence Selection

In the last chapter we discovered that the hop channel set and the sequence of hops through the channel set are determined by the lower 28 bits of a device's BD_ADDR, and the hop phase is determined by the 27 most significant bits of CLK. These two values are sent to the hop generator, and the output of this generator goes to the Bluetooth radio's frequency synthesizer (see Figure 4-16). For two FHSS devices to communicate, they must use the same hop channels, the same hop sequence from channel to channel, and the same phase so that they hop together. Also, one device must transmit while the other receives on the same frequency and vice versa.

The number of hop channels and the hop period are two different entities. The former is the number of different carrier frequencies used by the FHSS devices to communicate, which is 79 for the Bluetooth piconet. The latter is the number of hops that occur before the specific hopping sequence through the channels repeats. Clearly, the hop period must be equal to or greater than the number of hop channels. For the Bluetooth piconet, the hop period is 2^{27}. This means that, although a Bluetooth radio might hop through all 79 channels in a few dozen milliseconds, the hop pattern changes each time through. These hopping characteristics make Bluetooth relatively immune to interference, but also make it extremely difficult to establish communication between devices in the first place. For example, suppose a Bluetooth receiver is tuned to Channel 3 (2,405 MHz), and it hears a packet arrive, but it doesn't know the phase of the hopping sequence. On which channel will the next packet be sent? It's impossible to know because the long hop period means that the next packet could occur on any of the remaining 78 channels with about equal probability.

The challenge for paging and inquiry, then, is for the p-master and p-slave to start communicating on a particular hop frequency and then know which channel to go to next. To simplify this process Bluetooth devices use four sets of 32-hop sequences, each with a period of 32, for pages and inquiries. Because the period is equal to the number of channels, the hop pattern through the channels doesn't vary, and aligning the hop phase between two devices is greatly simplified. Continuing with our previous example, if a Bluetooth receiver hears a packet on one of these 32 frequencies, then it knows where to hop next and thus becomes aligned to the hop phase automatically.

A total of five hop sequences and periods are defined to cover inquiry, page, and connect activity for Bluetooth devices. These are

- **Channel hop sequence** Used for normal piconet communications between master and slave(s). Channels: 79. Period: 2^{27}.
- **Page hop sequence** Used by a p-master to send a page to a specific p-slave and to respond to the slave's reply. Channels: 32. Period: 32.
- **Page response sequence** Used by a p-slave to respond to a p-master's page. Channels: 32. Period: 32.
- **Inquiry hop sequence** Used by a p-master to send an inquiry to find Bluetooth devices in range. Channels: 32. Period: 32.
- **Inquiry response sequence** Used by a p-slave to respond to a p-master's inquiry. Channels: 32. Period: 32.

Each page hop frequency is associated with one page response frequency, and each inquiry hop frequency is associated with one inquiry response frequency.

Now we can provide more detail on the inputs to the hop generator used during transmit for the various states in Figure 5-3, which are given in Table 5-1. (Each CLKN in the Hop Phase Source column belongs to the particular device listed in the Device column.) The actual processing within the hop sequence generator is surprisingly complex and requires several pages in the Bluetooth specification for explanation.

Table 5-1

Hop generator inputs for device TX operation

Hop Sequence	Device	Channels/ Period	Hop Sequence Source	Hop Phase Source
Channel	Master	$79/2^{27}$	Master BD_ADDR	CLKN = CLK
Channel	Slave	$79/2^{27}$	Master BD_ADDR	CLKN + Offset = CLK
Page	P-master	32/32	P-slave BD_ADDR	CLKN + Offset = CLKE
Page scan	P-slave	32/32	P-slave BD_ADDR	CLKN
Inquiry	P-master	32/32	GIAC	CLKN
Inquiry scan	P-slave	32/32	GIAC	CLKN

Notice that, for each hop sequence pair (channel/channel, page/page scan, and inquiry/inquiry scan), the hop sequence is derived from the same source. The master determines the hop sequence for active piconet operation, and the p-slave determines the hop sequence for the page and response processes. For all inquiries, either general or dedicated, the hop sequence is determined by the general inquiry access code.

Notice also that the hop sequence input to the hop generator requires 28 bits, but the GIAC is only 24-bits long. Where do the other four bits come from? There's an 8-bit number called the *default check initialization* (DCI) that is placed into the HEC generator prior to calculating the HEC sequence for transmission of a FHS packet. (I'm not making this up.) The four least significant bits of the DCI become the four most significant bits of the hop generator sequence for inquiries. So what's the value of the DCI? It's 0x00. Talk about a circuitous way to get there! Anyway, the hop sequence input to the hop generator is 0x09E8B33 for inquiries.

Both the master and slave hop phase source generate identical numbers when using the channel hop sequence to participate in active piconet communication. As a result, the master and its slave(s) hop together from channel to channel, and their transmit/receive cycles are coordinated as well. For paging, the p-master may know the CLKN of the p-slave as a result of a previous inquiry, so an estimate CLKE of the p-slave's CLKN can be derived by adding an offset to the p-master's CLKN value. However, when issuing an inquiry the p-master may know nothing at all about any nearby devices, so CLKN is used by all devices participating in inquiry or inquiry scan.

As we'll discover shortly, pages and inquiries are expedited by the p-master, doubling its normal hop rate to 3,200 per second. Therefore, the

LSB of the appropriate clock source (CLKN or CLKE) is input to the hop generator to enable it to change to a new hop frequency every 312.5 μs. The LSB of the clock source is not used for the channel hop sequence because the hop rate is nominally 1,600 per second, and a new hop frequency is selected every 625 μs.

Paging Other Bluetooth Devices

The page process is used by a p-master for establishing a connection with a specific Bluetooth device. For a page to occur, the p-master must already know the BD_ADDR of the p-slave, which can be obtained either through the inquiry process or by another means such as the user manually entering the value. Furthermore, the p-master may know the approximate value of the p-slave's CLKN, which is also transmitted by the p-slave responding to an inquiry. Knowledge of the p-slave's CLKN is not required, but will speed the page procedure.

General Description of the Page Process

Several different page schemes are given in the Bluetooth specification, but we'll begin by discussing the mandatory page process that all Bluetooth devices are required to implement. For a page to be successful, the p-master must be in the PAGE state, and the p-slave must be in the PAGE SCAN state at the same time. The following tasks associated with the page are summarized here, and then described in greater detail in the next section:

1. The p-master transmits two p-slave DACs on two consecutive hop frequencies from the 32-hop page sequence during an even-numbered time slot.
2. The p-master listens on the two corresponding page response hop frequencies in sequence during the next odd-numbered time slot.
3. Meanwhile, the p-slave listens for its DAC on one of the page hop frequencies for about 11 ms each time it enters the PAGE SCAN state.
4. Upon hearing its DAC, the p-slave then transmits the same DAC 625 μs later on the corresponding page response hop frequency.

5. If the p-master hears the DAC returned on the corresponding page response frequency, then it sends its FHS packet to the p-slave.
6. If all is well, then the p-master and p-slave connect as master and slave.

As we've pointed out before, the page and page response hop sequences are both derived from the p-slave's BD_ADDR, but the hop phase hasn't yet been coordinated precisely, except for the p-master's estimate of the p-slave's CLKN. Bluetooth uses a "tortoise and the hare" technique for bringing the two devices together; that is, a fast-hopping master is trying to catch a slow-hopping slave.

By sending two DACs on different frequencies in a single time slot and then listening on two different response frequencies in the following time slot, the p-master has effectively doubled its nominal hop rate from 1,600 to 3,200 per second. Can its frequency synthesizer keep up with that blistering pace? We learned in Chapter 3 that the synthesizer has about 259 μs to change to a new frequency after finishing a packet transmission or reception under normal piconet conditions, so we can use that time as a benchmark for synthesizer switching speed. When paging, the p-master is required to send one 68-bit p-slave DAC and switch its synthesizer to a new frequency within a half-slot time of 312.5 μs. This gives the synthesizer about 244.5 μs to change frequencies, so a design that works with normal piconet operation should also work with pages.

Device Interaction in Frequency and Time To gain further insight into the p-master and p-slave interaction occurring during the page, we'll begin by taking a look at the activity as a function of frequency (see Figure 5-4). When a p-slave enters the PAGE SCAN state it begins listening for a nominal time of about 11 ms on one of the 32 frequencies associated with the page sequence. This task is labeled "A" on the p-slave's RX line.

Meanwhile, the p-master transmits two p-slave DACs 312.5 μs apart in time and on consecutive page frequencies during an even-numbered time slot (1 and 2 in Figure 5-4). Next, the p-master switches on its receiver and listens on the two matching page response frequencies (3 and 4) in the following odd-numbered time slot. In Figure 5-4, the p-slave isn't listening on either of the first two page frequencies, so there's no response on the associated page response frequencies.

Now the p-master selects the next two frequencies in the page hop set and sends two more p-slave DACs as before (5 and 6). The second of these frequencies happens to be the one that the p-slave is monitoring, so the p-slave transmits its own DAC on the corresponding page response frequency 625

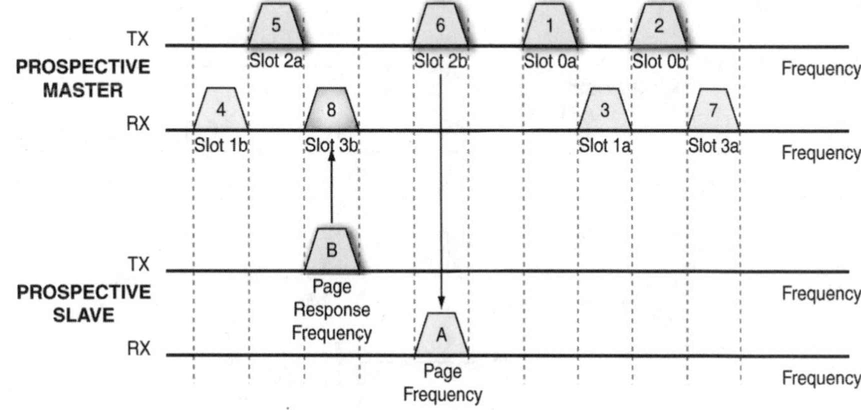

Figure 5-4 The beginning of the page process shown as a function of frequency

μs later (B). Because the p-master is programmed to be listening on that response frequency during that time (8), it will (hopefully) hear the slave's response, and it can proceed with the remaining part of the page process. In this way the p-master and p-slave become (mostly) hop-synchronized (hop phase aligned) to the page and page response hop sequences.

Once this initial hop phase synchronization occurs, it's perhaps easier to see what happens using a time plot, which is shown in Figure 5-5. The first p-master transmission in this example is during the first half-slot on frequency $f(k)$, which is also the frequency on which the p-slave happens to be listening for its page, so the p-slave hears its own DAC at this time. Now the p-slave changes its synthesizer to the associated page response frequency and waits 625 μs after the start of the page. Meanwhile, the p-master transmits another page DAC on frequency $f(k + 1)$, which isn't heard by the p-slave, and then switches to frequency $f'(k)$ to listen for a possible response to its page on $f(k)$.

Because the p-slave heard its page on $f(k)$, it transmits its response on $f'(k)$ while the p-master is there to hear it. (Note the prime mark on $f'(k)$, designating it as the page response frequency associated with $f(k)$.) At this time the p-master and p-slave are almost synchronized in phase. The only uncertainty at this point is with the p-slave: Did the page occur during the first or second half of the master's even-numbered time slot? The p-slave doesn't know, so it assumes worst-case, which is the later second half-slot. As a result, 312.5 μs later the p-slave is ready for the p-master's next transmission on frequency $f(k + 1)$.

Now the p-master switches its radio to $f(k + 1)$ and, at the beginning of its next even-numbered time slot as determined by its own CLKN, it

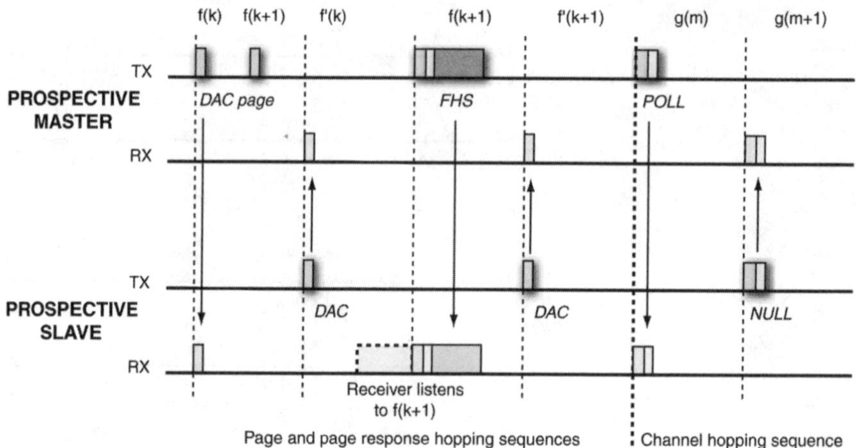

Figure 5-5 The completion of the page process shown as a function of time when the p-slave responds in the first half-slot

sends its FHS packet to the p-slave. The beginning of the FHS packet transmission is used by the p-slave to determine the start of a master-to-slave transmission slot. Following this, both p-master and p-slave switch to page response frequency $f'(k + 1)$, and the p-slave sends its DAC to the p-master if the FHS packet was received successfully.

Following successful FHS reception and response, the p-master becomes the master; the p-slave becomes the slave; and both devices switch their frequency synthesizers to the first frequency $g(m)$ in the channel hop sequence. That's the sequence that uses all 79 channels and has a period of 2^{27}. (The time index m in the channel hop sequence isn't related to the time index k in the page and page response sequences.) Notice that the p-slave must be able to decode the FHS packet and enter the new master's BD_ADDR and piconet CLK values into its hop generator within about 875 μs after the end of the FHS packet transmission. Finally, the master tests the link by sending a POLL packet to the slave, then both devices hop to the next channel $g(m + 1)$, and the slave responds to its poll with any packet, usually a NULL packet if there's nothing else to say. At this point both devices breathe a sigh of relief and congratulate each other on a job well done.

Figure 5-6 depicts a timing detail when the p-slave receives the p-master's page during the second half-slot instead of the first. In this case the p-master's FHS packet arrives 312.5 μs after the p-slave responds to its page.

Now that we've completed a general description of the page process, it's time to dive into some deeper subtleties associated with the activities of the p-master and p-slave both during the page and when either is occupied with other tasks between their respective PAGE and PAGE SCAN states. In the

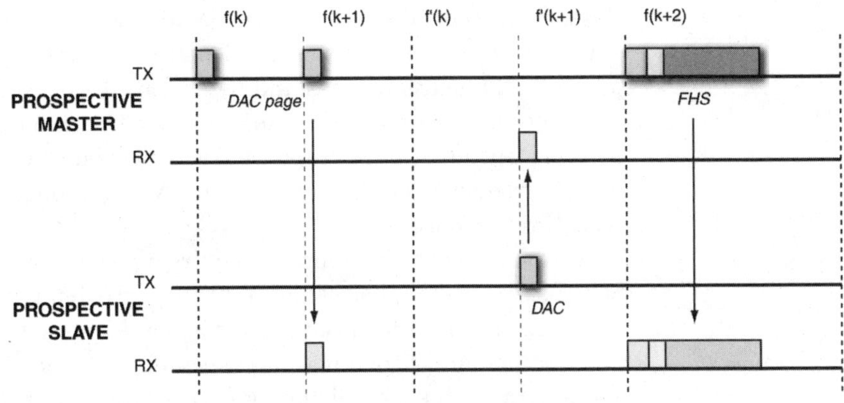

Figure 5-6 Page timing detail when p-slave responds to its page in the second half-slot

following section we'll address questions that I'm sure you're eager to have answered:

- How many page frequencies can the p-master probe during the time the p-slave is in the PAGE SCAN state?
- How often will a p-slave enter the PAGE SCAN state?
- How much time is required for a successful page to occur?
- What are some of the optional page modes?

Prospective Slave: PAGE SCAN and PAGE RESPONSE States

A p-slave can enter the PAGE SCAN state from either the STANDBY or CONNECT states. If it enters from the CONNECT state, then it must have scatternet capability because, if it detects its page, it must be able to enter a second piconet.

There are two major timing values that are associated with the page scan process. One, called $T_{w\_page\_scan}$, is the time duration (window) that the device spends in the PAGE SCAN state. The other is the T_{page_scan}, which is the time between successive page scans. For shortest response time, of course, the two times should be the same; that is, the p-slave is always scanning for its page. That situation won't be feasible, though, if the p-slave is already participating in another piconet. Furthermore, if the p-slave is battery powered and needs to conserve energy, then it may not be able to afford having its receiver on all the time.

Page Scan Window How long should a p-slave stay in the PAGE SCAN state once it enters? First, let's calculate how long a p-master will take to page all 32 hop frequencies and listen for a response. The p-master sends two pages in one time slot on different hop frequencies, then listens on the two corresponding page response frequencies in the next time slot, covering two frequency pairs every 1.25 ms. All 32 frequencies require 20 ms to cycle through once.

At first it may make sense to require a p-slave, once it enters the PAGE SCAN state, to stay there at least 20 ms to give a p-master a chance to page all 32 frequencies. However, if the p-slave re-enters the PAGE SCAN state every second or so, then it will be spending a rather large amount of time and energy in this state listening for pages that may actually arrive only a few times per hour or day. As such, the Bluetooth specification states that the minimum $T_{w\_page\_scan}$ is equal to the time for the p-master to probe only 16 frequencies in the page hop set or about 10 ms. The actual time in the PAGE SCAN state should be a bit longer (the default is 11.25 ms) to allow for timing differences between p-master and p-slave.

If the p-slave is already a member of another piconet and wants to enter the PAGE SCAN state, then it should put its ACL link into either the hold or park modes (see Chapter 6) before checking for pages. In this way the master of that piconet won't expect the p-slave to be available when it's otherwise occupied.

If the p-slave is participating in a SCO link in another piconet, then SCO packets have a higher priority and will periodically interrupt the page scan. Therefore, $T_{w\_page\_scan}$ should be extended accordingly. Of course, if HV1 packets are being exchanged, then the device is fully occupied in its present piconet and cannot exit the CONNECT state for other activities.

When in the PAGE SCAN state, the p-slave uses its own BD_ADDR to determine the hop sequence, and the hop phase is derived from its $CLKN_{16-12}$ (bits 16 to 12 of CLKN; see Figure 4-6). These five bits determine which of the 32 frequencies the p-slave listens on, but notice that the LSB of these five bits changes every 1.28 seconds. The p-slave, then, changes to a new page hop frequency every 1.28 s regardless of how often it enters the PAGE SCAN state.

Time Between Page Scans The p-slave has three options for setting the time between successive page scans, which is the length of T_{page_scan}. For continuous page scans, $T_{w\_page\_scan} = T_{page\_scan}$. If the p-slave is already occupied in a piconet or if it needs to conserve energy, then T_{page_scan} can be set to a maximum of either 1.28 s (which is the default value) or 2.56 s. This infor-

Table 5-2

Scan repetition (SR) values

SR Mode	FHS Packet SR Field	Time Between Successive Page Scans	N_{page}
R0	00	Continuous	≥1
R1	01	≤ 1.28 s (Default)	≥128
R2	10	≤ 2.56 s	≥256

mation is given in the *scan repetition* (SR) field in the device's FHS packet (see Table 5-2). The last column in the table will be discussed in the next section.

Prospective Master: PAGE and PAGE RESPONSE States

As we have already learned, when the p-master wants to page a p-slave, it enters the PAGE state and repeatedly transmits the p-slave's DAC on the page hop channels, two per even-numbered time slot, and listens to the two corresponding page response channels in each following odd-numbered time slot. Because the p-master doesn't know precisely when the p-slave enters its PAGE SCAN state, it must continue to transmit the DACs until receiving a response or experiencing a timeout. For a successful page to occur, the p-master must be transmitting a p-slave's DAC when that particular device is in the PAGE SCAN state and listening on the correct frequency.

Because the p-slave may be in the PAGE SCAN state only long enough for the p-master to page on 16 frequencies, the p-master should attempt to select those 16 frequencies to include the one on which the p-slave will be listening. However, on which frequency will the p-slave be listening? At worst the p-master can simply guess. However, if the p-master can estimate the value of the p-slave's CLKN, then it can examine bits 16 to 12 for the hop phase, and hence the correct hop frequency, which the p-slave will use when it next enters PAGE SCAN. The p-master could have received a copy of the p-slave's CLKN as part of its FHS transmission in response to the inquiry, or it may have asked for the device's CLKN value while connected to it in a previous piconet. If the p-master time stamped the arrival of the FHS packet with its own CLKN, then (provided neither device has been powered-down in the meantime) it can easily calculate an offset to generate CLKE, its estimate of the p-slave's present CLKN value.

To determine which 16 frequencies the p-master should use for paging, bits 16 to 12 of CLKE are extracted to estimate the frequency $f(k)$ on which the p-slave will be listening when it next enters the PAGE SCAN state. Now the p-master builds a set of 16 hop channels called *train A*, consisting of frequencies $f(k - 8), f(k - 7), \ldots, f(k), \ldots, f(k + 6), f(k + 7)$. Because the p-slave's page scan frequency changes every 1.28 seconds, train A changes every 1.28 seconds as well. As long as the p-master's CLKE is accurate from $-8 \times 1.28 = -10.24$ to $+7 \times 1.28 = +8.96$ seconds of the p-slave's CLKN, then the p-slave is guaranteed to be listening on a train A frequency when it next enters the PAGE SCAN state. Two Bluetooth clocks can drift relative to each other by up to 40 ppm (20 ppm in one direction and 20 ppm in the other direction), so train A should include the p-slave's page scan frequency for at least 2.6 days after the p-master last retrieved the p-slave's CLKN. The 16 page hop frequencies that aren't part of train A become train B.

While paging, the p-master cycles through train A, using each frequency in turn, over and over until the p-slave responds or a timeout occurs. If there's no response after a certain time, then the p-master switches to train B and continues paging. If there's still no response, then it returns to train A and so on until it gets *really* tired and another timeout occurs. (This latter timeout is often set by the user.) Then it gives up and tells the user that the page was unsuccessful.

How many times should the p-master cycle through train A before switching to train B? Suppose the p-master is having a bad day and begins the page process just as the p-slave exits its PAGE SCAN state. According to Table 5-2, the p-slave looks for pages every T_{page_scan} seconds, which depends upon the SR mode. For example, for SR mode R1, the p-slave will look for its page every 1.28 seconds. Because the p-master requires 10 ms for each trip through one of the page frequency trains, it may need to cycle through that train as many as 128 times before the p-slave returns to the PAGE SCAN state and its page is heard. These figures are given in the last column of Table 5-2. Of course, if the p-master doesn't know the p-slave's SR mode, then it should assume R2 and cycle through each train for 2.56 seconds.

Time Required for Completing the Page The time needed for the p-master to successfully page a p-slave can vary considerably, depending upon the accuracy of CLKE, the p-slave's SR mode, and the channel BER. Generally, though, if the BER is low and CLKE is accurate, then the average time required to complete a page will be about half the T_{page_scan} value.

This, of course, assumes that the p-master dedicates its entire attention to the page and isn't trying to support multiple operations such as maintaining a SCO link and paging.

If the p-master knows nothing about the paged device except its BD_ADDR, then the probability is one-half that it will select the wrong 16 frequencies for its first attempt at paging. Furthermore, it should use each frequency train for 2.56 seconds because the SR mode of the p-slave is also unknown. Page times in this situation can vary from nearly instantaneous to about 5 seconds. The latter time is fairly long by human standards, so developers of Bluetooth software packages should try to anticipate pages and enact them before being requested by the user.

Optional Paging Schemes

Up to now we've examined the mandatory-paging scheme that must be supported by all Bluetooth devices. The specification can support up to three optional paging schemes, of which one has been described. It's important to remember that the goal of the paging process is to facilitate establishing a connection between two Bluetooth devices without placing an excessive burden on the paged device (p-slave).

The main characteristic of the optional scheme is that the p-master can send p-slave DAC packets in both odd- and even-numbered time slots, rather than in even slots only, to speed the page process. The same 32 page hop frequencies are split into the A train and B train as before, but page response frequencies are designated differently. This page process takes place in the following way:

1. The p-master transmits two p-slave DAC packets in each of eight consecutive time slots, encompassing one 16-frequency hop train.
2. The p-master sends two marker packets in the ninth time slot.
3. The p-slave responds in the tenth time slot on the same frequency used by the p-master for transmitting the marker.

By bending the usual transmission rules and enabling the p-master to send pages during several consecutive time slots, the average time to completion of the page is reduced. Perhaps even more importantly, a p-slave only needs to enter the PAGE SCAN state for about 6 ms (rather than about 11 ms) to cover the time required for the p-master to transmit pages on an entire train of page hop channels.

Table 5-3

Mandatory scan period values

SP Mode	FHS Packet SP Field	Time Required for Mandatory Page Scan Following an Inquiry Response
P0	00	≥20 s (Default)
P1	01	≥40 s
P2	10	≥60 s

The two-bit *scan period* (SP) field in the FHS packet (see Chapter 4, "Baseband Packets and Their Exchange") specifies the amount of time that the device will support the *mandatory* PAGE SCAN state after responding to an inquiry. These times are given in Table 5-3. This is important because the mandatory and optional paging schemes are incompatible with each other. A p-master may not be able to support the optional scheme; therefore, all p-slaves are required to implement the mandatory page scheme for at least 20 seconds after responding to an inquiry.

Paging Timers

Several timers are defined at the link controller level to prevent the dreaded infinite loop while awaiting the response of another device that may never arrive due to channel errors or a naughty user shutting off the power or walking away.

The timer *page timeout* (pageTO) defines the number of time slots that the PAGE state can last before exiting if there's no response to the page. This value is transferred from the host to the Bluetooth module via the host controller interface (see HCI, Chapter 10, "Host Interfacing"). The time defaults to 5.12 seconds, but can often be set by the user.

The *page response timeout* (pagerespTO) timer has a different meaning for the paging and paged devices. For the p-slave, it is the maximum number of time slots that the device will wait between responding to the page and the arrival of the p-master's FHS packet. During the wait period the p-slave selects the next page response hop frequency every 1.25 ms. If this timer expires, then the p-slave returns to the PAGE SCAN state for one scan period ($T_{w\_page\_scan}$). After that it returns to whatever state it was in prior to PAGE SCAN. The pagerespTO value is eight slots for the p-slave.

For the p-master, pagerespTO defines the number of slots it will wait for the p-slave to respond to the FHS packet. While awaiting the p-slave's response, the p-master retransmits the FHS packet (with an updated CLK field) on a new page hop frequency at the start of each even-numbered time slot. If the timer expires, then the p-master begins the page process over again. The pagerespTO value is also eight slots for the p-master.

After the p-slave acknowledges the p-master's FHS packet, they both switch to the channel hopping sequence and the new master polls the new slave. If the master doesn't receive a response by expiration of *new connection timeout* (newconnectionTO), then the master and slave return to the PAGE and PAGE SCAN states, respectively. The newconnectionTO value is 32 time slots.

Finding Other Bluetooth Devices

Due to the dynamic, ad-hoc nature of Bluetooth, there will be many times when a device needs to discover the existence of other Bluetooth devices that are within range. The inquiry process enables a p-master to search for these other devices without necessarily knowing anything about them. This is equivalent to someone in a darkened room shouting, "Is anyone there?"

You would probably think that the inquiry procedure has much in common with paging, and you'd be mostly correct. The big challenge with the inquiry, though, is selecting the hop sequence and achieving hop phase alignment between p-master and p-slave. When paging a particular p-slave the hop sequence was based upon that device's BD_ADDR, and the p-master hopped rapidly in an attempt to catch a slow-hopping p-slave. Because a device in the INQUIRY state may know nothing about the devices it's searching for, a common hop sequence needs to be used for all inquiries. Not a bad idea, yes? What happens if two or more p-slaves respond to an inquiry on the same hop frequency and at the same time? A packet *collision* occurs, which may result in both responses being lost. The inquiry process, then, needs a mechanism for preventing collisions. A device responding to the inquiry will send its FHS packet from which the p-master can extract the BD_ADDR and the CLKN values, along with the p-slave's class and page scan modes, for possible use in a subsequent page.

General Description of the Inquiry Process

Like the page process, the p-master hops at a nominal 3,200 per second, sending two short inquiry packets in each even-numbered time slot. The p-slave listens for a relatively long time on a single hop frequency for an incoming inquiry packet. For an inquiry to be successful, the p-master must be in the INQUIRY state and the p-slave must be in the INQUIRY SCAN state at the same time. The following tasks associated with the inquiry are summarized here, then described in greater detail in the next section:

1. The p-master transmits two *inquiry access codes* (IACs) on two consecutive hop frequencies from the 32-hop inquiry sequence during an even-numbered time slot.
2. The p-master listens on the two corresponding inquiry response hop frequencies in sequence during the next odd-numbered time slot.
3. Meanwhile, the p-slave listens for the IAC on one of the inquiry hop frequencies for about 11 ms each time it enters the INQUIRY SCAN state.
4. Upon hearing the IAC, the p-slave delays a random time, then listens for the IAC again on the same frequency.
5. Upon hearing the IAC again, the p-slave transmits its FHS packet 625 μs later on the corresponding inquiry response hop frequency.
6. The p-master doesn't respond to the FHS packet, but simply stores it for future use if needed.

As we've pointed out before, the inquiry and inquiry response hop sequences are both derived from the GIAC preceded by 0x0. The phase, however, can't be estimated by the p-master under most conditions, so devices entering either the INQUIRY or INQUIRY SCAN states will use 0x09E8B33 for the sequence and their own CLKN for the phase in their hop generators. As with the page, a fast-hopping p-master is attempting to catch a slow-hopping p-slave.

The inquiry process enables the p-master to be somewhat selective in who responds to the inquiry. There are three levels at which the p-master can operate its inquiry, each with its own access code(s):

- ***General inquiry access code*** **(GIAC)** Used for the discovery of any Bluetooth devices within range. The GIAC has a value of 0x9E8B33.
- ***Dedicated inquiry access code*** **(DIAC)** Used to discover Bluetooth devices in range that have a specific capability. If the p-master is looking for a printer, for example, it can avoid responses from devices that have no printing capability. The DIAC is sometimes called the *device-specific inquiry access code*. There are 62 DIAC values that range from 0x9E8B01 to 0x9E8B32 and from 0x9E8B34 to 0x9E8B3F.

- ***Limited inquiry access code*** **(LIAC)** Finds another Bluetooth device that has been placed in a special response mode for a limited time, typically one minute. For example, if two users in a crowded Bluetooth environment want to connect quickly, then one device is placed into a limited INQUIRY SCAN state, and the other transmits a limited inquiry. The p-master will then find the p-slave and connect to it (through a subsequent page) without having to sort through a large number of responses to its inquiry. The LIAC has a value of 0x9E8B00.

Prospective Slave: INQUIRY SCAN and RESPONSE States

A p-slave can enter the INQUIRY SCAN state from either the STANDBY or CONNECT states. As with page scan, there are two times associated with the inquiry scan process. One, called $T_{w\_inquiry\_scan}$, is the time duration (window) that the device spends in the INQUIRY SCAN state (default: 11.25 ms). The other is the $T_{inquiry\_scan}$, which is the time between successive inquiry scans (default: 1.28 s). Because the p-master is using only 16 hop frequencies from the 32-hop inquiry set, $T_{w\_inquiry\_scan}$ should be greater than 10 ms and longer if the scan is interrupted by SCO packets. The time between successive inquiries can be at most 2.56 seconds.

When a p-slave receives an inquiry, it will reply with its own FHS packet (preceded by the same access code sent by the p-master along with a Bluetooth header) on the associated inquiry response frequency. However, this reply shouldn't occur in the next time slot because another p-slave may also be listening on the same inquiry hop frequency. If they both reply at the same time and on the same frequency, then the p-master could lose both responses in the resulting collision. Because there are only 32 channels in the inquiry hop sequence, the probability of collision increases significantly as more devices enter the INQUIRY SCAN state simultaneously. Therefore, any p-slave that hears an inquiry will enact a *random backoff* procedure before responding. This procedure uses the following steps:

1. Upon hearing an inquiry access code on frequency $f(k)$, the p-slave fetches a 10-bit uniformly distributed random number RAND, which has a value between 0 and 1,023.
2. The p-slave returns to its previous state (CONNECT or STANDBY) for RAND number of time slots. This translates to a delay of 0 to 639 ms.

3. After RAND expires, the p-slave returns to the INQUIRY SCAN state on the same frequency $f(k)$ and waits for another (identical) IAC.
4. Upon receiving this second IAC, the p-slave responds with its FHS packet 625 μs later on the corresponding inquiry response frequency $f'(k)$.
5. After sending its FHS packet, the p-slave returns again to the INQUIRY SCAN state, but now listens on the next inquiry hop frequency $f(k + 1)$ in the sequence.

This sequence of events is shown in Figure 5-7. Notice that the random backoff time averages about 320 ms. If, for example, the p-slave enters its INQUIRY SCAN state while 1.28 seconds (half) of an inquiry probe remains, then it will respond with its FHS packet an average of four times, each on a different inquiry hop frequency.

There are now three independent random events that help prevent collisions between different p-slaves responding to an inquiry. First, different p-slaves will enter their INQUIRY SCAN states at independently selected times. Second, different p-slaves will monitor independently selected inquiry hop frequencies, based upon their respective CLKN values. Finally, different p-slaves will generate independent RAND backoff numbers to (hopefully) separate their responses if they happen to be listening for an inquiry at the same time and on the same frequency.

Prospective Master: INQUIRY State

The p-master performs the inquiry process in a manner very similar to paging. The 32-hop inquiry sequence is divided into a 16-hop train A and

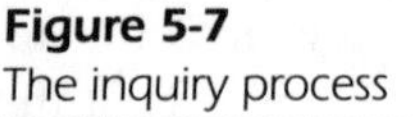

Figure 5-7
The inquiry process

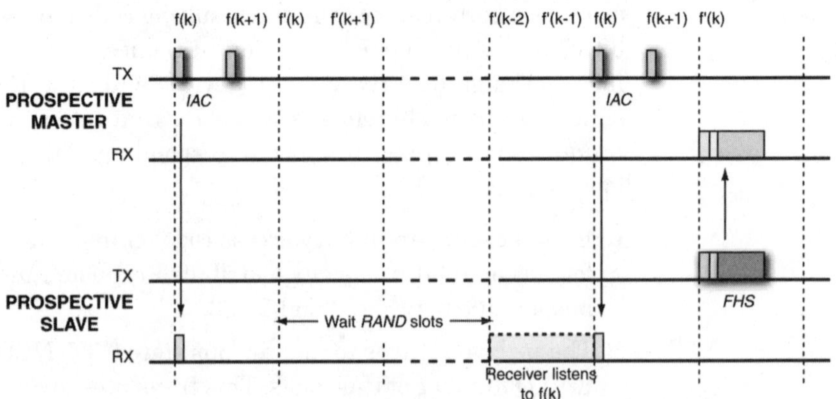

Inquiry and inquiry response hopping sequences

16-hop train B as before, but this division can be arbitrary because the p-master doesn't usually have any CLKN information from the devices it's trying to find. The appropriate IAC is sent on two consecutive inquiry hop frequencies in an even-numbered time slot, then the associated inquiry response frequencies are monitored in the next odd-numbered slot.

A single train of hop frequencies must be sent for at least $N_{inquiry} = 256$ times before switching to the other train to cover the 2.56 seconds between p-slave successive INQUIRY SCAN states. (If a SCO link periodically interrupts the p-master's inquiry process, then $N_{inquiry}$ should be lengthened accordingly.) One cycle through 16 inquiry (and their respective inquiry response) frequencies requires 10 ms to complete.

Upon receiving a FHS packet in response to the inquiry, the p-master stores it in a database for possible use in a later page. The inquiry is completed either when the p-master has received enough responses or a timeout occurs.

Time Required for Completing the Inquiry If a p-slave enters its INQUIRY SCAN state near the end of a p-master's 2.56 second cycle for a 16-hop train, then its RAND may not expire before the p-master switches to the other train. In that case, the p-slave won't hear the IAC a second time and won't respond with its FHS packet. Instead, after a timeout called *inquiry response timeout* (inqrespTO) (128 time slots, see the next section), the p-slave will return to its previous STANDBY or CONNECT state and re-enter the INQUIRY SCAN state within the next 2.56 seconds. As in the page hop sequence, the p-slave selects the next frequency in the inquiry hop sequence every 1.28 seconds. The result is that, under some conditions, the inquiry can take several seconds to complete. A subtle, detailed, infuriatingly obscure analysis follows.

When first entering the INQUIRY state, the p-master uses train A for 2.56 seconds. Any p-slaves entering their respective INQUIRY SCAN states on one of the train A frequencies will hear the p-master's inquiry. The p-slave will then send its FHS packet on a train A inquiry response frequency up to about 639 ms later, which is the maximum backoff time associated with RAND, plus up to an additional 10 ms, which is the time the p-master needs to cycle through a train. Therefore, as long as the p-slave hears an IAC for the first time prior to $2.56 - 0.649 = 1.91$ seconds into the inquiry probe, then its response will be heard by the p-master (assuming, of course, that the channel's BER is low).

If a p-slave's backoff time takes it beyond the p-master's first 2.56 second inquiry window, then the p-master will have already switched to inquiry train B and will miss the opportunity to trigger the p-slave's FHS response.

However, (stay with me, now), after another 2.56 seconds have elapsed, the p-master will switch once again to train A, and the p-slave will have another chance to respond.

The real problem occurs when the p-slave happens to switch from a frequency in train A to a frequency in train B for its INQUIRY SCAN state when the p-master is switching back to train A for its inquiry probe. It's clear (it *is* clear, isn't it?) that another 2.56 seconds must elapse before the p-master switches trains again and thus can complete the inquiry with that particular p-slave.

The bottom line is that it may take as long as four complete 2.56 second inquiry probes (train A, train B, train A, and train B) before the p-master hears a response from all devices within range of its inquiry. That equates to 10.24 seconds, which can seem like an eternity to a busy Bluetooth user. The best software implementations, then, will anticipate a user's needs and complete the inquiry process ahead of time. For example, the search for a suitable printer should occur upon opening an application, not when the user selects the Print command. Alternatively, the Bluetooth unit can enter the INQUIRY state periodically (see Chapter 10) in order to keep its database of in-range devices current.

Inquiry Timers

Two timers are defined at the link controller level to monitor the progress of an inquiry.

The timer *inquiry timeout* (inquiryTO) defines the maximum number of time slots that the INQUIRY state can last before the p-master exits, regardless of whether there are any responses. This value is transferred from the host to the Bluetooth module via the host controller interface (see Chapter 10, "Host Interfacing"). The time is usually about 10 seconds, but can often be modified by the user.

The inqrespTO timer is the length of time a p-slave remains in the INQUIRY SCAN state after the RAND has expired. If a second IAC hasn't arrived by the expiration of inqrespTO, then the p-slave returns to its previous CONNECTION or STANDBY state. The inqrespTO value is 128 time slots.

Summary

Bluetooth has implemented elaborate methods for finding (inquiry) and connecting with (paging) other devices. A dedicated channel for this purpose cannot be used because interference on this one frequency would effectively shut down the capability to connect, and this interference could even come from several piconets trying to form at the same time.

To facilitate reasonable interference immunity, both inquiries and pages use abbreviated 32-hop sequences for transmitting the request and corresponding 32-hop sequences for the response(s). The page sequence is derived from the p-slave's BD_ADDR, and the inquiry sequence is derived from the GIAC.

To save energy, a p-slave will periodically enter either the INQUIRY SCAN or PAGE SCAN states only long enough for the p-master to probe 16 frequencies, so both pages and inquiries can be delayed if the p-slave's receiver isn't tuned to one of the 16 channels selected by the p-master. Success will eventually occur, however, when the p-master switches to the other train of 16 frequencies.

A device responding to an inquiry sends its FHS packet to the p-master, which can use the information in it to page that device for connection into a piconet. We will examine the basic piconet operation in the following chapter.

CHAPTER 6

Advanced Piconet Operation

When paging has been successfully completed the master and slave are connected and packets can be exchanged between them. All members of the piconet use the master's BD_ADDR and clock values in their hop generators and they all hop together from channel to channel. The master begins its transmissions when the least two significant bits in the CLK are 00, marking an even-numbered time slot, and the previously addressed slave begins its transmission back to the master when those bits are 10, designating an odd-numbered time slot.

As we learned in the last chapter, the CONNECT state begins with a POLL packet to the paged slave to check that it has successfully transitioned to the channel hopping sequence, as verified by the slave responding with any packet, usually a NULL packet if it has nothing else to send. Now the master and slave can exchange link setup information and then begin sending user data to each other.

In this chapter we'll present some of the advanced operations available to piconet members. Low-power modes can be used to save energy and enable a device to perform additional tasks such as paging or inquiry when already an active piconet member. Link supervision timers are implemented to cause timeouts to occur if piconet members rudely depart without giving notice. Finally, we look closely at scatternet operation and its alternatives.

Low-Power Modes

Bluetooth devices connected in a piconet can be put into one of three different low-power modes called *sniff*, *hold*, and *park*. Although these are often referred to as *low-power modes*, they are useful in several different applications such as

- Enabling more than seven slaves to be in one piconet
- Giving the master time to bring other slaves into its piconet
- Providing a means for a device to participate in multiple piconets (scatternet)
- Conserving energy

Most of the energy conservation methods in these low-power modes are directed at reducing the time a device's receiver remains on. In the higher-power wireless devices of yesteryear, the transmitter needed far more power than the receiver. Indeed, my 1-kilowatt amateur radio transmitter dims the lights in the house! Systems like Bluetooth, with their extremely low transmit powers, often exhibit their greatest current drain during receiver operation. Furthermore, although only one transmitter is on at any one time during normal piconet operations, a great amount of energy is expended if the receiver in every slave must come on at the beginning of each master-to-slave transmission time slot. The main purpose of the low-power modes is to remove that requirement.

For a point-to-multipoint topology, the master must bring the slaves into its piconet one by one. This means that the master needs to exit the CONNECT state and enter either the PAGE or INQUIRY states. Of course, during the page or inquiry/page processes, the remaining slave piconet members won't be communicating at all; they can't talk to each other, and the master is temporarily unavailable. The salves may as well enter a low-power mode during the time the master is otherwise occupied. The slaves can thus reduce their power consumption and prevent some of those finicky timers from triggering and causing them to disconnect from the piconet.

Sniff

When involved in normal piconet operation the slave must turn its receiver on at the start of each even-numbered time slot, as determined by CLK, to check for the master's transmission. The exception to this rule is when it receives a packet header designating the transmission of a multislot packet for another slave; then our slave can leave its receiver off for the duration of the packet. The *sniff* mode gives a slave a chance to reduce its receiver duty cycle still further by activating it only at regularly spaced sniff intervals called T_{sniff}. Under normal piconet operation $T_{sniff} = 2$ time slots, but because the value is stored as an unsigned 16-bit integer, it can be as high as $2^{16} = 65{,}536$ slots or about 41 seconds between sniffs. The slave keeps its AM_ADDR while in the sniff mode.

One difficulty with these regular sniff time intervals is that the master could have a SCO packet or a higher priority ACL packet for another slave scheduled for the slot that our slave is scheduled to sniff. Therefore, another time, called $N_{sniff\_attempt}$, is the number of consecutive receive (even-numbered) slots the slave is required to monitor at each sniff interval. If a packet arrives during this time that has the slave's AM_ADDR in its header, then the slave will continue to monitor for an additional $N_{sniff\_timeout}$ receive slots or for the remaining $N_{sniff\_attempt}$ receive slots, whichever is greater. Why bother complicating matters with these two times? When a packet addressed to a particular slave arrives, there's a greater chance that another packet for that same slave will also arrive within a reasonably short time; therefore, $N_{sniff\_timeout}$ should be selected accordingly so that the slave doesn't go dormant before all packets have arrived. Both of these quantities are also 16-bit unsigned integers.

Finally, the quantity D_{sniff} is the number of time slots until the first sniff occurs. All of these quantities are passed between devices using *link manager protocol* (LMP) commands (see Chapter 7, "Managing the Piconet"). Together, they define the entire sniff process so that the master knows when the sniffing (don't you just love that word?) slave is available to receive a packet.

Figure 6-1 shows a slave in the sniff mode and how its receive timing is modified from normal piconet operation. The sniff mode affects only ACL packets, so the slave must still be available for any SCO links that may exist with the master.

Hold

Unlike sniff, which is periodic in nature, the *hold* mode is a one-time exit from the obligations of the piconet (see Figure 6-2). The slave keeps its AM_ADDR but suspends ACL packets for the *hold timeout* (holdTO) value, which is a number of slots to hold expressed as a 16-bit unsigned integer. Therefore, hold can last up to about 41 seconds. The *hold instant* specifies the CLK value when the hold begins. During hold a device can perform various page or inquiry duties, attend another piconet, or simply do nothing and save power.

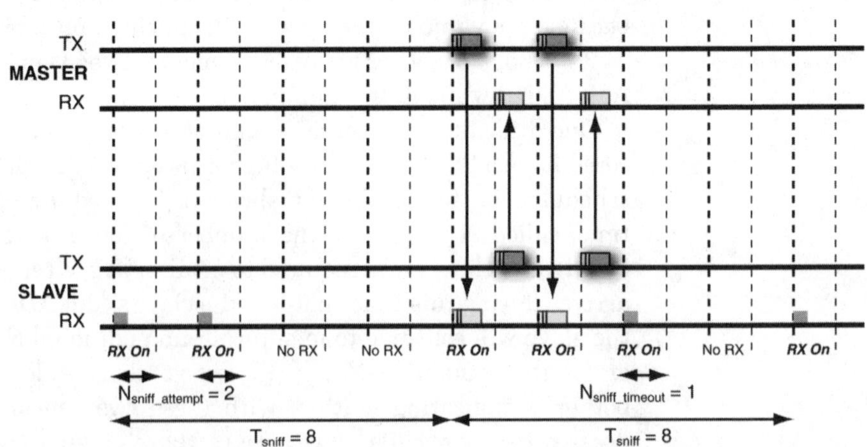

Figure 6-1 Sniff operation with parameter examples

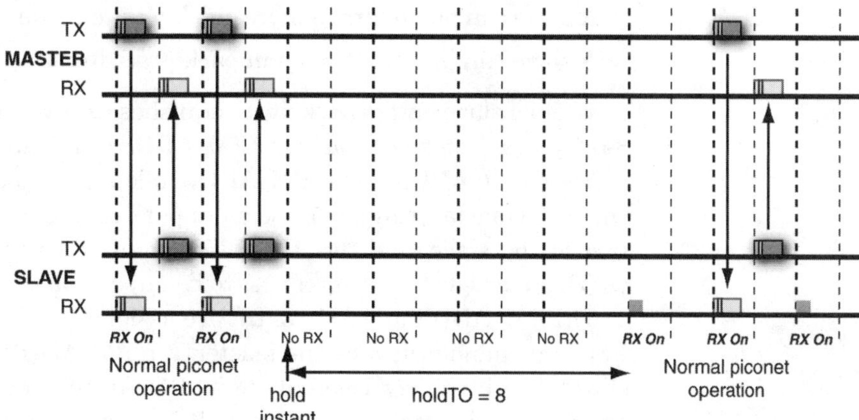

Figure 6-2 Hold operation with parameter examples

If a master needs to exit the piconet temporarily, then it can place all of its slaves into the hold mode. This has the effect of enabling the master itself to hold from the time the last slave enters hold until the first slave exits hold. Like the sniff parameters, hold parameters are exchanged as LMP packets, which is discussed in Chapter 7.

During normal piconet operation, the slave has a ± 10 μs uncertainty window over which it searches for the channel access code from a master's transmission. Even if that particular packet is addressed to another slave, all slaves can resynchronize their respective CLK values each time the master transmits a channel access code. When a slave returns from hold, though, the uncertainty window could be much greater than ± 10 μs due to clock drift between the master and slave, so the slave needs to increase the length of time its receiver is on when searching for the first master-to-slave transmission. The time required for resynchronization can be shortened if the master uses single-slot packets for a few time slots after each slave returns from hold. This will insure that a packet is actually transmitted on the hop frequency to which a returning slave's receiver is listening.

Park

The *park* mode is by far the most powerful of the low-power modes and as luck would have it, the most complex. In exchange for this complexity, the park mode presents several advantages. Among these are

- Low power consumption is in effect for parked devices.
- The master can bring more than seven slaves into the piconet.

- A slave can be unparked more quickly than it can be paged.
- A slave can initiate its own unpark procedure.

When a slave enters park, it relinquishes its AM_ADDR and receives an 8-bit *parked member address* (PM_ADDR) and an 8-bit *access request address* (AR_ADDR). The PM_ADDR takes the place of the AM_ADDR when the master wants to issue an unpark command or other communication to the slave, and the AR_ADDR determines a time window during which a parked slave can request to be unparked.

The PM_ADDR and AR_ADDR are assigned to the slave during a park LMP command issued by the master. If a PM_ADDR of 0x00 is given to a slave, then it will respond only to its BD_ADDR when being unparked by the master; otherwise, the slave will respond to either its PM_ADDR or BD_ADDR. As a result there's no limit to the number of slaves that can be parked, although I'm often amused thinking about how feasible it would be to have more than 255 Bluetooth devices located within a 10 meter radius. Communication between the master and parked slaves is accomplished through broadcast packets because there's no AM_ADDR assigned to any of the parked slaves. A slave involved in a SCO link cannot be parked.

Beacon Packets Slaves can remain parked for long periods of time, so a mechanism must be established to keep parked slaves synchronized to the master. When a slave is parked, the master makes a promise to transmit *beacon packets* periodically for the parked slave to use for resynchronizing its piconet timing. Beacon packets are any type of packet, either ACL or SCO. If the master has no information to convey when it's time to send a beacon, then it will transmit a NULL packet. The beacon packets have four purposes:

- Resynchronizing slaves to the master
- Carrying general broadcast messages to the parked slaves
- Unparking slaves
- Carrying messages to change beacon parameters to the parked slaves

Figure 6-3 shows how these beacon slots are organized. A beacon train consists of N_B packets spaced Δ_B apart over an interval of T_B slots. These parameters are passed from master to slave during the LMP park command (see Chapter 7) and should be selected such that a slave can resynchronize to the master periodically in error-prone conditions. The first beacon in the train begins at a time called the *beacon instant*. Resynchronization occurs when the slave successfully detects the master's channel

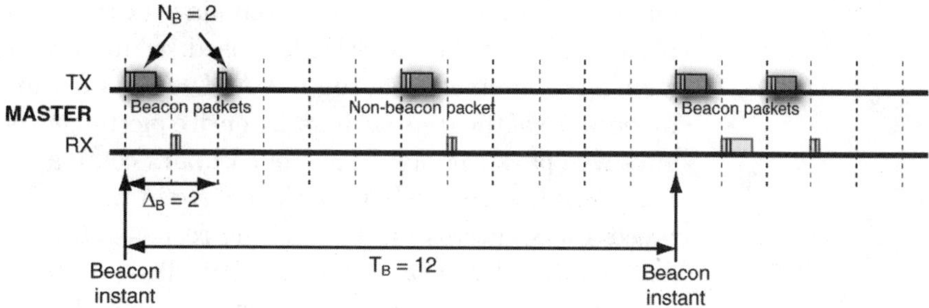

Figure 6-3 Beacon packets and examples of their timing parameters. Packets received by the master are from active slaves and play no role in the beacons.

access code with its correlator, so it's not necessary for N_B to be particularly large, but it needs to be at least 1.

A slave can sleep for extended periods while parked, and it can even skip entire sets of beacon slots. This sleep interval N_{Bsleep} is an integer multiple of T_B and is one of the parameters given to the slave when the park command is issued. Another parameter called D_{Bsleep} is the number of multiples of T_B before the slave first checks the beacon train. As before, knowledge of these parameters allows the master to know when the parked slave will be listening to beacon packets.

When the parked slave wakes up for resynchronization, then it should plan to listen for the first beacon in a beacon train. Because the master communicates with parked slaves via broadcast packets, these slaves should decode both access code and header. If the AM_ADDR = 000, then the parked slaves should also receive the packet payload.

Although beacon packets aren't required to be broadcast packets, a broadcast message to parked slaves must be placed in a beacon slot for them to hear it, and the message should be repeated during the entire beacon train unless interrupted by a SCO packet. Notice that broadcast messages can fall into two categories: those intended for all piconet members (including parked slaves) and those intended only for the active slaves. Messages to all piconet members must be broadcast during beacon trains in which all parked slaves are listening, but broadcast messages to only the active slaves can be sent during other times. Of course, the latter messages must take into account timing associated with active slaves that are in the sniff and hold modes.

Master-Activated Unpark A master can unpark a slave by sending it a LMP unpark command during a beacon slot in which that slave is listening. The unpark command includes the slave's new AM_ADDR. The slave will respond to either its PM_ADDR or its BD_ADDR. As we'll discover in Chapter 7, up to seven slaves can be unparked with a single command when the PM_ADDR is used, or up to two can be unparked with a single command using the BD_ADDR. This feature can be quite powerful because of the time savings; an entire piconet of slaves can be exchanged via seven park commands and one unpark command. However, the master must POLL each newly unparked slave in turn to insure that the unpark process was successful. If there's no response from an unparked slave for *new connection timeout* (newconnectionTO) (32) time slots after the end of the beacon train in which the unpark command was issued, then the master will try again during the next beacon train in which that slave is listening. If the slave doesn't receive a POLL packet within newconnectionTO slots after being unparked, then it returns to the PARK state.

One of the obvious applications of the park process is for a master to build a piconet of many slaves that will all share in a particular task such as parallel processing or data acquisition. Slaves are each given a computation task and then parked. The master can subsequently unpark the slaves periodically and check their progress in a manner similar to a CPU polling its peripherals periodically for service requests.

Slave-Activated Unpark In a rare digression from normal Bluetooth protocols in which the master has total control over the piconet, parked slaves can actually initiate their own unpark procedure by transmitting their desire to the master under carefully controlled conditions. This is done by the master creating a series of *access windows* after a beacon train during which parked slaves can call for an unpark command. An access window starts D_{access} slots after the beacon instant and lasts for T_{access} slots (see Figure 6-4). The access window train starts after the beacons are finished, and multiple access windows of number M_{access} can be used to improve the chance that the master hears the access request.

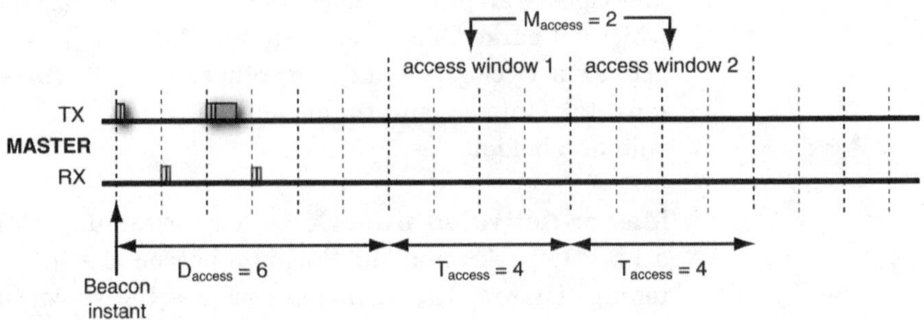

Figure 6-4 Access windows and their relationship to the beacon instant

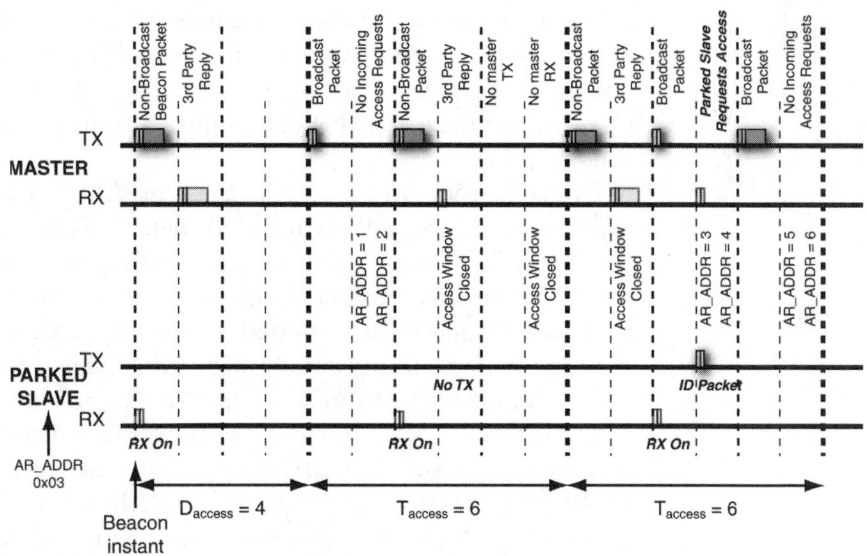

Figure 6-5 Using the access window to request unpark. The access windows are open only if the preceding slot contains a broadcast packet. The parked slave's AR_ADDR determines the time it transmits the master's ID packet to request an unpark command.

Each slave-to-master time slot within an access window is divided into two 312.5 μs half-slots, as shown in Figure 6-5. A parked slave uses its AR_ADDR to determine when it transmits the master's device access code (ID packet) as an unpark request to the master. The DAC is the same as the channel access code without the 4-bit trailer (see Chapter 4, "Baseband Packets and Their Exchange"). The master knows which parked slave requests the unpark by the timing of the DAC transmission. For example, the slave having AR_ADDR = 3 will transmit the DAC as an unpark request during the first half-slot of the second access request slot.

To avoid collisions with active piconet slaves, the parked slave is allowed to send an unpark request only if the previous master-to-slave slot contained a broadcast packet. Because active slaves never respond to a broadcast packet, the following slot becomes available for access request transmissions. If no packet at all appears in the previous master-to-slave slot, then parked slaves aren't allowed to transmit their access requests. Therefore, if the master has no broadcast information to send, then it can put a broadcast NULL packet in the slot. The duration of an access window must be sufficient to enable all parked slaves a chance to request unpark.

Because the AR_ADDR contains 8 bits (with 0x00 not used), the master must assign duplicate access request addresses if more than 255 slaves are parked. The specification is silent on how to resolve possible access request collisions, but a few solutions come to mind. Perhaps the simplest is for the master to insure that two parked slaves with the same AR_ADDR never use the same access window; that is, N_{Bsleep} should be at least $2T_B$, and slaves with duplicate AR_ADDR values alternate their monitoring of the beacon trains.

Even if two or more slaves with the same AR_ADDR use the same access window, a collision will occur only if both devices request unpark at the same time. To avoid a hopeless situation of repeated collisions, each unsuccessful access request could be accompanied by a random backoff before the next access request is transmitted. Although this solution is not in the specification, there's nothing prohibiting its implementation. (Are we spending too much time trying to answer a question no one asked?)

The slave-activated unpark procedure is useful because of its similarity to the *interrupt request* (IRQ) used by peripherals attached to a CPU. When the IRQ flag is activated by the peripheral, the CPU will service that device, just like the Bluetooth master will service a parked slave that activates its access request. For example, dovetailing with our earlier parallel-processing example, a host computer that needs to invert a large matrix can form a Bluetooth piconet with many slaves, each of which receives a portion of the matrix and is then parked. Because the associated host computers have varying speeds and efficiencies performing their task, each one can effectively interrupt the master through an access request when its task is completed. This removes from the master the burden of repetitively polling each parked slave individually; although in its present form, the access request process does require the master to effectively poll two parked slaves at a time using broadcast packets. The Bluetooth specification mentions the possibility of incorporating other access request procedures in the future.

Link Supervision

Because Bluetooth is a dynamic, ad-hoc network (where have we heard that before?), it's imperative that devices recover gracefully when various disruptions occur. For example, devices can move out of range after connecting to a piconet, or they could be turned off by the user. Regardless of whether the affected device is a master or slave, the others must be able to recover from its sudden disappearance.

Link supervision is the process of monitoring the possibility of link loss through a timer called $T_{supervision}$. The value *supervision timeout* (supervisionTO) is negotiated by the two link managers (see Chapter 7) and represents the maximum number of time slots each device (master and slave) will wait between successive packets from the other. The timer is reset whenever a packet arrives that has the correct CAC, AM_ADDR, and passes the HEC. (Aren't these acronyms great?) If $T_{supervision}$ reaches the supervisionTO value, then the connection is reset, and (if applicable) the user notified. Of course, supervisionTO should be greater than the longest hold or sniff periods. The timer also applies to parked slaves, so each slave should be periodically unparked and reparked by the master before expiration of its respective supervisionTO. This timeout has a default value of 20 seconds, but the value can be modified through a LMP command (see Chapter 7).

Although not officially part of link supervision, a poll interval called T_{poll} is established by the master as the maximum number of slots between consecutive transmissions from the master to a particular slave. The default value is 40 slots, but changes can be sent to the slave as a LMP packet (see Chapter 7). A NULL or POLL packet is used if the master has nothing to say to the slave, giving the slave an opportunity (mandatory for the POLL packet) to respond to the master in the following time slot. The purpose of T_{poll} is to enhance *quality of service* (QoS) by presenting the slave with guidelines for bandwidth allocation and latency. The master can violate T_{poll} times if they would conflict with page, page scan, inquiry, or inquiry scan routines.

Device Entry into an Existing Piconet

The setup procedure for a point-to-multipoint piconet can actually be quite complex due to a number of Bluetooth restrictions. First, the radio can only do one thing at a time, either transmit or receive, on a single frequency. If a master successfully polls a slave and begins a point-to-point communication link, bringing another slave into the piconet requires that the master reenter at least the page state and perhaps both inquiry and page states. Meanwhile, what does the original slave do? Inquiries can last more than 10 seconds in a perfect channel, and pages can take more than 5 seconds each.

Suppose a device wants to initiate its own entry into an existing piconet. The device initiating the link is always the master, but the existing piconet already has a master that may not be aware of the new slave's desire to enter. How should that be handled? We'll examine both of these issues next.

Master Brings in New Slave

Building a point-to-multipoint piconet requires a master to bring slaves into the piconet one-by-one. Existing slaves in the piconet perform no communicating while the master is busy with subsequent inquiries or pages. To free the slaves temporarily for other tasks (such as performing their own inquiries/pages for scatternet operation or power conservation), the master should place them in hold or park while it is occupied bringing other slaves into the piconet. If a SCO link has been established in the existing piconet, then the master must multiplex its SCO obligations with the inquiry/page process. The master may also need to temporarily leave the page or inquiry processes to transmit beacons to parked slaves as well.

Slave Initiates Its Own Entry

A device that wants to enter an existing piconet must perform several operations in sequence. They are

1. Page the existing master.
2. Establish a point-to-point piconet with the master as a slave.
3. Request a *master-slave* (MS) switch.

For this procedure to be successful, the existing piconet's master must periodically enter the page scan mode; that is, it must have scatternet capability. Upon completion of the master-slave switch, discussed in the next section, the initiating device becomes just another slave in the existing piconet.

The Scatternet and Its Implications

If a single Bluetooth device is a member of two or more piconets simultaneously, then the piconets form a *scatternet*. We will call the participating device the *scatternet member*. The scatternet member can be a master in one piconet and a slave in the other, or it can be a slave in both. Almost every article written on Bluetooth baseband over the years has pointed out that a device cannot be a master in two piconets because all slaves synchronize on the master for hop and packet timing, so the two piconets are indeed only one piconet. So we won't make that point yet again here, wasting valuable space.

The scatternet concept has several uses. Among these are

- A device initiates its own entry into an existing piconet by forming a scatternet with the master.
- Cross-piconet communication can take place without requiring devices to periodically disconnect from one piconet and reconnect to the other.
- Devices can participate in *store-and-forward* messaging, effectively removing any Bluetooth range limits if enough devices are available to relay data.

Accomplishing the first item requires scatternet capability on the part of the original piconet master because it must also temporarily act as a slave and respond to the initiating device's page. The scatternet demands on the original master are quite minimal, however, requiring only a short-lived point-to-point link to be established followed immediately by a master-slave switch. All of this could conceivably occur while the other members of the original piconet are in the hold state.

Implementing the scatternet also has some significant disadvantages, especially for the last two realizations, which are more complex than the first. Among these are

- Scatternet members must duplicate baseband timers and registers and manage both properly.
- Throughput is reduced substantially due to timing offsets between two independent piconets.
- Higher-layer protocols must track routing and error recovery during store-and-forward operations.

Scatternet Timing Issues

Scatternet operations require complex timing because a scatternet member must gracefully multiplex its operations between two or more piconets, keeping careful watch on various timers to prevent link disruption from timeouts.

ACL Timing When a scatternet member is involved only in ACL traffic across piconets, then it can take advantage of the sniff, hold, and park low-power modes to divide its attention between the different piconets. These methods are especially applicable to a scatternet member that operates as a slave in multiple piconets, but it should be noted that if a master puts all of its slaves into a low-power mode, then it has effectively put itself into the same low-power situation as well, liberating time for it to participate as a slave in another piconet.

Of course, a slave scatternet member might simply ignore each piconet in turn without informing the respective masters of its temporary exit, and as long as the various timeout values aren't exceeded, then the member won't lose any of its links. A master could do the same as long as it's available for beacon obligations and periodic POLL packets to reset each slave's $T_{supervision}$ timer. However, this stealth scatternet operating technique has the potential to waste bandwidth because the master in one piconet could begin sending packets to the (slave) scatternet member while it's participating in the other piconet. The result could be several consecutive no-response situations, each of which the master treats as a NAK with subsequent retransmission. Also, the master may conclude inaccurately that the channel is highly error-prone, causing it to reduce the size of its packets and/or add error control, further reducing throughput even when the scatternet member returns to that piconet. Efficiency is improved if a scatternet member arranges its temporary piconet departures through one of the low-power modes instead.

If the scatternet member is a slave in two or more piconets, then ACL scatternet operation can take place by using offset sniff intervals to multiplex traffic between piconets, as shown in Figure 6-6 for two piconets. The sniff interval T_{sniff} enables symmetric switching, so the scatternet member has the potential to divide its time equally between the two piconets.

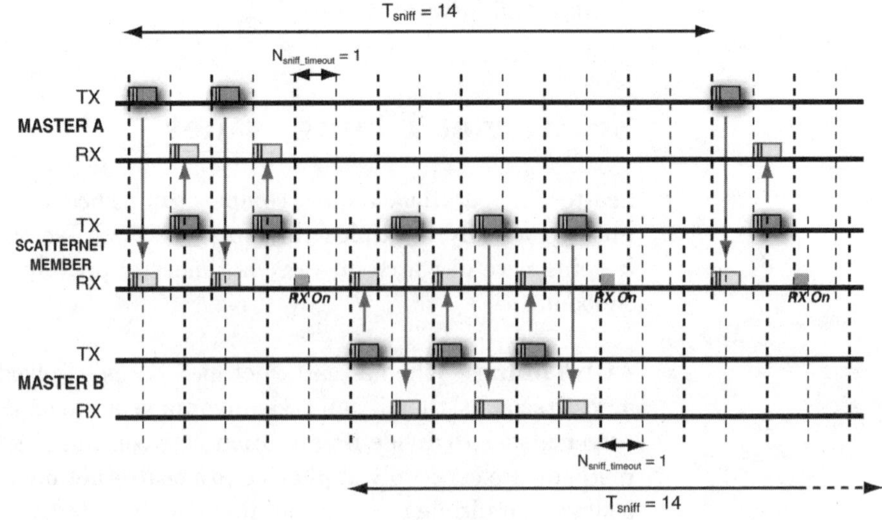

Figure 6-6 Scatternet ACL operation using offset sniff intervals. When traffic exists on one piconet, then the scatternet member may miss sniffs on the other piconet unless a hold time is negotiated with the active piconet.

Timing could be upset if the master in one piconet sends several consecutive packets to the scatternet member. In this case, the member is required to continue sniffing every even-numbered time slot for $N_{sniff\_timeout}$ slots after the last packet arrives, and this timeout value is reset for each new packet sent to the scatternet member. It's quite possible, then, that the member will miss the next sniff interval with the other piconet; in fact, there's no limit to the number of sniff intervals the member could miss in the other piconet when a large number of packets are addressed to it by the current piconet master.

For efficient operation, it makes sense for the scatternet member to remain in the piconet that has traffic for it rather than arbitrarily exiting, but there's a risk of timing out with the other piconet unless provisions are made for checking it periodically. This can be done by negotiating a hold time with the active piconet, during which the scatternet member can check-in with the other piconet. Of course, this hold time should coincide with a sniff period associated with the other piconet.

Instead of using sniff intervals to multiplex between two piconets, the scatternet member can use the hold or park modes instead. Efficiency is reduced for the hold mode because each entry into hold requires an exchange of link manager packets. Furthermore, as we'll discover again in Chapter 7, a slave can force the master to let it hold, but the hold cannot begin until at least $6*T_{poll}$ slots beyond the time the request reaches the master. Using the default T_{poll} value of 40 slots, then, the slave must wait at least 240 slots before it can even enter hold. The result is that the switching rate between piconets for the scatternet member is fairly slow when using hold times to multiplex. As before, packet traffic in one piconet could overrun the hold time in the other piconet. The solution is the same: The scatternet member should enter hold in the active piconet, and then renegotiate another hold in the other piconet before returning to exchange more ACL packets.

Perhaps the best way for a scatternet member to divide its time between two or more piconets is through the park mode. A device in park has much more versatility for monitoring a piconet for unpark commands and other broadcast packets, and it can skip several beacon trains by increasing N_{Bsleep} to a higher integer multiple of T_B beacon interval lengths. By offsetting beacon monitoring times, the scatternet member can operate in a manner similar to the arrangement of sniff intervals in our previous example.

A master attempting to unpark a slave will wait for newconnectionTO (32) time slots for a response after sending the unpark command, and if there's no response, then it will try to unpark the slave again. This gives a scatternet member some additional timing versatility because the master will continue to try and unpark the slave in this manner until supervisionTO (20 seconds default time) expires.

Suppose the master in *both* piconets wants to exchange ACL traffic with the scatternet member at the same time. The member can multiplex operations between the two masters or stay with one master until all packets are transferred, and then move to the other piconet and complete the packet exchange there. The method selected may depend upon timing constraints required by a higher-level protocol. For example, if two different streaming audio files are being played in real-time, then the scatternet member may be required to rapidly multiplex between the two piconets. On the other hand, the most efficient method of packet transfer is the one that requires the least amount of overhead by completing all traffic on one piconet before moving to the other (provided, of course, that no timeout violations occur).

SCO Timing Operating SCO links on a scatternet has some subtle timing implications beyond ACL packet traffic that aren't particularly obvious. Scatternet SCO requires that the scatternet member alternately switch between piconets for each pair of SCO packets exchanged.

Suppose a scatternet member just completed an exchange of two SCO packets in piconet A. How much time is needed for the member to exchange the next pair of SCO packets in piconet B? The random timing offset between the two piconets means that the scatternet member requires at least three full time slots to be available in piconet B to guarantee that a master-to-slave slot is followed by a slave-to-master slot. If the first full slot in piconet B, for example, is slave-to-master, then the SCO packet exchange occurs during the second and third slots.

After the piconet B exchange, there must be enough time for the scatternet member's synthesizer to return to the correct master-to-slave slot frequency in piconet A prior to receiving another packet from that master. Consequently, real-time two-way voice communication throughput is reduced from three HV3 links, if all were in the same piconet, to only two in the scatternet. Also, no scatternet capability exists for a piconet member that is passing HV2 or HV1 packets because three consecutive timing slots are never available with these two modes.

Figure 6-7 gives an example of timing and drift when a scatternet member is supporting two SCO HV3 links. The scatternet member is a slave to Master M and a master to Slave N. If the Master M CLKN runs at a slightly slower rate than the scatternet member's CLKN, then there is a gradual drift of Master M's slots to the right. To avoid eventual overlap, the scatternet member must periodically delay a SCO HV3 packet exchange by one pair of slots.

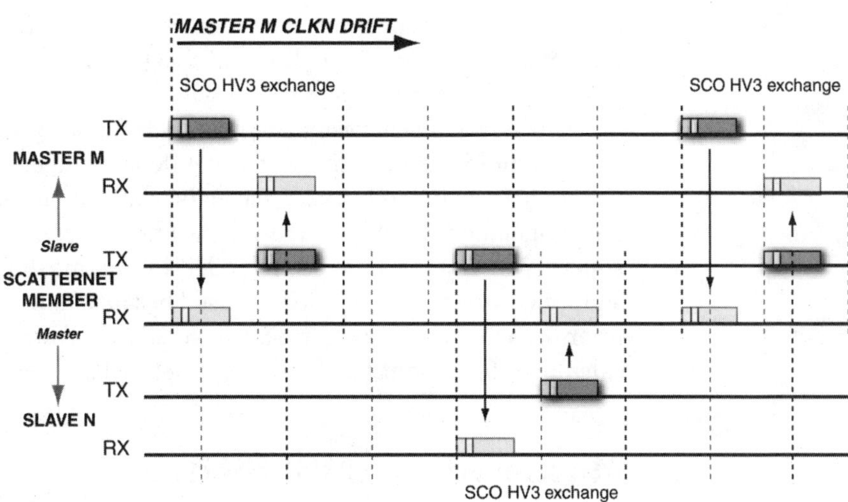

Figure 6-7 Two HV3 SCO links can be supported across two piconets. Drift between the independent piconet CLK values will eventually require a shift in the timing for one of the SCO links, so the scatternet member must be a master for this link.

Clock Drift and Scatternet Longevity As we pointed out before, Bluetooth devices are required to have a clock drift of 20 ppm or better relative to a true clock. The actual clock drift can be discovered by other devices through an LMP command, but for our calculations we'll assume that this worst-case value of 20 ppm occurs for each piconet in opposite directions, for 40 ppm total.

The most stringent requirements on drift are dictated by scatternet SCO operation because drift over a fraction of a time slot between piconets could disrupt the ability of the scatternet member to exchange consecutive HV3 packet pairs in both piconets (refer to Figure 6-7). In a worst-case situation, for example, the scatternet member could switch from piconet A to B just in time to receive the HV3 packet in the master-to-slave slot. If piconet B drifts into piconet A, then the scatternet member soon won't be able to switch fast enough to catch the first SCO packet in piconet B. If the combined drift is in the other direction, then the two piconets could shift nearly 1.25 ms before SCO slot overlap occurs, and this will happen in 1250/40 = 31.25 seconds. On average, then, the scatternet member can support one SCO link in each piconet for about 15.6 seconds at a time before SCO slot overlap occurs.

If a scatternet member has an active HV3 link in each piconet, then it should monitor the drift between them. When drift takes the cross-piconet SCO slots too close to each other, then the member can recover by establishing a new set of SCO slots in one of the piconets. Unfortunately, the master is the only entity that can specify when to begin the SCO packet pair exchange, so the scatternet member must be the master in one of the piconets for support of two different SCO links to be feasible.

Is all of this really worth the hassle?

Anyway, for a scatternet member participating in ACL links only, the effect of timing drift on the scatternet depends upon how often the scatternet member switches between piconets when using sniff or park. If the scatternet member is a slave, then the parameters can be adjusted during link supervision when the device is polled by the master. The effect of drift is simplified when the hold mode is used because a new hold time is specified prior to each temporary exit. The start time and duration of the hold can be calculated to account for relative piconet drift.

Master-Slave (MS) Switch

As we mentioned before, a device that initiates its own entry into an existing piconet begins communicating as a master and then requests a MS switch soon after connecting. These situations are likely to occur when a telephone handset first accesses a community base station or when a laptop first contacts a LAN gateway. Because both the telephone base station and the LAN gateway are often manipulating multiple connections, their tasks are greatly simplified if they are masters of their respective piconets. The initiating device begins as the master in a point-to-point piconet, then after the MS switch occurs, the device becomes a slave in the previously established point-to-multipoint piconet.

A more complex MS switch situation occurs when a slave in an existing point-to-multipoint piconet is to become the new master. Aside from the usual process of exchanging roles with the former piconet master, the new master must also send its timing information to each piconet slave in turn, so they can seamlessly transfer their allegiance to the new device. Such an arrangement enables the piconet to survive even when a master departs.

Either master or slave can initiate the MS switch process. Both devices stop transferring user data and stop any encryption processes before proceeding. We will examine the simple point-to-point MS switch first, followed by the more complex switch, where a slave becomes the new master in an established point-to-multipoint piconet.

Point-to-Point MS Switch The procedure for exchanging roles between a master and single slave in a point-to-point configuration begins with a LMP request, originating with either the master or slave. The current slave also sends a packet with the offset between the start of its own even-numbered time slot (used when it becomes the new master) and the present even-numbered time slot (used by the current master), accurate to 1 μs. This will help the new slave adjust its slot timing quickly once the new master takes over. If the request is accepted, then the MS switch takes place in several stages:

1. The current master and slave switch their TDD slots.
2. The new master sends the new slave its FHS packet and receives acknowledgment.
3. Both master and slave change to the new hopping sequence and slot timing.
4. The new master Polls the new slave for verification of the MS switch.

To avoid getting lost by using terms like *former master* and *current slave*, we'll just call the device that wants to become the new master *device A* and the current piconet master *device B*.

TDD switch The MS switch process begins with a TDD switch. Current slave A and master B simply reverse the slots on which they normally transmit; that is, device A begins sending packets during even-numbered time slots and device B during odd-numbered time slots. Device A's AM_ADDR is now used by device B. At this point, device A has become the new master and device B the new slave, but they're still using device B's timing and hop sequence.

FHS packet and acknowledgment Device A sends B its FHS packet (in which it assigns device B its old AM_ADDR) and receives its own ID packet in response. This is similar to the FHS packet transfer during a page, except that it takes place on the channel hopping sequence determined by device B (the old master) instead of the page hopping sequence. Also like the page process, newconnectionTO (32 slots) is used to monitor progress. If a timeout occurs during either FHS transmission or acknowledgment, then both devices return to their original roles.

New hop sequence and slot timing Because the FHS packet has a slot resolution of only 1.25 ms, device B combines this information with device A's previously transmitted slot offset to calculate an accurate starting time of A's next transmission slot.

Now the two units change to the hop sequence and slot timing determined by device A, the new master. (Why, you may ask, wasn't slot offset information needed during a page? The answer is that the p-master's transmission of the p-slave's ID packet established the p-master's half-slot timing, and complete slot synchronization was established when the subsequent FHS packet was transmitted. No such transmissions occur during the MS switch, so the new slave must determine the new master's timing by using a combination of slot offset and FHS information.)

MS switch verification with a POLL Following once again the example of a page, the new master sends a POLL packet to the new slave, which is acknowledged by a NULL if all is well. The newconnectionTO timer tracks its progress; if it times out, then the two devices return to the old piconet settings. At this point, it will be a miracle if they can still communicate.

Point-to-Multipoint MS Switch If a point-to-multipoint piconet exists, then the MS switch process begins with devices A and B exchanging roles. Next, each slave in the new piconet must be changed over to the new hop sequence and timing associated with the new master. Parked slaves must be unparked before the MS switch can be completed. Each old slave (with the exception of device A) becomes a slave in the new piconet through the following procedure:

1. The new master sends a slot offset LMP packet to the slave.
2. The new master sends its FHS packet to the slave and receives acknowledgment.
3. The new master and slave change to the new hop sequence and timing.
4. The new master POLLs the slave and receives acknowledgment.

Most of these events are similar to those used for the point-to-point MS switch with a couple of exceptions. First, a TDD switch isn't needed because the piconet's slaves are already using the correct odd-numbered time slots for their transmissions. Also, the FHS packet the new master sends to each slave contains that slave's old AM_ADDR in its header, followed by the new AM_ADDR in its payload (see Chapter 4). Of course, the old and new AM_ADDR values can be identical. Finally, any slaves that are out of range of the new master won't be able to participate in this procedure and will simply disconnect from the old piconet when their respective supervisionTO values are reached.

Alternatives to Scatternet Operation

Implementing a Bluetooth scatternet requires a lot of overhead and, for all the trouble, provides an aggregate throughput well below that possible within a single piconet. Therefore, it makes some sense to use the scatternet only when absolutely necessary. As we mentioned earlier, a temporary scatternet must be established for a slave to initiate its own entry into a piconet, so this is one of the necessary uses of the scatternet. Fortunately, at least one of the piconets in this type of scatternet is only point-to-point and lasts a very short time.

There are two general alternatives to other, longer-term scatternet operations that may be easier to implement and provide increased efficiency. They are

- Having the device disconnect from one piconet and connect to another as needed
- Installing multiple Bluetooth devices within a single host

The first option is the cheapest and probably the most practical, but it could prove to be quite time consuming, especially if the device jumping between piconets is a slave in both. As we've already learned, a slave initiating its own entry into a piconet must first join as a master, and then request a master-slave switch. Because the master in the established piconet is busy running its own show, it is probably using page scan mode R1 (1.28 seconds) or R2 (2.56 seconds) as the time between successive page scans. The average time for a successful page and the master-slave role switch in these situations could average more than a second. Clearly, this mechanism is satisfactory only if the jumping slave stays in one piconet for at least several seconds at a time.

For other applications that require normal Bluetooth throughput in a scatternet environment, it may be best to pursue the second option and install more than one Bluetooth device within a single host. Bluetooth *is* supposed to be cheap, yes? Because the devices have independent hardware, they can each achieve almost normal throughput figures provided one device's transmitter doesn't significantly block the other colocated receivers. If a single CLKN common to all devices on the host is used and if all Bluetooth devices on that host are masters in their own respective piconets, then the transmitters will activate together and the receivers will also activate together, and no receiver blocking occurs from host-based transmitters. Of course, the situation is more complex when one or more devices serve as slaves because their transmit/receive timing cannot be so easily coordinated.

Summary

The sniff, hold, and park modes each have different characteristics. A slave in the sniff mode can check master-to-slave transmission slots less often than every even-numbered slot. The hold mode enables a slave to stop ACL transmissions for a period of time before returning. The park mode is the most versatile, enabling a slave to resynchronize on the master's transmissions through beacon packets every now and then, but otherwise, it has no obligations in that piconet. A parked slave can be unparked by the master during a beacon slot, either because the master wants to bring the slave back into the piconet or as a result of an unpark request by the slave itself. SCO packet transmission and reception still take place during sniff and hold, but the SCO link must be terminated before a slave can be parked.

Although they're called low-power modes, sniff, hold, and park are also extremely useful for providing a device time to participate in activities outside the present piconet. For example, a device can become a member in another piconet, creating a scatternet. Scatternet operation is required for a unit to initiate its own entry into an existing piconet, but this short-lived point-to-point piconet disappears through the master-slave switch.

More complex scatternets have various timing requirements that must be carefully monitored, and their throughput to/from scatternet members is generally much lower than separate piconets could provide because of the time-multiplexing and lack of cross-piconet synchronization. Alternatives to scatternet operation include periodic connecting and disconnecting from separate piconets or installing multiple Bluetooth devices within a single host.

CHAPTER 7

Managing the Piconet

Up to now we've studied parts of the Bluetooth protocol stack that are predominately hardware such as the radio and a state machine in the form of a *link controller* (LC). As we move farther up the Bluetooth protocol stack, the layers gradually transition from hardware to firmware, and then, as we move over to the host, to software implementations.

The *link manager* (LM) resides between the *host controller interface* (HCI) layer (if it exists) above and the LC below within the Bluetooth protocol stack (see Figure 7-1). The LM is arguably the lowest layer in the protocol stack that begins to exhibit some software characteristics. As we'll discover in Chapter 12, "Module Fabrication, Integration, Test, and Qualification," manufacturers of Bluetooth radio and baseband ICs usually place the LM in firmware, either physically on the baseband chip or in a separate memory unit.

The LM communicates with three different entities during a Bluetooth session: the local host through HCI, the local LC, and the remote LM, as shown in Figure 7-2. The local LM usually resides on the module as part of a complete host-module Bluetooth implementation, and the remote LM is defined as the LM at the other end of a Bluetooth communication link. As an example, the host may direct the local LM to connect to another device,

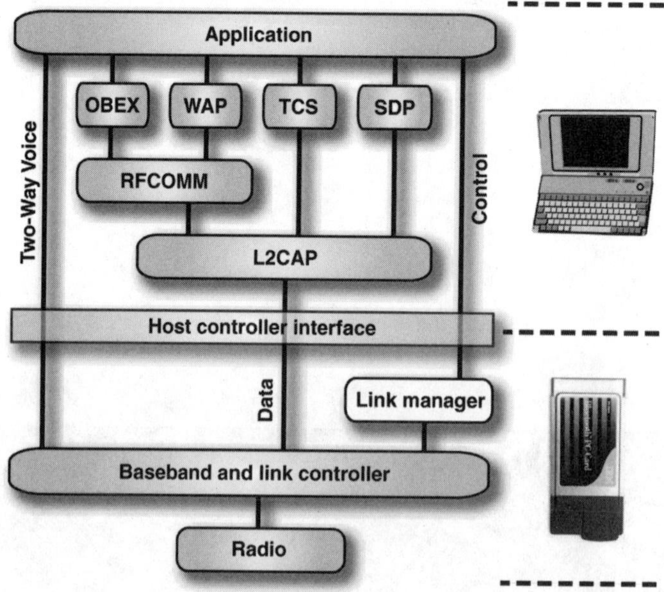

Figure 7-1 The LM communicates with the application through HCI and works directly with the LC.

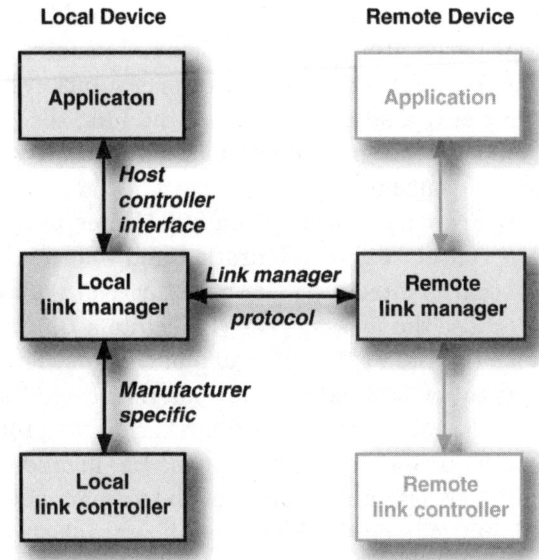

Figure 7-2 The local LM communicates with the host through HCI, with the LC for local operations and with the remote LM for link management.

in which case the local LM works with the local LC to carry out the required page. When the page is successful, the local LM begins communicating with the remote LM to set up and configure the link to prepare it for user data exchange. This communication path is depicted as directly connecting the two LMs in Figure 7-2, but, of course, they actually communicate through their respective Bluetooth LCs and radios. The Bluetooth specification presents a very structured method for LM-to-LM communications, but the method used by the local LM to interact with the local protocol stack may be specific to each Bluetooth device manufacturer if HCI isn't used.

In this chapter, we will describe the various tasks that the LM performs and reinforce the descriptions with some exceptionally exciting examples. However, a complete list of the commands and responses are in the Bluetooth specification so they won't be repeated here. Instead, we'll point out the subtleties associated with some of the LM tasks.

Link Manager Protocol (LMP)

Packets traded between LMs generally take the form of a *command* and *response* if the other LM is asked to enter a particular mode of operation, or a *request* and *response* if information is needed from the other LM. Information that is volunteered to the other LM usually requires no response. These *Link Manager Protocol* (LMP) operations are quite primitive and often involve only a single function of the link, such as entering the hold mode. If you enjoy programming in assembly code, a detailed study of the LM will warm your heart.

The LMP uses messages within a DM1 (or in some instances a DV) packet that contain link configuration and information, piconet management, and security directives. These messages are called *protocol data units* (PDUs). Because the messages are built into a Bluetooth packet, they are intended for the LM at the other end of the link. These special DM1 payloads, or the equivalent data portion of a DV payload, are "captured" by the destination LM and are not passed farther up the protocol stack. As usual, the Bluetooth specification tells *what* the LM must do, but not *how* to do it. If you really like assembly code, enough to dream of writing an assembler, then you'll find yourself thinking about how to take each LMP message and turn it into a set of actions by the LC.

LMP packets have a higher priority than user data between two devices, so while PDUs are being exchanged, the higher layers in the protocol stack may not see anything happening for several time slots. Furthermore, the LMP packets may also disrupt a SCO link if there are no unused ACL slots available because some PDUs have a longer data field than the DV packet can support. In these cases, a SCO packet pair will be dropped and replaced by a DM1 LMP PDU and its response. (Wow, three abbreviations in a row. Are we speaking Bluetooth or what?)

The LM relies on the LC and its implementation of error correction, error detection, and ARQ for reliable transmission of the PDU. This means that the LM itself has no procedure for acknowledging reception of a PDU, and it assumes that each PDU will eventually reach its destination error free. However, as with any ACL packet on the *user asynchronous* (UA) channel, maximum latency cannot be specified because poor channel conditions could result in many retransmissions of packets that were corrupted during reception. Also, a master's LC can reasonably guarantee that communication will take place with a particular slave only once every T_{poll} (default 40) slots. Consequently, link management tasks could require a significant time to complete. To prevent infinite waiting periods, the time between receiving a baseband PDU and sending a response must be less than 30 seconds, which is the LMP response timeout value. If this timeout is reached, the device assumes that the affected PDU was terminated unsuccessfully.

General Link Session

The life of a Bluetooth link between a master and single slave is shown in Figure 7-3. We've already studied the inquiry and page processes, where the devices find each other and connect. After this initial connection, the LM performs various link configuration tasks and then keeps watch while data is exchanged (see Chapter 8, "Transferring Data and Audio Information"), looking for any LMP packets and intercepting them for processing. Finally, when the data exchange is complete, either LM can perform the detach.

As expected, the LM plays its greatest role between the time the page is successful and when the devices begin exchanging user data. The link configuration process can often involve several LMP PDU exchanges between the master and slave LMs. Some PDU arrivals require the application and perhaps also the user to be notified and make some response, whereas others can be handled by the LM itself or perhaps by some other software/firmware kernel higher in the protocol stack. We'll examine the link configuration part of this diagram in more detail toward the end of this chapter.

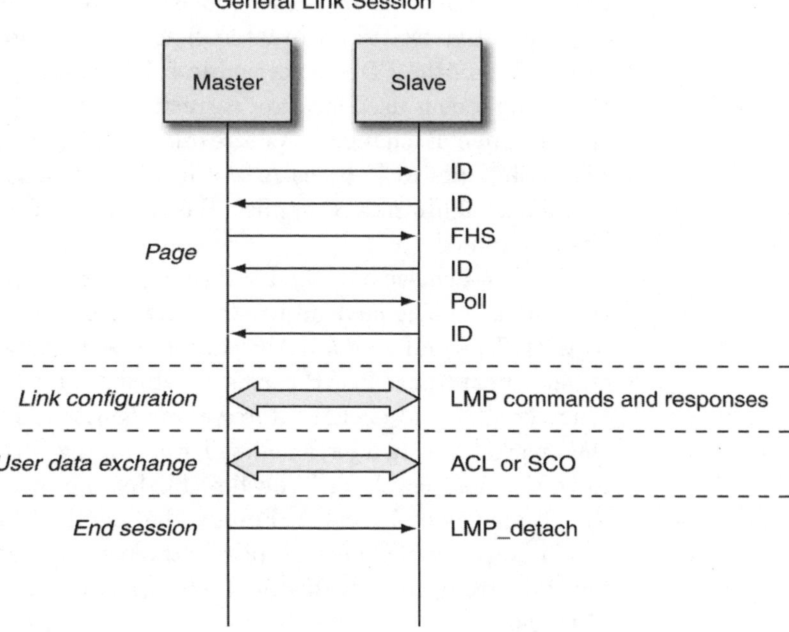

Figure 7-3 General link session from page to detach—either the master or slave can initiate the detach.

In general, the aspects handled by the LM fall into three categories:

Link configuration and information When the page is successful and the master and slave are connected into the piconet, they need to discover what link features (for example, support for multislot packets and RSSI) are available in the other device. LMP PDU packets also exist for setting QoS, power control, and other configuration functions during any time the link is active.

Piconet management Management of the piconet includes attaching and detaching slaves, using the master-slave switch, establishing SCO links, and handling the sniff, hold, and park low-power modes.

Security management The LMP also handles most of the implementations associated with authenticating and encrypting the Bluetooth link (see Chapter 9, "Bluetooth Security").

LMP Packet Structure

The payload of an LMP packet consists of the usual 8-bit payload header, data (the PDU itself), and 16-bit FCS normally found in any single-slot ACL packet. When the PDU is carried in a DM1 packet, the (15,10) shortened Hamming code is used for error correction (see Chapter 4, "Baseband Packets and Their Exchange"). As a result, the maximum possible length of a PDU is 136 bits, or 17 bytes. After the 8-bit payload header and 16-bit CRC are added and the FEC is applied, the maximum total payload length of 240 bits is obtained.

Figure 7-4 shows how the LMP DM1 packet appears before it is encoded with the FEC. The payload header consists of three fields: *logical channel* (L_CH), *flow*, and *length*. LMP packets always have L_CH = 11 to differentiate them from L2CAP packets that contain user data (see Chapter 8). Therefore, the LM packet interception algorithm is simple: Just grab any DM1 packet having a payload header beginning with binary 11 (after error correction and checking). The flow bit has no meaning for the LM so it should always be binary 1. Finally, the length field gives the size of the PDU in bytes, which doesn't include the payload header itself or the CRC. Five bits are needed in this field to designate a maximum PDU length of 17 bytes.

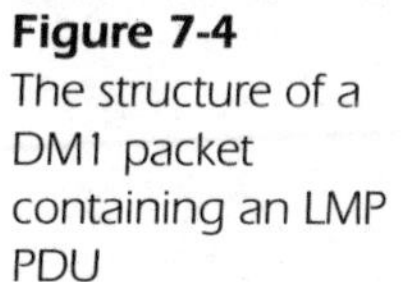

Figure 7-4 The structure of a DM1 packet containing an LMP PDU

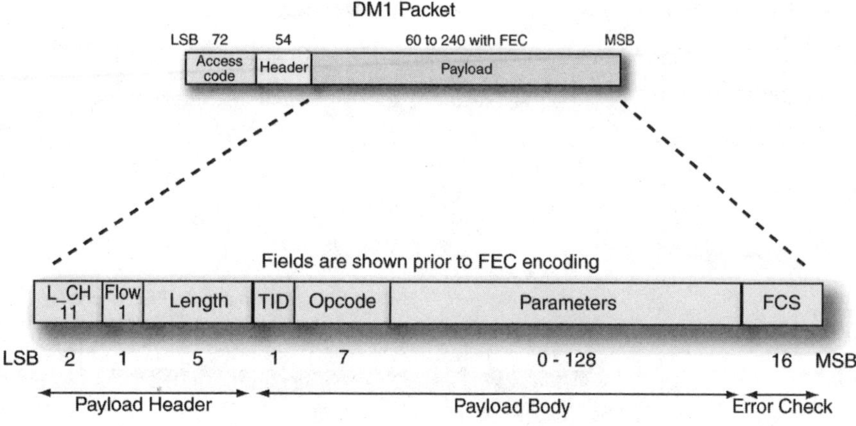

Within the DM1 packet's *payload body* field is the PDU itself, consisting (prior to FEC encoding) of a *transaction identifier* (TID), an *opcode*, and the *parameters*.

The 1-bit TID is in the LSB position of the payload body so it is transmitted first. The TID is binary 0 when the PDU originates with the master and is binary 1 when the PDU originates with the slave. This bit is needed because sometimes the master and slave exchange several LMP packets as part of a parameter negotiation process and the TID lets both LMs keep track of who started it all. The specific slave involved in the LMP packet exchange is designated by its AM_ADDR in the Bluetooth packet header.

The opcode is a 7-bit PDU designator, followed by a field that holds the parameters, if any, associated with that opcode. If a parameter exists, it can be as short as a single byte or as long as 16 bytes, but (with a few exceptions) each parameter has a length that is an integral number of bytes.

The shortest PDU contains just the 8-bit TID and opcode so the associated Bluetooth packet payload prior to FEC encoding consists of the 8-bit payload header, 8-bit PDU, and 16-bit FCS. This string is padded with 8 binary zeros so the resulting 40 bits form an integral number of 10-bit blocks. These in turn are encoded with the (15,10) shortened Hamming code to produce a 60-bit Bluetooth packet payload. Thus, LMP payloads in DM1 packets range from 60 bits to 240 bits.

The DV packet data payload looks similar to the DM1 payload, except the maximum PDU length is only 72 bits (9 bytes) to allow room for 80 bits of additional SCO information in a single-slot packet (see Figure 7-5). The DV packet data section also contains an 8-bit payload header structured as

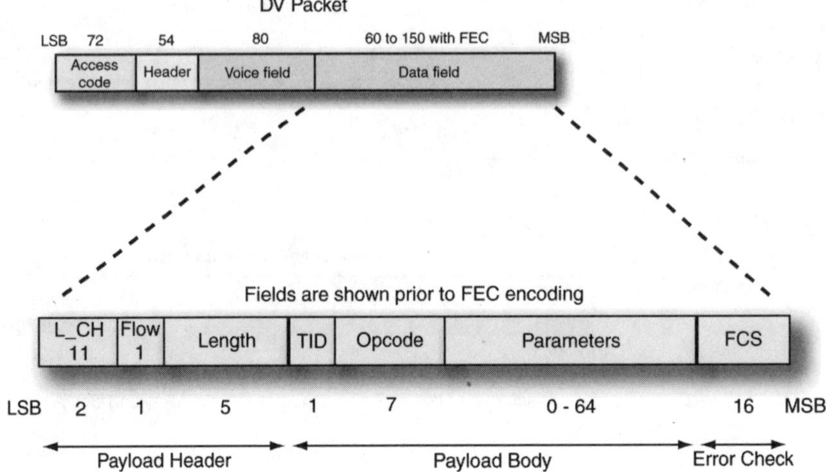

Figure 7-5 The structure of a DV packet containing an LMP PDU

before and the usual 16-bit FCS. The entire data payload prior to FEC application can be up to 96 bits long. This field is then padded with four binary 0s after the FCS so it contains an integral number of 10-bit blocks. After the (15,10) shortened Hamming code is applied, the data field becomes a maximum of 150 bits long for transmission. Isn't that simple?

General Response Messages

There are two general response messages: LMP_accepted and LMP_not_accepted. Both of these are used to successfully or unsuccessfully terminate many sequences of LMP packet exchanges. The former contains the opcode of the message accepted and the latter contains both the affected opcode and a reason for not accepting it.

Support for a particular PDU within an LM can be either mandatory or optional. The LM need not be programmed to transmit an optional PDU, but it must at least be able to recognize an incoming PDU as one that it cannot comply with and, if necessary, return an LMP_not_accepted with a "0x1A, unsupported LMP feature" error-code response. (That reminds me of the old "bad command or filename" reply for an unrecognized DOS command.) Some of the PDUs listed as optional in the LMP part of the Bluetooth specification may become mandatory for certain profiles, and vice versa. The basis for these changes will be discussed in Chapter 11, "Bluetooth Profiles." As an example, the LMP lists SCO connection establishment and termination as optional, but for a Bluetooth headset, support for these PDUs are clearly mandatory.

If both the master and slave begin the same LMP procedure together, LMP collision occurs even though the respective LMP packets are transmitted in adjacent time slots. As you might expect, the master seizes control of the situation. It does so by sending LMP_not_accepted with the "0x23, LMP error transaction collision" error code and then it completes the transaction on its own terms.

Figure 7-6 shows some typical command-response PDU exchanges between LMs, which are used to place the target LM into a new mode of operation such as hold or park. In the PDU mnemonic, *req* means "request." Although there are some exceptions to the basic rules, in most cases either the master or slave can initiate the command. If the master initiates the command, and if no negotiation takes place (see the following section), the slave either accepts or rejects the command. If the slave initiates the command, the master usually repeats the same command to the slave, which in turn accepts it.

A different PDU exchange can occur if one LM wants information from the other. The master or slave LM can ask for information, and if the responding LM has the capability to answer, it provides a return PDU with an ending mnemonic of *res*, which means "response." If the responding LM

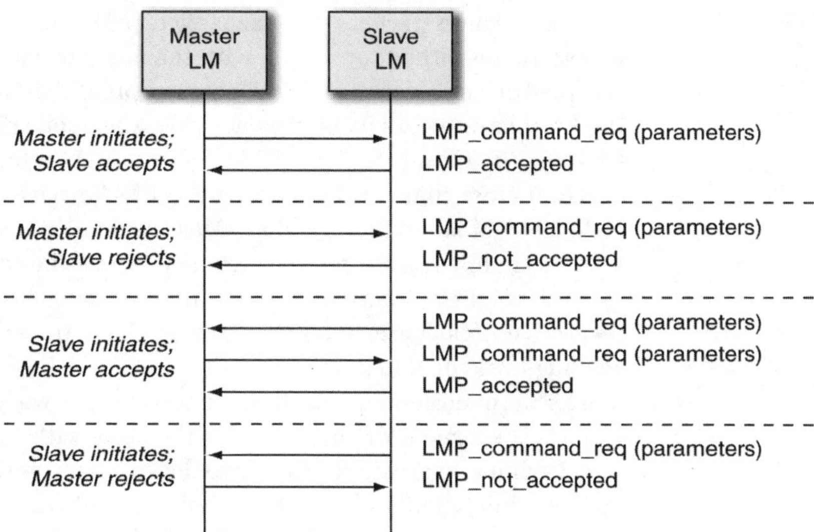

Figure 7-6 LM command-response PDU exchange examples

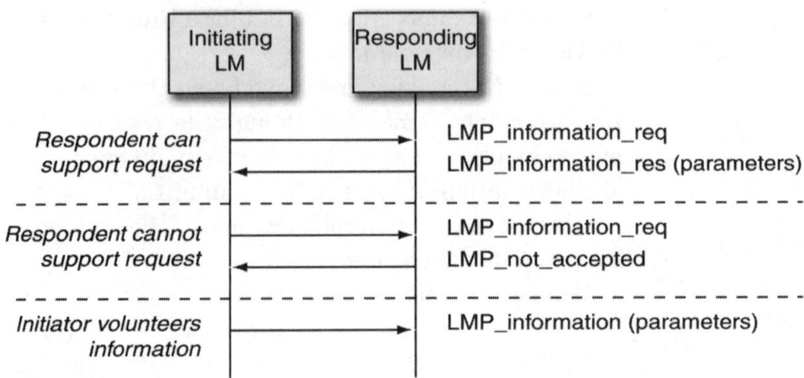

Figure 7-7 LM information request-response PDU exchange—the master or slave can initiate or respond.

cannot interpret the request, it returns LMP_not_accepted. This PDU exchange is shown in Figure 7-7. Sometimes information is volunteered by one LM to the other, and this often requires no response from the other LM because an ACK at the LC level indicates reception of the packet.

LMs can also negotiate the parameters for some of the PDUs, such as the hold time. This process is demonstrated in Figure 7-8. Notice that no additional PDUs are needed to implement negotiation. Instead, if a device wants to negotiate a parameter instead of just giving up and sending LMP_not_accepted, it repeats the same PDU with a different parameter attached. These packets are exchanged perhaps several times until the negotiation is either successful, with the last parameter being used (LMP_accepted), or unsuccessful (LMP_not_accepted). Authorization for negotiating LMP parameters usually comes from the application through HCI (see Chapter 10, "Host Interfacing"), and many HCI commands include a range of parameters that can be used by the LM for negotiations. For example, HCI_Hold_Mode contains a Hold_Max_Interval and a Hold_Min_Interval.

Sometimes the LMP packet exchange process has some unusual aspects to it, depending upon what particular function is worked between the respective LMs. For example, a slave can force the master to let it hold by sending LMP_hold to the master with the proper hold parameters in it. The master is required to accept this request, but, as we learned in Figure 7-6, it sends the same LMP_hold PDU to the slave with the same hold parameters. Perhaps surprisingly, the slave doesn't reply with LMP_accepted, but instead simply enters hold at the hold instant. Likewise, the master can force the slave to hold with the same LMP_hold PDU to which the slave doesn't reply. These situations are shown in Figure 7-9. It therefore goes

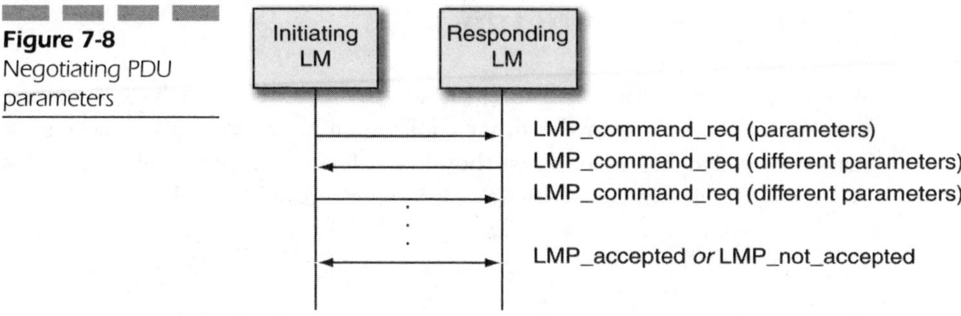

Figure 7-8 Negotiating PDU parameters

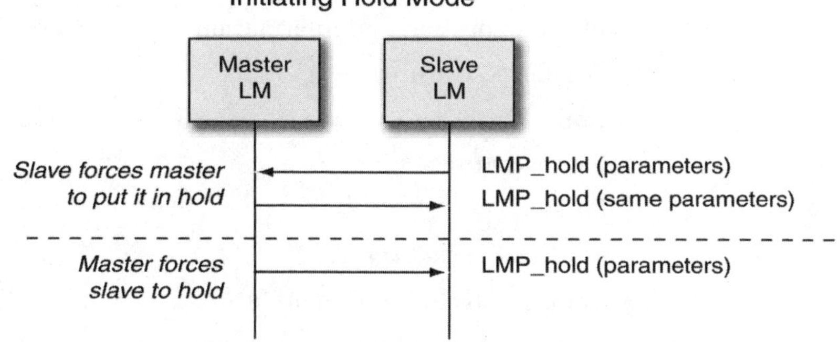

Figure 7-9 The master or slave forces slave to hold.

without saying (doesn't it?) that developers of Bluetooth LM firmware must carefully read the Bluetooth specification for such subtleties.

Requesting Host Connection

Another special PDU called LMP_host_connection_req is used to request a connection to the remote host. This request is usually sent by the master, who presumably already has permission from its host for the connection because that host initiated the page. Upon receiving this PDU, the slave LM passes the request through HCI to the host application, from which either a yes or no response is expected. The remote LM then replies with LMP_accepted, or LMP_not_accepted followed by LMP_detach. A complete treatment of this process is given in Chapter 10.

Completing the Setup

When the connection setup process is complete, each LM sends LMP_setup_complete to the other. Neither of these messages receives a response at the LM level since they are information only. Following this exchange, the devices can begin sending user data if desired.

Opcodes and Their Functions

There are 53 LMP opcodes designated in Specification 1.1. These opcodes can be grouped within the three major PDU categories as follows:

- Link configuration and information
  - Supported features
  - QoS
  - Power control
  - Packet types
  - Multislot packet control
  - Timing and clock information
  - LMP version and Bluetooth name
- Piconet management
  - Sniff, hold, and park modes
  - SCO link establishment and termination
  - Paging scheme
  - Link supervision
  - Master-slave switch
  - Link detach
  - Test mode
- Security management
  - Link key management
  - Authentication and pairing
  - Start and stop encryption

It could be argued that some of the previous PDUs could fit into different categories than those listed, so feel free to change them around if you want. We will examine some of these categories in detail, and sometimes we'll give example PDUs as well. We won't do them all because that would simply repeat what's in the Bluetooth specification, and why be any more boring than we already are? Also, most of the PDUs associated with Bluetooth security will be covered in Chapter 9. It's probable that more opcodes will be added in future specifications and the present 7-bit opcode field can support up to 128 of them.

Link Configuration and Information

When two Bluetooth devices first connect with each other, each device may know very little about the other. Can it support SCO connections? Does it have power control capability? Before user data can be exchanged in a meaningful and efficient way, the link must be configured so that neither device is surprised by something the other does or doesn't do. Many of the PDUs in this category are used for this initial probe of capabilities and link configuration process. Some of these are also used as the link progresses to retrieve information such as timing information in preparation for a master-slave switch.

Supported Features Before two Bluetooth devices are allowed full access to the many packet types and other LC capabilities, they must discover each other's supported features using LMP_features_req. Prior to this, the two devices are limited to exchanging only ID, FHS, NULL, POLL, DM1, and DH1 packets. Boring. The request for supported features can be done immediately after the master receives a successful response to its POLL upon completion of the page.

The response PDU, LMP_features_res, has a 3-byte parameter field that represents a bit map of the possible features being covered. The responding device sets the corresponding bit if the feature is supported; otherwise, the bit is cleared. The features that can be selected are

- Three-slot packets
- Five-slot packets
- Encryption
- Slot offset values available

- Timing accuracy values available
- Master-slave switch
- Hold mode
- Sniff mode
- Park mode
- RSSI available
- Channel quality driven data rate capability
- SCO link
- HV2 packets
- HV3 packets
- μ-law audio encoding
- A-law audio encoding
- CVSD audio encoding
- Alternate paging schemes available
- Power control
- Transparent SCO data
- Flow control lag (3 bits)

The meaning of many of the listed items may not be clear now, but these will be covered in later chapters. At this point, though, it may be useful to ponder Bluetooth device functionality and how the associated supported features map will look. For example, a Bluetooth-capable printer will probably have bits associated with SCO links cleared and multislot packet bits set, whereas Bluetooth headsets will do just the opposite. Also, the existence of a feature doesn't necessarily imply permission for its use. For example, multislot packet authorization must be obtained before these packets may be exchanged.

Quality of Service (QoS) The LMP_quality_of_service packet contains the poll interval T_{poll} specified in slots and the number of broadcast packet repetitions N_{BC} in its parameter field. Until modified through this PDU, T_{poll} defaults to 40 slots. At the HCI level, the value for N_{BC} defaults to 1.

Power Control As we learned in Chapter 3, "The Bluetooth Radio," a class 1 transmitter must use power control for transmit power levels above 4 dBm. Each power increment, up or down, must be between 2 and 8 dB. An LM can request the other device to increase or reduce its power by one increment. These commands should only be used if the other LM has power control implemented, which is part of the supported features PDU.

If an initiating LM requests the other LM to reduce power when its power output is already at minimum or to increase power when its power output is already at maximum, the other LM can inform the requestor that its power is already at the limit.

Packet Types An LM has tremendous flexibility in determining what types of ACL packets it prefers for data exchange. The PDU has a parameter that lets the LM choose between DH and DM packets, and whether it prefers packets using one, three, or five slots, or has no preference. Furthermore, an LM can ask the other LM to use only certain types of packets, or it can defer preference to the other LM instead. The responding LM isn't required to use the types of packets specified because only a preference is given. Not every LM can support a request for packet types; if it can, it will set the appropriate bit in the *supported features* bitmap. Devices that are primarily used for SCO packets, such as a Bluetooth headset, could forego this capability.

Multislot Packet Control Before a device is allowed to use packets longer than one time slot, it must obtain permission from the other LM. For example, a local LM enables the remote device to use multislot packets by sending LM_max_slot with the maximum number of slots (either three or five) in its parameter field. Because this is an information PDU, it is not acknowledged at the LM level.

If the local LM wants to send multislot packets to the remote device, it must first ask permission by sending LM_max_slot_req, with the maximum number of slots as the parameter. The other device responds with either LMP_accepted or LMP_not_accepted.

Because multislot packet transmissions are not needed during channel setup, these PDUs can be sent any time after the setup process is completed —that is, after each LM sends LMP_setup_complete, as described in the section "Connection Setup Procedure."

Timing and Clock Information Remember the clock offset? That was a value calculated by a slave to use in its frequency hop generator to match the phase of the master's hopping sequence (see Chapter 4). The master can request a slave's clock offset at any time during the life of a link, and it can use this offset to calculate the slave's CLKN value for future use in pages. That is, after the slave has disconnected from the piconet, the master can use the slave's CLKN to determine when the slave enters the page scan state, and on what frequency it will be listening, in order to speed the paging process.

The slot offset request and response is used by devices to perform the master-slave switch. This tells the present master when to expect transmissions from the new master after the switch takes place. This process was covered in detail in Chapter 6, "Advanced Piconet Operation."

Finally, both LMs can request the other device's timing accuracy information as a drift and jitter value. A device having the capability to respond to such a request will have in its memory the actual drift and jitter values measured by the manufacturer during the device test. A slave can use these values, for example, to determine when to begin listening for the next master's transmission after the slave exits from hold or to narrow the search window for beacon packets while parked. Until actual drift and jitter values are discovered through LMP packet exchange, each device must assume that the other conforms to a worst-case drift of 250 ppm and jitter of 10 μs.

LMP Version and Bluetooth Name The last two PDUs associated with link configuration are information parameters. The LMP version can be retrieved from the other LM to check compatibility, and the user-friendly Bluetooth name (for example, "Jim's cell phone") can be retrieved by the other device. The Bluetooth name can be up to 248 bytes long so LMP_name_req may generate several LMP_name_res packets in response.

Piconet Management

After the piconet is established, the respective LMs continue to actively manage piconet operation. The obvious example would be when a device wants to enter one of the various low-power modes to liberate time for scatternet operation or for inquiries or pages. Also, because all piconet links start as ACL-only, setting up a SCO link requires the exchange of PDUs containing the SCO parameters.

Sniff, Hold, and Park Modes The three low-power modes—sniff, hold, and park—are managed via PDU exchange between master and slave. The associated LM PDUs were developed as a result of pondering several questions, not all of which have obvious answers:

- Should the master be able to force a slave into a low-power mode?
- Can the master be forced to accept a slave's request for a low-power mode?

- Under what conditions can the other device's request be refused?
- How should the low-power parameters be negotiated?
- Does it matter which device originates the request for a low-power mode?

For example, the first item in the previous list was answered in an earlier Bluetooth specification by allowing the master to force a slave into the sniff mode. Specification 1.1 has removed this capability because the master's operation within the piconet isn't affected by a slave's refusal to enter sniff at the master's request. The master can always act as if the slave entered sniff, even if it didn't.

Sniff Mode Either the master or slave can request that the slave enter the sniff mode. The parameters can be negotiated in a manner similar to that shown in Figure 7-8 using the LMP_sniff_req PDU until the process is terminated with LMP_accepted or LMP_not_accepted. During a sniff slot, the master can force the slave out of sniff or the slave can force itself out of sniff by sending LMP_unsniff_req, which is always accepted by the other LM. Notice that, although the slave can initiate its own unsniff request, normal LC protocols are still followed; that is, the slave sends its LMP_unsniff_req packet only when specifically addressed by the master in the previous time slot.

Hold Mode The hold mode, which enables a slave to exit from piconet responsibilities for a period of time, is handled very differently from the sniff mode at the LM level. Whereas sniff can begin almost immediately after the parameters are negotiated, hold cannot begin until $6 \times T_{poll}$ intervals after the hold PDU has been sent ($9 \times T_{poll}$ intervals if the parameters must be negotiated). Furthermore, although the master or slave can request hold, they each can force hold as well. The latter makes sense because the master or slave must have the capability to liberate time for page/page scans or inquiry/inquiry scans, or perhaps scatternet operations, without having the other device(s) refuse to cooperate. When hold is forced, though, the maximum hold time cannot exceed what was previously negotiated.

Park Mode The park mode is by far the most complex of the three low-power modes. Either device can request the slave to enter park, but it cannot be forced to park. That's a change from earlier Bluetooth specifications in which the master could force a slave to park. What if the master needs to assign that particular slave's AM_ADDR to another device, but that slave refuses to park? The master can force the slave to disconnect from the piconet instead. So there.

A slave is parked by using the LMP_park_req PDU, which has so many parameters that its length is the maximum 17 bytes. The PDU includes

- Beacon parameters
- Beacon monitoring parameters
- Parked member address (PM_ADDR)
- Access request address (AR_ADDR)
- Access window characteristics

When a slave requests to be parked, it can suggest values for all of the previous parameters except PM_ADDR and AR_ADDR. If the request is approved, the master returns the same LMP_park_req PDU with all parameters included (some of which may be different from those requested by the slave) along with the PM_ADDR and AR_ADDR. The slave responds with LMP_accepted, followed by entering park, or LMP_not_accepted. Figure 7-10 shows the various ways that the park PDU can be implemented.

Once a slave is parked, the master communicates with it only through broadcast packets during beacon intervals. Because a parked slave isn't required to monitor all beacon trains (see Chapter 6), the master must use the correct beacon train for contacting a particular parked slave. The master is allowed to send only the PDUs associated with changing beacons and other broadcast parameters and unparking the slave during these broadcast times. These broadcast-only PDUs are special in that their parameter lengths aren't integral byte multiples. This increases their reliability in the

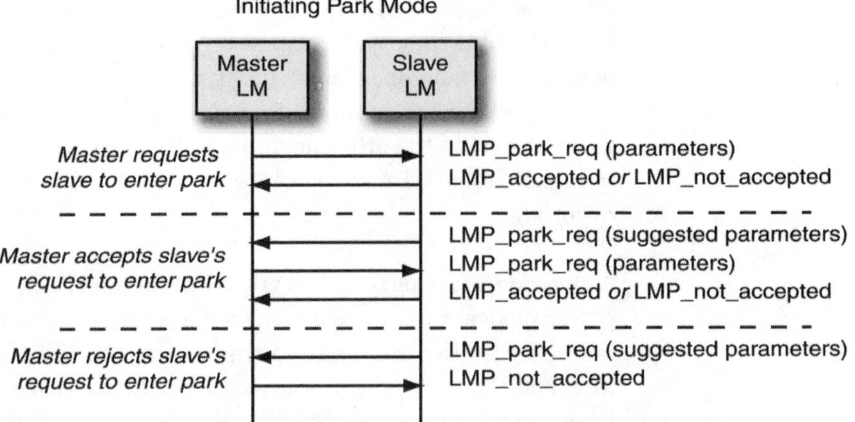

Figure 7-10 A slave can enter the park mode in various ways.

beacon environment by keeping them as short as possible—at least in theory.

A parked slave's former AM_ADDR can be assigned by the master to another device. However, the master needs to make sure that the first slave is actually parked or AM_ADDR duplication (collision) could occur with another slave. This duplication could be disastrous since the master loses the capability to differentiate between the two slaves, and any attempt to communicate with them will result in a packet collision when both slaves respond together in the following slave-to-master slot. Link timeout would eventually occur, both slaves would exit the piconet, and the master could then try to bring them in again. To avoid AM_ADDR duplication, the master can reassign a slave's AM_ADDR only after $6 \times T_{poll}$ slots have elapsed since receiving that slave's LMP_accepted response to the master's LMP_park_req.

One of the powerful aspects of the park procedure is that multiple slaves can be unparked with a single PDU. For slaves responding to their PM_ADDR, up to seven can be unparked together. The master can unpark up to two slaves with a single command when using their BD_ADDR. The unpark PDU contains each parked slave's PM_ADDR (or BD_ADDR) followed by its new AM_ADDR.

Just for fun, let's calculate the time required for a master to park seven slaves and then unpark seven other slaves. In other words, how long will it take to swap out an entire piconet of slaves? Assuming that no packet retransmissions are required and that the respective LMs respond as quickly as possible to incoming PDUs, the following happens:

1. The master sends LMP_park_req and receives LMP_accepted from each of seven slaves in turn (14 slots).
2. The master waits $6 \times T_{poll}$ slots before it can reassign the AM_ADDR from the parked slaves (default 240 slots).
3. The master unparks seven slaves with a single LMP_unpark_req command (2 slots, with no response in the slave-to-master slot).
4. The master polls each new slave in turn and receives LMP_accepted as a response (14 slots).

The total time to swap out an entire piconet is thus 270 slots, or about 170 ms. This time could be shortened significantly if T_{poll} were reduced below its default value of 40 slots. The master could even temporarily shorten T_{poll} by sending an LMP_quality_of_service PDU to each slave before initiating the piconet swap and then lengthen it again for the new slaves if necessary. Notice that it would probably require several seconds to detach the current slaves and page seven others for a piconet swap without using park.

SCO Link Establishment and Termination When a master-slave link is first established, it is, by default, an ACL link. To establish a SCO link, the originating LM sends the LMP_SCO_link_req to the other, which in turn responds by accepting or rejecting the request. There is no provision for negotiating SCO parameters in the traditional way—that is, by exchanging a PDU several times, if necessary, with new parameters. Instead, a slave receiving a SCO request with unacceptable parameters replies with LMP_not_accepted with an error code pointing to the problem areas. The master then has the option of repeating the request with new SCO parameters. On the other hand, if the slave initiates a SCO request with unacceptable parameters, the master simply repeats the LMP_SCO_req PDU with different parameters for the slave to consider.

The SCO link can be completely characterized by the following parameters located within the PDU:

- **SCO handle** An unsigned 8-bit number that identifies this particular SCO link
- **Timing control flags** A set of 8 one-bit flags affecting SCO timing details
- **D_{SCO}** The time in slots until the first SCO packet exchange occurs
- **T_{SCO}** The time between SCO packet exchanges
- **SCO packet** Identifies whether HV1, HV2, or HV3 packets will be used
- **Air mode** Identifies whether μ-law, A-law, CVSD, or just transparent data is used

The LMs can also change SCO parameters while the SCO link is operational. This enables, for example, the modification of the packet type to better match the channel error rate or timing requirements of the devices. Finally, a PDU also exists for removing the SCO link. Either the master or slave can initiate this request, and the other is required to accept it. More information on SCO communications can be found in Chapter 8.

Paging Scheme In Chapter 5, "Establishing Contact Through Inquiry and Paging," we discussed the mandatory and optional paging schemes, and we made the point that both p-master and p-slave must use the same scheme for a page to be successful. The mandatory page scan scheme must always be used for at least 20 seconds after a device responds to an inquiry. After the link is established, however, the master and slave can negotiate the paging scheme to be used the next time the slave is paged because subsequent pages are usually not preceded by an inquiry. This is done with two different PDUs, depending upon the situation.

The LMP_page_mode_req is initiated by device A (either the master or slave) and contains the paging scheme that device A will use when it next pages device B. The LMP_page_scan_mode_req is a proposal by device A to device B for use when B pages A.

Link Supervision The link *supervision timeout* (supervisionTO) in Chapter 6, is used to guard against an infinite waiting period if the device on the other end of the link should disappear for some reason. The default time is 20 seconds, but the LMP_supervision_timeout PDU contains an unsigned 16-bit integer that specifies the number of slots in the timeout that can be used to change the default. The 20-second default time is equal to 32,000 (0x7D00) slots. A timeout value of 0xFFFF equates to about 41 seconds. A value of 0x0000 specifies an infinite timeout—using this one is risky!

Master-Slave Switch The LMP_switch_req PDU is preceded by the slave sending the master an LMP_slot_offset packet with its BD_ADDR and slot offset timing information in preparation for the role switch. After accepting the LMP_switch_req PDU, the master and slave exchange roles, as outlined in Chapter 6. One bit of trivia here involves the TID in the LMP packet. If the master initiates the switch, then all LMP packets exchanged during the process have TID = 0, and if the slave initiates the switch, then they all have TID = 1. The LMP_switch_req PDU contains only one parameter, and that's the switch instant specified in terms of the current master's CLKN, which is also the CLK value for the piconet. To avoid timing problems, the switch instant must be at least $2 \times T_{poll}$ or 32 slots into the future, whichever is greater.

Link Detach The link can be detached any time by either the master or slave. The only parameter in LMP_detach is the reason, which is taken from the list of 8-bit LMP error codes. Perhaps the most common reason for a routine link detach will be "0x16, connection terminated by local host." The detach PDU is always accepted, and there's no response at the LM level to LMP_detach.

There are a few rules that govern when the AM_ADDR for the detached slave becomes available for reuse by the master. Remember that if two slaves somehow respond to the same AM_ADDR, the master can no longer separate them with LMP packets or any other packets for that matter. Furthermore, the two slaves will then collide in the following slave-to-master slot anyway, so the master has effectively lost the ability to function with these slaves.

If the master initiates the link detach, the affected AM_ADDR becomes available for reassignment $3 \times T_{poll}$ slots after the slave sends a baseband ACK packet in response to LMP_detach. If the slave never responds with an ACK, the AM_ADDR becomes available after the supervisionTO expires.

If the slave initiates the link detach, its AM_ADDR is available for the master to reassign $6 \times T_{poll}$ slots after that master receives the slave's LMP_detach PDU. Of course, if the master never receives the detach PDU and the slave disappears from the piconet, the master's supervisionTO for that slave will eventually expire and the master can reassign the AM_ADDR at that time.

Test Mode LM packets can be used to place a slave into the test mode so the master can perform various device tests required for Bluetooth certification. The test process itself will be covered in detail in Chapter 12. One PDU, called LMP_test_control, is sent from master to slave to enable a particular test scenario and other test criteria. Next, the master activates the test mode by sending LMP_test_activate to the slave.

The slave's host has control over whether the test mode can be entered by the slave. Because certain test modes can violate FCC or other government rules for frequency-hopping devices, such as disabling the hop generator, a normal Bluetooth user shouldn't be able to activate test mode.

Security Management

Numerous LM PDUs are designated for use in both authentication and encryption parts of Bluetooth security implementations. These will be covered in excruciating detail in Chapter 9.

Connection Setup Procedure

Finally we are ready to expand a bit on Figure 7-3 and include some of the details for setting up a link (see Figure 7-11). All of this occurs before the devices exchange user data or SCO packets.

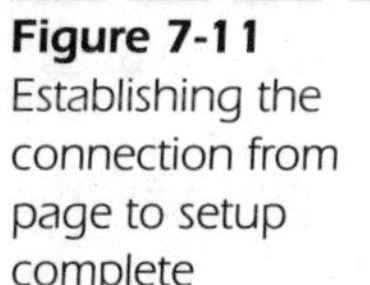

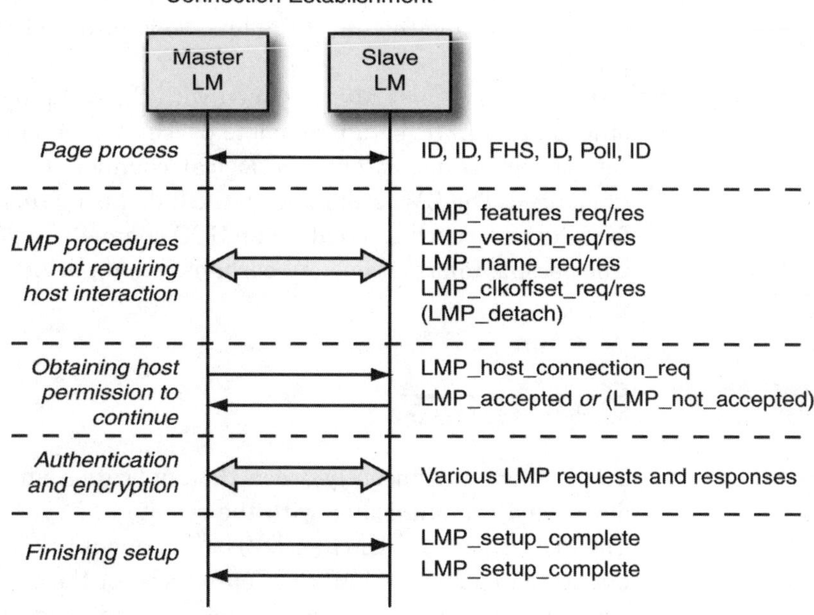

Figure 7-11 Establishing the connection from page to setup complete

After the paging process is completed at baseband, the respective LMs begin communicating with each other through LMP packets. The exchange begins with PDUs that don't require notification of the host, which are

- Supported features request and response
- LMP version request and response
- Bluetooth name request and response
- Clock offset request and response

Notice that the specification allows a Bluetooth device to effectively log in and extract the previous data without the host's (and hence, the user's) knowledge. Aren't you just a bit uneasy knowing this? I am.

Following these preliminary steps, one LM (usually the master because it initiated the connection) sends an LMP_host_connection_req to the other LM, which in turn informs the host that a connection has been requested. If the connection is accepted, then the latter LM returns LMP_accepted to the former; otherwise, LMP_not_accepted followed by LMP_detach is usually the result.

Assuming the connection is accepted by the remote host, the two LMs then work out any required authentication and encryption procedures, which are covered in Chapter 9. As we'll discover in that chapter, the authentication and (if needed) encryption procedures can occur during various times during the life of the link, so Figure 7-11 only shows one of the possibilities.

Finally, when an LM is satisfied with the setup procedure, it sends the other LM an LMP_setup_complete packet. When both have completed the setup, packets using another logical channel (UA, UI, or US) can be exchanged. The LMs still manage the link during this time as needed, performing tasks such as creating an SCO channel, implementing power control, passing information, and perhaps reconfiguring the link.

Summary

The LM acts as liaison between the application and the LC on the local device and communicates with the remote LM via PDUs using the LMP. The PDU is part of a DM1 (or DV) packet and is acknowledged at baseband but acted upon by the LM. PDU opcodes cover the categories of link configuration and information, piconet management, and security management.

A connection begins after the completion of a successful page. The two LMs first exchange packets that don't require intervention by the host in order to discover some basic characteristics of each other so communication efficiency can be enhanced. Next, the slave's host is notified of the requested connection. If the host responds favorably, then authentication and encryption security issues are processed by the LMs, followed by any other required configuration steps. Finally, each sends a setup complete message to the other to open the link to user data exchange. The two LMs continue to watch the link for LMP packets, which they intercept and act upon.

CHAPTER 8

Transferring Data and Audio Information

Once the ACL channel has been established and configured, the applications can finally get down to business and begin exchanging data. At least they are almost ready. If SCO packets are to be exchanged, then the LM must use LMP packets to set up the SCO channel. As discussed in the preceding chapter, the LMs will agree on the method of voice encoding used, whether HV1, HV2, or HV3 packets will be exchanged and their timing. In order to reduce latency, real-time two-way voice is usually communicated to the lower Bluetooth protocol layers directly by the application (see Figure 8-1). We will take a look at some of the details of audio encoding and transmission in this chapter.

If the users are exchanging ACL data, then the *Logical Link Control and Adaptation Protocol* (L2CAP) plays an important role in keeping track of the exchange, also shown in Figure 8-1. L2CAP is a middle manager, acting as a liaison between the application and the Bluetooth link controller. In fact, several applications can be communicating over a single Bluetooth RF link, and L2CAP has the responsibility of maintaining order in data streams. We will see how this is done next.

Now that we've entered the realm of the Bluetooth protocol stack as it appears on the host as part of a software package, it's important to be reminded of a few points we made at the beginning of the book. First, for

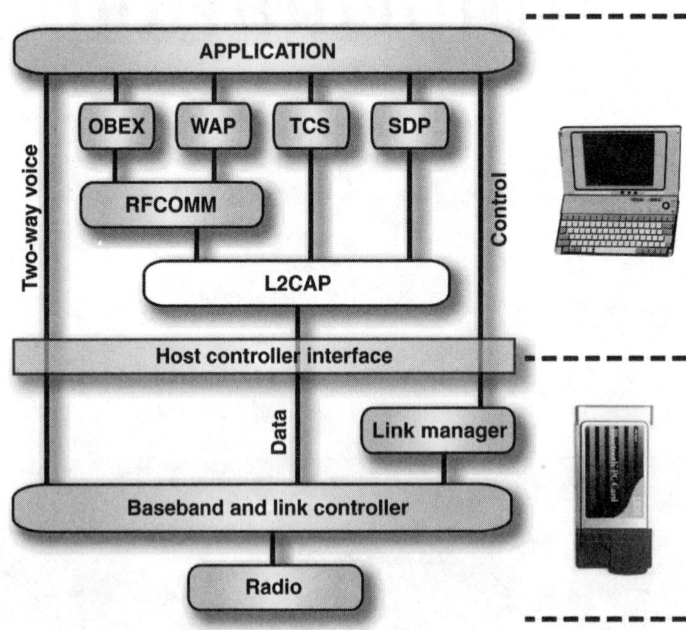

Figure 8-1 Bluetooth two-way voice takes a direct path from the application to the lower protocol layers. The logical link control and adaptation protocol provide user data services, acting as a liaison between the application and Bluetooth link controller.

software implementations, the Bluetooth specification usually enables manufacturers great flexibility in developing their own code, especially for communication between various levels of the protocol stack on one host. Indeed, the code is a source of pride (and intellectual property protection) for many developers. As such, it would be impossible to address each implementation approach here. Also, higher levels of the protocol stack become increasingly complex because they can appear to be applications themselves. Finally, we avoid including the entire specification here in order to reduce the slumber factor while reading the book. Our goal, then, is to explain the operation of the protocols by showing some common examples without necessarily covering every aspect of their operation.

In the Bluetooth specification, the L2CAP layer is described as if there were no master-slave relationship between piconet members. Indeed, the words "master" and "slave" are not found anywhere in the specification's L2CAP chapter. We don't take that approach here, though, because the way some L2CAP functions are accomplished depends upon whether the device is a master or slave. However, keep in mind that the goal of L2CAP is to make the master-slave relationship irrelevant and treat the communicating devices as peers whenever possible.

Logical Link Control and Adaptation Protocol (L2CAP) for Data

L2CAP is a software module that normally resides on the host and provides connection-oriented (master to one slave and slave to master) and connectionless (implied to be from a master to multiple slaves) data services. This means that L2CAP doesn't provide any real-time two-way voice (SCO) capability. It may, however, communicate with the LM for channel setup assistance. The main purpose of L2CAP is to act as a conduit for data on the ACL link between Bluetooth baseband and host applications.

Like the LM, L2CAP relies on the baseband packet exchange process for data integrity through ARQ and possibly, FEC as well. Therefore, no AUX1 packets are allowed for L2CAP communication because these have no CRC implementation and consequently, no ARQ capability. With all these abbreviations, isn't it nice that they're spelled out in the back of this book?

Two forms of user data can be exchanged through the services of L2CAP. One, called *user asynchronous* (UA), places reliability above all else and requires that packets be retransmitted until successful. UA service is needed when transferring data files from one device to another. The other form of data exchange is called *user isochronous* (UI), in which data integrity is important, but can be occasionally superceded by latency requirements. An example is found in one-way real-time audio or video (multimedia) streaming, in which a buffer at the receiving end must not be enabled to empty. A bad packet that cannot be successfully transferred after a certain number of attempts is flushed, and the system moves on to the next packet in the stream.

L2CAP Functions

Figure 8-2 shows some more detail of how L2CAP fits within the Bluetooth protocol stack. Host applications work through several different Bluetooth-dependent software kernels, which in turn communicate with L2CAP in a multiplexed fashion. Real-time two-way voice applications bypass L2CAP altogether and use a SCO link for their transfer, but audio streaming can use the UI channel through L2CAP instead. Also, L2CAP can communicate with its LM, if necessary, for ACL channel setup and other tasks.

The philosophy behind the development of L2CAP included the requirements of simplicity and low overhead. Devices having little memory and low computation speed should not be overwhelmed by L2CAP at the

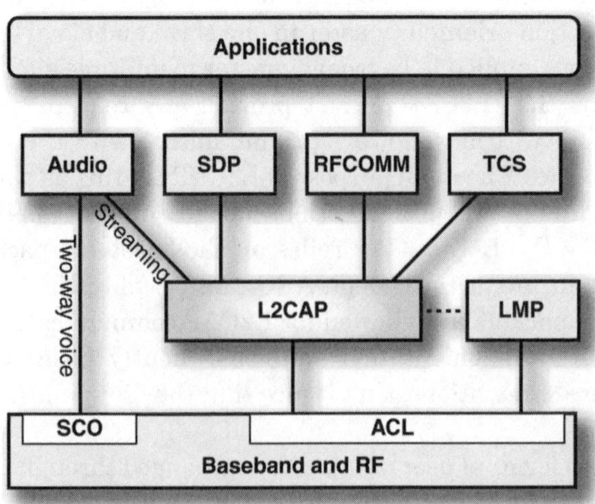

Figure 8-2 L2CAP must track several different application interfaces above it and work through baseband ACL packets to provide a data communication link. Coordination with the LM may be needed. Real-time two-way voice applications bypass L2CAP altogether.

expense of other operations, and the protocol shouldn't consume an excessive amount of power. Furthermore, L2CAP was designed for implementation on a large number of diverse units such as personal computers, *personal digital assistants* (PDAs), cordless and cellular phones, and interactive toys.

The functions of L2CAP can be divided into four categories: protocol multiplexing, packet segmentation and reassembly, quality of service (QoS), and group management. These can be defined as follows:

Protocol multiplexing You'll notice from Figure 8-2 that, unlike the lower protocol layers, L2CAP must communicate with several different layers above it. The communication is two-way; that is, the protocol takes packets from these higher layers and turns them into Bluetooth-size ACL payloads, and it takes payloads from the link controller and directs them to the correct higher layer. In other words, L2CAP provides protocol multiplexing in a way that is transparent to the higher protocols. As we'll see later, L2CAP implements protocol multiplexing by establishing virtual channels between devices that include a destination field within each L2CAP packet.

Segmentation and reassembly Packets used by the Bluetooth baseband protocol are small by most standards, having user payloads limited to 339 bytes at most. A higher-level protocol isn't hindered by a communication channel that has the reliability problems of wireless, so the size of its *maximum transmission unit* (MTU) is often significantly greater, sometimes as long as 64K bytes. These large packets must be segmented into several Bluetooth baseband packets for transmission, and incoming baseband packets are taken by L2CAP and reassembled into the large pieces of data that the higher level expects to receive. This process is also transparent to the higher protocol layer.

QoS L2CAP can implement a QoS level for each protocol. This includes such items as bandwidth requirements, how fast successive packets can arrive, maximum latency, and delay variation. QoS defaults to something called *best effort*, which is another way of saying that it will do the best it can under the circumstances.

Group management Many higher protocols on the host require the capability to manage a group of addresses. The Bluetooth LM manages a group called the *piconet*, consisting of both active and parked slaves. L2CAP takes this concept a step further and enables the mapping of protocol groups within the piconet itself. An example would be sending an MP3 music file for real-time listening to two or three slaves within a seven-slave piconet.

L2CAP Operating Environment

For L2CAP to operate properly, several assumptions are required. These include

The baseband link provides orderly packet delivery. An ACL link is set up between two devices using the LMP. Although an L2CAP command may initiate the creation of a baseband link between two devices (depending upon the software manufacturer's implementation of the protocol stack), L2CAP plays no role in the inquiry, paging, and configuration process. L2CAP assumes that Bluetooth baseband provides orderly delivery of packets. Delivery is orderly because the baseband always sends packets in consecutive order from start to finish. In other words, a later packet in a file is never sent before an earlier packet. This means that L2CAP need not number packets for correct arrangement after reception as is required for TCP/IP communication over the Internet.

The baseband link is full duplex. L2CAP assumes that each baseband link between devices can be treated as full duplex. Of course, not every L2CAP communication is bidirectional, but the capability is there for such functions as L2CAP channel setup and data exchange.

The baseband link is reliable. The baseband channel is reliable using the methods available to it. Reliability is guaranteed through error checking, where packet transmissions are repeated until they either arrive intact or a timeout occurs. No such reliability can be guaranteed for broadcast packets, so these are prohibited if L2CAP requires a higher QoS than what broadcast can provide.

Items Outside the Scope of L2CAP At this point it would be useful to list the items that are outside the scope of L2CAP. These mostly involve tasks that are performed by other layers in the protocol stack, but they also reflect the limitations inherent in wireless communications in general and Bluetooth in particular:

- L2CAP doesn't transport audio for SCO links.
- L2CAP doesn't enforce reliable data through its own error checking or retransmissions.
- L2CAP doesn't support a reliable multicast (broadcast) channel.
- L2CAP doesn't support the concept of a global group name.

L2CAP Channel Concept

As a data conduit, L2CAP operates with the concept of *channels*. A *connection-oriented* (CO) channel means that the data is exchanged between an application at the master and another application located at a single slave. A *connectionless* (CL) channel is normally used to send data from the master to a number of slaves simultaneously.

The endpoint of each data pipe from one host application to another is designated by a *channel identifier* (CID) within L2CAP. For example, a master may have four slaves in its piconet, as shown in Figure 8-3. Slave 1 has two CO channels to the master, and slave 2 has one CO channel. Slaves 3 and 4 are a group, and each has a CL channel for the master to use. The 16-bit value for each CID is determined by the local device (the device on which the CID is located) for use by the remote device when it sends the local device a L2CAP packet on that particular channel. When a local device is referring to one of its own CID values, it uses the term *local channel*

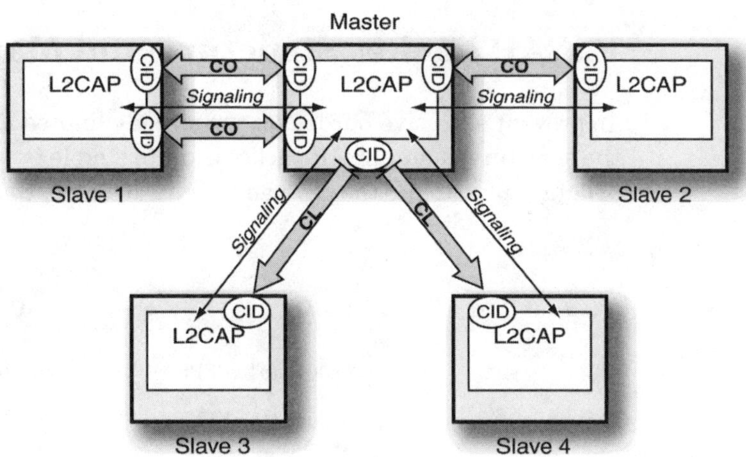

Figure 8-3 Several L2CAP channels can exist between master and slaves in a piconet. The signaling channels exist by default and are used to set up connection-oriented and connectionless channels.

identifier (LCID), also called the *source channel identifier* (SCID) if the L2CAP packet originates at that device. When it sends the packet across the L2CAP channel to the remote device, it uses the *destination channel identifier* (DCID), which is the CID belonging to the endpoint for that particular packet.

Each slave has one L2CAP *signaling channel* between it and the master. This channel exists by default as soon as an ACL link is established between two devices, and is used by L2CAP to establish the other channels for the exchange of user data.

The master has only one ACL link at baseband between each slave, so multiple L2CAP channels must have their data multiplexed on the ACL channel. Furthermore, the piconet itself is constrained by the maximum throughput figures that were given in Chapter 4, "Baseband Packets and Their Exchange," and this value cannot be exceeded by the sum of the throughput requirements for all slaves in the piconet.

The CID values that correspond to the various L2CAP channels are given in Table 8-1. The signaling channel is always present and has an assigned value of 0x0001. Any L2CAP packets sent to multiple slaves simultaneously (as a group) will have an associated DCID of 0x0002. Each CO channel is set up individually using the signaling channel, and these are assigned a DCID within the range of 0x0040 to 0xFFFF. The local device must assign a unique LCID to each of its L2CAP channel endpoints. There are sufficient CID values so that each Bluetooth device can support up to 65,472 unique L2CAP CO channels. Does that sound like overkill? It does to me as well (see Figure 8-4).

L2CAP Packet Structure and Mapping

To prevent excessive burden on the already limited Bluetooth throughput, the structure of an L2CAP packet is quite simple (see Figure 8-5). A chunk of data is placed into the payload, and a 2-byte length field and 2-byte DCID

Table 8-1

CIDs

Channel Type	SCID	DCID
CO	0x0040 to 0xFFFF	0x0040 to 0xFFFF
CL	0x0040 to 0xFFFF	0x0002
Signaling	0x0001	0x0001

Figure 8-4
There are how many unique CID values? (Source: Bill Lae)

"Got a little carried away again, eh Henderson."

Figure 8-5
The composition of the L2CAP packet

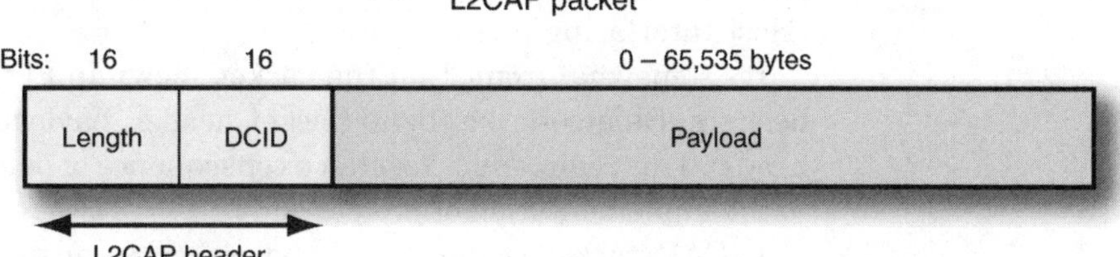

field are attached to the beginning of the packet. The length field is the length in bytes of the L2CAP packet payload. For the maximum possible data length of 64K bytes, the 4-byte length and DCID fields represent an overhead of only 0.006 percent. The actual maximum payload length is controlled by the MTU capability of the application communicating with its local L2CAP software module.

Segmentation Because the L2CAP packet is usually significantly longer than that supported by the Bluetooth baseband, it must be broken up into several segments and sent as a sequence of baseband ACL packets. Figure 8-6 shows an example of a DH5 baseband packet that contains the first segment of an L2CAP packet. As usual, the baseband packet contains an

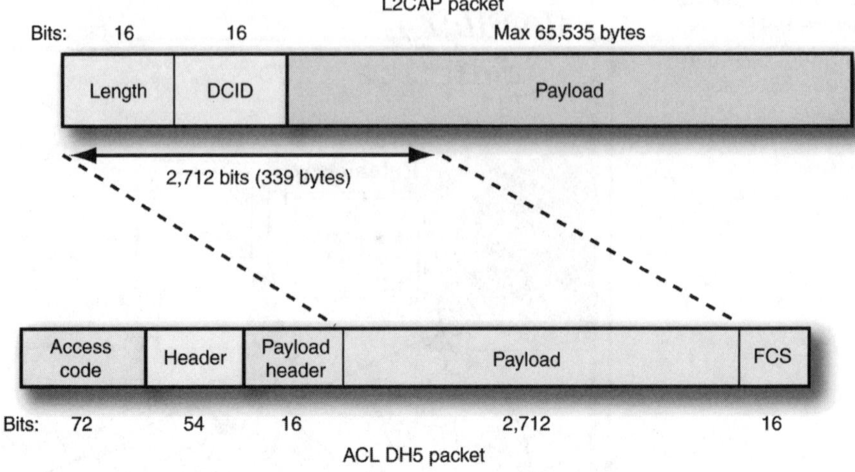

Figure 8-6 Placing the first part of a L2CAP packet into a DH5 baseband packet for transmission to another piconet member

access code, header, and payload. Within the payload is a payload header, L2CAP segment, and CRC code for error checking. (The astute reader is probably wondering at this point where the *host controller interface* (HCI) fits in because L2CAP typically resides on the host and the link controller resides on the Bluetooth module. This will be explained in Chapter 10, "Host Interfacing.")

It's somewhat ironic that the packet shown in Figure 8-6 now has *three* headers: Bluetooth baseband packet header, payload header, and L2CAP header. This redundancy is often a consequence of using a layered approach to the implementation of protocols.

L2CAP packet segments can be placed in any ACL baseband packet except AUX1. A single-slot ACL packet has an 8-bit payload header, but the length of multislot packets requires the payload header to be extended to 16 bits to incorporate a longer length field. The fields in payload headers belonging to single- and multislot Bluetooth ACL packets are shown in Figure 8-7.

The segmentation and reassembly processes require that the starting segment of a L2CAP packet be identified differently than a continuation segment. This is accomplished through the *logical channel* (L_CH) field of the baseband packet's payload header, as shown in Table 8-2. As we stated in Chapter 7, "Managing the Piconet," the LM intercepts all baseband packets having L_CH = 11 and sends the others to L2CAP. The start of an L2CAP segment has L_CH = 10. These bits are listed in Table 8-2 with the MSB first, as they usually are when presented in written form. However, remember that the LSB is sent first in a Bluetooth baseband packet, so

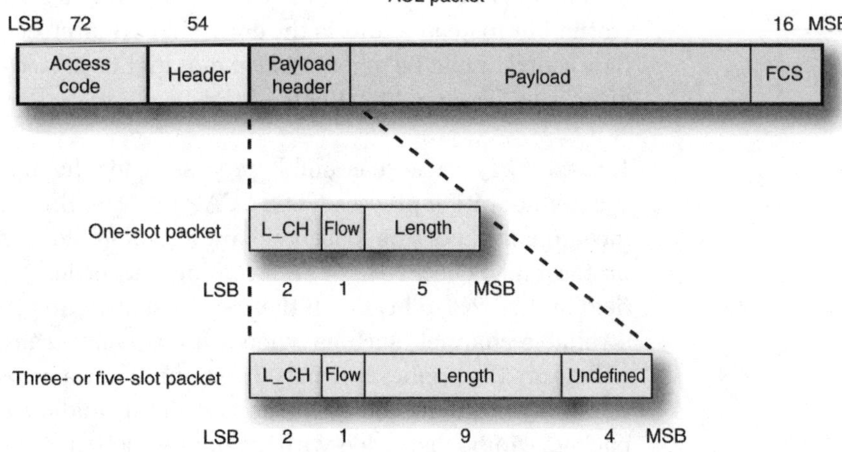

Figure 8-7 Payload header fields of Bluetooth ACL single- and multislot packets

Table 8-2 Logical channel fields

L_CH	Logical Channel	Protocol	Type
00	Reserved	Reserved	Reserved
01	UA or UI	L2CAP	Continuation of L2CAP segment
10	UA or UI	L2CAP	Start of L2CAP segment
11	LM	LMP	LMP packet

their order will be reversed when transmitted. Notice that there's no need to identify the final segment of a L2CAP packet because that information can be deduced from the packet's length field.

Once the starting segment of a L2CAP packet is transmitted, all of the remaining packet segments must be sent before a new L2CAP packet is begun. In other words, multiple L2CAP packets cannot be segmented and interleaved on the baseband channel because there's no information in the L2CAP packet to enable the destination to perform the deinterleaving process.

The FLOW bit in the ACL payload header is different from the FLOW bit in the baseband packet header, but its function is similar. This bit is set to binary 1 to enable the flow of L2CAP traffic and is set to 0 to stop L2CAP traffic. The payload header FLOW bit applies only to the L2CAP layer, so SCO traffic and LMP packets aren't affected. Unfortunately, the payload header's FLOW bit cannot be used to stop only a single channel of L2CAP traffic, but instead it affects the entire L2CAP protocol layer. Single-channel flow control must be accomplished at a higher protocol layer, which we will discuss in Chapter 11, "Bluetooth Profiles."

Reassembly The reassembly process is, for the most part, the reverse of the segmentation process. The L2CAP layer on the receiving end takes the incoming packets from baseband and rebuilds the original long packet. The protocol also checks that the incoming packet length matches the length field in the L2CAP header. If there's a mismatch and the application expects a reliable channel (such as when a file is being transferred), then L2CAP will notify the application that the channel has a reliability problem. If full reliability isn't needed (such as streaming audio), then the bad L2CAP packets can be discarded without further action.

Communicating with the L2CAP Layer

A request for L2CAP processing usually begins from a higher protocol layer on the initiator end of the Bluetooth link and proceeds down to the radio, across the channel, and up to the higher protocol layer at the responder end of the link, as shown in Figure 8-8. An example might be when the initiator asks that a L2CAP channel be created between local and remote applica-

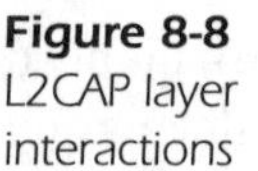

Figure 8-8 L2CAP layer interactions

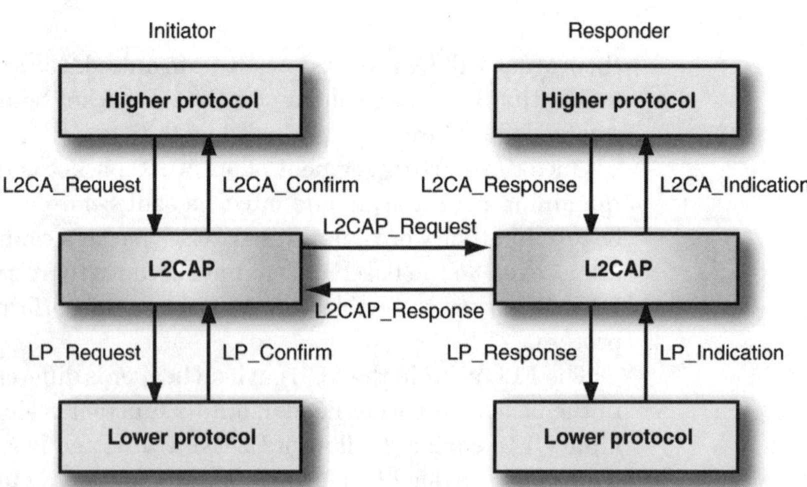

tions. Although this diagram looks complicated, the activity is actually quite straightforward once the hierarchy is understood. In our diagram, the two devices simply represent the initiator of, and responder to, a particular request, so their roles could swap several times during the life of a Bluetooth link.

Notice that communication paths exist that are *vertical*, occurring between protocol layers within one Bluetooth-equipped host, and *horizontal* between the two L2CAP layers in a manner similar to the way LMs communicate (see Chapter 7). These paths are labeled using prefixes in the following way:

- **L2CAP** Communication horizontally between L2CAP layers using the wireless link
- **L2CA** A vertical communication path between L2CAP and a higher protocol layer within the host software
- **LP** A vertical communication path between L2CAP and a lower protocol layer, either within the host software or across the HCI to the Bluetooth module

The last item in the previous list could be termed LM, for example, if L2CAP were communicating with the LM, asking it to create an ACL channel for L2CAP to use. L2CAP-to-L2CAP communication is completely defined in the specification, but the method for communicating between vertical layers is implementation-dependent. Also, the specification often refers to the layer itself as L2CA and the protocol as L2CAP, sort of like treating the LM and LMP as separate entities. On the other hand, sometimes L2CAP is used in place of L2CA because it's easier to say. Confused? Me too.

Tasks leaving the initiator's application and proceeding down the protocol stack are called *requests* (Req). When they cross over to the responder, they then proceed up its protocol stack in the form of *indications* (Ind). The responder's application then forms one or more *responses* (Rsp) that run down the stack, across the Bluetooth channel, and then up the initiator's stack in the form of *confirmations* (Cnf). All of these can be written as *service primitives*, which are simply mnemonics in fractured English. We'll give an example of a string of service primitives used to set up a L2CAP channel after defining events and actions.

Events *Events* are all incoming messages to the L2CA layer. These can come from a higher protocol layer, lower protocol layer, or from the other device's L2CA layer, as shown in Figure 8-8, and from timers that are not shown in the figure. Events can be divided into five different categories:

- Lower layer to L2CA Ind and Cnf
- Signaling from the other device's L2CA layer via L2CAP packets
- Incoming data from the other device's L2CA layer via L2CAP packets
- Higher layer to L2CA Req and Rsp
- Timer expirations

Actions *Actions* are all outgoing messages from the L2CA layer. Again referring to Figure 8-8, actions can go to higher or lower protocol layers, across the Bluetooth channel to the other L2CA layer, or consist of setting timers. The five major action categories are

- L2CA to lower layer Req and Rsp
- Signaling to the other device's L2CA layer via L2CAP packets
- Outgoing data to the other device's L2CA layer via L2CAP packets
- L2CA to higher layer Ind and Cnf
- Setting timers

L2CAP Channel Setup Example As an example of how these events and actions progress from one device to another, and through the various protocol layers, see Figure 8-9. This shows a step-by-step process of estab-

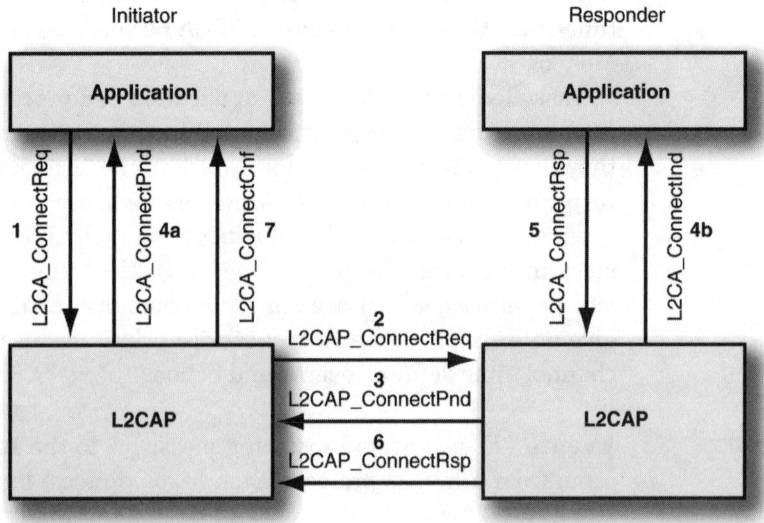

Figure 8-9 The flow of service primitives for establishing a L2CAP channel

lishing a L2CAP channel for use as a data pipe by an application. We assume that the two devices already have an ACL link established and configured. However, the first command in the following list will direct the local LM to establish an ACL channel if it doesn't already exist. To improve clarity, the role played by HCI is not shown.

1. L2CA_ConnectReq is sent by the initiator's application to its own L2CA layer to request that a L2CAP channel be created between the requester and responder.
2. L2CAP_ConnectReq is sent from the initiator's L2CA across the ACL link to the responder's L2CA to request that the channel be created.
3. L2CAP_ConnectRspPnd is returned across the ACL link by the responder's L2CA layer to indicate that the request is being processed (response pending).
4. L2CA_ConnectPnd indicates to the initiator's application that a connection response is pending at the responder. Meanwhile, L2CA_ConnectInd indicates to the responder's application that a request for L2CAP channel setup has been received from the initiator.
5. L2CA_ConnectRsp tells the responder's L2CA layer that the request has been approved.
6. L2CAP_ConnectRsp is then returned across the ACL link by the responder's L2CA layer, indicating to the initiator's L2CA layer that the channel request has been approved.
7. L2CA_ConnectCnf confirms to the initiator's application layer that the L2CAP channel has been established.

If the responder's host in step 5 had disapproved of the request to establish the channel, then it would have sent a L2CA_ConnectRspNeg (negative response) to its L2CA layer, which would have in turn sent L2CAP_ConnectRspNeg to the initiator. Also, various security measures may be required by a host before the L2CAP channel is enabled (see Chapter 9, "Bluetooth Security").

As we will discover in the next section, the initiator's CID is contained in its L2CAP_ConnectReq packet, and the responder's CID is included in its L2CAP_ConnectRsp packet. All of the L2CAP-to-L2CAP communication takes place on the signaling channel.

L2CAP Timers The L2CA layer has its own timers to prevent infinite wait times for command responses from the other device. The *response timeout expired* (RTX) timer is used for request retransmissions or L2CAP channel termination when signaling requests go unanswered. This timer is implementation dependent, but should expire between 1 and 60 seconds after each signaling request. Because multiple requests can be pending, one timer is needed for each. The software manufacturer can decide how to implement a signaling retransmission scheme. The scheme should be based upon the flush timeout option (discussed later), which determines how many baseband packet retransmissions are allowed for a given L2CAP packet. A higher flush timeout value means greater baseband reliability and reduced need for initiating retransmissions at the L2CAP level. If this timer expires, then a duplicate request can be sent or the channel disconnected. If a duplicate request message is sent in response to a RTX, then the timer must be reset to a new value that is at least double the previous value.

Another timer called the *extended response timeout expired* (ERTX) is used in place of the RTX if the local device suspects that the remote device requires additional processing time for a particular L2CAP request. This timer is started when the remote device returns L2CAP_ConnectRspPnd, for example, proving that it's at least alive. The ERTX timer expires between 60 and 300 seconds after the pending service primitive is received and no further response occurs. This provides time for human interaction, if needed, in response to an incoming request to form a L2CAP channel. If this timer expires, then a duplicate request can be sent or the channel disconnected. If a duplicate request is sent, then the ERTX timer disappears and a new RTX timer is started.

Incidentally, timers in protocol layers above the LM are set in seconds, not baseband time slots, so host applications need not be aware of how the lower Bluetooth protocol layers operate in order to communicate with L2CAP.

L2CA Service Primitives The L2CA primitives that are used by higher protocol layers to communicate with their local L2CA layer are listed in the specification, but the actual means by which this communication takes place is up to the software developer. These primitives contain the parameters required by the L2CA layer for creating any associated L2CAP command packets for transfer to the other device's L2CAP layer over the RF link.

Several primitives are used for communication between a higher protocol layer and L2CAP. Although these are listed in the Bluetooth specification, we've categorized them and added to their descriptions to hopefully make their use a bit more clear. Remember, Req and Rsp travel from the higher protocol to the L2CA layer as events, and Ind and Cnf travel from the L2CA layer to the higher protocol as actions.

- Establishing the L2CAP channel
  - The L2CA_ConnectReq event initiates a channel connection. This primitive also includes the destination device's BD_ADDR for use by the LM in a page if the ACL link hasn't been established yet.
  - The L2CA_ConnectInd action indicates to the higher protocol layer that a L2CAP channel connection request has been received from another device.
  - L2CA_ConnectPnd confirms to the higher protocol that a connection response is pending at the remote device.
  - The L2CA_ConnectRsp event tells the local L2CA layer to approve establishing the channel, while L2CA_ConnectRspNeg disapproves establishing the channel.
  - The L2CA_ConnectCnf action tells the higher protocol that the requested L2CAP channel has been established, while L2CA_ConnectCnfNeg indicates that the L2CAP channel was not established.
- Configuring the L2CAP channel
  - The L2CA_ConfigReq event initiates the channel configuration process.
  - The L2CA_ConfigInd action indicates to the higher protocol layer that another device wants to configure the L2CAP channel.
  - The L2CA_ConfigRsp event is the higher protocol layer's response to the configuration request indication from the other device's L2CAP layer.
- Reading and writing data to the L2CAP channel
  - L2CA_DataWriteReq requests the transfer of data from the higher protocol to the local L2CA layer and from there across the RF link to the other device.
  - The L2CA_DataWriteCnf action confirms to the higher protocol the number of bytes transferred and whether or not the transfer was successful.
  - L2CA_DataRead is used by the higher protocol layer to request the transfer of data from the local L2CA layer to the higher protocol.

- Group management
  - L2CA_GroupCreate and L2CA_GroupClose create or close a local CID representing a logical connection to multiple devices. This local CID enables L2CAP to identify different groups and their respective members. The remote CID for a group is always 0x0002.
  - L2CA_GroupAddMember and L2CA_GroupRemoveMember add or remove a device with the specified BD_ADDR from a group identified by the local CID.
  - L2CA_GroupMembership requests a report of group members associated with a particular local CID.
- Miscellaneous channel activity
  - L2CA_Ping checks a device with a specified BD_ADDR to see whether the L2CA layer is alive. Note that the ping doesn't check a particular L2CAP channel, but only the L2CA layer itself.
  - L2CA_GetInfo retrieves various information from a target device with a specified BD_ADDR. The information retrieved is implementation specific, and the only field defined in the specification is for retrieval of the CL MTU.
  - L2CA_EnableCLT and L2CA_DisableCLT enable and disable the capability to receive connectionless data traffic. Connectionless traffic might be disabled if a host can't accept the QoS limitations placed on receiving data via broadcast packets.
  - The L2CA_TimeOutInd action indicates to the higher protocol that a RTX or ERTX timer has expired. The implementation determines the number of times this timeout occurs before L2CAP gives up and sends L2CA_DisconnectInd.
  - The L2CA_QoSViolationInd action indicates to the higher protocol that the QoS agreement during L2CAP channel configuration has been violated.
- Disconnecting the L2CAP channel
  - L2CA_DisconnectReq initiates the channel disconnect.
  - The L2CA_DisconnectInd action indicates to the higher protocol layer that a disconnect request has been received from the other device's L2CAP layer.

A similar set of primitives exist for communication between L2CAP and the lower protocol layer. These are listed in the Bluetooth specification and won't be repeated here.

L2CAP Signaling

The L2CAP channel setup procedure given in the last section includes several instances of communication that take place between the two devices' L2CA layers in the form of L2CAP_Connect service primitives. This communication occurs on the L2CAP signaling channel that is automatically established whenever two devices form an ACL link.

Figure 8-10 shows the general format of a L2CAP signaling command packet. This is simply the L2CAP packet in Figure 8-5 with signaling commands in its payload. Multiple commands can be included in a single signaling packet, but unless negotiated otherwise, the total packet length cannot exceed a signaling MTU (MTU_{sig}) of 48 bytes. Therefore, an entire signaling command packet using its default payload length will fit into one 3- or 5-slot ACL packet.

The command itself consists of a command code, a DCID = 0x0001 corresponding to the signaling channel on the remote device, a 2-byte length field, and data such as command parameters. Remember when we referred to L2CAP as a middle manager? That job is characterized by a high level of responsibility, but little authority, and L2CAP signaling is no different. Table 8-3 lists the entire set of signaling commands, some of which you should recognize from the channel setup example given in Figure 8-9.

Because L2CAP signaling occurs between two L2CA layers, each can be considered as an action or event, depending on whether the L2CA layer sends or receives the command. For simplicity, the Bluetooth specification lists these as events. Furthermore, all of these signaling commands must actually pass from the L2CA layer to the lower protocol layers for transmission to the other device, and then back up to the L2CA layer at that device.

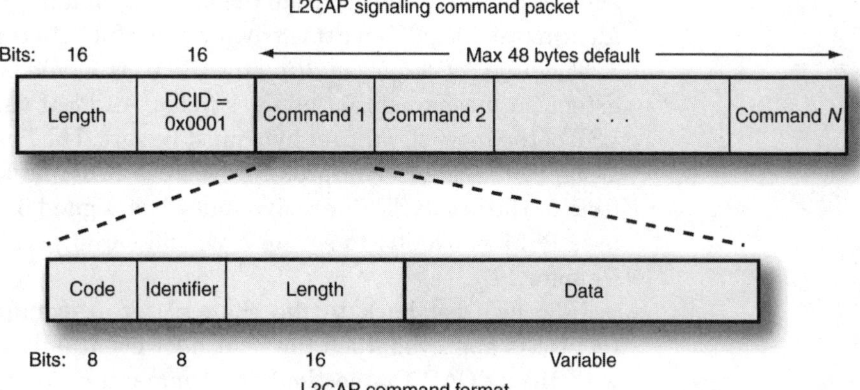

Figure 8-10 The format of the L2CAP signaling command packet and command format

Table 8-3

L2CAP signaling command set

Code	Command
0x00	Reserved
0x01	Command reject
0x02	Connection request
0x03	Connection response
0x04	Configure request
0x05	Configure response
0x06	Disconnection request
0x07	Disconnection response
0x08	Echo request
0x09	Echo response
0x0A	Information request
0x0B	Information response

As an example, the command service primitive L2CAP_ConnectReq (command code 0x02) is sent from the initiator to responder to request that a L2CAP channel be established. The format of the actual command represented by that service primitive is shown in Figure 8-11. The initiator's CID (0x0040 in the example) is part of this packet, thus telling the responder what DCID to use in L2CAP data packets that are later sent back to the requestor on that channel.

The identifier (0x01 in the example) is set by the requesting device so that both L2CA layers can keep track of the various commands because several may be pending at any given time. Identifiers can be reused after 360 seconds have elapsed from the initial command transmission. The 0x00 identifier is illegal. Isn't this trivia wonderful? There's more.

The *protocol service multiplexor* (PSM) is a field based on the ISO 3309 extension mechanism for address fields. All PSM values are odd, and the LSB of the most significant byte must be zero. The first few PSM values are assigned to higher layer protocols such as SDP and RFCOMM, and values from 0x1001 to 0xFEFF are dynamically assigned by L2CAP. Our example uses PSM = 0x0003 to set up a virtual serial port using RFCOMM (see Chapter 11).

Now let's get back to the channel setup example. If the connection request is approved, then the responder eventually replies to the initiator with the L2CAP_ConnectionRsp (command code 0x03) service primitive

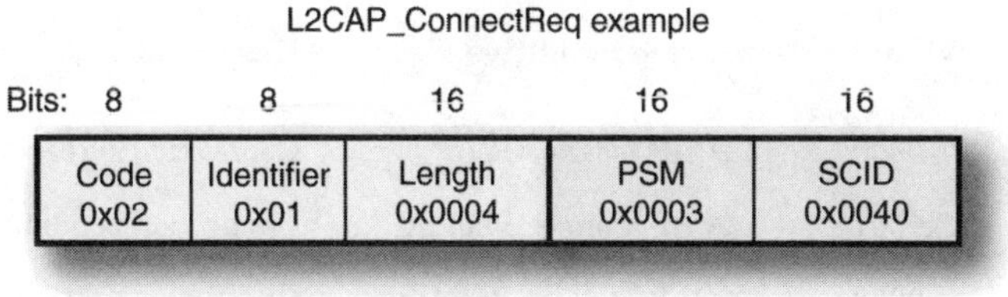

Figure 8-11 L2CAP_ConnectReq packet from initiator's L2CA to responder's L2CA. The initiating device's CID is 0x0040, and the PSM is 0x0003 corresponding to RFCOMM. The command is tracked using identifier 0x0001.

L2CAP_ConnectReq example

Bits:	8	8	16	16	16
	Code 0x02	Identifier 0x01	Length 0x0004	PSM 0x0003	SCID 0x0040

that takes the form of the command in Figure 8-12. Both the initiator's and the responder's CIDs (0x0040 and 0x0050, respectively) are contained in this response. Upon receipt of this response, both L2CA layers now have the information they require to construct L2CAP packets with the correct DCID field for that channel. The responder uses the initiator's CID, and the initiator uses the responder's CID; hence, both of these CIDs are the DCIDs for L2CAP data packets sent by the respective devices.

The connection response contains two additional fields. One is the *result*, with possible values given in Table 8-4. If the channel setup is successful, then 0x0000 is in this field, which identifies the service primitive as L2CAP_ConnectRsp. Finally, the status field (see Table 8-5) is defined only for L2CAP_ConnectPnd to indicate the reason for the wait.

Figure 8-13 shows how this process works with amazing clarity. In Step 1, the initiator sends its CID to the respondent, establishing the first L2CAP channel endpoint. Next, in Step 2, the respondent sends both CID values back to the initiator, and the L2CAP channel is established and ready for configuration.

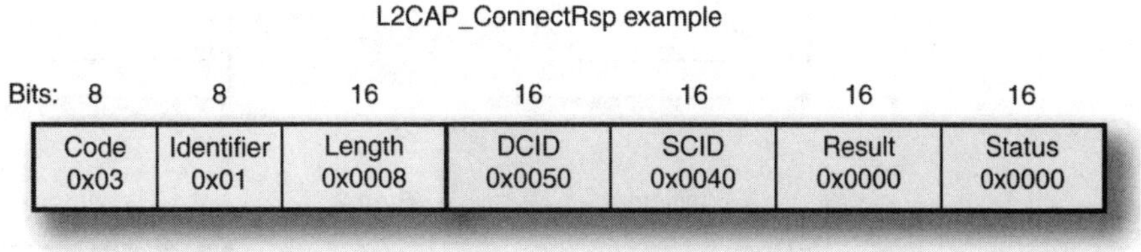

Figure 8-12 L2CAP_ConnectRsp packet from responder's L2CA to initiator's L2CA. The responding device's CID is 0x0050. The result field of 0x0000 means the connection is successful, and the status field doesn't apply.

L2CAP_ConnectRsp example

Bits:	8	8	16	16	16	16	16
	Code 0x03	Identifier 0x01	Length 0x0008	DCID 0x0050	SCID 0x0040	Result 0x0000	Status 0x0000

Table 8-4
Result values for connection response

Value	Service Primitive	Description
0x0000	L2CAP_ConnectRsp	Connection successful
0x0001	L2CAP_ConnectRspPnd	Connection pending
0x0002	L2CAP_ConnectRspNeg	Connection refused—PSM not supported
0x0003	L2CAP_ConnectRspNeg	Connection refused—security block
0x0004	L2CAP_ConnectRspNeg	Connection refused—no resources available
Other	Reserved	Reserved

Table 8-5
Status values for connection response

Value	Service Primitive	Description
0x0000	L2CAP_ConnectRspPnd	No further information available
0x0001	L2CAP_ConnectRspPnd	Authentication pending
0x0002	L2CAP_ConnectRspPnd	Authorization pending
Other	Reserved	Reserved

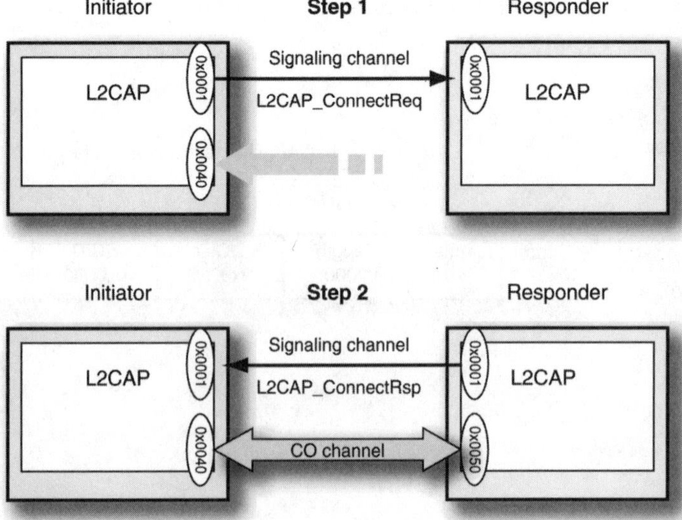

Figure 8-13 Setting up a L2CAP channel using signaling commands

Configuring the L2CAP Channel Once the channel is established, it must be configured before it can be used. This is done by sending a (surprise!) configuration request packet on the signaling channel. The packet contains the DCID for the channel being configured, and the configuration applies only in the direction in which the configuration request packet is sent. In other words, each end of the L2CAP channel must generate its own L2CAP_ConfigReq and receive a L2CAP_ConfigRsp from the other end.

The L2CAP_ConfigReq packet contains the MTU size, flush timeout option, and QoS option as described in the following list:

- **MTU size** This field specifies the largest L2CAP payload size that can be accepted. The default size is 672 bytes, which is roughly the L2CAP packet payload capacity in two DH5 baseband packets. The maximum possible MTU size is 65,535 (0xFFFF) bytes.
- **Flush timeout option** This informs the recipient of the amount of time in ms that the originator's baseband will attempt to transmit a L2CAP packet segment in a baseband packet before flushing the packet and moving on to the next one. The minimum time is 1 ms (actually 1.25 ms, the time between successive transmission slots at baseband), which means that the segment is sent only once before being flushed, that is, no retransmission. The maximum is 0xFFFF, indicating an infinite number of retransmissions. The latter value is termed a *reliable channel* and is the default. Fortunately, an infinite retransmission loop due to loss of the ACL link will not result from using the reliable channel because a link timeout will eventually occur at the LM level. However, a reliable channel flush timeout option could increase the effectiveness of a denial-of-service attack (see Chapter 9).
- **QoS option** The QoS option describes the outgoing traffic flow from the device sending the configuration packet if used in a L2CAP_ConfigReq and the incoming flow if used in a L2CAP_ConfigRsp. The default is *best effort*, which specifies no QoS guarantees. The other possibility is *guaranteed* QoS. There are several additional values that are included with the QoS portion of the configuration packet. These are taken as requirements for guaranteed QoS and as merely hints for best effort QoS.
  - **Token rate** This 4-byte value specifies the data rate that the application associated with the L2CAP channel wants to use. The default is 0x00000000, which means that no rate is specified. A value of 0xFFFFFFFF means that the maximum possible rate is desired.

The LM must insure that the polling rate is fast enough to service the selected token rate. If the maximum token rate is selected and the service type is guaranteed, then the LM should refuse any additional ACL connections and disable any periodic inquiry or page scans in order to support the fastest possible data rate.

- **Token bucket size** This 4-byte field is the size of the data buffer that can be dedicated to the L2CAP channel. A value of 0x00000000 means no token bucket is needed, which is the default. A value of 0xFFFFFFFF asks to use the largest bucket available. If the token bucket overflows, then a QoS violation occurs, and the sending unit must either hold or discard packets.
- **Peak bandwidth** This is another 4-byte field that represents the bandwidth (throughput) in bytes per second that packets can be sent back-to-back from applications. The default is 0x00000000, which represents an unknown bandwidth.
- **Latency** This 4-byte field specifies the maximum acceptable delay in microseconds between generating a bit of data and actually sending it over the air. The default is 0xFFFFFFFF, which is the don't care value. For other latency values, the LM should insure that the polling rate is fast enough to meet the requirement.
- **Delay variation** This 4-byte field represents the difference between the maximum and minimum delay that a packet will experience in microseconds. The value helps an application determine the amount of buffer space needed at the receiving side. The default is 0xFFFFFFFF, which is the don't care value. The LM can ignore this value if it can't comprehend the length field in the L2CAP packet.

All of these configuration options can be negotiated by both devices through use of the parameters in the L2CAP_ConfigReq and L2CAP_ConfigRsp exchanges. Once the channel has been configured, it is considered open and ready for user data exchange.

Sending Data Across the L2CAP Channel User data is passed across the L2CAP channel using packets formatted as shown in Figure 8-5 for CO channels. Data on the CL channel has a 2-byte PSM (shown in Figure 8-11) following the DCID field; otherwise, it is identical to the CO data packet. The DCID field enables the receiving L2CA layer to route the packet to the proper higher-level destination protocol. Incoming packets on a CO channel are routed to the DCID value that was established during channel setup, and incoming CL packets are always sent to DCID = 0x0002.

Managing Group Data Data over a group-oriented CL channel is sent in a best effort manner, but the channel has no QoS associated with it. In other words, the CL channel is unreliable. Although not specifically spelled out in the Bluetooth specification, it's obvious that CL communication can take place using the broadcast feature of the ACL link. Furthermore, the formation of groups only makes sense for the piconet master because slaves can't communicate with more than one piconet member. Because ARQ isn't allowed for broadcast packets, there's no guarantee that all slaves in a group successfully receive the data unless they're polled individually. At the L2CA level, any such polling activity must take place over a different (CO) channel.

If a broadcast packet is heard by all piconet members, how can more than one group be designated at the L2CA level? First of all, group communication is assumed to be nonexclusive, meaning that data intended for one group could also be received by any other slave in the piconet. If privacy within a group is required, then it can be achieved through encryption (see Chapter 9).

For large piconets some of the slaves will be parked, and parked slaves won't necessarily receive broadcast packets outside their beacon windows (see Chapter 6, "Advanced Piconet Operation"). Furthermore, even parked slaves can skip beacon trains, so further group differentiation between parked slaves can be obtained through the timing of broadcast packets. The sniff mode offers a similar method of differentiating between groups of active piconet members. Slaves in different groups can be given nonoverlapping sniff slots. Of course, implementation of these issues requires careful coordination between the L2CA and LM protocol layers.

Disconnecting the L2CAP Channel The disconnection process begins when a higher protocol sends a L2CA_DisconnectReq to its local L2CA layer, which in turn sends a L2CAP_DisconnectReq command packet to the remote L2CA layer. Both SCID and DCID are specified in the packet, and the response L2CAP_DisconnectRsp contains the same two CIDs for a validity check. Finally, L2CA_DisconnectRsp is passed from each L2CA layer to its respective application to finalize the disconnect.

L2CAP for Streaming Audio Applications

As we've already discovered, real-time two-way voice has stringent latency requirements that are met in Bluetooth by using SCO packets in a virtual circuit-switched configuration that bypasses L2CAP altogether. However, streaming audio applications are probably better served over the L2CAP channel. Because streaming audio is one way, delays of a few hundred milliseconds, or even a few seconds, for destination buffering are perfectly acceptable to the user as long as the audio stream is not interrupted once it begins. L2CAP using best effort communications with a flush timeout option to prevent excessive retransmissions is well suited for one-way audio.

One of the most common streaming audio applications is the transmission of CD-quality audio. Can Bluetooth support stereo CD? The CD standard requires sampling each analog audio channel at 44.1 kHz using 16 bits per sample. Stereo CD thus produces a bit rate of 44.1 kb/s × 16 × 2 = 1.41 Mb/s, which is higher than even an asymmetric DH5 channel can support. Therefore, the audio stream must be compressed. One of the most popular compression schemes is MP3, which can produce excellent stereo results using a 128 kb/s data rate. Any ACL packet except DM1 can support this rate in a perfect channel, but asymmetric multislot transmissions are best if the channel is somewhat error prone.

Real-Time Voice Communication

Real-time two-way voice communication isn't processed by the L2CA protocol layer because L2CAP is most suited for data exchange that places data integrity over latency. Voice communication between two individuals needs just the opposite: Low latency must take precedence over data integrity. Those of you old enough to remember when overseas telephone service was in its infancy will recall two characteristics of such calls. They required shouting into the phone to be heard over the significant noise on the line, and they had long latencies from the many delays along the link. The result was that the caller on each end would begin talking simultaneously, and then both would stop when each heard the others voice come through about a half-second later. Then they would both say "go ahead" together and, of course, because this request was also delayed, both would then begin again together as well. How frustrating!

To reduce latency to a tolerable level, Bluetooth uses a simulated circuit-switched environment manifested by dedicating slots to SCO packet exchange (see Chapter 4) without any retransmission. As a result, latency is not only low, but it's also deterministic; that is, no random processes affect the delay between the time a microphone at one end of a Bluetooth link hears a voice and the speaker at the other end reproduces it. Thus, the latency doesn't vary over time.

Figures 8-1 and 8-2 show the path that real-time two-way audio takes within the protocol stack, and you'll notice that most of the stack is bypassed. Once the SCO link is established through the respective LMs (see Chapter 7), the application communicates with the link controller for SCO packet assembly and transmission. Indeed, the audio portion of the application may be something as simple as a microphone and speaker within a Bluetooth headset. A Bluetooth manufacturer's chipset might have an analog audio input pin for a microphone connection, and an analog audio output pin for sending voice to the headset speaker via an audio amplifier. If a host controller interface is present, then the audio is digitized on the host and transferred to the Bluetooth module via HCI packets (see Chapter 10).

Methods of Digitizing Voice

Bluetooth has two different options for encoding voice into a bit stream. They are *pulse code modulation* (PCM) and *continuous variable slope delta* (CVSD) modulation. Both of these create a data stream at 64 kb/s, but they create the data in very different ways, and they are not compatible with each other. The data stream is then placed into HV packet payloads and transmitted. At the receive end, the payloads are stripped, concatenated, and decoded into an analog audio stream as shown in Figure 8-14.

Digitizing Voice Using PCM PCM is perhaps the most straightforward method for turning an analog waveform into a string of bits. For *toll quality* telephone coding, an audio waveform is sampled at a rate of 8k per second, and each sample is quantized using 8 bits, for a total bit rate of 64 kb/s.

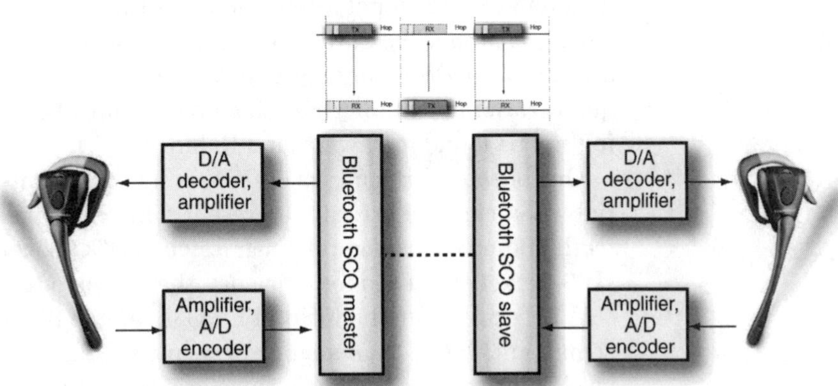

Figure 8-14 Real-time two-way voice communication using Bluetooth. The audio encoders and decoders may be part of the Bluetooth baseband chipset or separate devices depending upon the manufacturer.

Eight bits allow for a total of 256 quantization levels to be assigned to each sample. The 64 kb/s rate is commonly used in many commercial voice applications. For example, the digital hierarchy called *T1* in the United States encodes 24 speech channels at 64 kb/s using PCM for an aggregate data rate, including signaling, of 1.544 Mb/s.

The highest frequency in the analog waveform that can be accurately reconstructed at the receiving end depends on the sampling rate according to the Nyquist criteria. Nyquist states that the sampling rate must be at least twice that of the highest frequency component in the sampled waveform for accurate reproduction, so we can conclude that the highest frequency enabled in the audio signal for an 8k per second sampling rate is 4 kHz. Frequency components higher than this will appear as lower frequency distortion (aliasing) at the receiver's *digital-to-analog* (D/A) decoder, so it's important to filter the audio signal prior to *analog-to-digital* (A/D) encoding to remove all frequency components higher than 4 kHz.

The *International Telegraph and Telephone Consultative Committee* (CCITT) sets the standard for toll quality voice, and it requires a signal-to-noise ratio of about 34 dB over a range of input signal variations of about 30 dB. To meet this standard using PCM, a method called *companding* (compressing/expanding) is used. This is a means to reduce the spacing between quantization levels for signal levels near zero and increase the spacing for signal levels near maximum. In this way, quiet audio passages have reduced quantization noise for improved *signal-to-noise ratio* (SNR) at the expense of increased quantization noise for louder passages. This is not a problem because the signal level in a loud audio passage will mask the increased noise levels.

Two different companding schemes supported by Bluetooth are A-law and μ-law. The former is used mainly in Europe and the latter in North America and Japan. Both companding techniques are approximately linear for small analog signal levels and logarithmic for larger values. A-law provides a *companding gain* (the highest SNR improvement) of about 24 dB, and the μ-law companding gain is about 30 dB over the SNR for linear quantization. More information on the technical issues of speech companding can be found in Jayant[1] and Rabiner[2].

One of the shortcomings of PCM is its somewhat high distortion potential when channel errors are present. This can cause problems with Bluetooth audio because no retransmission of SCO packets are enabled, and if HV3 packets are used for efficiency, then no error correction exists either.

A bit error near the MSB of a PCM sample distorts the received audio signal much more than a bit error near the LSB. Various techniques can be used to interpolate the audio waveform when the incoming data appears to be incorrect given the trend of the analog waveform that's being generated. Suspect 8-bit blocks in the HV3 packet can be discarded and their values estimated at the receiver. HV2 packets containing more than one error in any 15-bit block will often trip the error alarm (see Chapter 4), so the parity check bits can simply be discarded and the remaining 10 bits used in the receiver's decoder and D/A converter if they appear to have a reasonable value. Finally, if an HV1 packet seems to have problems even after FEC decoding, then the bad sections can be removed and interpolated instead.

Digitizing Voice Using CVSD CVSD is a relatively new (1970) method for encoding speech that uses a predictive algorithm for generating a bit stream that represents an audio signal. CVSD is considered to be more robust than PCM under poor channel conditions, in part because the distortion on a reconstructed audio signal from a bit error is about the same regardless of where in the stream the error occurs. This is unlike PCM, where a bit error at the MSB position causes much more havoc than an error at the LSB position. CVSD is the encoding method of choice for *voice over Internet protocol* (VoIP). Random bit errors in a CVSD data stream are manifested by an increase in background noise at the receiver's decoder, but the signal itself usually remains intelligible in all but the worst BER conditions.

The concept of *delta modulation* is based upon the fact that if the sampling rate of an analog waveform is much faster than the Nyquist rate, then the sample-to-sample correlation is very high. In other words, if an audio signal is at a particular voltage level at a particular time, then a short time (perhaps 16 μs or 1/64,000 second) later, the value will have changed by a very small amount. This leads to the conclusion that perhaps only a single bit can be used to represent each sample: A binary 1 indicates that the new value is greater than the previous value, and a binary 0 indicates the opposite.

There are a number of problems with the foregoing method. One is that there's no way to represent how fast the analog signal is changing from sample to sample. This can be solved by using a *predictive algorithm*, where the encoder predicts where it expects the audio waveform to be when the next sample is taken, then it looks at the actual value and assigns a binary 1 if it's greater than, or a binary 0 if it's less than, the predicted value. If the receiver's decoder uses the same algorithm, then it can hopefully reconstruct an analog waveform that closely matches the one sampled at the transmitter.

Another problem with delta modulation is called *slope overload*. This happens when the audio waveform increases or decreases so quickly that the algorithm can't track it fast enough, and distortion occurs at the receiver's decoder output. CVSD reduces slope overload by using an algorithm that dynamically adapts to the varying slope of the audio waveform, so its reproduction accuracy is enhanced at the receiver. More details on CVSD and other adaptive delta modulation techniques can be found in Jayant[1] and Rabiner[2].

Latency Calculations Table 8-6 expands upon the information in Table 4-3 from Chapter 4. We've already discussed in that chapter the fact that a Bluetooth piconet is fully loaded with one HV1 channel, two HV2 channels, or three HV3 channels. Table 8-6 also enables us to roughly calculate latency from one end of the link to the other. For simplicity, we'll ignore the additional delay caused by the presence of an HCI (see Chapter 10). At the sender, delays fall into the following categories:

- Performing A/D encoding of the audio signal
- Assembling a HV packet
- Waiting for the transmission window to arrive
- Sending the HV packet

Delays at the receiver include

- Receiving the HV packet
- Extracting the payload
- Performing the D/A decoding and sending the signal to the amplifier and speaker.

The first two steps require a time at least as long as the number of milliseconds of digitized audio sent within a single HV packet. For the worst-case HV3 situation, 3.75 ms of audio is taken, so let's use 5 ms as the time needed to perform the A/D encoding and assemble the HV3 packet. Once the packet is ready, then the wait for the transmission window begins. Once

Table 8-6

Structure of SCO HV packets

Type	Bits of Digitized Audio	Milliseconds of Digitized Audio	FEC	Payload Size in Bits	Packets per Second for 64 kb/s
HV1	80	1.25	(3,1)	240	800
HV2	160	2.50	(15,10)	240	400
HV3	240	3.75	None	240	267

again, HV3 can require the longest wait because one transmission occurs only every six time slots. The worst-case situation is a six time slot delay of 3.75 ms. Propagation time is negligible, so the next delay is the 0.625 ms required to send/receive the SCO packet, followed by another 2 ms, or so, for the receiver to extract the payload and begin the D/A decoding. We'll assume that the D/A conversion can occur in real-time so that the listener begins hearing the audio associated with that particular HV3 packet when decoding begins. Total latency from one end of the SCO channel to the other is pessimistically estimated at just over 11 ms, which would be imperceptible to most users. Although quite subjective in nature, latencies can begin to interfere with two-way voice communications when they approach about 100 ms.

In Chapter 10, we'll include the additional latency that comes from using HCI to pass SCO data between host and module.

Establishing and Disconnecting the SCO Channel

When a Bluetooth link is first established, it is by definition an ACL link. After the LMs at either end finish the configuration process and LMP_setup_complete PDUs are exchanged (see Chapter 7), then a SCO channel can be established for two-way voice communications. The process is begun when either the master or slave LM sends LMP_SCO_link_req to the other LM. If the master initiates the SCO request, then it determines the values for the PDU parameters, which are

- **SCO handle** An 8-bit unsigned integer identifying the particular SCO link
- **Timing control flags** An 8-bit mapped value describing how timing initialization is to be accomplished relative to the master's Bluetooth clock
- **D_{SCO} (SCO offset)** Determines when the first SCO packet exchange takes place
- **T_{SCO} (SCO interval)** Specifies how often SCO packets are to be exchanged
- **SCO packet** Specifies whether HV1, HV2, or HV3 packets are to be used
- **Air mode** Specifies whether A-law, μ-law, CVSD, or transparent data is being sent

The slave responds with LMP_accepted or LMP_not_accepted. If the slave initiates the request, then the timing control flags, D_{SCO}, and SCO handle are invalid. The master returns another LMP_SCO_link_req PDU with these values filled in, and with other parameters changed if the master so desires. The slave LM can accept or reject the return request.

The ability to specify these parameters independently provides significant coding flexibility. For example, notice that a SCO link can be set up to send transparent data over the CO channel. This enables other audio encoders to be used and can even accommodate different error control methods and encryption of an audio bit stream.

Once the SCO channel is established, the specified time slots are reserved. The slave is allowed to transmit its SCO packet even if it doesn't successfully decode the master's packet in the previous time slot. Either master or slave can terminate the SCO channel by sending LMP_remove_SCO_link_req, which the other device is required to accept.

Summary

The L2CAP provides user data services over the Bluetooth ACL channel. These services consist of protocol multiplexing, packet segmentation and reassembly, QoS issues, and group management. L2CAP assumes that the Bluetooth baseband provides a reliable channel with full duplex capability and delivers packets in the same order as they are generated at the remote device.

L2CAP operates by setting up channels representing endpoints of each data pipe between devices. The signaling channel is used to designate a CID at each end, and then to configure the channel to meet either a best effort (the default) or guaranteed QoS. The L2CAP packet includes in its header the destination CID so that the remote L2CA layer can direct the packet to the correct higher-level protocol.

Communication between the L2CA layer and other layers on both the local and remote devices can occur through indications, confirmations, requests, and responses. Incoming messages to L2CA are called events, and outgoing messages are actions. These are designated in English using service primitives. Communication between L2CA layers on different devices is done with L2CAP primitives that are precisely designated in the Bluetooth specification. Vertical communication with the L2CA layer is part of the host software package, and its implementation is up to the software developer.

Real-time two-way voice communication is not part of L2CAP, but is instead set up by the respective LMs. To reduce latency to a few milliseconds, applications typically communicate directly with the link controller (through HCI if implemented). Aren't these summaries boring?

End Notes

1. Jayant, N. and Noll, P. *Digital Coding of Waveforms: Principles and Applications to Speech and Video*, Englewood Cliffs, NJ: Prentice-Hall, 1984.
2. Rabiner, L. and Schafer, R. *Digital Processing of Speech Signals*, Englewood Cliffs, NJ: Prentice-Hall, 1978.

CHAPTER 9

Bluetooth Security

Because Bluetooth is a wireless communication system there is a definite possibility that its transmissions could be deliberately intercepted or jammed, or false information passed to piconet members. To provide usage protection for the piconet, the system must establish security at several protocol levels. Unlike many network protocols that leave the issue of security up to attached software modules, Bluetooth offers built-in security measures at the link level.

As an ad-hoc network, Bluetooth devices are subject to security needs beyond what is normally required for either wired or centralized wireless networks. Wired links do radiate somewhat when information is sent from one end to the other, but the range is typically very short. As a result, there is usually more emphasis on insuring that users have been authenticated properly than on preventing remote interception of data. Centralized wireless networks, such as 802.11, generally operate with a central database of security entities such as passwords and keys. Of course, if this database is compromised, then the consequences can be disastrous, but protecting a single database is generally easier than guarding several different databases distributed throughout an ad-hoc network.

Threats in distributed networks can be roughly divided into three categories.[1] These are

- **Disclosure threat** Leakage of information from the target system to an eavesdropper that doesn't have authorization to access the information.
- **Integrity threat** Deliberate alteration of information in an attempt to mislead the recipient.
- **Denial of service (DoS) threat** Blocking of access to a service, making it either unavailable or severely limiting its availability to an authorized user.

These different threats will be discussed in more detail toward the end of this chapter.

Because ad-hoc networks have no centralized infrastructure, security issues must be distributed as well. There are some specific exceptions to this rule, such as an Internet access point that may initiate security from its centralized database for each connecting user, but for the most part Bluetooth devices connecting to each other will implement security without the intervention of any third-party manager.

Overview of Bluetooth Security

Security within Bluetooth itself covers three major areas: authentication, authorization, and encryption. The process of *authentication* proves the identity of one piconet member to another. The results of authentication are used for determining a client's *authorization* to access various services on a server. The process of *encryption* is used to encode the information being exchanged between devices such that eavesdroppers (even other members of the same piconet) cannot read its contents. These three security processes are implemented within several layers of the Bluetooth protocol stack, so security is often termed a *cross-layer function*. For example, the link controller has random number generation capability and includes methods for managing security keys and providing the mathematical operations for authentication and encryption. The LM (see Chapter 7, "Managing the Piconet") includes several commands for handling security issues, and L2CAP (see Chapter 8, "Transferring Data and Audio Information") may initiate security procedures when a channel connection attempt is made. The HCI (see Chapter 10, "Host Interfacing") handles security communication between host and Bluetooth module, and several higher protocols outline specific requirements for their implementation.

Security Levels

The Bluetooth *generic access profile* (GAP), (see Chapter 11, "Bluetooth Profiles") specifies how Bluetooth security is organized. Security begins when a user decides how a device will implement its discovery and connectability options. Different combinations of these capabilities can be divided into three general categories:

- **Silent** The device will never enter the PAGE SCAN or INQUIRY SCAN states, and thus will not accept any connections. The device simply monitors Bluetooth traffic.
- **Private** The device will periodically enter the PAGE SCAN state, but never enters INQUIRY SCAN, so the device cannot be discovered. Connections will be accepted if the device's BD_ADDR is known by the prospective master during PAGE.
- **Public** The device periodically enters both INQUIRY and PAGE SCAN, so it can be both discovered and connected to.

The GAP refers to the device's discoverability, connectability, and pairing modes in the context of whether it is set up for silent, private, or public access capability. These will be covered in more detail in Chapter 11.

Security levels are defined for many different scenarios involving public or private devices and the services they provide. The GAP defines three different security modes that a device can implement. They are

- **Mode 1 (nonsecure)** A device will not initiate any security measures, so communication takes place without authentication or encryption.
- **Mode 2 (service-level enforced security)** Two devices can establish an ACL link in a nonsecure manner. Security procedures are initiated when a L2CAP channel request is made.
- **Mode 3 (link-level enforced security)** Security procedures are initiated when the ACL link is being established.

A device using security level Mode 2 will initiate security procedures when a L2CAP_ConnectReq is received that is associated with a channel that requires security. Security measures are completed before L2CAP_ConnectRsp is returned. The security requirements of a particular channel could include authentication, authorization, and perhaps encryption, and different channels may have different security requirements. For Mode 3, security procedures are initiated when the LMP_host_connection_req is received and are completed before LMP_setup_complete is sent.

Mode 2 is probably the most flexible way to implement security on a Bluetooth device. For example, a client could connect to a server with an ACL link, followed by a L2CAP channel, to browse for services without the need for security. When an attempt is made to access a service, then authentication, followed by an authorization check, could be required before access is granted.

Limitations of Bluetooth Security Architecture

From a real-world point of view, security can be categorized into three different scenarios.[2] The first is a simple point-to-point link in which each device has its own set of applications and security requirements. The situation becomes more complex in the second scenario when a point-to-multipoint piconet is established with a master communicating with several slaves, any of which may have unique security implementations. Finally, the last scenario is exemplified by access to an infrastructure such as the Internet or a LAN via Bluetooth. Security issues are far more complex in this scenario because Bluetooth is only one of perhaps several links needed for access to services. In this case different access layers may require their own authentication and authorization processes.

Based on these three scenarios, the Bluetooth security architecture has the following limits:

- *Support for legacy applications may require a software module to act as liaison between the application and the Bluetooth security manager*. A legacy application is one that has no built-in knowledge of Bluetooth or how to support it.
- *Automatic Bluetooth authentication applies to the device, not the user*. Security could be severely compromised if a device is used by an unauthorized individual. To prevent this situation, user authentication could be set up at the application level, such as requesting a user-provided password before access is granted.
- *Although authentication can be required in either one or both directions on a Bluetooth link, once the connection is established, data flow is bidirectional*. The Bluetooth security architecture doesn't provide a way to enable flow on a L2CAP channel in one direction, for example, but restrict it in the other direction. That enforcement must occur at the application level.
- *End-to-end security issues in the last scenario (such as Internet bridge access) will probably require a security solution that has the capability to incorporate Bluetooth security into its operation*. If this isn't done, then several passwords may be required before the link is complete, increasing user frustration.

Summary of Bluetooth Security Operations

The security problem in general terms requires the capability of a device to verify the identity of another device (authentication) in a way that doesn't provide any useful information to a third party listener, nor does it permit a third party access via unauthorized authentication. Furthermore, data encryption should be possible, again without providing any information to a third party that would help it break the encryption code.

The philosophy behind Bluetooth security is to build a chain of events, none of which provides meaningful information to an eavesdropper, but all must occur in a specific sequence for security to be set up successfully. The following example applies to point-to-point security, where a master is communicating with a single slave. The devices begin with a common *personal identification number* (PIN), also called the Bluetooth *passkey*, upon which several 128-bit keys are built. The PIN is used to create an *initialization key*, and this key in turn is used to create a *link key*, which can be a *unit key* (for devices with limited resources) or a *combination key* (for most devices). The link key becomes part of the authentication process. If encryption is desired, then the link key is used to generate yet another key called the *encryption key*.

Figure 9-1 shows the sequence of events leading to the eventual generation of the encryption key. During key generation and authentication, the devices transmit random numbers to each other that provide no information to an eavesdropper. As long as both devices begin with the same PIN, they will normally generate the same initialization key, authentication (link) key, and encryption key.

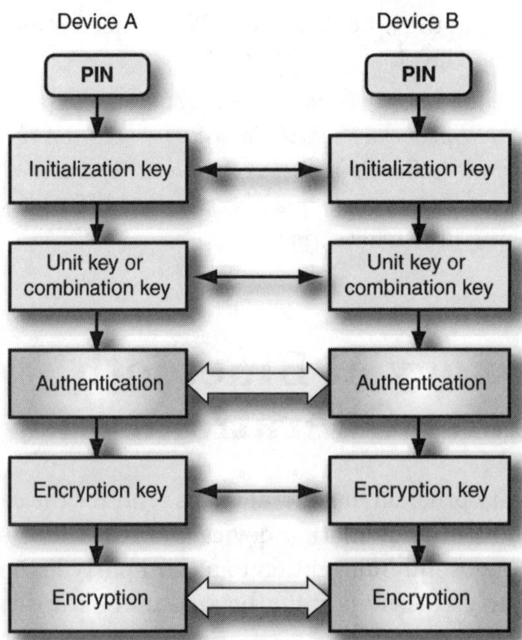

Figure 9-1 To enhance security, a Bluetooth device uses a chain of events to develop the various keys used for authentication and encryption. Each new key depends on the value of the previous key.

Authentication

Authentication is the process of verifying the identity of the device at the other end of a link. The *verifier* queries the *claimant* and checks its response; if correct, then authentication is successful. This technique is often seen in films on espionage, where each of two spies speaks a certain phrase that the other expects to hear for authentication. Of course, anyone else within earshot can receive the code words as well, so in this situation security is assured only for a one-time authentication.

Authorization

In a typical client-server scenario, the client links to the server and requests use of its services such as file information or Internet access. In this scenario, the server takes the role of verifier and the client of claimant. Once the client has successfully authenticated itself to the server, it is granted access to services based on its authorization level. Authorization can be implemented in a number of ways:

- Access is granted to all services.
- Access is granted to a subset of services.
- Access is granted to some services when authentication is successful, but further access requires additional authentication based on some user input at the client device.

The last item is usually implemented at the application layer and involves the intercession of an *external security control entity* (ESCE). This is usually a human operator who decides how to proceed with a security-related matter. People are sometimes given strange names, yes? Anyway, authorization can be implemented by a security manager that interacts with several protocol layers above HCI as we'll see shortly.

Pairing, Bonding, and Trust

The concepts of pairing, bonding, and trust are sometimes blurred in the various publications on Bluetooth security. This is often due to the fact that security methods differ, depending upon the situation and types of devices that are operating. Here are some guidelines on what is meant by the terms, although there will sometimes be areas of overlap between the categories.

Pairing Two devices become *paired* when they start with the same PIN and generate the same link key, and then use this key for authenticating at least the present communication session. The session can exist for the life of a L2CAP link (for Mode 2 security) or the life of the ACL link (for Mode 3 security). Pairing can occur through an automatic authentication process if both devices already have the same stored PIN from which they can derive the same link keys for authentication. Alternatively, either or both applications can ask their respective users for manual PIN entry.

Bonding Once devices are paired they can either store their link keys for use in subsequent authentications or discard them and repeat the pairing process each time they connect. If the link keys are stored, then the devices are *bonded,* enabling future authentications to occur using the same link keys and without requiring the user to input the PIN again. Figure 9-2 shows the bonding options included with a commercial software implementation. Bonding can expire immediately after the link is disconnected (not bonded), after a certain time period expires (temporarily bonded), or

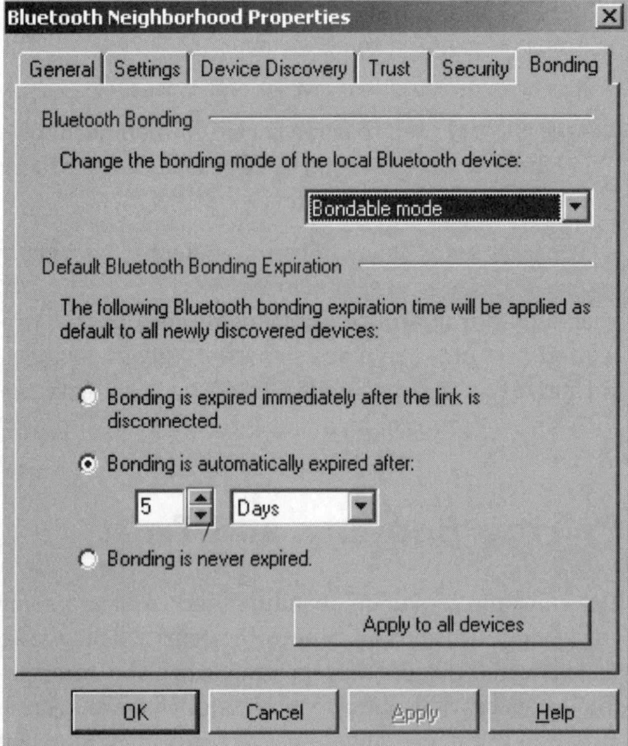

Figure 9-2 Bonding options as part of a host software package (Source: Digianswer)

never (permanently bonded). When bonding expires, the devices must go through the pairing process again.

The decision whether or not to bond devices is based upon the relative risk of storing a link key against repeating the pairing process each time the devices connect. When devices perform pairing, there's a possible vulnerable period when the PINs are used to form the initial keys, during which time a random number is sent over the air from one device to another. A third party could intercept the random number, guess the PIN, and then possibly derive one of the link keys. A PIN is often much shorter than the 128-bit link key and can be more easily guessed.

On the other hand, storing a link key, either on the host or on the Bluetooth module itself, has its own risks if the host and module aren't physically secured when not being used. An intruder with access to the host or module may be able to download the stored link keys and use them for later unauthorized authentication or decryption of files.

Trust The concept of *trust* applies to a device's authorization to access certain services on another device. A trusted device is previously authenticated and, based upon that authentication, has authorization to access various services. An untrusted device may be authenticated, but further action is needed, such as user intervention with a password, before authorization is granted to access services. The user intervention doesn't necessarily establish trust, so another password may be needed each time the service is accessed.

Figure 9-3 shows how trust is implemented in a commercial Bluetooth software package. This is a simple approach that gives the user the option of accepting all connections (trusting everyone), rejecting all connections (trusting no one), or being asked to accept or reject each incoming connection as it arrives (selecting whom to trust). This particular implementation of trust is global in nature, applying to incoming connections rather than access to specific services.

Encryption

From a security standpoint, one of the major shortcomings of wireless is the fact that its transmissions can be heard over relatively large distances away from the simple line-of-sight path between endpoints. Indeed, one of the characteristics of most ad-hoc wireless devices is their use of omnidirectional antennas so that their signals can be received by other devices, regardless of where they are relative to each other. Sending information in

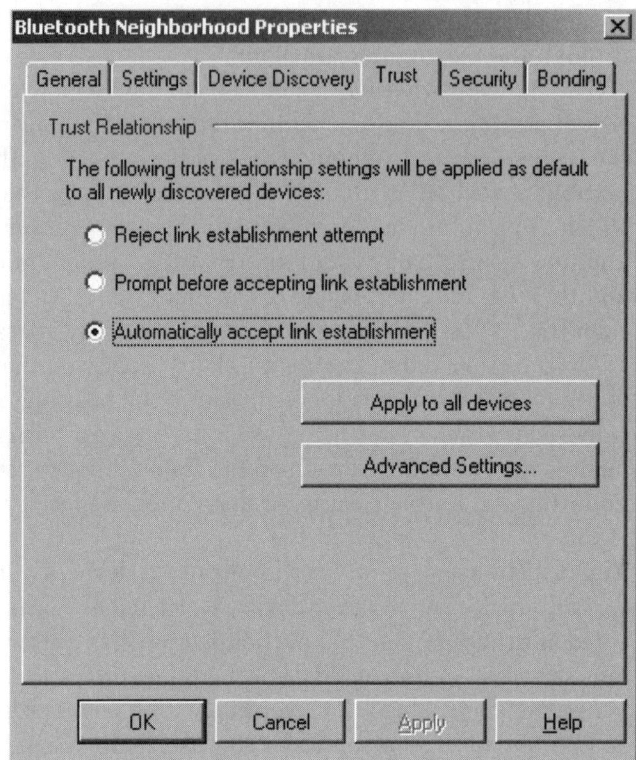

Figure 9-3 Establishing trust as part of a host software package (Source: Digianswer)

all directions facilitates eavesdropping as well, so sensitive data should be encrypted to prevent its unauthorized use.

Encryption is done by changing the message (*plain text*) into what looks like a random sequence of bits (*cipher text*) prior to transmission. Authorized users are able to change the cipher text back into plain text after reception. The cipher process uses an *encryption key* together with the plain text as an input to an encryption algorithm, out of which emerges the cipher text. At the destination, the cipher text and a *decryption key* are used in another algorithm to reconstruct the plain text. Usually, the algorithms themselves are made public, as is the case with Bluetooth ciphers. The strength of encryption is in the algorithm and key together, not in the algorithm itself.

Cryptographic algorithms can be either symmetric or asymmetric. A *symmetric algorithm* uses the same algorithm for encryption and decryption. Both ends of the link also use the same key, so the system is fairly simple. The danger, of course, is that if the single key is compromised (intercepted or otherwise determined) by an eavesdropper, then security is lost.

This danger increases in proportion to the number of devices possessing the key. Bluetooth uses a symmetric encryption/decryption algorithm.

The *asymmetric algorithm* is based upon a mathematical concept that enables one key to be used for encryption only, and another key is required for decryption. The keys are unrelated to each other, so knowing one will not provide any information for deriving the other. This algorithm lends itself to the *public key cryptography* method, in which one of the keys, called the *public key*, is given to anyone who wants it. This key can be used to create cipher text from plain text, but is useless for reversing the process. Only the *private key* can do that, and that key is kept carefully secret.

The main advantage of the public key cryptography method is that the decryption key is (ideally) kept in only one location, so it is more difficult to compromise than in the symmetric method, where the encryption/decryption key is possessed by both users of the link. Furthermore, unlike the symmetric algorithm, there's no requirement to negotiate or somehow transfer the encryption key from one device to another, during which time a compromise could occur. The main disadvantage to the public key system is its encryption and decryption speed, which is about 1,000 times slower than using a symmetric algorithm.[3] Because of this, real-world encryption tasks are often accomplished by using the public key system to securely distribute symmetric keys to the participating parties, which are then used for a single communication session and then discarded. Bluetooth has no built-in public key encryption system, but one can certainly be implemented at the application level.

Frequency Hopping and Whitening as Encryption Measures Because Bluetooth routinely incorporates frequency hopping and data whitening, will either of these provide protection from eavesdropping? The short answer is no. An eavesdropper will almost certainly use an easily obtained Bluetooth chipset as the basis for the intercepting receiver and processor. The hop sequence is known because the master's BD_ADDR and CLKN values are sent over the air as part of the FHS packet during the page, and when these values are put into the eavesdropper's hop generator, it will happily hop along with the rest of the piconet. Simple firmware modifications can enable it to receive in all slots, thereby intercepting every transmission.

For similar reasons, the data whitening and dewhitening algorithms don't enhance security either. Both of these algorithms are part of every Bluetooth-qualified chipset, and the dewhitening process automatically takes place on every received packet. The only reasonable encryption value inherent in frequency hopping and data whitening is protection from casual eavesdropping using nonBluetooth equipment.

Security Management

If any part of Bluetooth security is to take place automatically, then a security manager should be part of the host software package. Furthermore, for greatest flexibility, authentication and authorization should occur after determining the security level of the requested service, so security measures must be implemented after the ACL link is established. This implies that Mode 2 (link-level enforced security) takes place. Of course, another authentication could occur with initial establishment of the ACL link, but in many cases that authentication would be redundant.

Figure 9-4 shows a security manager residing on a Bluetooth host and communicating with L2CAP and with the LM through HCI.[2] A typical security scenario looks like this:

1. A connect request from another device arrives at the L2CAP level.
2. L2CAP requests evaluation from the security manager.
3. The security manager looks up the requested service in the database for security information.

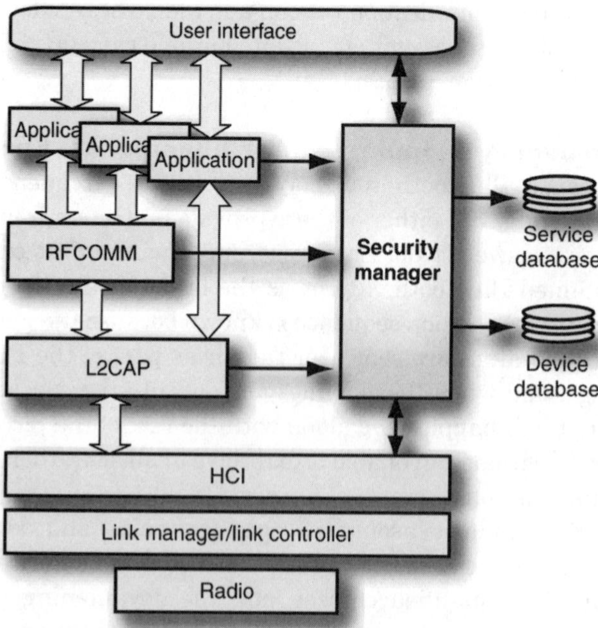

Figure 9-4 Possible realization of a Bluetooth security manager implementing Mode 2 security[2]

4. The security manager looks up the requesting device's BD_ADDR in the database for access authorizations.
5. The security manager begins the necessary authentication and (if needed) encryption procedures with the LM through HCI.
6. If all is well, then the LM provides a favorable response through HCI.
7. L2CAP finishes the connection setup process.

The security manager architecture in Figure 9-4 could be used to implement Mode 3 (link-level) security as well. One possible conflict occurs when link keys are stored in two different places: on the Bluetooth module for link-level enforced security and on the host for service-level enforced security. In order to prevent unwarranted trust being granted to devices whose link keys are stored on the module, the security manager can remove these keys, thus removing the bonding between the devices, by using HCI commands.

Bluetooth Security Implementation Process

Bluetooth implements basic security operations primarily at the LC and LM levels. The LC implements the key-generating algorithms, the random number processes, and the basic communication of the various security parameters between devices. The LM provides a set of commands that enable the formation of LMP packets containing the security parameters. HCI provides a means for the host to communicate security items to the Bluetooth module for use by the LC and LM.

Security Entities

At the link layer, there are several different entities that are used to maintain security. Table 9-1 lists these entities, their length in bits, and whether their values are assumed to be available to anyone listening (public) or secret (private).

PIN Selection and Management The PIN can be either a fixed number preprogrammed into the Bluetooth unit or a number that's entered by the user at the beginning of each secure session. The former situation

Table 9-1
Security entities

Entity	Description	Length (Bits)	Status
PIN	Personal identification number	8, 16, . . . , 128	Private
BD_ADDR	Bluetooth device address	48	Public
K_{init}	Initialization key	128	Private
K_A	Unit key	128	Private
K_{AB}	Combination key	128	Private
K_{master}	Master key	128	Private
K_C	Encryption key	8, 16, . . . , 128	Private
IN_RAND	Random number for generating K_{init}	128	Public
LK_RAND	Random number for generating K_{AB}	128	Private
AU_RAND	Random number for authentication	128	Public
EN_RAND	Random number for generating K_C	128	Public
SRES	Authentication result	32	Public
ACO	Authenticated ciphering offset	96	Private

commonly occurs if there's no *man-machine interface* (MMI) such as using a preprogrammed key card or when network managers program security entities into devices such as Internet bridges or printers with access control. The latter situation is encountered when a person uses a Bluetooth-enabled client to access a server or for interaction between two devices such as *personal digital assistants* (PDAs) being operated by humans.

There are several ways that two devices can be provided the same PIN which is, of course, never sent over the air. If two PDAs are being set up to exchange files, then each can ask its respective user for a password. The users look at each other and one might say, "Let's use 1234." This password becomes the common PIN from which the link keys are derived. In another scenario, suppose a Bluetooth-equipped printer is set up with user accounts consisting of a database of BD_ADDR values and associated PIN codes. The security manager can enter these into the printer over an encrypted Bluetooth link or through an ordinary cable hookup. When a user connects to the printer via Bluetooth, the user's application asks for a PIN (or retrieves one that was previously stored), from which the link keys are derived. If the user's PIN matches that in the printer's database, then both devices create the same link key, and authentication and, if needed, encryption can proceed successfully.

Perhaps the most secure approach to PIN management on a client is to associate the PIN with a user rather than with the device, which implies that the PIN shouldn't be stored on the device itself. This will prevent someone who obtains possession of the equipment from easily breaching security as long as the device isn't bonded to another. An ATM card is protected in this way. The downside to a PIN that's both defined and entered by the user is that it is usually fairly short (such as four digits) and can often be guessed by an intruder. If the PIN is programmable, the Bluetooth specification states that its default is all zeros, so that would be the first guess. Other guesses could be the user's birth month and year or last four digits of his/her social security number. One counter to these easily-guessed numbers is to have the network security administrator assign a PIN to a user, which is the approach taken by the ATM card providers. The danger here, of course, is that the user must either memorize the PIN or (as is more often the case) write it down somewhere, again compromising security.

Longer PIN values that are stored and periodically changed provide greater over-the-air security, but may require locking up the device when it's not being used to prevent a security breach from theft of the device. Insisting that a user manually enter a long PIN is probably not feasible due to the inconvenience and the fact that the user will almost certainly write it down.

BD-ADDR The BD_ADDR from one or both of the involved devices is used to create several link keys. This personalizes the key to a particular device, and it adds another variable into the security chain of events. The master's BD_ADDR is not a secure number because it is transmitted as part of a FHS packet. A slave's BD_ADDR isn't secure either because it can be discovered either by listening to an inquiry or performing an inquiry. Furthermore, a slave's BD_ADDR is sent as part of the LMP_slot_offset packet during a *master-slave* (MS) role switch, or when it is unparked with LMP_unpark_BD_ADDR_req.

Random Numbers Random numbers are used in several key-generation processes, and these numbers are intended to provide no information to a third party that doesn't have the PIN. Every Bluetooth device is equipped with a random number generator that can create a 128-bit random binary number on demand. These numbers are usually produced by a pseudorandom sequence generator with outputs that are required to be *nonrepeating* and *randomly generated*. Nonrepeating means that successive requests for random numbers should produce different results for a large number of requests, and randomly generated means that it should not be possible to predict the generator's output with a probability significantly greater than 2^{-L} for a L-bit random number.

If a random number is used for a specific purpose and transferred from one device to another, then we use the notation shown in Table 9-1. For example, the random number IN_RAND is used to help generate the initialization key K_{init}, as we'll discover later. General random numbers or those that are used only within the local device will usually be named RAND.

Link Key Types The *authentication key*, which is also a link key, is always 128 bits long and is used by one device to insure that the other device is indeed who it claims to be. The authentication key is often called the link key in the Bluetooth specification, and we'll do the same here. However, the specification also sometimes calls K_{init} a link key, even though it isn't normally used for authentication. The length and random nature of the authentication key makes it rather impossible to guess, and obtaining access via sequential or random link key searches by an intruder is impractical.

The link key can either be temporary, where it is used for one session only (devices not bonded) or semipermanent in which it is stored and used for several sessions or over a time period (devices bonded). Stored link keys are called semipermanent because they can be either changed or removed at a later time. As a result, paired and/or bonded devices can derive and store a new link key during each session if desired. The link key is also used to generate the encryption key.

Four different types of link keys have been developed for Bluetooth. These are

- **Initialization key (K_{init})** This is used as a link key when two devices first meet. It is normally created only once and used to protect the generation and transfer of other keys that are more secure than the initialization key.
- **Unit key (K_A)** This key is associated with a single Bluetooth device that has limited resources and can't store a large number of keys. This key is usually generated once and is almost never changed.
- **Combination key (K_{AB})** This key is more secure than K_A because it is derived from inputs provided by both devices on the Bluetooth link. Unlike K_A, K_{AB} is unique to a *pair* of devices, not just one device.

- **Master key (K_{master})** This key is temporary only and is used for the generation of an encryption key for broadcasting packets to multiple slaves.

Encryption Key The *encryption key* is used in a streaming algorithm to change plain text into cipher text and vice versa. The key can be as short as 8 bits and as long as 128 bits. The reason for the variable length is associated with import/export restrictions of devices from one country to another. Not all governments allow 128-bit encryption keys, so Bluetooth devices can accommodate these regulations by restricting the maximum encryption key length. The maximum encryption key length should be permanently programmed into the Bluetooth module without the possibility of user modification.

Key Generation Functions

We've mentioned many times that a link key is derived from different values such as a BD_ADDR and a random number. The derivation process, based upon an algorithm called SAFER+, is done though different *key generation functions*, depending upon the type of key being generated. These key generation operations simply use variations of our old friend the *linear feedback shift register* (LFSR) that is loaded with the input values, sometimes in stages, and clocked to produce the desired link key. The hardware designs for these shift registers are given in the specification, and these can be built into the LC state machine as-is or implemented under software/firmware control. A good key generation function produces a key that looks like a random number regardless of the input values given to it, and the same input values always produce the same 128-bit output value.

The different key generation, authentication, and encryption functions are designated by the letter E (because they're all technically encryption algorithms) with a numeric subscript that denotes the particular function being performed. Two versions of the E_2 function are used to generate the four link keys. The E_{21} function produces K_A and, indirectly, K_{AB}, where the former is associated with only device A and the latter with both devices A and B. The E_{22} function produces K_{init} and K_{master}. The output (the link key) of a particular key generation function can be expressed in terms of the function itself and its inputs; for example, $K_A = E_{21}(BD\_ADDR_A, RAND_A)$. The key generation function names and the formulas for the link keys derived from them are given in Table 9-2.

Table 9-2 Formulas for generating the link keys

Link Key Name	Symbol	Formula
Initialization key	K_{init}	E_{22}(PIN', L', IN_RAND)
Unit key	K_A	E_{21}(BD_ADDR$_A$, RAND$_A$)
Combination key	K_{AB}	E_{21}(BD_ADDR$_A$, LK_RAND$_A$) $\oplus$ E_{21}(BD_ADDR$_B$, LK_RAND$_B$)
Master key	K_{master}	E_{22}(RAND1, RAND2, 16)

Following our random number convention, the same IN_RAND at both devices is used to help generate the initialization key; device A's unit key is formed using a locally generated RAND$_A$; the combination key is generated with two LK_RAND numbers, one from device A and another from device B; and the master key is built from two random numbers, both created locally in the piconet master. Figure 9-5 shows how E_{21} is used to derive K_A, and Figure 9-6 shows how E_{22} is used to derive K_{init} and K_{master}.

When K_{init} is generated, the inputs to E_{22} are stamped with a device's identity by modifying the PIN and its length L into two different quantities, called PIN' and L', before sending them to the E_{22} algorithm. If the PIN is less than 16 bytes, then it is augmented by appending a device's BD_ADDR. Starting from the least significant byte, the BD_ADDR is appended to the PIN until PIN' either reaches 16 bytes total length or the entire BD_ADDR is appended, whichever comes first. This augmentation process doesn't really enhance PIN security because any BD_ADDR is assumed to be publicly known.

The device that provides the BD_ADDR for this operation is the one having a variable PIN, that is, one that can be programmed or entered by a

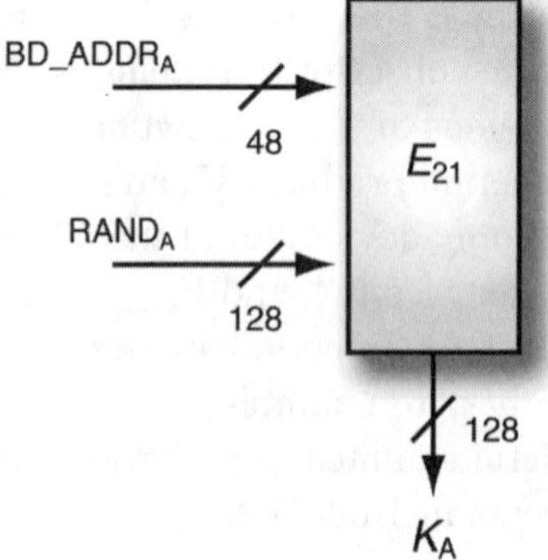

Figure 9-5 Deriving K_A from the key generation algorithm E_{21}

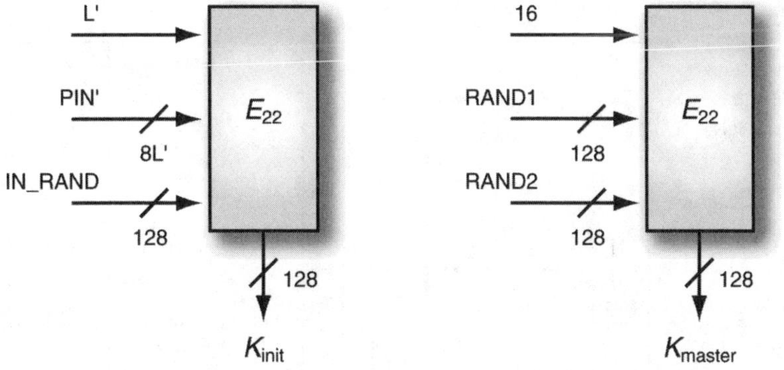

Figure 9-6 Deriving K_{init} and K_{master} from the key generation algorithm E_{22}

user. If both devices have a variable PIN, then the BD_ADDR is the one from the device that receives IN_RAND through a LMP_in_rand command during pairing. Notice that, before pairing occurs, both the master and slave link know each other's BD_ADDR. The master received the slave's BD_ADDR during INQUIRY, and the slave received the master's BD_ADDR during PAGE.

Pairing

Suppose two devices have never met before via Bluetooth. The pairing process begins when the initiator creates the initialization key $K_{init} = E_{22}(\text{PIN}', \text{L}', \text{IN\_RAND})$. The random number IN_RAND is passed to the responder with LMP_in_rand, so the responder can then calculate the same K_{init}, provided that it has the same PIN. This process is shown in Figure 9-7. K_{init} isn't normally used for authentication itself, but instead it provides a way to encrypt parameters that are exchanged during the next step in the authentication process.

Pairing must be accomplished when Bluetooth devices meet for the first time and require authentication, and it must be reaccomplished if their respective link keys have expired, been deleted, or don't match for some reason. If a device doesn't enable pairing and it receives a LMP_in_rand packet, then it responds with LMP_not_accepted with "pairing not allowed" as the reason. This doesn't necessarily imply that a connection cannot be made because a device with no authentication or encryption requirements has no need to implement pairing.

If one device has a fixed PIN (one that cannot be changed by a user), then the only way pairing can be successful is for the other device to have a PIN

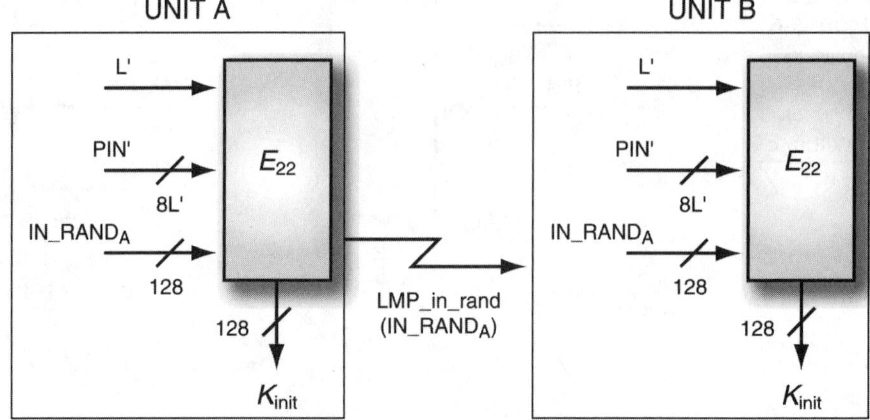

Figure 9-7 Pairing between two Bluetooth devices begins when each calculates the same K_{init}.

that can be made to match the fixed one. This is usually implemented by having a window appear for the user to enter a PIN for pairing. Of course, if both devices have different fixed PINs, then pairing is not possible.

After both devices create K_{init}, the pairing process continues through the creation of the link key and the use of it to perform an authentication.

Link Key Exchange

Once the link key(s) are generated by the two devices, they begin a process called *link key exchange*. This name is slightly misleading because the link keys themselves are never sent over the air. Instead, each device provides the other enough information that they will both generate the same link key, provided that they both began with the same PIN.

One of two different link keys can be used during the authentication process. The simpler one is called the *unit key*, designated K_A for device A. The unit key provides only a low level of security for devices that are typically inexpensive and may have limited resources, including memory. The *combination key* K_{AB} is functionally identical to K_A; that is, it is used for subsequent authentications and encryption key derivations. K_{AB} is much more secure, however, because its structure is based upon input from both devices, A and B. As such, K_{AB} is associated with the link between the two devices, not just one of the end points.

Unit Key Device A creates its own unit key, K_A, as shown in Figure 9-6 and generally uses the key for an extended period of time such as several sessions, or perhaps even for the life of the device. K_A is sent to another device, B, for use as a common link key by encrypting it with K_{init}, as shown in Figure 9-8. After K_A is transferred and decrypted, both devices can discard K_{init}. Notice that device B plays no part in the link key generation process other than decryption with its own K_{init}.

The unit key is then used for subsequent authentications between the two devices. A weakness in the unit key is that every device that wants to authenticate with device A uses the same K_A. We will discover the implications of this in a later section.

Combination Key Generating the combination key looks complex at first, but is actually straightforward and quite clever. Look again at Table 9-2, and you'll note that the combination key is just the two unit keys *exclusive ORed* (XORed) together. That is,

$$K_{AB} = K_A \oplus K_B \tag{9-1}$$

Because each device can create its own unit key and each device has the other's BD_ADDR, the only item that needs to be exchanged between the two devices is the respective random numbers used to create each other's unit key. This secure transfer is done by each device encrypting its respective LK_RAND with the current link key, which could be an initialization key, unit key, or combination key that was created earlier. The initialization key is used when the devices are creating a link key for the first time together. The unit key could be used if the link key is being upgraded to a

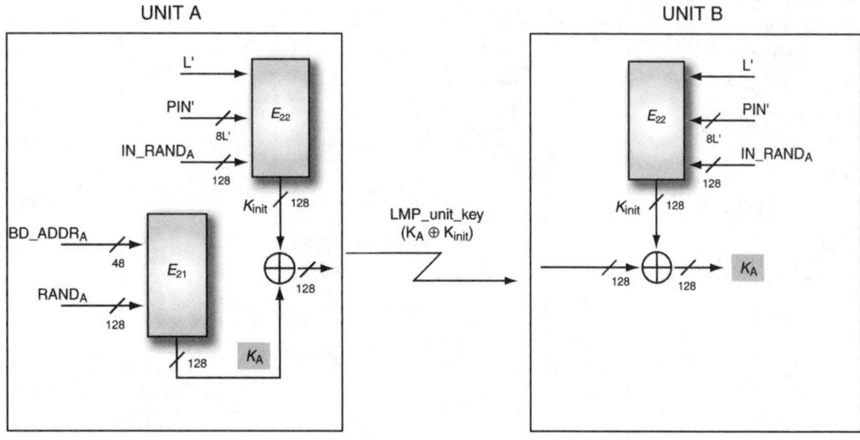

Figure 9-8 Transferring the unit key K_A from device A to device B by encrypting it with the initialization key K_{init}

combination key, and a combination key is used if the link key is being changed.

If the present link key is called K, then A sends $LK\_RAND_A \oplus K$ to B, and B sends $LK\_RAND_A \oplus K$ to A. Now each unit can decode the other's LK_RAND by performing a second XOR operation using the same link key, that is,

$$(LK\_RAND \oplus K) \oplus K = LK\_RAND \qquad (9\text{-}2)$$

The two devices are simply using a symmetric cipher. Clever, yes? Finally, both devices can create the other's unit key and hence K_{AB}. This entire sequence is shown in Figure 9-9. This figure is similar to one given in the Bluetooth specification, but we simplified the terminology. (This figure *does* look simple, doesn't it?) The exchange of the encrypted LK_RAND values is accomplished using LMP packets.

Selecting the Link Key to be Used for Authentication After two devices meet and begin the pairing procedure, both create a K_{init} for use in generating the link key for authentication. At this point, neither device knows if the two initialization keys are identical because they don't necessarily know if the PINs are identical.

Next, the link keys are created for use in authentication. If a device has unit key capability, then it will send LMP_unit_key to the other (see Figure 9-8), and if it has combination key capability, then it will send

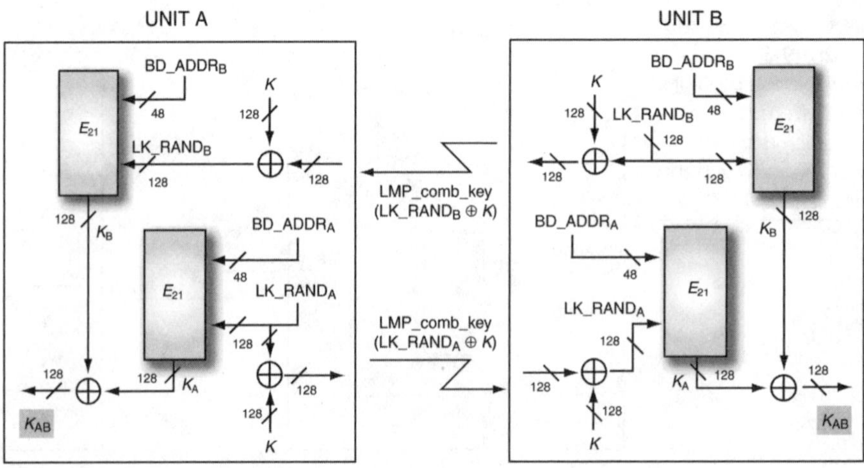

Figure 9-9 Creating the combination key K_{AB} from a previous key K. The previous key could be an initialization key, unit key, or old combination key.

LMP_combination_key to the other (see Figure 9-9). The manner in which the link key is generated follows this algorithm:

- If both units send LMP_unit_key, then the master's unit key will be the link key.
- If both units send LMP_combination_key, then the combination key will be the link key.
- If one unit sends LMP_unit_key and the other sends LMP_combination_key, then the unit key will be the link key.

At this point, each device has its respective link key, but they still don't know if the two link keys are identical. This is tested during the authentication process.

Authentication

Bluetooth authentication takes the form of a challenge-response action, where a verifier challenges the claimant to prove that it has the correct link key. If it does, then the verifier assumes that the claimant is who it claims to be. The problem, of course, is that the verifier cannot simply ask the claimant for its link key because this would cause a nasty security breach if an eavesdropper is present. Instead, the verifier sends the claimant a challenge whose response depends on the entire 128-bit link key value. It's important that neither challenge nor response provides any useful information to an eavesdropper.

The challenge is simply a 128-bit random number, and the response is produced by the output of an authentication function E_1, which is different from the algorithms used to generate the link key itself. Both verifier and claimant have the ability to run this algorithm and calculate a response to the challenge. This number is sent from the claimant back to the verifier using a LMP packet and checked against the verifier's own response for a match. If the two responses are the same, then the verifier can conclude that the claimant indeed possesses the correct link key.

Figure 9-10 shows how the challenge-response authentication is applied. The verifier places an authentication random number AU_RAND_A (the challenge), the BD_ADDR of the claimant, and the current link key into E_1, out of which comes a 32-bit result *signed response* (SRES), which is the desired response. The verifier sends AU_RAND_A to the claimant via a LMP packet, and the claimant performs the same operation using E_1, and sends its result SRES' back to the verifier using another LMP packet. If SRES = SRES', then authentication is successful.

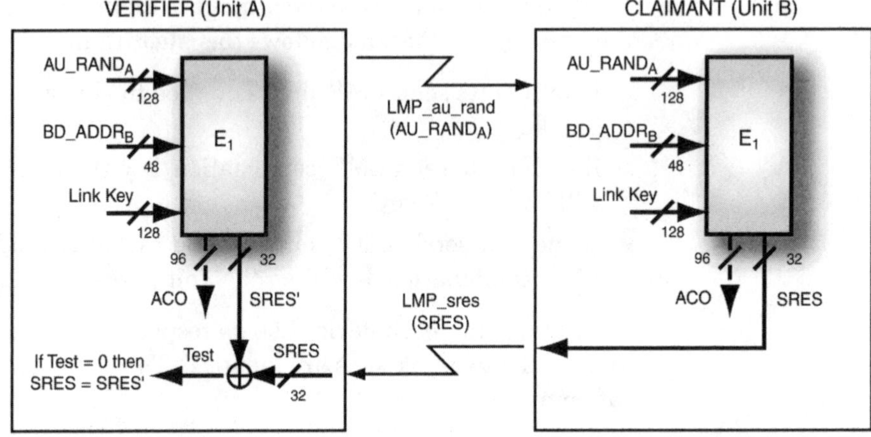

Figure 9-10 The challenge-response authentication process

Notice that SRES and SRES' are 32-bit numbers, not 128-bit numbers. This might seem odd at first because the chance of an intruder correctly guessing SRES is about one in 4 billion, significantly higher than if SRES were 128 bits long. However, if the authentication algorithm E_1(AU_RAND$_A$, BD_ADDR$_B$, Link key) always produced a unique SRES, then an intruder could conceivably intercept the various pairing and authentication parameters sent over the air and use them to discover the PIN, as we'll see in the section on Bluetooth piconet vulnerability. Limiting SRES to 32 bits provides reasonable protection against an intruder guessing its value, while reducing the chance that the PIN will be compromised if indeed SRES can be determined.

The authentication performed in Figure 9-10 is one way, and in some cases, that may be sufficient. However, if an intruder is attempting to gain access, then it will probably begin the communication session by posing as the verifier, and of course, it will always successfully verify the identity of the claimant. Therefore, the claimant shouldn't conclude anything about the verifier based only upon the fact that it asked for authentication. Instead, the claimant should also become the verifier and initiate its own authentication process.

If an authentication attempt fails, then the verifier should wait before initiating a new authentication attempt. The waiting interval should be extended for each failed authentication, up to some long maximum time. This prevents a fraudulent claimant from performing a sequential check of link keys or SRES values, or from forcing the verifier to devote significant time to the task of authentication. Each of these attacks will be examined more closely near the end of this chapter.

In addition to SRES, algorithm E_1 also produces a 96-bit result called the *authenticated ciphering offset* (ACO), which is used for creating an encryption key. This preserves the chain of events that we mentioned earlier, where each key is based upon knowledge of previous information that's (hopefully) not in an intruder's possession.

Encryption

If user information needs to be hidden from eavesdroppers, then encryption is employed. Three steps are needed to implement encryption. First, the same *encryption key* must be constructed at both ends of the link. Next, a *key stream generator* is loaded with the encryption key (and other information as well), and a sequence of what appears to be random bits are produced at its output. Finally, a *bit-by-bit XOR operation* is performed between the key stream and the plain text payload of each transmitted Bluetooth packet to obtain cipher text.

Only the payload of Bluetooth packets is encrypted, never the access code or header. Encryption is done after the FCS is attached to the payload, but before whitening or the FEC code are applied. Because the encryption algorithm is symmetric, decryption occurs at the destination simply by performing another bit-by-bit XOR operation between the cipher text and the key stream generator to produce plain text.

Generating the Encryption Key If the link key is either K_A or K_{AB}, then the encryption key is formed by placing a 128-bit random number called EN_RAND, the 96-bit ACO, and the 128-bit current link key into the encryption key generation algorithm E_3, which in turn produces a 128-bit encryption key K_C, as shown in Figure 9-11. Both master and slave possess all of these inputs to E_3 except EN_RAND, which is generated by the master and sent to the slave using a LMP_start_encryption_req packet. Every time encryption is activated by the LM, a new K_C is created. Of course, if the inputs to E_3 haven't changed, then the encryption key won't change either. The length of K_C out of algorithm E_3 is always 128 bits, but the key stream generator (see the following section) has the capability to shorten the key, if necessary, to the correct key length as given in Table 9-1.

If a combination key K_{AB} is the current link key, as will usually be the case if data security is really important, then each slave in the piconet will have a different K_{AB} and hence a different K_C. How then can the master encrypt broadcast packets? One method might be to derive another special key for use only with broadcast packets, but that would require slaves to

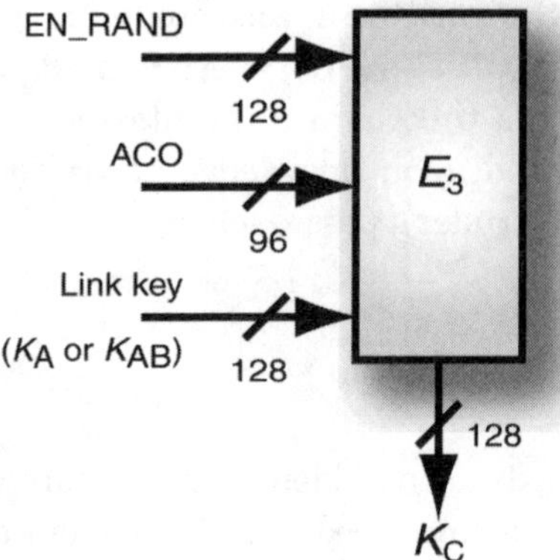

Figure 9-11 Generating the encryption key from the link key, a random number, and the ACO

have the capability to check the AM_ADDR in a packet header and quickly plug the correct key into the decryption engine. The severe timing restrictions placed upon the slave for such a task make this approach impractical.

Because each slave can have only one encryption key, the master can decide whether to send broadcast packets individually to each slave using different cipher streams from the different K_C or to replace the individual encryption keys with a common encryption key that is then used for all traffic, both broadcast and individually addressed. The particular encryption mode desired is communicated via a LMP_encryption_mode_req packet with the content field set to the encryption mode. These various scenarios are summarized in Table 9-3. Before a common encryption key can be generated, the master must first replace all of the different K_{AB} combination

Table 9-3 Encryption options for different link keys and traffic

Link Key	Broadcast Traffic Option	Individual Traffic Option	Encryption Mode
K_A or K_{AB}	No encryption	No encryption	0x00
K_A or K_{AB}	No encryption	Encryption with individual key	0x01
K_{master}	No encryption	No encryption	0x00
K_{master}	No encryption	Encryption with common key	0x01
K_{master}	Encryption with common key	Encryption with common key	0x02

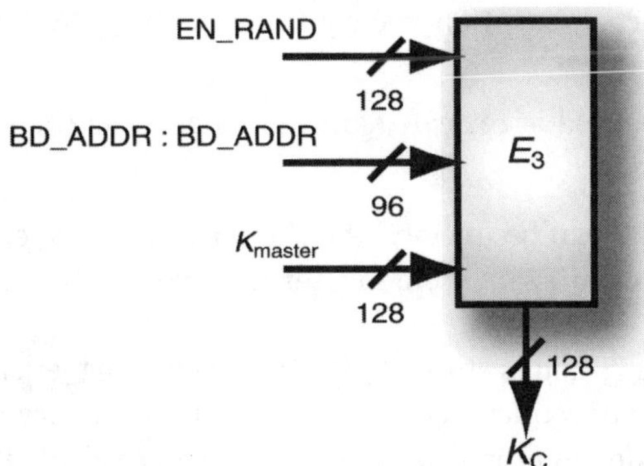

Figure 9-12 Generating the encryption key from the master key, a random number, and two concatenated copies of the master's BD_ADDR

keys with a common K_{master} key. When the link key becomes K_{master}, then the encryption key is derived according to Figure 9-12.

Generating and Distributing the Master Key The master key is generated according to Figure 9-6, where the inputs to key generation algorithm E_{22} are two different random numbers and a length field of 16, which is the length of the random numbers in bytes. Using two random numbers and the key generation algorithm, as opposed to having the master simply select a random number, is to mask possible correlations from a poorly designed random number generator. The master then transmits K_{master} to each slave in turn, encrypting it in a manner similar to the method for transferring $RAND_A$ and $RAND_B$ when K_{AB} was generated.

Next, the master and each slave perform a mutual (two-way) authentication to verify proper reception of K_{master}. The master then sends the same EN_RAND to each slave one at a time using LMP_start_encryption_req, and then encryption can begin for either *point-to-point* (individually addressed slaves) or *broadcast* packets. Note that, because every slave in the piconet has the same encryption key, there's no privacy guarantee by addressing a packet to only one slave.

Encrypting and Decrypting Data The master and slave initiate encryption through the exchange of several LMP packets. These usually occur in the following order:

1. LMP_encryption_mode_req is used to set the mode of encryption to none, point-to-point, or point-to-point and broadcast.
2. LMP_encryption_key_size_req negotiates the required size of the encryption key.

3. LMP_start_encryption_req contains the EN_RAND value and begins encryption.
4. LMP_accepted is returned, and this packet is the first one that's encrypted.

Steps 1 and 2 can be initiated by the master or slave, but the LMP packet in step 3 is always sent from master to slave and the response from slave to master.

After encryption is started with the LMP_start_encryption_req PDU, the payloads of subsequent Bluetooth packets are encrypted according to the selected mode, all of which are summarized in Table 9-3. Both master and slave place the master's BD_ADDR, the piconet's $CLK_{26\text{-}1}$, and K_C into key stream algorithm E_0. If K_C is required to be shorter than 128 bits, then its length is reduced using a special polynomial that is factory preset into E_0.

The key stream algorithm E_0 differs from the key generation algorithms in that E_0 produces a cipher bit stream that is used to encrypt or decrypt a packet payload bit by bit, as shown in Figure 9-13. Because the cipher is symmetric, the same algorithm and the same input values are used for both encryption and decryption.

The E_0 algorithm is reinitialized for each new packet, either sent or received. For each new packet to be encrypted or decrypted, the algorithm receives an updated $CLK_{26\text{-}1}$ so that the input parameters to E_0 are never identical for more than one packet. Thus, a new cipher stream is produced for each packet, and an eavesdropper will find it much more difficult to break the cipher.

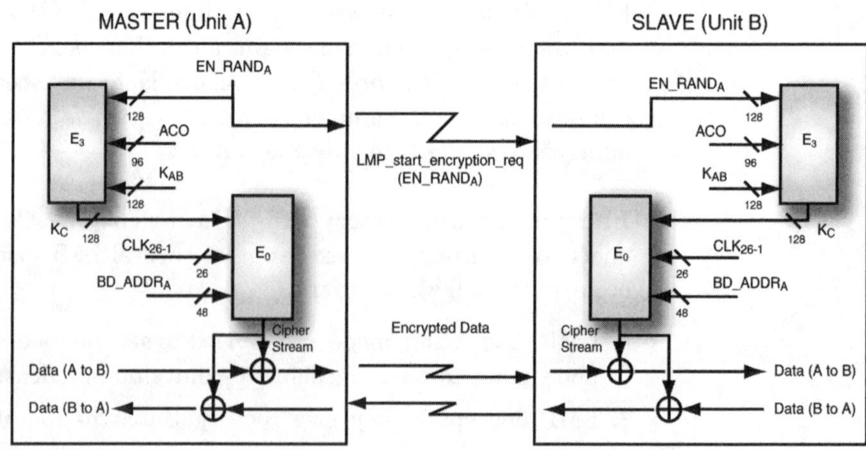

Figure 9-13 Encrypting and decrypting a Bluetooth packet payload. The encryption key can be derived from the unit key, combination key (illustrated), or master key.

In summary, Bluetooth encryption has the following strengths:

- The encryption key can be as long as 128 bits, compared to 40 or 64 bits for many other encryption implementations, greatly reducing the chance that an eavesdropper will be able to guess the key.
- The encryption key is built from several secret entities derived over a period of time, reducing the chance that it can be rederived by an eavesdropper.
- The cipher stream generator is reinitialized at the beginning of every transmitted packet, preventing an intruder from intercepting several copies of the same cipher stream encrypting different data.

Encryption is stopped when the master sends LMP_stop_encryption_req to the slave, and the slave acknowledges. The request to stop encryption PDU is itself encrypted, but the slave's LMP_accepted PDU is not.

Bluetooth Piconet Vulnerability

Now that we have a basic understanding of Bluetooth security, it's time to look at the various features from a critical point of view. Where are the weaknesses? How can they be exploited by an intruder? What countermeasures can be taken? These issues are extremely important because it's tempting to treat wireless like the older and more established wired communication systems, where eavesdropping and disruption are usually quite rare, at least for the smaller wired LAN. Intruding on a wired network almost always requires actual contact with the *physical layer* (PHY), and the proximity of an unauthorized person in such a context is often easy to detect.

Wireless is much different, however, due to the PHY encompassing a relatively large volume of space, even for short-range systems such as Bluetooth. We'll discover that the distance over which the Bluetooth network can be intercepted or disrupted can be significantly greater than that over which normal communication takes place. Also, an intruder can place a bug nearby to intercept and record Bluetooth activity over a long period of time. The bug can be retrieved later and its contents analyzed, perhaps using powerful computation techniques in an attempt to break the cipher on encrypted data.

Security within wireless networks has been a topic of great interest, especially because the encryption scheme used by IEEE 802.11b *Wireless*

Fidelity (Wi-Fi) has shown some weaknesses. Specifically, the Wi-Fi encryption key can be decoded by an intruder that merely listens to encrypted packets over a period of time. The irony is that the 802.11b encryption protocol is called *wired equivalent privacy* (WEP). Although the Bluetooth encryption algorithm has some significant differences from WEP, there's always the chance that a mathematical cracking code will be developed that can extract K_C or the cipher stream itself after a reasonable processing time.

We will examine Bluetooth security from the standpoint of the *disclosure threat*, in which an eavesdropper is present, the *integrity threat*, in which an unauthorized user is granted access to a target device, and the *DoS threat*, in which normal operation of the target piconet is deliberately disrupted. Incidentally, most of these activities are in violation of various laws in different countries. However, we will concentrate on the technical aspects, not the legal issues, of the threats and their countermeasures.

Disclosure Threat

When most people ponder the nature of security breaches in a communication system, they usually think of the disclosure threat first. This threat is especially serious; the presence of an eavesdropper is often not realized because of the physical separation of eavesdropping equipment from the communicating devices. Unencrypted transmissions are easy prey for an eavesdropper, but even encrypted packets can be recorded for later cryptographic analysis.

As we discovered in an earlier chapter, *infrared* (IR) transmissions have a built-in security feature in that their transmissions are usually confined to a single room, and their beamwidths are usually quite narrow as well. Interception of IR is difficult unless an eavesdropper has managed to penetrate the room itself to place a bug (hidden receiver). Obstructions, such as walls, are opaque to light but not to radio, so interceptions of RF energy can occur from much further away.

Range of Vulnerability to Eavesdropping Suppose a sensitive file is being transferred between two Bluetooth units. How far away can their transmissions be heard? We answer that question by determining the *range of vulnerability*, d_e, to eavesdropping. A receiver located within that range can intercept the transmissions from at least one of the communicating devices and most likely from the entire piconet as well. The range of vulnerability depends greatly upon path loss, transmitted power and antenna gain, and the eavesdropper's receiver sensitivity and antenna gain. All of these quantities are part of the link budget equation in the version first given as Equation 2-2 in Chapter 2, "Indoor Radio Propagation and Bluetooth Useful Range," which is repeated here as

$$P_{r(dBm)} = P_{t(dBm)} + G_{t(dB)} + G_{r(dB)} + 20\log\left(\frac{\lambda}{4\pi}\right) + 10n\log\left(\frac{1}{d}\right) \tag{9-3}$$

where $P_{r(dBm)}$ and $P_{t(dBm)}$ are the received and transmitted powers, respectively, in dBm, $G_{t(dB)}$ and $G_{r(dB)}$ are the transmit and receive antenna gains, respectively, λ is the carrier wavelength in meters, n is the path loss exponent, and d is the distance in meters separating the two antennas.

To adjust Equation 9-3 for applications to eavesdropping, we'll assume that the target piconet members are using antennas with 0 dBi of gain. Also, we'll include a term called *AF* (in dB) that can be used to adjust the link budget for an overall partition attenuation factor because most eavesdroppers will probably not be in the same room as the piconet itself. After performing substitutions for n and λ and rearranging terms in Equation 9-3, we also drop the dB subscripts and obtain

$$P_r = P_t + G_r - 40 - 10n\log(d) - AF \tag{9-4}$$

It should be obvious that the range of vulnerability for eavesdropping is the distance d between transmit and receive antennas, so we'll rename this variable d_e and again rearrange terms, giving

$$\log(d_e) = \frac{P_t - P_r + G_r - AF - 40}{10n} \tag{9-5}$$

It's probably reasonable to assume that an eavesdropper will use an ordinary Bluetooth chipset for intercepting transmissions, and these have a receive sensitivity of −70 to −85 dBm. However, a determined eavesdropper can improve both the receive sensitivity and receive antenna gain significantly. Receiver sensitivity levels can be enhanced by adding a *low-noise amplifier* (LNA) between antenna and receiver input. A LNA is commercially available that provides a gain of 16 dB at the input of a 2.4 GHz receiver. Also, a 2.4 GHz planar antenna can be readily obtained that has a gain of 20 dBi in the direction perpendicular to its surface. This antenna measures 33 × 33 × 7 cm, so it's a bit large for an eavesdropper to carry around unnoticed, but it could be used from a hidden, fixed location.

Recall that when various link budget assumptions were made in Chapter 2 for finding the useful range of two Bluetooth devices, we tried to be pessimistic, so that actual performance would hopefully exceed that which was calculated. In the case of range of vulnerability calculations, we would prefer to be optimistic instead to give us an idea of how far away an eavesdropper can be and still intercept our transmissions.

Eavesdropping Example Let's suppose Eve the eavesdropper wants to intercept Bluetooth transmissions occurring inside a building. To avoid detection, Eve works from her car in a nearby parking lot. Is the range of vulnerability sufficient for Eve to hear the transmissions? To find out, we'll assume that the path loss exponent is $n = 2.5$, corresponding to light clutter within the parking lot itself because Eve put her car in a favorable location, and there's 20 dB of additional attenuation on the signal as it exits the building ($AF = 20$ dB). Eve has at her disposal a commercial Bluetooth receiver with a sensitivity of -85 dBm preceded by a LNA with a gain of 15 dB, for a total receive sensitivity level of -100 dBm. For this application, Equation 9-5 simplifies to

$$\log(d_e) = \frac{P_t + G_r + 40}{25} \tag{9-6}$$

For the remaining two variables on the right side of Equation 9-6, let's assume that Eve has two antennas, one with a gain of 0 dBi (omnidirectional) and another with a gain of 20 dBi (directional). The omnidirectional antenna has an advantage because it doesn't need to be aimed. The directional antenna, while requiring precise aiming, has the advantage of both extending eavesdropping range and reducing interference from 2.4 GHz transmissions coming from undesired directions. We'll also assume that the target piconet members use a transmit power of either 0 dBm (class 3) or 20 dBm (class 1). Table 9-4 gives the range of vulnerability for these various combinations.

Table 9-4

Range of vulnerability for eve the eavesdropper on a target piconet

Eve's Antenna Gain	Target Piconet TX Power	Range of Vulnerability
0 dBi	0 dBm	40 meters
0 dBi	20 dBm	250 meters
20 dBi	0 dBm	250 meters
20 dBi	20 dBm	1,580 meters

It should be distressingly obvious that Eve will have no problem intercepting Bluetooth transmissions from a considerable distance away, especially if she uses a gain antenna. If Eve runs a commercially available software package intended for Bluetooth protocol analysis, she can see all the piconet activity on a laptop computer screen (see Figure 9-14). Do you think it's important to encrypt Bluetooth transmissions?

Defenses Against Eavesdropping Defenses against eavesdropping are based upon countering the various steps that an eavesdropper must take before the target piconet traffic can be intercepted and used. These steps are

- Placing a receiver within the range of vulnerability of the target piconet
- Synchronizing on the target piconet's hopping sequence
- Recording each packet sent by the piconet members and assembling the message

Some of these activities can be defended against better than others by the target devices.

Reducing the Range of Vulnerability The first defense against an eavesdropper is to prevent him/her from entering the range of vulnerability.

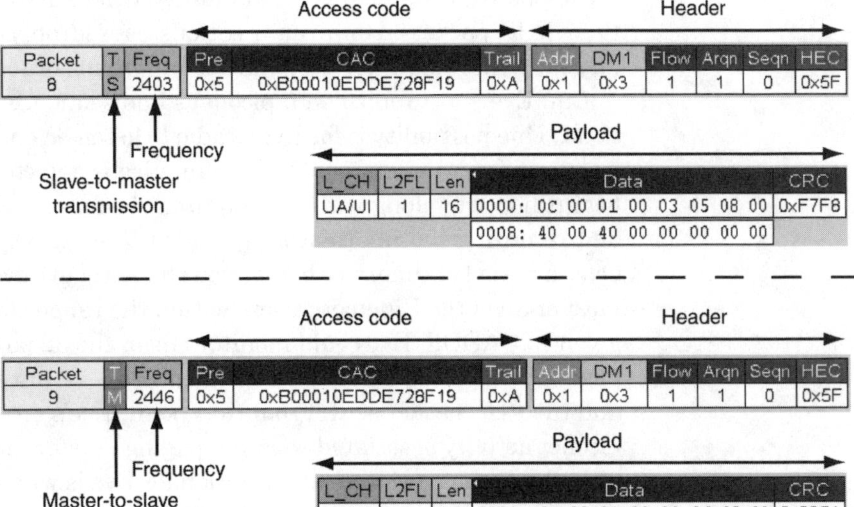

Figure 9-14 Bluetooth piconet activity can be displayed on a laptop screen. (Source: CATC)

Although the ranges given in Table 9-4 are discouragingly long, devices in the target piconet can decrease this range substantially by using only enough power to reach other piconet members. The Bluetooth specification enables transmit power control to be optionally employed to levels below 0 dBm for any transmitter class, with a recommended minimum of −30 dBm (one microwatt) for a nominal range of 1 meter. Even if Eve the eavesdropper uses her supercharged Bluetooth receiver and 20 dBi gain antenna, the range of vulnerability, under our other assumptions, is only 16 meters when target piconet TX powers are −30 dBm. Therefore, if Bluetooth is to be used for sensitive file transfers, make sure that automatic TX power control is employed down to −30 dBm, and place the two communicating devices as close to each other as possible.

Other more esoteric methods of reducing the range of vulnerability could be employed, such as confining the Bluetooth users in a room that's shielded from RF emissions, using the devices in a large area such as a field where any eavesdroppers can be spotted, or working in isolated areas where others aren't likely to be nearby. Perhaps we'll see more security-conscious Bluetooth users who are out standing in their field.

Hop Synchronization by an Eavesdropper Synchronizing on the target piconet can provide some challenges for the eavesdropper, but is not particularly difficult. Commercial testing and monitoring products for Bluetooth protocol analysis usually require a brief contact with piconet members to extract addresses and hop sequence information. This can be accomplished through a general inquiry, where all devices in range return their FHS packets. However, a serious eavesdropper will never transmit anything that might disclose his/her intentions or location.

Can Eve synchronize on a piconet's hop sequence without transmitting at all? One possibility is for Eve's radio to listen on one of the 32 inquiry hop frequencies for an inquiry. Once an inquiry is detected, then Eve's receiver begins hopping along with the inquirer, checking each response frequency and recording the information provided in each responder's FHS packet. Over a period of time, Eve has a good chance of discovering the identities of most or all of the Bluetooth users within the range of vulnerability. For any given BD_ADDR, Eve could monitor one of the 32 paging frequencies for a page and then hop along with the paging device until a response is heard, and the FHS packet sent. From that point, Eve's receiver can intercept all piconet activity associated with the paging master and paged slave.

Perhaps the most difficult situation for Eve is when no inquiries occur in the target area due to every target device being configured as *private*. This situation requires that Eve monitor traffic on various hop frequencies over a longer period of time, perhaps intercepting a page or FHS packet every now and then. A single page packet contains only the 24-bit LAP of the paged device's BD_ADDR, so there's not enough information there for Eve to derive the page hop sequence: She also needs the 4 LSBs of the UAP. However, a page (ID) packet informs Eve that a page process is taking place, a process that successfully concludes with the transmission of a FHS packet and response. If Eve's receiver happens to intercept the FHS packet, then the included BD_ADDR and CLK information can be exploited by Eve for monitoring the piconet itself.

A *very* determined eavesdropper can employ several receivers in parallel to increase the probability of intercepting useful information. Indeed, 79 receivers and associated processors will intercept all Bluetooth activity within the range of vulnerability. The bill of materials for such a network of receivers will be quite low due to the low cost of Bluetooth chipsets.

The point of this discussion is that an eavesdropper who is within the range of vulnerability for a long period of time has a good chance of finding the active Bluetooth devices.

Message Encryption Finally, users in the target piconet can encrypt their messages in an attempt to guard against a successful intrusion upon the piconet by an eavesdropper. Encryption can take place either with the methods built into Bluetooth itself or at the application level with any type of encryption desired. As we'll discover in the following section, Bluetooth encryption has a few weaknesses that could perhaps be exploited by an eavesdropper. Although many attacks on the encryption system could be categorized as disclosure threats, other attacks require connecting into the piconet and gaining unauthorized access by misleading another piconet member. We'll therefore postpone a discussion of Bluetooth encryption shortcomings until the following section.

Integrity Threat

Unlike eavesdropping, where an attacker simply listens for desirable information, the integrity threat involves unauthorized access by an attacker who actually connects into the piconet. Once access is granted, the attack can proceed in many different directions: Sensitive files can be retrieved, deeper access into the network can be attempted, or misinformation can be transferred into the network. In many cases the goal of such an attack is the retrieval of information, which one could argue is a disclosure threat. However, because we assume here that this attack is carried out by actually connecting to a target device rather than just listening, we will consider it to be an integrity threat instead.

One basic approach to gaining unauthorized access to another Bluetooth device is for the attacker to assume the identity of a legitimate user. This is often called *spoofing*, so our attacker will be Spiff the spoofer. Several methods can be employed by Spiff in his attempt to access the piconet. Among these are

- Exploiting a weakness in the PIN and initialization process
- Exploiting a weakness in the unit key
- Attempting to pair repeatedly with different authentication responses
- Using a *reflection attack* to obtain the correct authentication response
- Acting as the verifier in a one-way authentication
- Using a stronger RF signal to displace an active piconet member

Once access is granted, the intruder may be able to exploit weaknesses in Bluetooth encryption to retrieve sensitive information and perhaps even introduce damaging information or software into the piconet.

PIN and Initialization Weaknesses Recall that when two Bluetooth devices meet for the first time and begin the pairing process, they both create an initialization key K_{init} by running the E_{22} key generation function on PIN', L', and IN_RAND. The value for PIN' is the PIN augmented by a BD_ADDR, L' is the length in bytes of the augmented PIN, and IN_RAND is the initialization random number sent from the initiator to the responder. Out of these quantities, only the PIN is unknown to the eavesdropper. Once the PIN becomes known, PIN' and L' can be calculated because we assume that any BD_ADDR is public knowledge.

The weakness in the initialization process is rooted in PIN selection. Suppose Spiff the spoofer is able to intercept IN_RAND, the link key exchange parameters, AU_RAND, and the correct SRES value(s) from two target devices. Because he has access to the various key generation algorithms and both BD_ADDR values, Spiff can simply test different PIN values in his own E_{22} for a trial K_{init}, and then run through the link key generation and authentication sequences until an SRES match occurs. Note that SRES is only 32 bits (not 128 bits) long, so an SRES match doesn't necessarily guarantee that Spiff has discovered the PIN, but the chances are fairly high, especially if the PIN is short.

If Spiff discovers the PIN, then he can create any of the link keys assuming that he can intercept the respective random numbers used in their generation. That's a significant security breach! However, if Spiff isn't able to intercept IN_RAND, then the PIN becomes hidden behind a 128-bit random number K_{init} that's essentially impossible to guess.

Of course, if the PIN itself is a random 16-byte (128-bit) value, then it has the same protection from this attack as any other 128-bit random number. As we'll discover in the section on encryption weaknesses, that protection is considerable. Also, longer PIN values can significantly increase the chance that an incorrect PIN will produce a SRES match. On the other hand, if the PIN consists of merely four decimal digits, then there are only 10^4 possible PIN values (and, hence, only 10^4 possible K_{init} values for a given BD_ADDR), and Spiff can test those in a very short time, such as 10 seconds at a rate of 1,000 trials per second.

Defending against this attack is accomplished by selecting a longer PIN composed of alphanumeric characters instead of just numbers. For example, if a six character PIN is composed of case-sensitive alphanumeric characters, then even if Spiff knows the PIN length, he will need 1.8 years to test all possible PIN values at a rate of 1,000 per second.

Unit Key Weaknesses Unlike the combination key K_{AB}, which is derived from secret information provided by both devices A and B, the unit key K_A is derived from secret information at device A alone. Now suppose device B and C both possess K_A from prior communication with A. If A and B connect at some later time, C can intercept their AU_RAND and derive the ACO from E_1(AU_RAND, BD_ADDR, K_A). Next, C can intercept EN_RAND and derive the encryption key from E_3(EN_RAND, ACO, K_A). Finally, C can create the cipher stream from E_0(BD_ADDR, CLK, K_C) and decrypt all messages between A and B.[1] Obviously, a combination key, not a unit key, should be used for authentication and subsequent encryption whenever sensitive data is being exchanged.

Authentication Weaknesses The Bluetooth authentication process always uses a 128-bit link key in the E_1 authentication algorithm (see Figure 9-10), so a sequential search attack by an intruder posing as a legitimate claimant would require a very large number of authentication attempts before there can be any hope of success. However, sometimes such an attack is motivated by an attempt to occupy the verifier with processing so many false SRES values that its services are severely limited or denied altogether to legitimate users. Other variants of this so-called DoS attack will be examined in a later section.

Fortunately, it's relatively easy for a verifier to recognize such an attack due to the many incorrect SRES values emanating from the same

BD_ADDR. Both sequential authentication searches and DoS attacks using authentication can be countered if the verifier waits before requesting another authentication after a failed attempt. The waiting period should increase exponentially (such as twice as long) after each failed authentication, up to some maximum value (such as 30 seconds).

Another authentication attack on a verifier is called the reflection attack. This is accomplished in the following way:

1. An intruder claims to be legitimate device A (let's name the intruder A') and requests authentication from target device B, the verifier.
2. The verifier B sends claimant A' an AU_RAND.
3. Before answering B, device A' assumes the role of verifier for the second step in a two-way authentication and sends (reflects) to B the same AU_RAND.
4. Device B as the claimant replies with SRES, which is the correct response to the AU_RAND challenge.
5. Device A' then returns (reflects) the same SRES back to device B, which as verifier finds a match and successfully authenticates A'.
6. Device A' "authenticates" device B, completing the process.

There are two reasons that the reflection attack cannot work with Bluetooth authentication. First, there is no preemption in the Bluetooth authentication process; that is, step 3 cannot take place until the first authentication with A' as claimant and B as verifier has been completed. Second, recall from Figure 9-10 that SRES = E_1(AU_RAND, BD_ADDR, K), where the BD_ADDR is that of the claimant. Therefore, the SRES provided by B in step 4 isn't necessarily the SRES that A' needs for its attempted authentication to succeed unless A' also assumes the BD_ADDR of device B, in which case B is immediately alerted to the attack.

Access via RF Capture The goal of an eavesdropper is usually to obtain information that is being passed from one target device to another. However, the eavesdropper may need to monitor communications for a long time before the desired information appears over the Bluetooth link. The longer the eavesdropper remains in position, the more likely the attack will be discovered. The desired information may be obtainable more quickly through an integrity attack, where the intruder actually connects with the target piconet to actively obtain the desired information. However, as we've already discovered, Bluetooth employs several formidable authentication obstacles that must be circumvented by the intruder before access is granted. Can both of these attack mechanisms be somehow combined for the intruder to obtain the desired information without being hindered by authentication? The answer is yes, possibly, through using the method of *access via RF capture*.

Recall from Chapter 3, "The Bluetooth Radio," that the radio is required to meet a cochannel C/I specification of 11 dB (see Table 3-5). This means that if a desired signal is at least 11 dB stronger than an interfering signal on the same channel at the Bluetooth receiver, then the desired signal can be detected with a BER not greater than 10^{-3}. In other words, the desired signal has *captured* the channel. Unfortunately for the principles of truth and justice, Spiff the spoofer can exploit capture for his own purposes by transmitting a strong signal into the piconet with timing such that he can replace a legitimate piconet member during almost any time that member is active. The range at which Spiff can use this technique will be explored in the section on DoS attacks.

Here's how Spiff the spoofer uses access via RF capture to download a sensitive file:

1. Spiff eavesdrops until he can identify the BD_ADDR of a server S on which the target file F is located. He also identifies the BD_ADDR of a device A that seems to have the proper authorization to access that file.
2. Spiff eavesdrops long enough to learn how A accesses the files on S, and he gathers any other intelligence that he may need to eventually access file F himself.
3. When he is ready to carry the attack to the next level, Spiff waits until A connects to S and is properly authenticated. Device A may have no intention of downloading file F during this particular session.
4. At an appropriate time during the communication session between S and A, Spiff captures A's channel by assuming the identity of A and transmitting signals that are at least 11 dB stronger at S's receiver than what A was sending.
5. Continuing to pose as A, Spiff then requests the desired information from S.

There are several points to note about Spiff's attack. First, he doesn't necessarily need to perform an authentication with S because that was already accomplished by A. Second, although it's possible that A will think something is amiss when S begins its strange behavior in response to Spiff's capture, A will never hear Spiff because Spiff transmits only in A's transmit time slots. Furthermore, A may not be able to communicate any of its concerns to S due to Spiff's capture of the link. As a result, there's a good possibility that the attack won't be discovered until it's too late, and the target file has already been compromised. For example, if Spiff cleverly sends his first capture transmission in the same slot that A transmits LMP_detach, then even without a response from S, A may detach after $6 \times T_{poll}$ slots anyway according to the requirements of the Bluetooth specification, very possibly ignoring altogether the fact that S is now communicating with Spiff masquerading as A.

Several approaches can be taken by a Bluetooth device to defend against this type of attack. These include

- Encrypting all sensitive material
- Requiring reauthentication prior to every request for sensitive data, not just at the beginning of an ACL or L2CAP session
- Watching for a sudden increase in the RSSI, especially from devices that won't respond to LMP_decr_pwr_req (request to decrease TX power) and reauthenticating, disconnecting, or taking other appropriate action

Finally, we'll mention in passing that a particular transmitter usually has a unique RF signature, or fingerprint, consisting of various amplitude, frequency, and phase characteristics as it turns on and off, and during symbol modulation. If security is of critical importance, then the affected devices could be equipped with signal processing capability that can check for the proper RF signature from other devices attempting to connect and reject any attempts from units having alien fingerprints.

Encryption Weaknesses As we pointed out earlier, Bluetooth encryption has several strengths, but there are some weaknesses as well. Perhaps the most significant weakness occurs when an encryption key shorter than 128 bits must be used. When two devices negotiate the encryption parameters, the key length is restricted by the device with the shorter maximum value. Therefore, if one device can only support a 64-bit encryption key, then the other device must accommodate that length or encryption cannot be enabled.

Suppose an eavesdropper wants to perform a sequential search on the encryption key, hoping to eventually find its correct value. This can be done by recording several encrypted packets, and then attempting to decrypt them using cipher streams generated by the various trial keys. Although the cipher stream generator has the master's BD_ADDR and piconet CLK as additional inputs, both of these values are assumed to be known by the intruder. In particular, the intruder can stamp each arriving packet with the associated CLK before it is stored. If there's any plain text in the tested packet, then the correct cipher stream will unveil it.

Let's assume that all of the tested packets have plain text in them, so the intruder's key testing algorithm will realize its success simply by finding a string of words. How long will it take for the correct key to be found using a sequential search? The time required to find the correct key depends upon the key length, how quickly each trial key can be tested, and how many of the possibilities must be checked before success. For a key that is L bits long, the intruder must perform an average of 2^{L-1} tests before the correct key is discovered. Let's suppose that the eavesdropper can test $2^{10} = 1{,}024$ keys per second of processing time on one computer. The second column in Table 9-5 shows the average time required to find the correct key for several different key lengths. A clear advantage is obtained by using an encryption key that is at least 40 bits long.

For a more efficient sequential key search, the intruder can use a method called *parasitic computing*.[4] This involves embedding a computation problem into TCP/IP packets and exploiting the error-checking feature associated with web request messages on the Internet. When other computers receive these messages, they will return an acknowledgment if the correct result is included in the message. Although several hurdles must be overcome before this method can become practical for use in a sequential key search, for the sake of our pessimistic approach, let's assume that an intruder can employ parasitic computing to increase the key search speed

Table 9-5

Average sequential search times for discovering an encryption key

Key Length (Bits)	Average Search Time at 2^{10} Trials per Second	Average Search Time at 2^{30} Trials per Second
8	0.125 seconds	120 nanoseconds
16	32 seconds	30 microseconds
32	24 days	2 seconds
40	17.4 thousand years	8.5 minutes
48	4.5 million years	36.4 hours
64	293 billion years	272 years
128	A really long time	Also a really long time

to 2^{30} trials per second, which would be the equivalent to about one million computers, each performing about one thousand searches per second. The average time required to search keys of varying lengths is given in the third column of Table 9-5. In this case, a key that is at least 64 bits long has reasonable protection from this attack.

Because sequential search is impractical for any encryption key of 64 bits or longer, other techniques must be used instead. We discussed one of these, working backward from intercepting IN_RAND and other over-the-air transmissions to finding the PIN, in the "Bluetooth Piconet Vulnerability" section. Other methods involve exploiting mathematical weaknesses in the key generation or cipher stream algorithms themselves. This was the basis for the successful attack on the WEP encrypting key for 802.11b Wi-Fi. The Bluetooth cipher stream process E_0 apparently has a weakness in which the 128-bit key length can be discovered in about 2^{64} trials, but this requires access to different data encrypted with the same cipher stream, something that Bluetooth doesn't provide due to the reinitialization of E_0 with a different CLK prior to each baseband packet transmission.[1]

Another possible attack on the encryption process can occur if an intruder actually connects to a target device and successfully negotiates any authentication firewalls. Now the intruder asks for a sensitive file, which the target device encrypts using a cipher stream that is assumed to be unknown to the intruder. During encryption parameter negotiation, the intruder insists that the encryption key can be only 8 bits long. When the encrypted files arrive, the intruder can easily discover the encryption key by using an elementary sequential search. To prevent this attack, a device should refuse an encryption request, and hence the file transfer, if the other device's maximum key length capability is below some minimum value.

Even if an intruder cannot decrypt files, there is still information to be gained by merely intercepting Bluetooth packet access codes and headers from the target piconet. These are never encrypted, so simply discovering that two target devices are communicating may be of interest to the listener.

Finally, the intruder can acquire one of the target devices through theft and use it to access the desired information. Devices that provide automatic authentication and encryption using stored parameters are especially vulnerable to this attack. Defenses include requiring a user-provided PIN for authentication, and perhaps another for encryption, and physically securing devices when they're not in use.

Clandestine Software Modifications Up to this point we've discussed how an intruder could connect to the piconet with a misleading identity and retrieve sensitive information. This is one possible outcome of a successful integrity attack. Another possible outcome is the transfer of misinformation from the intruder to a target device. Although the type of misinformation sent by the intruder can span a wide range of possibilities, if the intruder has the ability to modify the Bluetooth protocol stack on another device, then the following nasty things could happen:

- Authentication and/or encryption could be disabled without user notification.
- Files could be accessed and transferred without user notification.
- The Bluetooth device could be rendered unable to power down, enabling prolonged attacks to occur.
- Safeguards, such as timers, could be modified or disabled.

The previous is only a partial list of disruptions, but the effect could be extremely serious. Some of these require access by an intruder to Bluetooth firmware, which is part of the LM on the Bluetooth module, and others can be accomplished by modifying the portion of the stack residing on the host. For reduced cost, it's likely that the LM and perhaps even the LC itself, will eventually reside on the host as part of a software protocol stack. Code residing in this location is arguably more vulnerable to clandestine modification. Can you say virus?

Denial-of-Service (DoS) Threat

The classic DoS threat involves disruption of the piconet such that legitimate throughput is either slowed considerably or shut down completely. This threat can be accomplished in several different ways and at different levels of the Bluetooth protocol stack. At the physical layer, an intruder can either capture the channel from a legitimate piconet member or jam the piconet entirely. Attacks on higher protocol layers try to exploit some of their characteristics in an attempt to occupy the attention of one or more members of the piconet such that they're unable to service other devices in a timely manner.

Disrupting the PHY Everyone who has used wireless devices has probably been inconvenienced by interference, whether on a cordless phone at home, cell phone on the road, or even manifested by poor television reception. Deliberate interference is called *jamming*, and this attack can be detrimental on any wireless device unless countermeasures are taken. Bluetooth already incorporates a modest antijam mechanism through

FHSS, so a few channels can experience interference without much detriment to throughput. Of course, the hopping pattern can be derived by anyone who can determine the master's BD_ADDR and CLKN through interception of its FHS packet, so a determined jammer can hop from channel to channel along with the piconet members, disrupting communications thoroughly if the jamming signal is strong enough. Alternatively, the jammer can transmit a wideband signal to disrupt the entire 2.4 GHz band at once.

As we learned earlier, service can also be penetrated and even disrupted when an intruder captures the channel used by one of the piconet members and spoofs that member's identity. We will examine both of these techniques and calculate how far an attacker can be from the target piconet and still carry out an attack.

Range of Susceptibility to Jamming Suppose a jammer wants to disrupt a target Bluetooth piconet by hopping with its members and sending random data in every time slot. If the jamming signal is strong enough to significantly interfere with the reception of desired signals, then various timeouts will trigger within the piconet members, and they will eventually disconnect without having accomplished their task. This is a classic DoS attack, but how close does the jammer need to be before its signal is strong enough?

We turn to Equation 9-4 as a starting point. This equation gives the received signal power in terms of several other factors. Unlike the eavesdropping situation where a range of vulnerability was calculated based upon the link budget from the target piconet to the eavesdropper, determining the range of susceptibility to jamming must consider the reverse link budget from jammer to target receiver. Assuming once again that the target piconet members are all using omnidirectional antennas with 0 dBi gain, we can solve for the target piconet's range of susceptibility d_j to the jammer as

$$\log(d_j) = \frac{P_t + G_t - P_r^{(j)} - AF - 40}{10n} \tag{9-7}$$

where P_t and G_t are the jammer's transmit power and transmit antenna gain, respectively, and $P_r^{(j)}$ is the minimum power necessary at a target piconet receiver for jamming to occur. The remaining quantities are identical to those in Equation 9-4.

According to Table 3-5 in Chapter 3, the Bluetooth radio must have a C/I of at most 11 dB for certification. This means that piconet disruption could begin occurring when the jammer's signal is 11 dB weaker than a desired signal at a target receiver. For the jammer to be effective, though, we will assume that the C/I must be at least 0 dB; that is, the jammer's power is at least equal to the desired signal's power at the target receiver. Although these desired signal levels aren't necessarily known by the jammer, it can increase the range of susceptibility by using a power amplifier with the Bluetooth transmitter. These amplifiers are commercially available with power outputs of 1 W and higher.

Jamming Example Suppose Jim the jammer wants to disrupt a piconet in a building from his car in the parking lot. We'll assume that Jim can intercept the master's FHS packet and extract the parameters necessary for him to hop along with the target piconet members, jamming their communications until they give up and disconnect. To find the target piconet's range of susceptibility to jamming, we'll assume that the path loss exponent is $n = 2.5, AF = 20$ dB, and Jim uses a directional antenna with a gain of 20 dBi. For this application, Equation 9-7 simplifies to

$$\log(d_j) = \frac{P_t - P_r^{(j)} - 40}{25} \tag{9-8}$$

For the remaining two variables on the right side of Equation 9-8, let's assume that Jim can transmit with either of two power levels: 20 dBm or 30 dBm. We'll also assume that the target piconet receivers all have a desired signal power level of either −60 dBm or −40 dBm. These power levels often fall within the "Golden Receive Power Range" discussed in Chapter 3 for Bluetooth links that have power control. Jim needs to create a jamming signal power at least that high at the target receiver to be successful. The range of susceptibility to jamming under each of these conditions is given in Table 9-6.

We can conclude from the values in Table 9-6 that the range of susceptibility to jamming is not nearly as high as the range of vulnerability for eavesdropping under similar path conditions. If jamming is to be successful from a reasonable distance, the jammer will require both a gain antenna and a power amplifier that can operate properly with a transmitter hopping at 1,600 channels per second.

Wideband Jamming Suppose Jim wants to jam the entire 2.4 GHz band at once using a wideband signal. In this way, Jim can shut down all piconets within the range of susceptibility, and he no longer needs to use a Bluetooth

Table 9-6 Range of susceptibility for Jim the jammer on a target piconet

Minimum Jamming Signal Power at Target Receiver	Jim's TX Power	Range of Susceptibility
−40 dBm	20 dBm	6.3 meters
−40 dBm	30 dBm	16 meters
−60 dBm	20 dBm	40 meters
−60 dBm	30 dBm	100 meters

chipset to generate hopping patterns. To cover all 79 channels with the same power density as before requires a transmit power level at least 79 times greater and probably more if the channels at the band edges are to be jammed as effectively as the center channels. Wideband jamming would be difficult to accomplish by a parking-lot jammer. Instead, the jammer will probably need to plant a device physically close to the target piconet.

RF Capture Finally, let's look at the feasibility of executing an integrity threat by using the RF capture as we discussed earlier. To do this, Spiff the spoofer will need to produce a signal at the target receiver that's at least 11 dB higher than the desired signal to effectively capture the channel. If Spiff uses a TX power of 30 dBm and a 20 dBi gain antenna, then he can capture a −60 dBm desired signal at a maximum range of only 36 meters. Still, this distance is far enough under many conditions to enable Spiff to conceal his location.

Defending Against PHY Disruptions The best means of defense against a disruption in the PHY is to ensure that adequate signal strength exists at each receiver in the piconet. When power control is implemented, the devices could set their RSSI near the upper end of the golden receive power range, at least over shorter distances. According to Figure 3-10 in Chapter 3, the upper threshold for this power is −30 dBm, a strong signal indeed. Jim the jammer, with his 1 W TX power and 20 dBi gain antenna, would need to be within 2.3 meters of the target receiver to jam it effectively under these conditions. However, when a piconet reduces its range of susceptibility to jamming through increased TX power, it also increases its range of vulnerability to eavesdropping.

Other defenses against jamming and capture include monitoring the area within the range of susceptibility for unauthorized personnel (that's military talk) and using a 2.4 GHz receiver to check for jamming signals. If such a signal is found, then *direction-finding* (DF) equipment, or perhaps even a physical search, could be used to pinpoint the location of the unwanted signal.

Disrupting Protocols above the PHY Aside from occurring at the PHY, DoS attacks can also take place within the framework of protocols within the data link layer and higher. Here is a list of possible attacks on some of the higher protocols and defenses against them:

- **Big NAK attack** An intruder can connect to the target device and request some information. When the transfer begins, the intruder always sends a NAK in response to the baseband packet, possibly putting the target device into an endless transmission loop and taking bandwidth away from other piconet users. The target's supervision timeout won't be triggered because the timer is reset each time the intruder sends the NAK. The intruder could prevent an L2CAP flush timeout from occurring by requesting a reliable channel, which designates the default infinite flush timeout. Defenses include disconnecting if excessive NAK responses occur and then placing the intruder's BD_ADDR into a database of suspicious devices.
- **BD_ADDR duplication** An intruder could plant a bug that spoofs the BD_ADDR of a target device. Now whenever users attempt to contact the target device, either the bug responds (if it hears the page first) or both units respond (if their respective CLKN and page scan parameters are identical) and jam each other, thus denying access to the legitimate device. Defenses include using a portable Bluetooth unit to find the area where the bug captures the channel, and then physically searching for the bug.
- **SCO attack** An intruder can set up a HV1 SCO link with a target device, thus removing it from service altogether because HV1 uses every time slot. Defenses include disabling SCO links on devices that don't require real-time two-way voice, and requiring authentication before a SCO link can be established on devices with SCO capability.
- **L2CAP guaranteed service attack** An intruder can request that the target device set up a L2CAP link with guaranteed service and highest possible token (data) rate. The LM under these conditions is supposed to refuse any further ACL channels and disable periodic scans to support the requested data rate, thus denying access to the target device by others. A similar attack could be mounted by requesting minimum latency, which requires the maximum possible

polling rate of every other time slot, again denying access by others. Defenses include requiring authentication for any L2CAP request, especially those that appear to be part of a DoS attack.

- **Battery exhaustion attack** Many of the previous attacks that occupy so much of a target device's time could also drain its battery rather quickly if that's how the target device obtains its power.[1] Defenses, of course, are the same as those listed for the different attacks.

Summary

Threats to distributed networks such as Bluetooth fall into three categories: disclosure, integrity, and denial-of-service attacks. Bluetooth security attempts to counter these threats. Devices can be either silent (no page or inquiry scans), private (no inquiry scans), or public (both page and inquiry scans). Security levels are Mode 1 (nonsecure), Mode 2 (service-level enforced via L2CAP), and Mode 3 (link-level enforced).

There are three general aspects to Bluetooth security: authentication, authorization, and encryption. Authentication attempts to verify the claimant's identity, authorization determines the level of access a device has, and encryption conceals information that is sent over the air. Two Bluetooth devices begin pairing by deriving an initialization key from their supposedly identical PINs, and then proceeding with the derivation of a link key. After authenticating with the link key, pairing is complete. The link keys can be stored for use in subsequent communication sessions, in which case the devices are said to be bonded. A trusted device is one that is granted access to services based upon a successful authentication.

The piconet can be vulnerable to a disclosure threat if an eavesdropper is located within the range of vulnerability. The intruder may be able to determine the PIN by intercepting the IN_RAND and other authentication parameters, and then testing different PIN values until the same authentication SRES is produced. An integrity threat is manifested when a spoofer assumes the identity of another device and gains access. A DoS threat can take the form of a jammer entering the range of susceptibility and attempting to prevent legitimate users from connecting.

End Notes

1. Vainio, J. "Bluetooth Security," Helsinki University of Technology White Paper, May 2000.
2. Muller, T. "Bluetooth Security Architecture," Bluetooth SIG Document 1.C.116/1.0, July 1999.
3. Persits, P. "15 Seconds: Crash Course in Cryptography," White Paper, December 1999.
4. Peterson, I. "Sneaky Calculations," *Science News*, November 17, 2001.

CHAPTER 10

Host Interfacing

One of the most common physical implementations of Bluetooth is in the form of a card or module that plugs into a host computer. When configured in this manner, the host must have some way of communicating with the module, passing commands to it, and obtaining results. The host also requires a method for sending the module data packets to be transmitted and for accepting from the module data packets that have been received from another Bluetooth device. The Bluetooth specification includes an entity called the *host controller interface* (HCI) that provides such communication between the host and module. Within the Bluetooth protocol stack, the module usually contains the radio, link controller, and link manager, and the rest of the protocol stack from the L2CAP and above reside on the Bluetooth-enabled host. HCI thus provides communication between the application and LM for piconet management, two-way audio packet transport, and user data packet transport between host and module (see Figure 10-1).

Implementing HCI in a Bluetooth design isn't required for certification, but is definitely a good idea if a module is being designed that is supposed to have compatibility with a wide variety of hosts. Also, if the module is intended to operate with a third-party host software package (the protocol stack), then some type of HCI will probably be required. On the other hand, if Bluetooth is part of a self-contained design, such as a radio-

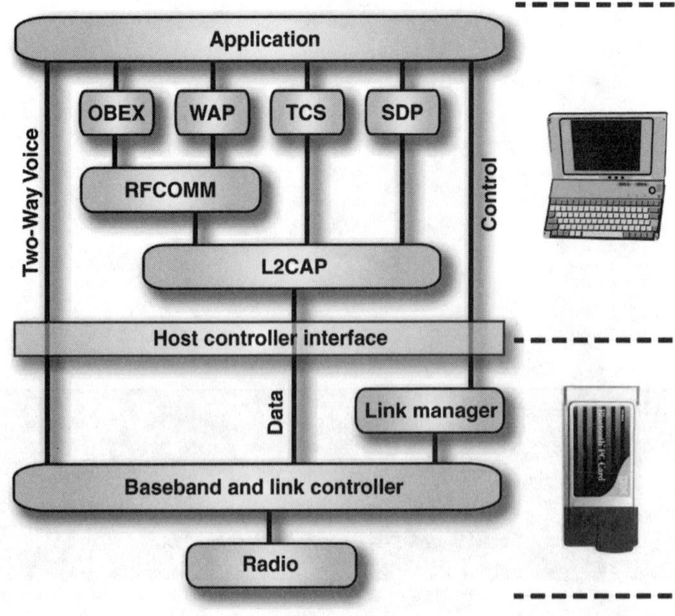

Figure 10-1 HCI resides between the Bluetooth module containing the radio, LC, and LM, and the Bluetooth host containing L2CAP and higher protocol layers. Control signal and both data and real-time two-way voice communications must pass through HCI.

controlled toy, then HCI may not be needed at all. Instead, the Bluetooth radio and (perhaps) baseband controller chipset will most likely communicate directly with a microcontroller into which the unit's functionality has been programmed.

In this chapter, we will examine the various methods that can be used to incorporate Bluetooth into a host device, followed by the role of HCI. Next, we'll examine some of the details of HCI itself, beginning with the structure of command and event packets, followed by data packets proceeding either to or from the Bluetooth module's ACL and SCO channel implementation. Finally, we give some examples of how HCI can be treated as a high-level language.

Host Controller Interface (HCI) Functionality

Because the purpose of HCI is to communicate between the host and Bluetooth module, it must make provisions for any communication that may be needed between these two entities. This includes control, data, and two-way real-time voice. HCI simply sets up the protocol for such communication, but the actual physical bus on which the communication takes place between host and module can be over a *universal serial bus* (USB), a *Personal Computer Memory Card International Association* (PCMCIA) or compact flash card, RS-232 serial communications, or a *universal asynchronous receiver/transmitter* (UART). Full implementation of HCI, then, requires four additional entities to the Bluetooth protocol stack: an HCI software driver on the host, a physical bus driver on the host, a physical bus driver on the module, and an HCI firmware on the module. The goal of the HCI is to ensure that the particular physical driver being used is irrelevant to communication between the host's HCI driver and the module's *host controller* (HC) firmware. As such, changing the transport mechanism from, say, USB to RS-232 requires no changes to HCI and hence no changes to the host software stack or to the module's firmware.

Figure 10-2 shows how these entities are arranged on two Bluetooth-enabled hosts that are communicating with each other. Because HCI is located on both the host and the module, it's important to make a distinction between the two so they can be discussed separately. As such, we will follow the Bluetooth specification and refer to the host entity as the *HCI driver* and the module entity as either the HC or the *HCI firmware*. Isn't it ironic that, instead of one cable connecting the two hosts together for a

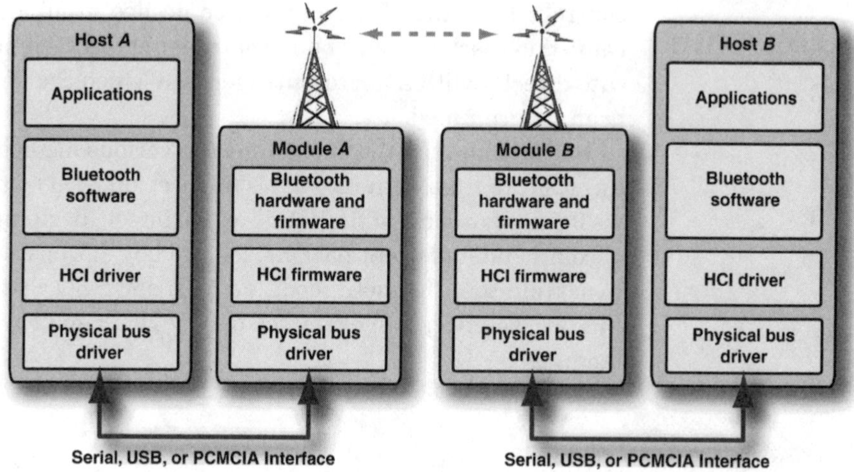

Figure 10-2 Two Bluetooth-enabled hosts communicating with each other through their Bluetooth modules. This configuration requires each host to have an HCI and a physical bus driver and each module to have a physical bus driver and HCI firmware.

wired link, we have two cables (or bus attachments) connecting the hosts to their respective Bluetooth modules for a wireless link?

Now that there are two hosts, two modules, two wired links, and one wireless link, confusion can reign supreme unless we're careful to specify which of these entities is being referred to in our many descriptions. Therefore, we focus our attention on one host-module pair and call these the *local host* and *local module*. That's the device that we're using. The other host-module pair, called the *remote host* and *remote module*, is being used by someone else. The goal is to establish and operate a Bluetooth wireless link between the local and remote modules so the two hosts can communicate with each other.

Commands, Events, and Data over HCI

There are three general types of communication that can take place between a host and its Bluetooth module over HCI. These are depicted in Figure 10-3. HCI *commands* are sent from the host to the module to direct the module to perform a task such as connect to another device with a particular BD_ADDR. HCI *events* are responses to commands or requests originating from a remote Bluetooth-enabled host (such as a connection request) that are passed from the local module to the local host. Finally, *data* is passed in either direction over HCI. Data sent from a local host to a module will be transmitted to the remote Bluetooth-enabled device, and

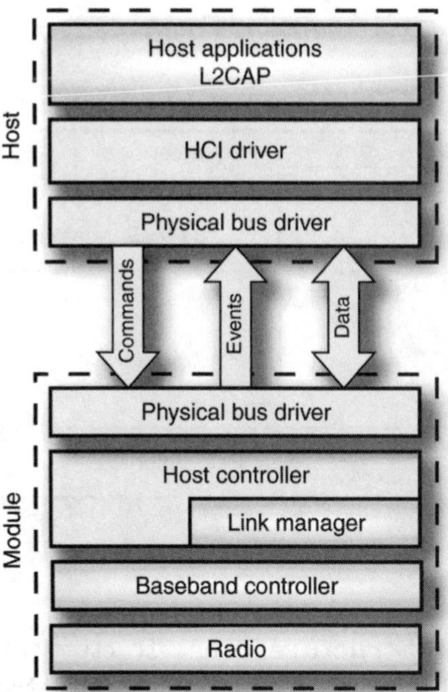

Figure 10-3 The host sends commands to the module, and the module sends events to the host via HCI as part of the control process. Both ACL and SCO data are also communicated between host and module. Data originating on the host is passed to the module for transmission, and data received by the module's radio is passed to the host.

incoming data is passed over HCI to the local host. This data can be either asynchronous, isochronous, or synchronous.

The HCI firmware (HC) in the Bluetooth module implements the HCI commands from the host by accessing the LM, LC, and various hardware status, control, and event registers on the module. As we'll discover later in this chapter, HCI firmware on the module can act as a form of compiler, taking the relatively high-level HCI commands and turning them into a sequence of operations at the LM and LC levels.

In a manner similar to the link manager protocol, HCI commands have their own set of mnemonics, each of which is preceded by "HCI," just like LMP mnemonics were preceded by "LMP." The mnemonics are worded to make their meaning as obvious as possible; for example, HCI_Hold_Mode is used to place a particular ACL link into the hold mode. On the other hand, HCI events are distinguished by a plain mnemonic so they won't be confused with HCI commands; for example, the Connection Complete event tells the host that the requested connection has been established.

Packets used at the HCI level in the protocol stack have their own specific composition defined for commands, events, and ACL and SCO data. Packets are sent over the HCI transport layer using the Little Endian format (LSB first). Negative values use 2's complement representation.

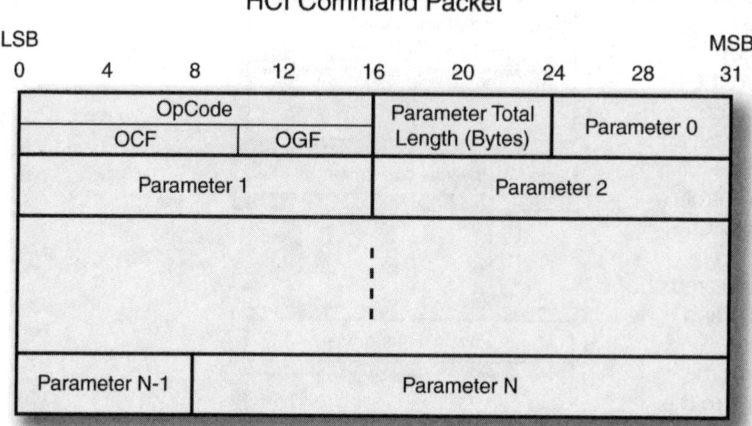

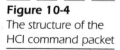

Figure 10-4 The structure of the HCI command packet

Command Packet Structure The structure of the HCI command packet is shown in Figure 10-4. Due to its length, the packet is depicted a bit differently than previous packets. Instead of a linear representation (for example, Figure 8-5), we show these in stacked form. The various fields are shown from left to right and from top to bottom as if you're reading a page of text.

The HCI command packet begins with a 16-bit OpCode that's divided into a 6-bit *OpCode Group Field* (OGF) and a 10-bit *OpCode Command Field* (OCF). This organization enables the HC on the module to infer information from the OGF without having to decode the entire OpCode. The OGF of 0x3E is reserved for Bluetooth logo testing, and 0x3F is reserved for vendor-specific debugging.

The next field is 8 bits long and designates the number of bytes occupied by the parameter field. This is not necessarily the total number of parameters themselves. Each command packet has a specific number of parameters associated with it that may vary in length, but each is an integral number of bytes in size.

Event Packet Structure The event packet is used by the HC on the Bluetooth module to notify the host of the occurrence of various events. The host must be able to accept event packets with up to 255 bytes of data, excluding the event packet header. The event packet structure is shown in Figure 10-5. The 8-bit event code identifies the event, the 8-bit parameter length field specifies the total parameter length in bytes, and each parameter is an integral number of bytes.

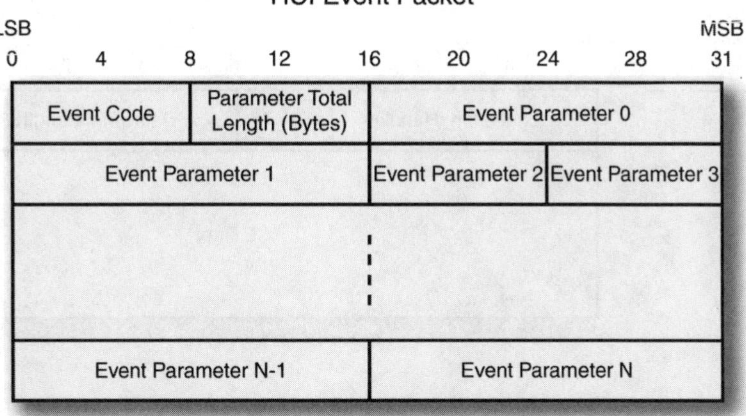

Figure 10-5 The structure of the HCI event packet

ACL Data Packet Structure Data that is to be used in an ACL link, either asynchronous or isochronous, also has its own HCI packet structure as shown in Figure 10-6. The Connection_Handle field consists of 16 bits, but only 12 bits are significant. Each ACL link is identified by this Connection_Handle, which can have a value between 0x0000 and 0x0EFF.

The 2-bit *Packet_Boundary_Flag* (PB) is used to identify the first packet from a higher-layer message such as the start of a L2CAP packet (PB = 10) or the continuing fragment (PB = 01). The 2-bit *Broadcast_Flag* (BC) is used in the following way:

BC _ 00 The packet is point-to-point.

BC _ 01 Active broadcast. The packet is broadcast to all active slaves. This packet, which may require several ACL packets to be transmitted, may or may not be sent during parked slave beacon slots, and slaves in the sniff mode may or may not receive this data. Slaves in the hold mode will not receive the data.

BC _ 10 Piconet broadcast. The packet is intended for all active and parked slaves in the piconet, so it will be broadcast during parked slave beacon slots in which the parked slaves are listening.

BC _ 11 Reserved for future use.

ACL packets that are meant for broadcast have their own unique Connection_Handle and, of course, are sent from the host to the HC only if it is the master of the piconet.

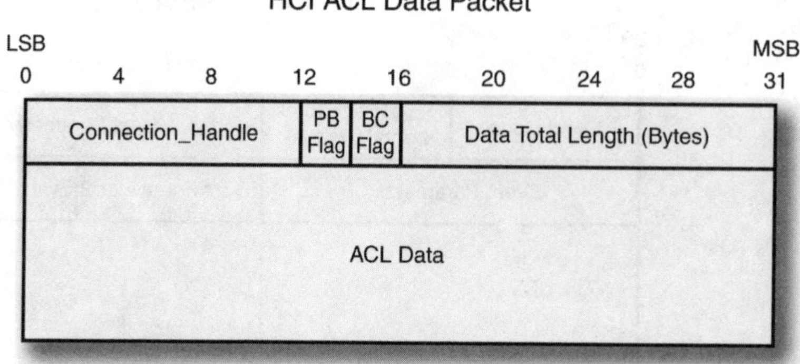

Figure 10-6 The structure of the HCI ACL data packet

The 16-bit Data_Total_Length field specifies the length of the HCI ACL data field in bytes, enabling a maximum possible data field of 65,535 bytes, the same as the maximum L2CAP data field. However, it's probable that the module may not be able to accept data fields this large. The actual maximum size of the module's data field can be read by the host using the HCI_Read_Buffer_Size command.

SCO Data Packet Structure and Latency The HCI SCO data packet composition is shown in Figure 10-7. SCO links are also identified with a unique 16-bit Connection_Handle with values ranging from 0x0000 to 0x0EFF. The next four bits are reserved, followed by a 8-bit Data_Total_Length field that specifies the total length of the HCI SCO data field in bytes. Unlike the HCI ACL packet, the maximum SCO packet data field is limited to only 255 bytes to prevent latency problems from waiting to encode long segments of audio.

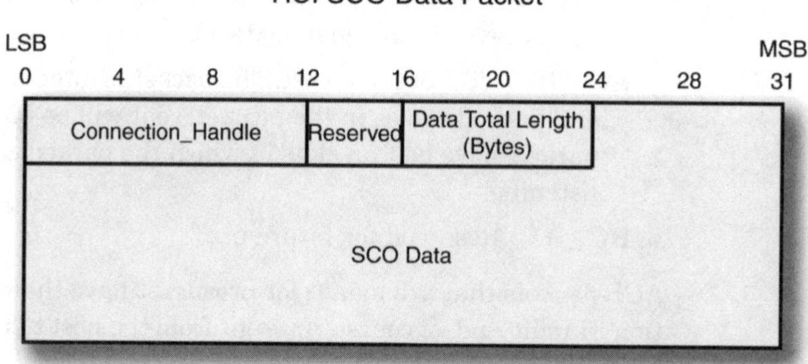

Figure 10-7 The structure of the HCI SCO data packet

In Chapter 8, "Transferring Data and Audio Information," we determined that the overall SCO latency without accounting for HCI delays was about 11 ms. Latency was calculated from the time the voice was spoken at one end of the Bluetooth link until it was heard at the other end. The delay experienced when HCI is used depends upon various encoding times and the size of the HCI SCO packet.

Let's assume that voice encoding and decoding, along with HCI packet manipulations, add about 5 ms of delay. We can determine the maximum HCI delay by assuming that 255 bytes of voice data is included in each HCI packet. This is equivalent to $^{255}/_{8} = 32$ ms of audio. Once the module has this packet, it can send the data over the SCO link, which takes another 32 ms at 64 kb/s. Finally, the SCO packets are received at the remote location, and another HCI packet is assembled and passed from the remote module to its host. Because analog voice can begin to be heard at the destination shortly after the HCI packet's arrival at the host, we need not include any additional latency beyond the 5 ms decoding time that we assumed earlier. Total latency is thus $5 + 32 + 32 + 5 = 74$ ms, which is still less than the 100 ms latency that many consider to be the maximum that can be tolerated.

Example Command: Creating a Connection An example of a commonly used HCI command packet is HCI_Create_Connection. This packet is a member of the link control group, which has an OGF of 0x01, and the command itself has a OCF of 0x0005. (Although the OCF is only a 10-bit field, it is represented in the specification as a 2-byte hexadecimal word.) The packet has six parameters (see Figure 10-4):

BD_ADDR (6 bytes) This is the BD_ADDR of the remote device to which the host wants to connect.

Packet_Type (2 bytes) This determines the type of packets that the module's LM is enabled to use on the ACL link after it is formed. The parameter enables different bits to be set that represent DM1, DH1, DM3, DH3, DM5, or DH5 ACL packets. This approach enables packet types to be selected individually. At least one must be selected.

Page_Scan_Repetition_Mode (1 byte) This parameter lists the remote device's page scan repetition mode (R0, R1, or R2) that was discovered during the inquiry process (see Chapter 5, "Establishing Contact Through Inquiry and Paging").

Page_Scan_Mode (1 byte) Also discovered during inquiry, this parameter states whether the remote device uses the mandatory or one of the optional page scan modes (see Chapter 5).

Clock_Offset (2 bytes) This is the difference between the local and remote module's $CLKN_{2-16}$ for possible use in timing calculations. The host may not know this information, in which case the Clock_Offset_Valid_Flag in this field is reset to 0.

Allow_Role_Switch (1 byte) This parameter lets the remote device know if the local device will accept a master-slave switch during connection setup. Clearly, the local device starts as a master because it is initiating the connection through HCI_Create_Connection, so the role switch parameter determines whether the local device can accept a slave role.

When the HC receives the HCI_Create_Connection command, it immediately returns a Command Status event telling the host that the command was received, and its status is pending. Next, the HC causes the Bluetooth module to begin paging the BD_ADDR listed in the command. If the page is successful, then a Connection Complete event is sent to the host with several parameters, including the 4-byte (12 bits significant) Connection_Handle. Although the HCI_Create_Connection command specified a BD_ADDR for the connection, subsequent references to this particular link are made using the Connection_Handle. This enables the host to avoid using Bluetooth-specific addressing any more than necessary. The details of creating a connection will be given later in this chapter.

It's easy to see how HCI can be treated as a high-level language. Upon receiving HCI_Create_Connection, the HC communicates with the LC to complete the paging process and then works with its LM to, for example, pass information in the Packet_Type and Allow_Role_Switch fields to the remote LM as part of a LMP_features_res packet.

HCI Flow Control

One of the reasons that HCI exists is to relieve the processor on the Bluetooth module from the burden of running the higher-level protocols in addition to managing the radio, LC, and LM. In this way, the module can have a relatively inexpensive processor, and the host can devote a portion of its considerably greater processing resources to L2CAP and above. When the host devotes its attention to the Bluetooth module, however, it's possible that during a relatively short period of time a large amount of data could be sent to the module for transmission, exceeding the module's buffer capacity. Also, because there's no latency guarantee for ACL packets, there may be relatively long delays (and low throughput) when the communication channel is particularly poor. Consequently, there may be periods of time during which the module cannot accept any more data from the host. Several HCI commands and events are implemented for the module to communicate its buffer status to the host, but flow control to the module is the responsibility of the host.

When the host and module initialize together, the host is required to issue a HCI_Read_Buffer_Size command before it can send any HCI data packets to the module. The module will return to the host the maximum size of HCI ACL and SCO packets that it can accept from the host, and the total number of each of these packets that the module can have in its buffers. Also, when there's at least one connection to another Bluetooth-enabled device (or when in a local loopback mode), the HC on the module periodically sends the host a Number of Completed Packets event that tells the host the number of packets that have been successfully transmitted from its Bluetooth module to the destination since the last time this event was sent to the host. The manufacturer decides how often this event occurs. The host can use this information to calculate actual buffer usage and decide when to send more packets to the module for transmission.

When the host sends a command packet to the module, the module returns an event that includes the parameter Num_HCI_Command_Packets. This is the number of additional HCI command packets that the HC can accept from the host. The host can refer to this parameter and possibly send additional commands to the module, if desired, before action on a previous command has been completed. Upon power-up, the host must assume that the HC can accept only one HCI command packet until the host can retrieve Num_HCI_Command_Packets from the module.

It's also possible in certain conditions that flow control may be necessary in the direction of module to host. Flow control in this reverse direction can be enabled with an HCI_Set_Host_Controller_To_Host_Flow_Control command (whew!) to the module, and the host sends its buffer size to the module using the HCI_Host_Buffer_Size command. Now the host can send an HCI_Host_Number_Of_Completed_Packets command periodically to the module, which uses this information to control its transfer of ACL and SCO data packets to the host in a manner similar to that described earlier. HCI therefore enables flow control to be implemented in both directions.

HCI Commands

The HCI command packets provide a standardized method to access the local Bluetooth module and direct it to perform certain tasks. Some of these tasks are performed locally and some of them require the module to communicate with the remote module across the Bluetooth radio link. In the latter situation, the local module's LM responds to the HCI command by forming a set of LMP packets that are exchanged with the specified LM at the remote end of the Bluetooth link.

Commands that don't require access to the Bluetooth wireless link can be completed immediately, such as when the local host asks its own module for its BD_ADDR. In these cases, the local module's HC returns a Command Complete event containing the requested parameters. Other commands may take some time for completion, such as asking the module to form a remote connection. In this situation, the HC on the module responds to the request with a Command Status event, which is equivalent to saying "I'm working on it, so be patient." When the module later completes the command, either successfully or unsuccessfully, it generates a completion event that is associated with the command, which is usually something different than the generic Command Complete event. This completion event contains the parameters that may be needed by the host. For example, suppose the host sends an HCI_Create_Connection command. The HC on the module responds with a Command Status event, and then proceeds to page the requested BD_ADDR. When the connection has been formed, the HC sends the Connection Complete event to its host.

To avoid infinite delay situations at the host, there should be a timeout if the module fails to respond to a host command. The recommended default timeout is one second, but it may be longer if there are several HCI commands in the HC's queue. The timeout may, for example, be triggered if the user removes the module from the host at an inopportune time.

Many of the HCI commands are written to be independent of whether the particular Bluetooth module is the master or a slave in a given piconet. For example, HCI_Hold_Mode simply requests that a particular ACL link, identified by its Connection_Handle, be placed into hold for a certain duration of time, regardless of whether the module is a master or slave. This simplifies the HCI instruction set, and because it's already over 200 pages long in the specification, it needs all the simplification it can get. Furthermore, in many cases, it's irrelevant to the host whether it is participating in the piconet as a master or slave, especially if the connection is point-to-point, so HCI shields the host from always having to track this Bluetooth-specific information.

In the next few sections, we'll summarize the HCI commands that belong to each respective OGF. Although these summaries appear in longer form in the specification, we've arranged them in subgroups to show the versatility of HCI. Furthermore, we've included some extremely useful gems of information within some of these commands, so maybe you shouldn't skip over

Figure 10-8 Being a Bluetooth expert can have its disadvantages. (Source: Bill Lae)

Biff was unequivocally correct for the very last time.

them after all. Once you understand how the HCI commands operate, you will be near the pinnacle of Bluetooth expertise (see Figure 10-8).

Link Control Commands

The Link Control commands are used by the host to control connections to other Bluetooth devices. As such, their execution may take some time, so the HC will respond to a command in this OGF with a Command Status event. Each of these commands causes the local LC and LM to perform several actions in response. These commands are summarized in the following section.

Inquiry Commands

HCI_Inquiry Causes the local module to enter the inquiry state. The *lower address part* (LAP) for the inquiry is included as a parameter, and this determines the type of inquiry (general, limited, or dedicated) to be performed. Other parameters specify the length of time the inquiry will take place and how many replies can occur before the inquiry is terminated.

HCI_Inquiry_Cancel Cancels the inquiry process.

HCI_Periodic_Inquiry_Mode This configures the local module to enter the inquiry state periodically, with the maximum and minimum times between successive inquiries given as two parameters in this command.

HCI_Exit_Periodic_Inquiry_Mode Cancels the periodic inquiry process.

Connect and Disconnect Commands

HCI_Create_Connection Causes a connection to be attempted with the BD_ADDR given as a parameter.

HCI_Disconnet Terminates an existing connection.

HCI_Add_SCO_Connection Causes an SCO link to be established on the link given by the Connection_Handle parameter. The type of SCO packets enabled (HV1, HV2, and HV3) is also specified. The local LM decides which packet type is to be used if given a choice. If successful, the SCO connection will be assigned a Connection_Handle by the HC as part of the Connection Complete event packet returned to the host.

HCI_Accept_Connection_Request When a remote device initiates a connection with the local device and the local host approves the connection, then it sends this command to the module's HC. Parameters include the remote device's BD_ADDR and whether a role switch is required, that is, whether the local host insists on becoming the master of the resulting connection.

HCI_Reject_Connection_Request Used by the host to reject a connection request from a remote device. Parameters include the remote device's BD_ADDR and a Host Reject Error Code.

HCI_Change_Connection_Packet_Type The host can change the packet types that can be used on a connection that has already been established with this command. The command applies to either an ACL or SCO link as determined by the Connection_Handle parameter and contains a bit map of the enabled ACL or SCO packet types.

Security-related Commands

HCI_Link_Key_Request_Reply If a link key is stored on the host and requested by the HC using the Link Key Request event, then the host uses this command to pass the link key value to the module.

HCI_Link_Key_Request_Negative_Reply Used by the host to reject a request from the local HC for a link key.

HCI_PIN_Code_Request_Reply The host sends the *personal identification number* (PIN) to the local HC using this command. The HC uses the PIN for pairing with another Bluetooth device and subsequent encryption if needed (see Chapter 9, "Bluetooth Security").

HCI_PIN_Code_Request_Negative_Reply Used by the host to reject a request from the local HC for the PIN.

HCI_Authentication_Requested Authentication for the particular Connection_Handle is requested. This causes the local LM to initiate the authentication process as the verifier (see Chapter 9).

HCI_Set_Connection_Encryption Used to enable and disable link layer encryption for a particular Connection_Handle.

HCI_Change_Connection_Link_Key Forces the devices on both ends of the Connection_Handle to form a new link key.

HCI_Master_Link_Key This command forces the master of the piconet to use either the master key for broadcast packet encryption or the semipermanent link key for point-to-point encryption.

Remote Device Information Commands

HCI_Remote_Name_Request Used to obtain the user-friendly Bluetooth name of the device specified by the BD_ADDR parameter in the command.

HCI_Read_Remote_Supported_Features This command requests a list of the remote device's supported features corresponding to the Connection_Handle parameter.

HCI_Read_Remote_Version_Information This command obtains the version information of the remote device specified by Connection_Handle.

HCI_Read_Clock_Offset Used to obtain clock offset information from the remote device specified by Connection_Handle. This information can be used to speed the paging process by enabling the local device to determine the page channel on which the remote device will be listening for future pages (see Chapter 5).

Link Policy Commands

The Link Policy commands are used by the host to affect how the local LM manages the piconet. The ACL link is assumed to already exist between the two devices. Notice that, for the most part, whether the local device is a master or slave is irrelevant to the HCI commands themselves, although execution of the commands by the local LM will differ, depending upon whether the device is a master or slave. These commands are summarized in the following section.

Setup and Configuration Commands

HCI_QoS_Setup Specifies the *quality of service* (QoS) parameters for Connection_Handle. These are the same parameters as those in the L2CAP QoS (see Chapter 8). The local LM therefore has access to all the QoS parameters being requested by the host and can use LMP packets to request QoS parameters from the remote (master) device or request the remote (slave) device to accept the QoS parameters.

HCI_Role_Discovery Determines the role (master or slave) that the local device in an ACL link specified by Connection_Handle has.

HCI_Switch_Role Initiates a role (master-slave) switch with a remote device given by BD_ADDR.

HCI_Read_Link_Policy_Settings / HCI_Write_Link_Policy_Settings Reads/writes the link policy settings for the ACL link specified by Connection_Handle. The policy settings determine whether the local device's master-slave switch, hold, sniff, or park modes are enabled.

Low-Power Mode Commands

HCI_Hold_Mode Causes the local LM to place the ACL link specified by the Connection_Handle parameter into the hold mode. Two other parameters called the Hold_Mode_Max_Interval and the Hold_Mode_Min_Interval determine the maximum and minimum hold intervals that are acceptable to the local host. These two parameters can be used by the local LM when negotiating the hold interval (see Chapter 6, "Advanced Piconet Operation"). The device exits the hold mode automatically when the hold time expires.

HCI_Sniff_Mode Causes the local LM to place the ACL link specified by the Connection_Handle into the sniff mode. Two other parameters called the *Sniff_Max_Interval* and the *Sniff_Min_Interval* determine the maximum and minimum sniff intervals that are acceptable to the local host. The Sniff_Attempt and Sniff_Timeout parameters complete the information that the local LM needs to execute the sniff (see Chapter 6). These two parameters are called $N_{sniff\_attempt}$ and $N_{sniff\_timeout}$, respectively, at the LM level.

HCI_Exit_Sniff_Mode Causes the local LM to exit sniff on the ACL link specified by the Connection_Handle.

HCI_Park_Mode Causes the local LM to place the ACL link specified by the Connection_Handle into the park mode. Two other parameters called the *Beacon_Max_Interval* and the *Beacon_Min_Interval* determine the maximum and minimum beacon intervals that are acceptable to the local host. For this command to execute successfully, there must not be a Connection_Handle corresponding to a SCO link with the remote device. One fairly powerful feature associated with park is the capability of the HC to automatically unpark the ACL link when the host sends a packet to the HC for transmission on a link represented by a parked Connection_Handle. After successful receipt of the data, the HC will return the ACL link to the park mode.

HCI_Exit_Park_Mode Returns the ACL link to the active mode. If the local device is the master, then its LM issues LMP_unpark_BD_ADDR_req or LMP_unpark_PM_ADDR_req in a subsequent beacon slot. If the local device is a slave, then the LM performs an access request (see Chapter 6).

HC and Baseband Commands

The HC and baseband commands provide access to and control of the local Bluetooth module itself. Because none of these commands use the radio link for their execution, the module's HC responds to them with a Command Complete event containing the requested parameters, if any. Most of these commands are paired as both read (from the module) and write (to the module), so we'll try to put these together whenever possible to save space. When parameters are written, they are part of an HCI command packet, and when they are read, they are part of an event packet provided by the HC. These commands are given in the following section.

Setup and Configuration Commands

HCI_Set_Event_Mask Controls which events are generated by the HC for return to the host. This enables the host to control how often it's interrupted. Every event can be masked by the host except Command Complete, Command Status, and Number of Completed Packets. All events are enabled as default.

HCI_Reset Resets the HC and LM, but doesn't affect the HCI transport layer (such as USB). After being reset, the local Bluetooth device enters the standby state.

HCI_Set_Event_Filter Used to determine how the local Bluetooth module responds to various link activities. For example, incoming connections can be enabled from all devices, or only from devices having a specific class, or only from a specific BD_ADDR. Inquiry responses can be similarly filtered. A remote connection request can be automatically accepted or require user intervention before it is accepted, and the master-slave role switch can be enabled or disabled. This command can be used to enhance Bluetooth security in many situations.

HCI_Read_Local_Name / HCI_Change_Local_Name Used to read/modify the user-friendly Bluetooth name stored on the local device. It usually makes more sense to attach a Bluetooth name to a host rather than to a module, so when the host first communicates with its local module upon power-up, this name will be written to the module.

HCI_Read_Class_of_Device / HCI_Write_Class_of_Device Reads/writes the 6-byte class of device parameter, which is used to indicate the capabilities of the local host to other devices. Device classes are listed in *Bluetooth Assigned Numbers*.[1]

HCI_Read_Voice_Setting / HCI_Write_Voice_Setting Reads/writes the voice connection settings, which must be the same for all voice connections. Inputs to the Bluetooth module can be linear, μ-law, or A-law PCM; 1's or 2's complement or sign magnitude format, and for linear PCM either 8- or 16-bit samples with selected MSB positions. The coding for transmission over the air can be A-law, μ-law, or CVSD. The default input coding is linear PCM using 2's complement representation, with 16-bit sample size and the sample MSB at the input word's MSB position. The manufacturer can set any air mode as its default.

HCI_Read_Num_Broadcast_Retransmissions / HCI_Write_Num_Broadcast_Retransmissions Reads/writes the number of broadcast packet retransmissions to increase the probability of successful receipt. The number can range between 0 (0x00) and 255 (0xFF) and defaults to 1 (0x01).

Flow Control Commands

HCI_Set_Host_Controller_To_Host_Flow_Control Used by the host to turn flow control on or off for data and/or voice sent from the HC to the host. Either ACL or SCO packets can be flow-controlled independently by the host though periodically sending a Host_Number_Of_Completed_Packets command to the HC. Host flow control is turned off as default.

HCI_Host_Buffer_Size Used by the host to notify the HC of the maximum data portion of HCI ACL and SCO packets that can be sent in the direction of the HC to the host, along with the total number of these packets that can be stored in the host.

HCI_Host_Number_Of_Completed_Packets Used by the host to notify the HC of the number of HCI data packets that have been completed for each active Connection_Handle since the last such command was sent to the HC. This command is only used if flow control from the HC to the host (or local loopback) is enabled.

HCI_Read_SCO_Flow_Control_Enable / HCI_Write_SCO_Flow_Control_Enable Reads/writes whether or not the HC will send Number Of Completed Packets events for SCO connection handles. SCO flow control is disabled as default.

HCI_Flush Discards all data pending for transmission in the HC for the ACL link specified by Connection_Handle. The entire current HCI ACL data packet is affected by this command, even if some of it still resides on the host itself. This command works independently of the L2CAP automatic flush timers (see Chapter 8).

Timeout Commands

HCI_Read_Connection_Accept_Timeout / HCI_Write_Connection_Accept_Timeout Reads/writes the maximum time duration enabled between the HC sending a Connection Request event to the host, and if there's no response from the host, the connection is refused by the HC. A 16-bit parameter specifies the number of baseband slots, and the timeout can range between 0.625 ms (0x0001) and 29 seconds (0xB540), with a default of 5 seconds (0x1FA0). If user intervention is required to determine whether or not an incoming connection is accepted, then this value places an upper limit on how long the user can wait before responding.

HCI_Read_Page_Timeout / HCI_Write_Page_Timeout Reads/writes the maximum time that the local LM will wait for its page to be responded to by a remote device. If this timeout expires, then the HC considers the connection attempt to have failed. A 16-bit parameter specifies the number of baseband slots, and the timeout can range between 0.625 ms (0x0001) and 40.9 seconds (0xFFFF), with a default of 5.12 seconds (0x2000). This default corresponds to the maximum time that paging could require if the prospective master begins paging on the wrong 16-hop frequency train (see Chapter 5).

HCI_Read_Automatic_Flush_Timeout / HCI_Write_Automatic_Flush_Timeout Reads/writes the flush timeout value for the ACL link corresponding to Connection_Handle. This determines the amount of time before all segments of the current L2CAP packet are flushed by the HC for isochronous data support (see Chapter 8). The 4-byte timeout value defaults to 0x0000 for infinite time (no automatic flush), but it can be changed to designate a number of time slots between 0.625 ms (0x0001) and 1.28 seconds (0x07FF).

HCI_Read_Link_Supervision_Timeout / HCI_Write_Link_Supervision_Timeout Reads/writes the Link_Supervision_Timeout for the ACL link designated by Connection_Handle. This timeout is used by the master and slave to monitor link loss. If this timeout expires, then the link is disconnected. This 2-byte value can be 0x0000 for an infinite (disabled) timeout; otherwise, the value is the number of slots and ranges between 0.625 ms (0x0001) and 40.9 seconds (0xFFFF). The default value is 20 seconds (0x7D00). If the timer is disabled prior to parking a slave, then that slave no longer needs to be unparked and reparked periodically. However, the parked slave runs the risk of remaining in park indefinitely, even if its master disappears.

Inquiry and Page Scan Commands

HCI_Read_Scan_Enable / HCI_Write_Scan_Enable Reads/writes whether or not the Bluetooth module will periodically scan for pages and inquiries. Either can be enabled or disabled independently. No scans are enabled as default.

HCI_Read_Inquiry_Scan_Activity / HCI_Write_Inquiry_Scan_Activity Reads/writes the time between successive inquiry scans (Inquiry_Scan_Interval) and the time that the module spends in an inquiry scan (Inquiry_Scan_Window). Each of these parameters is a 4-byte value expressed as a number of time slots and can vary between 11.25 ms (0x0012) and 2.56 seconds (0x1000). The Inquiry_Scan_Window must be less than or equal to the Inquiry_Scan_Interval. For inquiry scans to occur, they must be enabled through HCI_Write_Scan_Enable. The default Inquiry_Scan_Interval is 1.28 seconds (0x0800), and the default Inquiry_Scan_Window is 11.25 ms (0x0012). Despite these default values, the prospective master usually probes a particular 16-hop inquiry train 2.56 seconds before switching to the next train (see Chapter 5). This insures that a prospective slave will hear an inquiry even if its Inquiry_Scan_Interval is set to the maximum 2.56 seconds. (However, the prospective master may not hear the prospective slave's response until several seconds later; see Chapter 5.)

HCI_Read_Number_Of_Supported_IAC Reads the value for the number of *inquiry access codes* (IAC) that the local module can listen for simultaneously during an inquiry scan. All devices are required to support the *general inquiry access code* (GIAC), but others may be supported as well. There are a total of 64 IAC values, so the number of possible supported IACs can range between 1 (0x01) and 64 (0x40).

HCI_Read_Current_IAC_LAP / HCI_Write_Current_IAC_LAP Reads/writes the current LAP used to create each IAC value that the module can simultaneously search for during inquiry scans. Each parameter is a 3-byte (24-bit) value and ranges from 0x9E8B00 to 0x9E8B3F.

HCI_Read_Page_Scan_Activity / HCI_Write_Page_Scan_Activity Reads/writes the time between successive page scans (Page_Scan_Interval) and the time that the module spends in a page scan (Page_Scan_Window). Each of these parameters is a 4-byte value expressed as a number of time slots and can vary between 11.25 ms (0x0012) and 2.56 seconds (0x1000). The Page_Scan_Window must be less than or equal to the Page_Scan_Interval. For page scans to occur, they must be enabled through HCI_Write_Scan_Enable. The default Page_Scan_Interval is 1.28 seconds (0x0800), and the default Page_Scan_Window is 11.25 ms (0x0012). These values correspond to Scan Repetition Mode R1, with a scan window long enough for a prospective master to page 16 frequencies in the 32-hop set (see Chapter 5).

HCI_Read_Page_Scan_Period_Mode / HCI_Write_Page_Scan_Period_Mode Reads/writes the period in which the device enters the mandatory page scan mode after responding to an inquiry (see Chapter 5). The choices are 20 seconds (P0), 40 seconds (P1) or 60 seconds (P2). The default is mode P0. If an optional page scheme is available, then the module can either replace it with the mandatory page scan mode during this period, or it can operate the two modes simultaneously if it has the capability.

HCI_Read_Page_Scan_Mode / HCI_Write_Page_Scan_Mode Reads/writes the default page scan mode used by the local device. Specification 1.1 defines one mandatory and three optional page scan modes (see Chapter 5). The default is set to the mandatory page scan mode.

Low-Power Mode Commands

HCI_Read_Hold_Mode_Activity / HCI_Write_Hold_Mode_Activity Reads/writes what activities should be suspended while the device is in the hold mode. Page scans, inquiry scans, and periodic inquiries can be independently enabled or disabled while the device is in the hold mode. The default is to enable all of these activities.

Security-related Commands

HCI_Read_PIN_Type / HCI_Write_PIN_Type Reads/writes whether the host supports variable or fixed PIN types (see Chapter 9).

HCI_Create_New_Unit_Key Initiates the creation of a new unit key (see Chapter 9).

HCI_Read_Stored_Link_Key / HCI_Write_Stored_Link_Key Enables the host to read or write any of the 128-bit link keys that may be stored in the module. Note that these commands can pose a security risk if an intruder can obtain momentary physical access to a Bluetooth module containing link keys. Aside from reading the link keys, the intruder can also write link keys that can be exploited later to gain unauthorized access to a host over a Bluetooth link.

HCI_Delete_Stored_Link_Key Enables the host to remove one or more link keys from the module. If an intruder can gain physical access to the module and delete all link keys, then the pairing process may need to be reaccomplished by the legitimate user, during which the intruder can intercept IN_RAND and perhaps use it to assist in discovering the PIN (see Chapter 9).

HCI_Read_Authentication_Enable / HCI_Write_Authentication_Enable Reads/writes whether the local module requires authentication prior to establishing an ACL link (link level authentication; see Chapter 9). This authentication occurs between the HCI_Create_Connection command and the corresponding Connection Complete event (if the local host initiated the connection) or between accepting an ACL connection and the Connection Complete event (if the remote host initiated the connection). Authentication is disabled as default.

HCI_Read_Encryption_Mode / HCI_Write_Encryption_Mode Reads/writes whether or not encryption is required at ACL connection setup. Encryption can be enabled for only point-to-point or for both point-to-point and broadcast packets. Encryption is disabled as default.

Transmit Power Commands

HCI_Read_Transmit_Power_Level Reads the values for the Transmit_Power_Level parameter for an ACL link represented by Connection_Handle. This is the power level at the local transmitter, and it can be read as either the current power or the maximum power. The 1-byte signed integer value returned by the HC ranges from −30 to +20 dBm.

Informational Parameters

The informational parameters are fixed by the manufacturer of the specific Bluetooth hardware being used by the host. The host can read these parameters from the module, but cannot modify any of them. Because none of these commands requires access to the Bluetooth radio link, the module responds with a Command Complete event containing the requested parameters.

HCI_Read_Local_Version_Information Reads the values for the local module's HCI version and revision, LMP version and subversion, and manufacturer's name.

HCI_Read_Buffer_Size Reads the packet length and number of packets that the HC can accept. Each of these two quantities can be designated independently for HCI ACL or SCO packets. These values apply to all Connection_Handles combined, and the host determines how the buffers are to be divided between the different Connection_Handles. The host isn't enabled to send any HCI data packets to the module until this command is executed. Both the host and HC must support command and event packets with data lengths of 255 bytes, excluding the header.

HCI_Read_Country_Code Reads the country code that applies to the local Bluetooth module. This code will determine, among other factors, the segment of the 2.4 GHz band on which the module operates. The 1-byte value is defined as 0x00 for North America, Europe (except France), and Japan, all of which use the 79-hop sequence. A value of 0x01 is defined for France, which (as this is being written) requires use of the 23-hop sequence in at least some situations. (We don't discuss the 23-hop sequence in this book because I have great confidence that the 79-hop sequence will eventually be used worldwide.) Other country codes are reserved.

HCI_Read_BD_ADDR Reads the 48-bit Bluetooth Device Address on the local module.

Status Parameters

These status parameters provide information about the state of the HC, LM, and LC, and are modified by the HC. Other than specific resets, the parameters cannot be changed by the host.

HCI_Read_Failed_Contact_Counter Reads the value in the 16-bit Failed_Contact_Counter for a particular ACL Connection_Handle. This counter is incremented by one each time the L2CAP flush timeout expires. The counter is reset to zero if an L2CAP packet is subsequently acknowledged for that connection, a new connection is established, or a Reset_Failed_Contact_Counter command is issued by the host.

HCI_Reset_Failed_Contact_Counter Resets the 16-bit Failed_Contact_Counter to 0x0000.

HCI_Get_Link_Quality Reads the value for the Link_Quality for the specified ACL Connection_Handle. The quality is assigned a value between 0 (0x00) and 255 (0xFF), which is a vendor-specific value that measures the relative quality of the Bluetooth link between two devices. Higher values designate higher link quality.

HCI_Read_RSSI Reads the difference between the actual RSSI and the limits of the Golden Receive Power Range (see Chapter 3, "The Bluetooth Radio") for a particular ACL Connection_Handle. A positive RSSI value indicates how many dB the actual RSSI is above the upper limit, and a negative value indicates how many dB the actual RSSI is below the lower limit. A value of zero indicates that the actual RSSI is inside the Golden Receive Power Range. Note that there's no provision within HCI to read the RSSI itself. The 8-bit signed integer can range between −128 and 127. For each Connection_Handle, the HC should have an 8-bit RSSI register in which it stores the RSSI value from the latest incoming transmission.

Testing Commands

The testing commands are used to provide access to functional tests on the Bluetooth module. Because many of these test modes can violate either government regulations or the Bluetooth specification, these commands should not be accessible by the end user. Further information on testing will be given in Chapter 12, "Module Fabrication, Integration, Testing, and Qualification."

HCI_Read_Loopback_Mode / HCI_Write_Loopback_Mode Reads/writes the value for the setting of the HC's loopback mode. The *nontesting mode* is used for normal Bluetooth operations, which is the default setting. In the *local loopback* mode, the HC simply returns every ACL and SCO data packet and every command packet back to the host with no modifications to enable the host to test HCI transport. In *remote loopback mode*, the HC will cause the module to simply retransmit every incoming baseband packet that arrives over the air from a single remote device. In this way, the remote device can check the channel BER.

HCI_Enable_Device_Under_Test_Mode This command puts the local Bluetooth module into the test mode so it will accept the various LMP commands associated with testing the module. Upon receiving this command, the HC will respond with a Command Complete event, but continues to operate normally until receiving the correct LMP command from the remote device to place it into the *device under test* (DUT) mode. This mode is exited when the local host issues the HCI_Reset command.

HCI Events

Communication over the HCI transport layer that is from the module to the host takes the form of events. Events can occur as a result of a host command, or they can be triggered by an incoming packet from a remote device. Event packets have the structure given in Figure 10-5 and are uniquely identified by the 1-byte event code.

We've already discussed the contents of many events already in the previous section because an event is generated any time a HCI_Read_Something command is sent from the host to the module. Therefore, instead of describing all of the events in the specification, we'll simply give some examples of how events are used in different situations.

Event Process: Command with Immediate Execution

When the host issues an HCI command that can be executed locally by the HC, then the response is the Command Complete event that includes as its parameters the desired information.

For example, if the local host wants to read the BD_ADDR of its Bluetooth module, then the following occurs:

1. The host sends HCI_Read_BD_ADDR to the module. This particular command contains no parameters.
2. The local module's HC sends a Command Complete event to the host. The parameters in this event packet are: Num_HCI_Command_Packets, designating the number of additional HCI command packets that the HC can accept, a reflection of the Command_Opcode so the host can determine what the corresponding command was, and the Return_Parameters, in this case, the requested BD_ADDR.

Event Process: Command with Delayed Execution

If a host command requires the local HC to access another device over the Bluetooth link, then there will be an indeterminate delay until the local host can be given the return parameters. In these cases, the Command Complete event isn't used at all; instead, the local HC responds immediately with a Command Status event that lets the host know that the command was received and is being acted upon. After the HC has obtained the desired response from the remote module, then it notifies the local host with an event packet that corresponds to the particular command previously given.

For example, suppose the local host wants to create a connection to a remote device. The following sequence occurs:

1. The host sends HCI_Create_Connection to the module. This command contains the following parameters: the BD_ADDR of the remote device, the Packet_Type field of enabled baseband packets, the

Page_Scan_Repetition_Mode, the Page_Scan_Mode, the Clock_Offset of the remote device discovered during inquiry, and the Allow_Role_Switch parameter designating whether or not the local device can later accept a slave role in the piconet.

2. The local module's HC sends a Command Status event back to the host containing the following parameters: Status, which is set to 0x00 (command pending), Num_HCI_Command_Packets, which is the number of additional command packets that the local HC can accept, and a reflection of the Command_Opcode.
3. The HC works with the local LM and LC to page the requested remote device and set up a connection.
4. If the connection is successful, then the local HC sends a Connection Complete event to the host with these parameters: Status, which is set to 0x00 (connection successfully completed), the Connection_Handle for that particular link, the BD_ADDR of the device connected to, the Link_Type designating whether the connection is ACL or SCO (ACL in this example), and the Encryption_Mode to designate whether encryption is disabled, enabled for point-to-point only, or enabled for both point-to-point and point-to-multipoint communication.

Event Process: Triggered by an Incoming Baseband Packet

If an incoming baseband packet triggers an HCI event, then the host is notified without having necessarily issued a previous command. The host typically responds to this event with a new command.

For example, if a remote device requests a connection with the local device via LMP_host_connection_req, then, assuming that the local host will approve such a connection, the following sequence occurs:

1. The local module's HC sends a Connection Request event to the host with these parameters: BD_ADDR of the requesting remote device, the Class_of_Device, and the Link_Type (ACL or SCO) being requested.
2. If the connection is accepted, then the local host sends HCI_Accept_Connection_Request to the HC. The parameters in this command are the BD_ADDR of the remote device, and the Role, which tells the HC to either remain as a slave or initiate a master-slave switch for this connection.

Event Status and Errors

The Bluetooth specification contains a list of particular errors that can occur in response to an HCI command packet from the host. An accompanying error code can be put into the Command Status Event Status field to indicate a failure of the command. A common example would be 0x01 "Unknown HCI Command." If the status parameter is 0x00, then the status is pending. Likewise, the event associated with completing a command will have a Status parameter in which either 0x00 (command successfully completed) or an appropriate error code will be placed by the HC. These error codes are all listed in the specification, and it would be agonizingly boring to list them here as well.

HCI as a High-Level Language

As we mentioned earlier, a single HCI command that involves the Bluetooth radio link will often cause several actions to occur in sequence at the LM and LC levels within the Bluetooth module. As such, HCI commands can be thought of as a high-level language, sometimes called an *abstraction*, with the HC operating as the compiler.

We will now give several examples of link activity that occurs when stimulated by a HCI command packet. These are, in order of typical occurrence

- Requesting the remote device's name, assuming that the ACL connection doesn't yet exist
- Establishing an ACL connection
- Authenticating an ACL connection that has already been established

Device Name Request

The Bluetooth device name is a user-friendly name that can convey certain useful information about the Bluetooth-enabled host. For example, a printer might be named, "Color laser printer, room 3F." This name can be retrieved from the remote module without its HC necessarily interacting with the remote host. In other words, the remote host and its user may not know that their Bluetooth module passed this information to another Bluetooth device.

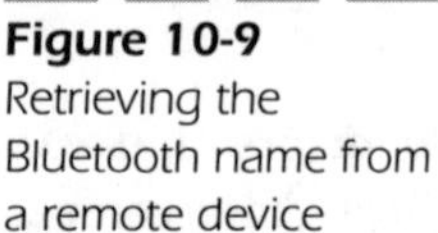

Figure 10-9 Retrieving the Bluetooth name from a remote device

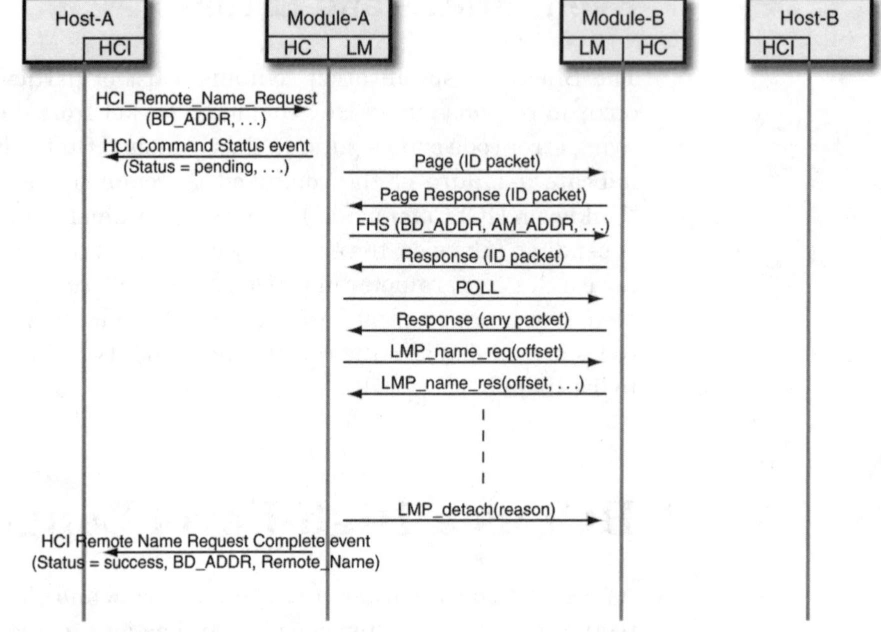

Figure 10-9 shows how the remote Bluetooth name is retrieved by using a single HCI command. Host-A and Module-A are connected via USB, as are Host-B and Module-B. The two modules communicate with each other over the Bluetooth RF link. Time progresses from top to bottom. The local host issues HCI_Remote_Name_Request that contains, among other parameters, the remote BD_ADDR associated with the name.

Upon receiving HCI_Remote_Name_Request(BD_ADDR, [other parameters]), the HC in Module-A returns a HCI Command Status event that lists the Status as pending (0x00). Next, the HC works with the LC and LM to perform several tasks over the Bluetooth RF channel. First, the remote device having the specified BD_ADDR is paged and then POLLed to check for a successful ACL connection. Next, the LMP_name_req(offset) is sent to request the remote device's Bluetooth name. Because this name had already been written into Module-B when it first communicated with Host-B, it's not necessary to contact the host to retrieve the name.

The LM on Module-B responds with its LMP_name_res(name offset, name length, name fragment) to the LM on Module-A. As soon as all the name fragments have been retrieved using these LMP packets, the LM on

Module-A sends LMP_detach(reason), with the reason probably listed as 0x16 "Connection Terminated by Local Host." Now the local HC sends an HCI Remote Name Request Complete event with the Status as "success" (0x00), along with the remote BD_ADDR and the remote device's Bluetooth name as additional parameters.

Establishing an ACL Connection

Figure 10-10 shows the interaction between local and remote devices when an ACL connection is being established between them. The initial ACL connection is started with an HCI_Create_Connection(BD_ADDR, [other parameters]), after which the local HC returns the usual Command Status event with Status pending. The local LM pages the desired BD_ADDR and checks for a successful ACL connection with a POLL. Next, the two LMs trade their supported features, and the local LM sends LMP_host_connection_req to the remote LM. This triggers a remote HCI Connection Request event to the remote host.

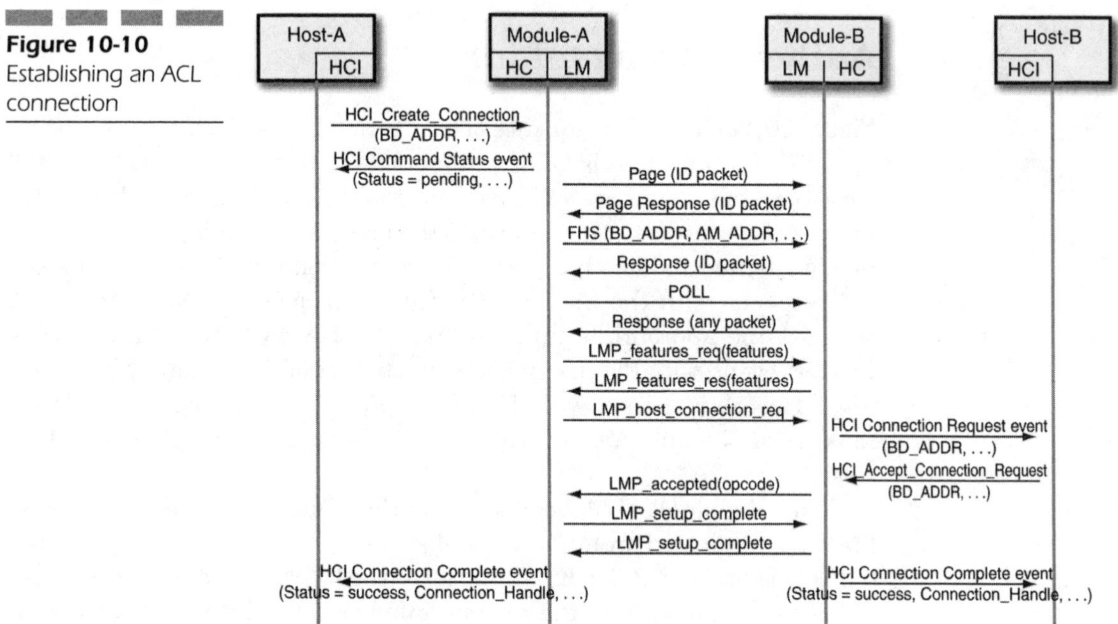

Figure 10-10 Establishing an ACL connection

Assuming that the connection is approved, the remote host returns a HCI_Accept_Connection_Request (BD_ADDR, Role) command. If the remote host is content with being the slave in the piconet, which is the situation shown in Figure 10-10, then the remote LM sends LMP_accepted to the local LM. If the remote host requires a role change, then the two LMs initiate the master-slave switch at this point. Finally, each device sends LMP_setup_complete to the other, which triggers an HCI Connection Complete event to each host.

The request to establish an ACL connection could originate when the local host's application wants to open an L2CAP channel. The local host begins by sending L2CA_ConnectReq to its own L2CAP layer (see Chapter 8). L2CAP then communicates this request to HCI on the local host, and the process shown in Figure 10-10 begins. Now when the remote host receives the HCI Connection Request event, this request is elevated to L2CAP and then to the application using L2CA_ConnectInd. The application responds with L2CA_ConnectRsp if the request is approved, and the remote HCI continues by sending the HCI_Approve_Connection_Request command to the remote module. After the connection is completed, then the local L2CAP layer sends the application L2CA_ConnectCnf, completing the establishment of the L2CAP channel on top of the new ACL link.

Authenticating a Link

Figure 10-11 shows the sequence of events that takes place when an established ACL connection is being authenticated. Host-A is the verifier, and Host-B is the claimant. We'll assume that the link keys at both ends are stored on the host. Host-A begins the process by sending HCI_Authentication_Requested for that particular connection handle, and Module-A's HC responds with the Command Status event pending. Next, Module-A requests the appropriate link key from Host-A with an HCI Link Key Request event, with the BD_ADDR equal to that of the verifier (Module-B). After Host-A responds with HCI_Link_Key_Request_Reply(BD_ADDR, Link Key), Module-A's HC sends HCI Command Complete event to Host-A.

Next, Module-A's LM begins the authentication process by sending LMP_au_rand(random number) to Module B's LM. Because the required link key is on Host-B, Module-B retrieves the link key and uses it to calculate a result SRES from the random number it received from Module-A. This result is sent to Module-A in an LMP_sres(authentication result) packet, and if the result is correct, then Module-A's HC sends an HCI

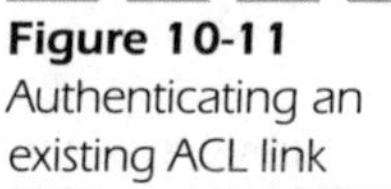

Figure 10-11 Authenticating an existing ACL link

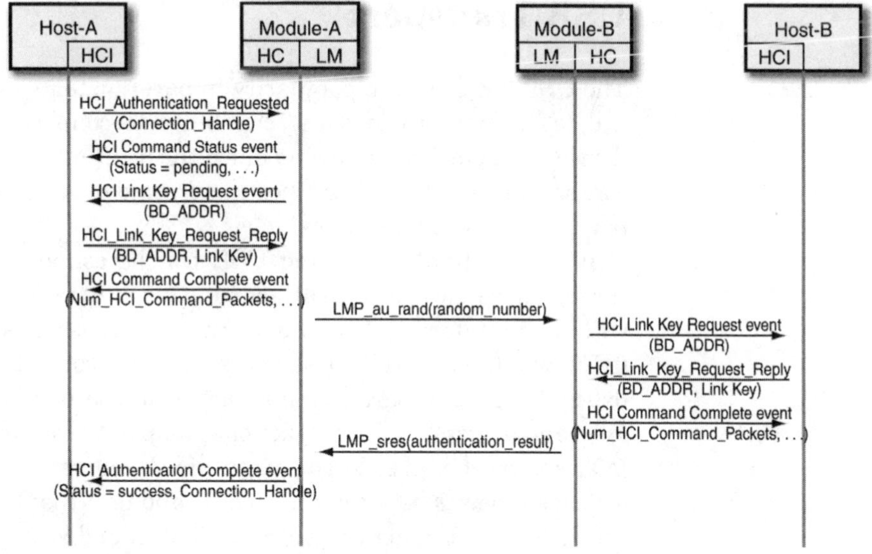

Authentication Complete event, with a Status = 0x00 success and the affected Connection_Handle, to Host-A.

HCI Transport Mechanisms

The Bluetooth host incorporates an HCI driver between its L2CAP layer and the physical bus driver used to send the information across to the Bluetooth module, as shown in Figure 10-2. Also, the module's HC firmware resides between the LM/LC layers and the physical bus driver. This architecture presumably enables any module with a compatible bus and driver to be attached to the host, but it complicates the interface by adding an extra transport layer. Also, an HCI packet has no type field that designates whether the packet is a command, an event, ACL data, or SCO data. Therefore, other methods must be used within the HCI transport layer to specify these packet types.

The Bluetooth specification defines three HCI transport layers: USB, UART, and RS-232. A PC-card transport layer also exists,[2] but it isn't part of the specification because the method of indicating what type of HCI packet (command, event, ACL data, and SCO data) is implementation dependent. Consequently, we'll focus on the three transport layers listed in the specification itself.

USB Transport

The USB has grown in popularity, in part due to its simple four-wire physical layer that provides the capability to connect and disconnect USB devices without first powering down the equipment. Multiple USB devices can be attached to a single host, and some devices can even obtain their power over the USB physical layer.

Bluetooth hardware using USB can be connected to the host either through a dongle (external attachment) or integrated onto the motherboard with a USB interface. The Bluetooth SIG has assigned a USB class code of 0xE0 (wireless controller), subclass 0x01 (RF controller), and protocol code 0x01 (Bluetooth programming). Each 1-millisecond USB frame is 64 bytes long, and a single USB transaction (group of frames in response to a single I/O request) should collectively contain an entire HCI packet. USB defines logical endpoints between the module and host that can be used to identify the HCI packet being transferred. Control endpoint 0x00 is used for HCI commands. HCI events are interrupt driven on endpoint 0x81.

The USB bulk endpoint 0x82 is used to transport HCI ACL packets. The bulk endpoints have built-in error control, so ACL packets are assumed to be transported error free. The isochronous endpoint 0x83 is used to transport HCI SCO packets, and these packets could be corrupted across the USB transport. However, the raw bit error rate over USB is only 10^{-13}, so SCO audio data is far more likely to be corrupted over the Bluetooth link than over USB.

There are a couple of power issues that the designer and user of USB Bluetooth modules should be aware of. First, if the module is powered over USB, then removing the module will erase volatile memory unless backup power is provided on the module. Second, when a module is installed, it will check with the host once every millisecond for tasks, which may prevent the host from entering any of its low power modes. If the host is a laptop, then its battery may be exhausted much sooner when the USB module is attached.

Packet Flow from Host to Module In Chapter 8, we showed how a L2CAP packet can be segmented and a portion of it transmitted in a baseband Bluetooth packet (refer to Figure 8-6). By now you've no doubt suspected that things are probably not that simple in reality, and you would be correct. Figure 10-12 shows what really happens to a L2CAP packet when HCI is used over USB between a local host and module. The L2CAP packet is first segmented into the payload of (usually) several HCI packets, and these in turn are segmented into several USB packets for transport to

Figure 10-12 Packet activity on the Bluetooth host and module. The L2CAP packet is segmented into several HCI packets, each of which is in turn segmented into several USB packets for transport to the module. Incoming USB packets to the module are reassembled into HCI packets that are then segmented into baseband packet payloads for transmission over the Bluetooth RF link. The process runs in reverse on packets arriving over the Bluetooth RF link.

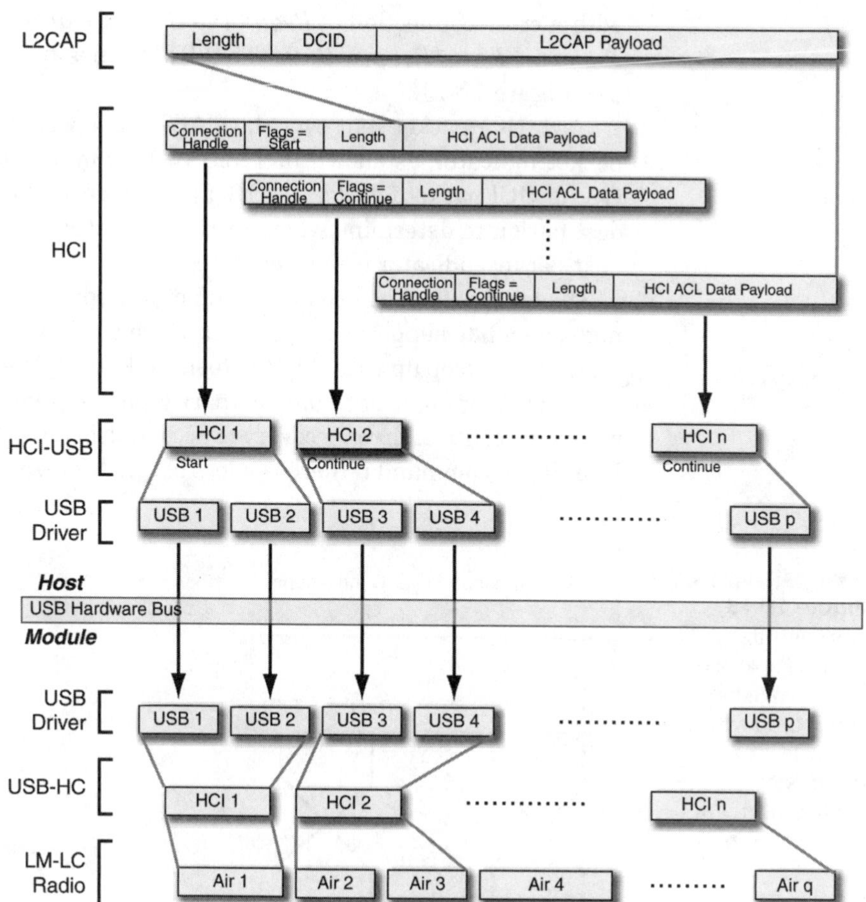

the Bluetooth module. At the module, the USB packets are reassembled into their HCI packets, which in turn are segmented into the payloads of several baseband packets for transmission over the air. At the remote end, the process is reversed until the L2CAP packet is completed at the remote host. It's a good thing that the segmentation and reassembly processes are computationally efficient!

UART Transport

The UART transport layer is perhaps the simplest, providing a serial interface between two UARTs on the same printed circuit board. The two UARTs are connected together in the classic null-modem configuration, with a cross-connection of *Request to Send / Clear to Send* (RTS/CTS) and *Transmit Data / Receive Data* (TXD/RXD) lines between the two devices (see Figure 10-13).

Each HCI packet sent over the UART transport is preceded by a 1-byte packet indicator, as shown in Table 10-1, to inform the destination what type of HCI packet follows. The HCI packet length field can be used by the destination to determine when an HCI packet is completed and when the next packet indicator is to be expected.

The baud rate to be used is manufacturer specific, but clearly, it must be high enough to support the speeds at which the host and module must communicate to keep up with the Bluetooth link itself. The serial port should be configured for eight data bits, no parity, one stop bit, and RTS/CTS hardware flow control. Error recovery is accomplished by the host performing a HCI_Reset command if there is a loss of synchronization. The local HC will

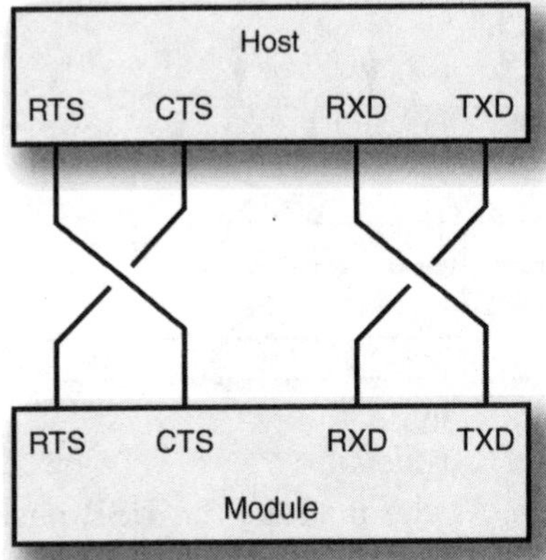

Figure 10-13 The UART transport layer is implemented in hardware by connecting two UART devices in a null-modem configuration.

Table 10-1 UART packet indicators

HCI Packet Type	HCI Packet Indicator
Command	0x01
ACL data	0x02
SCO data	0x03
Event	0x04

realize an error situation has occurred when an incorrect packet indicator has been detected or a length field is out of range.

It's clear that the UART transport of HCI packets is extremely simple, but the downside is that no error control is provided, and recovery from either errors or loss of synchronization is accomplished by the somewhat drastic reset operation.

RS-232 Transport

The RS-232 transport layer, also called the *serial port interface* (SPI), is another method of connecting the host to the module using serial communications, but, unlike UART, RS-232 is more sophisticated in its approach to error recovery and link parameter negotiation. RS-232 transport will support devices connected in a four-wire null-modem configuration with hardware RTS/CTS flow control, or as a two-wire configuration using only the TXD and RXD lines with special framing delimiters for synchronization and error recovery.

Each packet over RS-232 transport is preceded by a 1-byte packet indicator, as shown in Table 10-2, and a 1-byte sequence number that is incremented by 1 each time a new packet is sent. The sequence number remains unchanged if the packet is retransmitted as part of an error recovery procedure. The first four entries in Table 10-2 are the same as for UART, but the last two are used for reporting errors and supporting the negotiation of communication settings and protocols.

The RS-232 link is initiated with a baud rate of 9600, 8 data bits with no parity, and 1 stop bit. The default baud rate will, of course, need to be increased if the RS-232 transport layer is to support maximum Bluetooth data rates.

Table 10-2

RS-232 packet indicators

HCI Packet Type	HCI Packet Indicator
Command	0x01
ACL data	0x02
SCO data	0x03
Event	0x04
Error message packet	0x05
Negotiation packet	0x06

Summary

The HCI provides a standard interface between a module and a host. HCI packets can be commands from the host to the module, events from the module to the host, ACL data in either direction, and SCO data in either direction. In other words, all Bluetooth-related communication between the host and module takes place over HCI.

Some HCI commands can be executed locally by the module itself, and these are responded to with a Command Complete event. Others require communicating across the Bluetooth radio link, and these result in a Command Status event to acknowledge the command, followed by a completion event tied to the command itself once the command has been completed.

HCI commands that involve communication across the Bluetooth radio link can be treated as a high-level language, where a command from the host is compiled by the module's HC into several sequential activities by the link controller and LM.

The Bluetooth specification lists three transport layers for HCI: USB, UART, and RS-232. Either can be used for external or internal Bluetooth modules. The UART is simple, but has a very inelegant error recovery mechanism provided by resetting the module. Both USB and RS-232 are more sophisticated and include their own respective error control mechanisms.

End Notes

1. *Bluetooth Assigned Numbers,* Version 1.1, published by the Bluetooth SIG, February 22, 2001.
2. *Bluetooth PC Card Transport Layer,* Version 1.0, published by the Bluetooth SIG, August 25, 1999.

CHAPTER 11

Bluetooth Profiles

The function that a Bluetooth-enabled device is supposed to perform is based upon its usage model, which is a real-world model of what a customer would expect from this device. As we mentioned in Chapter 1, "Introduction," the usage models were originally conceived for the following categories:

- Three-in-one phone
- Ultimate *headset* (HS)
- Internet bridge
- Data access point
- Object push
- File transfer
- Automatic synchronization

Other usage models have been added as the capabilities of Bluetooth are matched to other areas where a wireless link would be a logical means for communication. Among these are

- *Human interface device* (HID)
- Audio/video distribution
- Audio/video remote control
- Basic printing
- Basic imaging
- Hardcopy cable replacement
- *Personal area network* (PAN)
- Operating a phone via an in-car device

You'll notice that as the usage models proliferate, they are branching out into several diverse areas, even those that have in the past been predominately serviced by infrared transmission and reception.

These usage models aren't part of the Bluetooth specification. Instead, a set of *profiles* has been established that gives a Bluetooth-enabled device its personality. Do you want the device to be an HS? Use the HS profile. A mouse? Use the HID profile. Profiles associated with the original usage models are included as a compilation in the *Bluetooth Profiles* document that was released as part of Specification 1.1. Profiles associated with subsequent usage models are being released over time as separate documents.

The purpose of a Bluetooth profile is threefold:

- It reduces the number of options and sets parameter ranges within the protocols.
- It specifies the order in which procedures are combined.
- It provides a common user experience across devices from different manufacturers.

The manufacturer of a Bluetooth HS, for example, won't need to implement all of the features listed in the specification but only those directly related to HS operation as detailed in the HS profile.

Overview of Bluetooth Profiles

Several profiles exist in Specification 1.1, and their interaction with each other is shown in Figure 11-1. For example, if a Bluetooth device is to have the capability to perform automatic file synchronization, then the Generic Access profile, Serial Port profile, Generic Object Exchange profile, and Synchronization profile will all play a role in the device's capabilities. Profiles can be envisioned as a vertical slice through the Bluetooth protocol stack, in which a subset of capabilities in each layer is selected for the particular Bluetooth function being developed. Automatic file synchronization, for example, doesn't require the use of two-way real-time audio, so implementing audio isn't necessary for that application. Each usage model has its own corresponding profile(s).

The profiles themselves use the various Bluetooth protocols as building blocks. Indeed, some of the higher-level protocols, such as *radio frequency communication* (RFCOMM), *object exchange* (OBEX), the *Wireless Application Protocol* (WAP), the *telephony control protocol specification* (TCS), the *service discovery protocol* (SDP) (see Figure 11-2), and others not shown, such as the *Bluetooth Network Encapsulation Protocol* (BNEP), are fully encompassed by one or more profiles. Therefore, we'll summarize the functions of some of these protocols when we discuss their use in a profile.

During our descriptions of the upper-level protocols and profiles, we'll deviate from our usual practice of in-depth analysis and instead summarize what these entities can do without spending a lot of time on implementation details. There are a number of reasons that we've taken this approach. First, the means by which the protocols and profiles perform

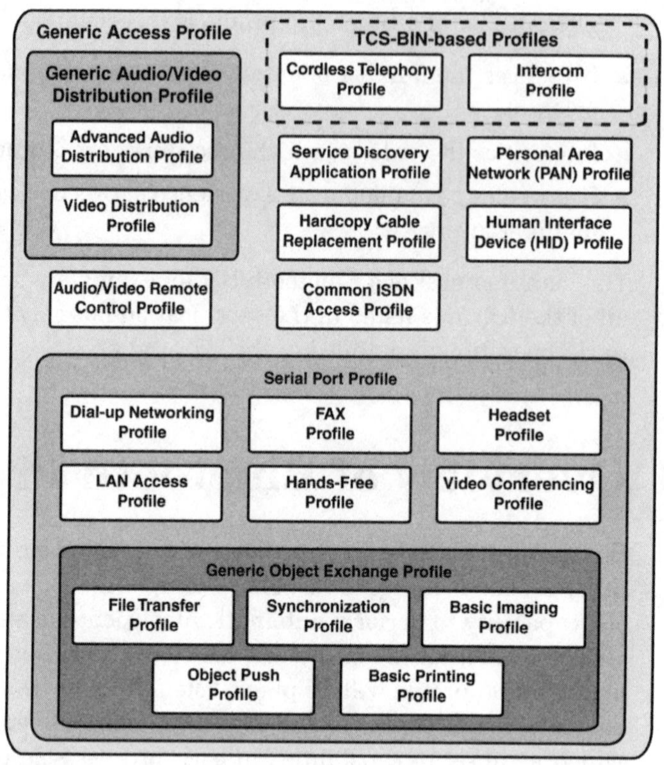

Figure 11-1 Bluetooth profiles implement the various usage models by taking a vertical slice through the protocol stack to incorporate only those protocols and features that are needed to realize the profile. Some profiles build upon others, as shown in this nested diagram. More profiles will be added as they are developed.

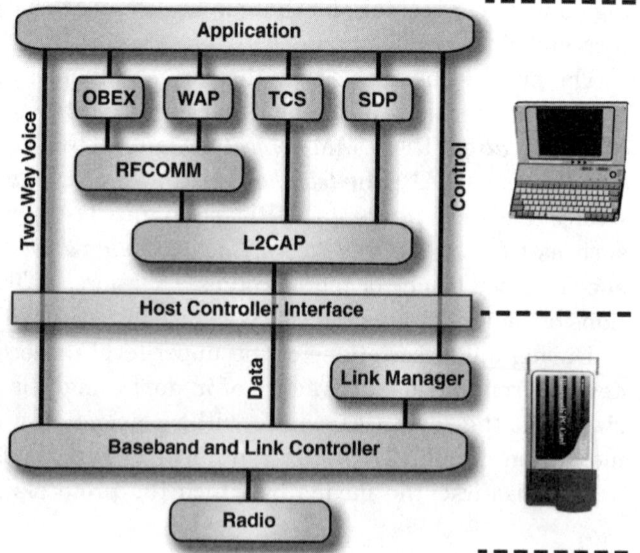

Figure 11-2 The profiles also encompass the higher-level protocols in the stack.

their various tasks are determined by the software developer and form the base for much of that developer's intellectual property. Second, development details that are given in the specification mostly consist of a plethora of charts and tables, and it would be somewhat boring (for both reader and author!) to duplicate them. Finally, most developers of Bluetooth-enabled devices require an in-depth analysis of perhaps one or two profiles, not the whole lot. It therefore makes little sense to add hundreds of pages of profile details here.

Profile Descriptions

Most of the profile descriptions in the specification include a section that examines the protocol stack layer by layer and specifies how the various features in each layer are to be implemented in the context of that profile. The features are assigned *conformance requirements* that fall into the following categories:

- **M** Implementation of that feature is mandatory to conform to the profile.
- **O** Implementation of that feature is optional.
- **C** Implementation of that feature is conditional, based on some subset of performance criteria.
- **X** The feature may be supported in the unit, but it is never used (excluded) in that profile.
- **N/A** The capability is not applicable to that particular profile.

Some of these requirements will have a number after the letter, corresponding to a note; for example, C1 means "Implementation is conditional based upon information in Note 1 given elsewhere."

As an example, a portion of the Cordless Telephony profile's LM functions is shown in Table 11-1. The cordless telephone implementation consists of a base station called the *gateway* and a number of telephone handsets called *terminals*. The LM features needed for these two entities can differ, and these in turn can be different from the support required in the LM part of the specification. To keep all of this straight, Table 11-1 shows how some of the LM features are listed within the profile description.

Each feature's expected support (M or O) within the LM chapter in the specification is given in the second column; then the terminal and gateway support is specified in the third and fourth columns if it's different. For

Table 11-1

Cordless telephony profile requirements for the LM (partial list)

Procedure	Support in LMP	Support in Terminal	Support in Gateway
Authentication	M		
Pairing	M		
Encryption	O	M	M
Switch of master-slave role	O	M	C1
Hold mode	O		
Sniff mode	O		
Park mode	O	M	C1
SCO links	O	M	M

Note: If multiterminals are supported, then M; otherwise, N/A.

example, authentication and pairing are mandatory in the LMP, and because there's no entry in the terminal and gateway columns, the features remain mandatory in this profile. Support of SCO links, although optional in the LMP, becomes mandatory in this profile because the telephone devices use real-time, two-way voice communication. Finally, the master-slave role switch must be supported in the terminal because the gateway insists on being the master if multiple terminals are connected to it, and yet the terminal may initiate a connection to the gateway. In that light, gateway support of the master-slave role switch is conditioned upon whether it has multiple-terminal capabilities. Also, if a gateway supports multiple terminals, then it must be able to park those in the piconet while performing page or inquiry scans or while brining another terminal into the piconet.

A Paradigm Shift*

Up to this point, we've concentrated on the operation of Bluetooth itself, often referring to a Bluetooth device. We shift gears now and call such a device *Bluetooth-enabled* to emphasize that the purpose of Bluetooth is to enable a device to perform some useful task that Bluetooth can support.

*I read somewhere that the word *paradigm* must be included in any book that is to become successful.

Profiles focus on the use of a Bluetooth-enabled device and how to make it perform in a consistent and (hopefully) straightforward way. In other words, if a customer is given a Bluetooth-enabled HS, it should be obvious how that HS should be used, regardless of who manufactured it. To obtain Bluetooth certification, a manufacturer's device must conform to the profile associated with the usage model being addressed.

Some manufacturers may balk at the requirement that a particular profile be followed when designing a particular Bluetooth-enabled device. After all, how can a manufacturer gain a significant market share if the devices from all manufacturers operate alike? Actually, the profiles allow the manufacturer enough latitude that other performance criteria, such as data rate and operating convenience, may be very different for Bluetooth-enabled devices that conform to the same profile. The purpose of profiles is not to eliminate competition in areas other than price and marketing prowess, but instead they exist to reduce the possibility of customer frustration while using complex and somewhat unpredictable wireless technology.

It's also important to keep in mind that usage models and their respective profiles are in a state of dynamic change and expansion. As an example, profiles associated with the second group of usage models listed previously were added after the Bluetooth specification was first published, despite the fact that many of these involve obvious uses for the Bluetooth wireless link. New profiles are being added at a rapid pace, encompassing such diverse areas as interactive games, automotive applications, local positioning, and secure access. Although Figure 11-1 shows the relationship of the profiles that existed as of this writing, more profiles are certain to be approved and existing profiles could change as Bluetooth applications expand.

We begin by examining the profiles that aren't necessarily associated with any particular usage model. These include the *generic access profile* (GAP), the *service discovery application profile* (SDAP), and the *serial port profile* (SPP). After that, we summarize the remaining profiles that tend to be associated with specific usage models.

General Profiles

Several Bluetooth profiles have been defined that aren't associated with any one-usage model. The GAP, for example, applies to all the usage models. If a Bluetooth-enabled device is developed for a custom application different from any of the approved usage models, it must still conform to the

GAP. The SDAP defines the features and procedures required to perform in-depth searching and browsing of services supported by another Bluetooth-enabled device. Finally, the SPP works with RFCOMM to define the emulation of a serial cable. We will highlight the characteristics of each of these profiles in turn.

Generic Access Profile (GAP)

The GAP defines generic procedures related to the discovery, link management, and use of security levels for Bluetooth-enabled devices. Also specified in the profile are format requirements for some of the parameters that the user may access. This profile must be included with all Bluetooth-enabled devices that don't claim conformance to any other profile, or if the device is to perform a custom application. If the Bluetooth-enabled device conforms to another profile, then modified portions of the GAP will be listed in that profile.

A large part of the GAP is dedicated to the definition of terms to be used as part of the Bluetooth vocabulary, including those terms employed at the *user interface* (UI) level. For example, the A-party is defined as the paging device when the link is being established or the initiator of a procedure on an already-established link. The B-party is the paged device or acceptor of the procedure. Even if the two devices have no operation in common and cannot possibly function together (such as HS and printer), they should be able to at least connect via a Bluetooth link and discover their incompatibility. Careful definition and use of the various UI terms should reduce the chance that a poorly written instruction manual will enter the marketplace (see Figure 11-3).

The GAP profile fundamentals are

- Stating the requirements for the definition and use of names, values, and coding schemes
- Defining the general procedures for discovering the identity, name, and basic capabilities of another discoverable Bluetooth-enabled device
- Defining general bonding procedures
- Describing the general procedures used for establishing a connection to another connectable Bluetooth-enabled device

As we mentioned in Chapter 9, "Bluetooth Security," the GAP defines the various *operational modes* of the Bluetooth-enabled device. These define the discoverability, connectability, and paring modes of the device as follows:

Excerpts from poorly written user instruction manuals.... *"Finally, push the red button. But first, pull down on the big lever."*

Figure 11-3 Badly written instruction manuals could be detrimental toward the success of Bluetooth. (Source: Bill Lae)

- **Discoverability modes**
  - **Nondiscoverable** The device never enters the inquiry scan state.
  - **Limited discoverable** The device responds to the *limited inquiry access code* (LIAC) and the *general inquiry access code* (GIAC).
  - **General discoverable** The device responds to the GIAC but not to the LIAC.
- **Connectability modes**
  - **Nonconnectable** The device never enters the page scan state.
  - **Connectable** The device periodically enters the page scan state.
- **Pairing modes**
  - **Nonpairable** The device will respond to LMP_in_rand with LMP_not_accepted with the reason "pairing not allowed."
  - **Pairable** The device will respond to LMP_in_rand with LMP_accepted, or its own LMP_in_rand if it has a fixed PIN.

Notice that these definitions are quite precise and give the developer specific instructions on the operating states that the device is or is not enabled to enter. Furthermore, minimum recommended time periods for each of these modes are also listed in the profile. Following these definitions, the GAP then addresses the three security modes (nonsecure, service-level-enforced security, and link-level-enforced security) that we covered in Chapter 9, "Bluetooth Security."

Next, the GAP defines the *idle mode procedures* as follows:

- **General inquiry** This mode is used by devices that make themselves discoverable through the GIAC either continually or based upon no specific condition. When a general inquiry is initiated, the device shall be in the inquiry state for at least 10.24 seconds. This is called *Bluetooth device inquiry* at the UI level.
- **Limited inquiry** This mode provides the initiator the capability to discover devices that are made discoverable for a limited time, at least 30.72 seconds, using the LIAC either simultaneously with, or in sequence with, the GIAC. The inquiring device shall be in the inquiry state for at least 10.24 seconds. This is also called *Bluetooth device inquiry* at the UI level.
- **Name discovery** This mode provides the initiator with the Bluetooth device name of connectable devices. This is called *Bluetooth device name discovery* at the UI level.
- **Device discovery** This mode combines the inquiry and name discovery processes into a single sequence of operations. This is called *Bluetooth device discovery* at the UI level.
- **Bonding** This mode creates a relationship between two Bluetooth-enabled devices through the creation and exchange of link keys (refer to Chapter 9). *Dedicated bonding* is performed when the link is formed to create and exchange a Bluetooth link key between devices. Bonding can also involve higher-level initialization procedures in a process called *general bonding*. Either of these is called *Bluetooth bonding* at the UI level.

The various modes (operational, idle, and so on) specified in the GAP are addressed in each Bluetooth profile, in which a table lists their implementation requirements (M, O, C, X, and N/A). Finally, the GAP defines and gives the procedures for establishing an ACL physical link, a L2CAP channel, and a connection between applications. In other words, the profile describes the complete setup process so two applications at different locations can communicate.

Service Discovery Application Profile (SDAP)

The Bluetooth specification includes a comprehensive means for one device to search for services that are available on another device. SDAP works through the SDP, which provides the capability for another device to discover which services are available and determine the characteristics of those services. The protocol is based upon the classic client/server model, where the client wants to access services that are provided on the server. The client queries the server in the form of requests, and the server responds to the requests with information.

Most Bluetooth profiles have a relatively narrow SDP scope, such as retrieving information required to set up a transport service or usage scenario. SDP is often initiated automatically in these cases, and the required information is retrieved without user interaction. In contrast, SDAP is intended to be initiated by a human user and enables extensive searching for services on a client device. An example would be accessing a library card catalog or set of financial databases.

Overview of the SDP SDP can be used to access a specific device (such as a digital camera) and retrieve its capabilities or to access a specific application (such as print job) and find devices that support that application. The former task requires paging a single device and forming an ACL link to retrieve the desired information, and the latter involves connecting to and retrieving information from several devices that are discovered via an inquiry.

SDP supports the following:

- Browsing for services on a particular device
- Searching for and discovering services based upon desired attributes
- Incrementally searching a device's service list to limit the amount of data to be exchanged

An L2CAP channel with a *protocol service multiplexer* (PSM) of 0x0001 is used for the exchange of SDP-related information. SDP has both client and server implementations, with at most one SDP server on any one device. However, if a device is client only, then it need not have an SDP server.

SDP also has some negative characteristics:

- Third-party participation in SDP is not required.
- Access to services is not provided by SDP.
- No warnings are given to client devices by SDP when services become unavailable.
- SDP does not provide a means for negotiating service parameters or availability.

Each service is listed in the device's SDP database as a *service record* having a unique ServiceRecordHandle, and each attribute of the service record is given an attribute ID and an attribute value. Attributes include the various classes, descriptors, and names associated with the service record. For example, a Bluetooth HS would have SDP attributes specifying that it's a member of the HS and generic audio classes and that it requires the use of L2CAP and RFCOMM protocols for setup. Attributes are structured in a hierarchy to give the client the ability to probe as deeply as necessary until the requirements are satisfied without having to sift through a large amount of irrelevant information. Service inquiries can take the form of searches by service class, searches by service attribute, and service browsing. Each profile in the Bluetooth specification has a section in which the associated SDP requirements are listed.

Service Discovery Using SDAP The SDAP specifies the functions to be performed by both devices that participate in service discovery. The first is the *local device* (LocDev), which is the device that initiates the service discovery procedure and contains both SDAP and the client portion of SDP. The second is the *remote device* (RemDev) that responds to the LocDev service inquiries through its server SDP. Depending upon the context, the LocDev and RemDev may eventually exchange roles or assume different roles, depending upon their needs and capabilities. However, to qualify as a LocDev, the device should have a UI for entering service requirements and perhaps returning results of service searches.

Of course, before services can be searched or browsed, the LocDev and RemDev must first create an ACL link and accomplish any required authentication and encryption. However, no assumptions are made as to whether the LocDev or the RemDev is the master in the piconet, and the profile doesn't require authentication or encryption for its use; that's up to the particular implementation.

SDAP and the Protocol Stack The *service discovery application* (SrvDscApp) itself enables significant flexibility in how the services are discovered by the LocDev. For example, user input can be collected first, and the application can inquire and connect to several RemDevs, access their services, and display the results. Alternatively, because the inquiry process can take as long as 10.25 seconds in a perfect channel, SrvDscApp can perform the inquiry prior to or during the time user input is being collected in an attempt to speed the process. Connection to the remote devices can take place one at a time (point-to-point) or as a point-to-multipoint piconet. A point-to-multipoint piconet makes sense if the SDAP user wants to browse services on many devices in some indeterminate way. The SrvDscApp can react differently to the RemDev, depending upon whether it is trusted, new, or already connected to the LocDev.

The L2CAP layer is used by the LocDev to create a *connection-oriented* (CO) channel to support SDAP. This channel is configured to be reliable, so the default infinite flush timeout (0xFFFF) is specified. The *connectionless* (CL) channel isn't used because the (one-way) broadcast concept doesn't apply to service discovery. The LocDev usually terminates the L2CAP channel, but the RemDev must also have that capability in case the link is disrupted.

The LM layer has the usual capabilities except that authentication and/or encryption may be required, depending upon the use, and SCO links are excluded from participating in SDAP. No low-power mode use is specified in the profile, but its use is up to the individual LMs. At the LC layer, the AUX1 and all SCO packets are excluded from use in SDAP.

Serial Port Profile (SPP)

The SPP defines the protocols and procedures that are to be used by devices using Bluetooth for serial cable (RS-232 or similar) emulation. The goal of this profile is to make legacy applications think that Bluetooth is actually a wired serial cable with all of the expected hardware and software entities in place. SPP makes extensive use of the RFCOMM protocol layer.

This profile uses the convention put forth in the GAP by defining DevA as the *initiator* of the serial link and DevB as the *acceptor* of the link. This designation applies to link setup only and doesn't necessarily relate to the order in which the legacy applications access the link. The RFCOMM protocol is designed to be independent of which device is *data terminal equipment* (DTE) and which device is *data circuit-termination equipment* (DCE), and it supports either DTE-DCE or DTE-DTE (null modem) configurations.

SPP has mandatory support for one-slot packets, so data rates up to 128 kb/s can be used as a minimum. Support for multislot packets and their respective higher data rates is optional. Also, even though the profile is set up for only a point-to-point connection (consistent with a wired serial port), multiple executions of SPP can be used to support any configuration of point-to-multipoint piconet or even scatternet operation by one device.

Overview of RFCOMM The RFCOMM protocol provides the emulation of serial ports over L2CAP, based upon ETSI standard TS 07.10. Provisions are made for RFCOMM to emulate the nine circuits of the RS-232 serial port. These signals are

- Signal Common
- *Transmit Data* (TD or TxD)
- *Receive Data* (RD or RxD)
- *Request to Send* (RTS)
- *Clear to Send* (CTS)
- *Data Set Ready* (DSR)
- *Data Terminal Ready* (DTR)
- *Data Carrier Detect* (DCD)
- *Ring Indicator* (RI)

The protocol supports either a direct connection between two Bluetooth-enabled devices (DTE-DTE) or a wireless connection between a Bluetooth-enabled device and a DCE that, in turn, is connected to a non-Bluetooth device over (say) a wired link.

RFCOMM can provide up to 60 open emulated serial ports on one Bluetooth-enabled device, but this number may be smaller due to implementation limitations. These ports can be multiplexed into one RFCOMM implementation for a point-to-point piconet (see Figure 11-4[a] or in a point-to-multipoint piconet by dedicating one RFCOMM realization to each ACL link (see Figure 11-4[b]).

Unlike a wired serial port, where flow control has a simple hardware (RTS/CTS) or software (XON/XOFF) implementation, flow control through RFCOMM is much more complex because the serial port is being emulated through many layers of the Bluetooth protocol stack. L2CAP and ACL flow control have been described in Chapters 4, "Baseband Packets and Their Exchange," and 8, "Transferring Data and Audio Information," respectively. Flow control between RFCOMM and L2CAP on a single host is implementation dependent. Aggregate flow between two RFCOMM entities on

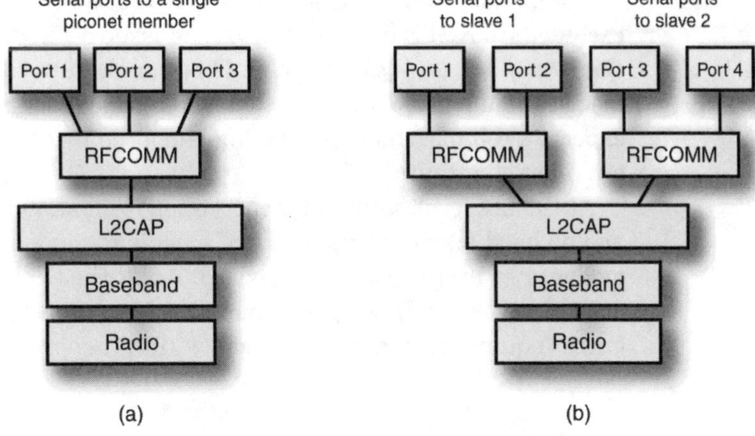

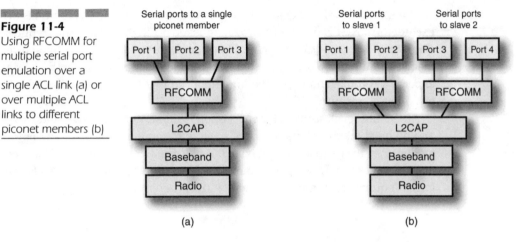

Figure 11-4 Using RFCOMM for multiple serial port emulation over a single ACL link (a) or over multiple ACL links to different piconet members (b)

different devices can be controlled using special FCon and FCoff commands, but for enhanced versatility, each *data link connection* (DLC) within RFCOMM should be provided with its own flow control mechanism.

Wireless emulation of wired flow control is also made more difficult by delays inherent in the Bluetooth link. Whereas a wired link responds to hardware flow control within perhaps a few hundred nanoseconds, the Bluetooth link may respond much more slowly, especially in an error-prone channel. To accommodate these delays, a *credit-based flow control* is implemented on each DLC, where each device knows how many RFCOMM frames the other is able to accept before its buffer is full. Frame credits are passed from one device to the other, which decrements the credit count each time it sends a RFCOMM frame. When the count reaches zero, it cannot send any more frames until more credits are obtained.

SPP and the Protocol Stack To use the Bluetooth serial port, the application at DevA must be able to establish the link and set up a virtual serial connection, and DevB's application must be able to accept the link. Furthermore, DevB must also be able to register its application service record in the SDP database for access by DevA. To initiate the serial port, DevA uses SDP to discover the existence of the RFCOMM capability and its parameters (baud rate and so on). While performing any necessary authentication and encryption, a new L2CAP channel is established and configured, and an RFCOMM session is initiated on that channel. Finally, a serial data link is started on the RFCOMM session.

At the RFCOMM protocol level, DevA must be able to initiate, and DevB to accept, a session, but either device must be able to terminate the session.

At the L2CAP level, CO channel support is mandatory, and CL channels are excluded from SPP use. Only DevA can issue a L2CAP connection request. *Quality of service* (QoS) support is optional, but a device in a low-power mode should be able to return to the active mode within 500 ms. RFCOMM expects a reliable channel to be implemented by L2CAP, so the flush time-out must be infinite.

At the LM level, support for encryption is mandatory for DevA and DevB, and no fixed master-slave roles are specified. Implementation of low-power modes is up to the respective device LMs. At the LC level, no AUX1 packets are allowed. Only DevA performs inquiries and pages, and only DevB performs inquiry scans and page scans within this profile.

Profiles Based upon Serial Port Communication

Once an RFCOMM port is set up between Bluetooth-enabled devices, their applications can begin transferring data using what appears to be a familiar serial port. Aside from the obvious object-exchange-based functions, such as file transfer, five other profiles are based upon having a serial link between devices. These are the Dial-Up Networking, Fax, HS, LAN access, and *hands-free* (HF) profiles, each of which will be summarized in this section. In the next section, we summarize the profiles that use the serial port for object exchange.

Dial-up Networking (DUN)

The DUN profile defines the protocols and procedures associated with the Internet Bridge usage model (refer to Chapter 1). This involves establishing a Bluetooth link between a computer and a cellular phone or modem, which in turn dials into the Internet or other service using a means other than Bluetooth. The profile also defines the procedures for the modem or cellular phone to receive calls and connect via Bluetooth to the computer.

The profile defines a *gateway* (GW) as the device that provides access to the public network and the *data terminal* (DT) as the device that uses the dial-up service. In RS-232-speak, the GW is DCE and the DT is DTE. The profile has the following characteristics:

- The profile requires support for only one-slot packets for data rates up to 128 kb/s.
- Only point-to-point single-call situations are supported.
- Authentication and encryption support are required.
- Standard modem AT commands are used.
- The GW isn't required to discriminate between different incoming call types.
- One SCO link is supported for call progress audio feedback.

LAN Access

The LAN access profile defines access to a LAN using the *Point-to-Point Protocol* (PPP) over RFCOMM. PPP is widely used and can support many networking protocols such as the *Internet Protocol* (IP); however, this profile doesn't mandate the use of any specific networking protocol. The device that provides access to the LAN is called the *LAN access point* (LAP). The connection between the LAP and the LAN itself can be via Ethernet, fiber, cable modem, USB, or any other means. The Bluetooth-enabled device that uses the services of the LAP is the DT.

The LAN access profile defines the support of PPP networking in three different situations:

- LAN access for a single Bluetooth-enabled device (see Figure 11-5[a]). In this situation, either the DT or LAP can be the piconet master.
- LAN access for multiple Bluetooth-enabled devices (see Figure 11-5[b]). In this situation, the LAP must be the master of the piconet, and aggregate piconet throughput is divided between the DTs.
- A computer-to-computer link using PPP over RFCOMM serial cable emulation (see Figure 11-5[c]). Either device can be the piconet master.

Communication between a DT and LAP proceeds in the following way:

1. DT finds a LAP within a range that provides PPP, RFCOMM, and L2CAP service.
2. DT forms an ACL link with the selected LAP.
3. DT establishes a L2CAP/RFCOMM/PPP connection with the LAP with authentication and encryption.
4. Non-Bluetooth (PPP) authentication, if needed, occurs from LAP to DT.

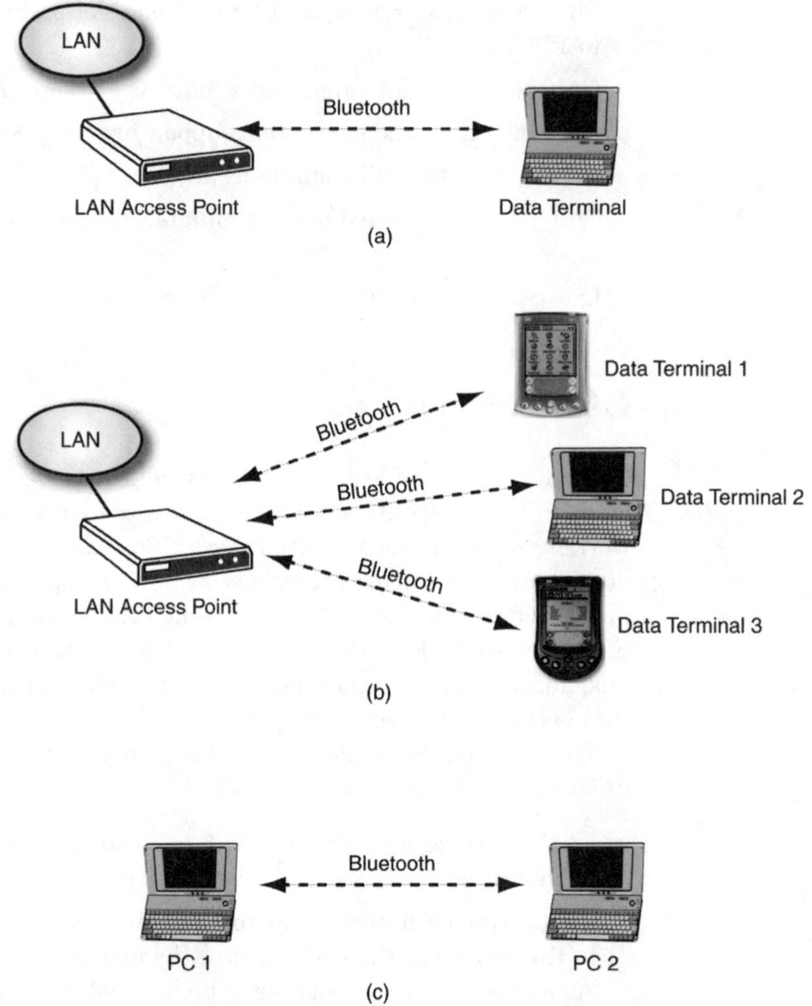

Figure 11-5 The LAN access profile enables three basic configurations using PPP: LAN access by a single data terminal (a), LAN access by multiple data terminals (b), and computer-to-computer communication (c).

5. A suitable IP address is negotiated between LAP and DT.
6. IP traffic can now be exchanged between LAP and DT.
7. Either device may terminate the PPP connection.

During the step 1 inquiry process, the DT may be provided a list of LAPs that are within the radio range and what services they provide, enabling the DT to select one of them. Alternatively, the DT user may be provided a list of services from all the LAPs within range, and the particular LAP eventually connected to depends upon the service selected. Another possibility is for the DT user to enter a service name, and the LAN access profile will then find and connect to the LAP providing the selected service.

Bluetooth baseband encryption is required any time PPP traffic flows across the link. Pairing and authentication must therefore take place between the DT and LAP. The default PIN is one byte set to 0x00. Note to hackers: you didn't notice the default PIN value. Note to users: the hackers noticed the default PIN value, so change it to a different value.

For RFCOMM, the DT is DevA, and the LAP is DevB. RFCOMM should make use of three- and five-slot packets for this profile to enhance throughput. Within the LM, implementation of pairing, authentication, and encryption are mandatory for both the DT and LAP, along with master-slave switch capability.

Fax

The Fax profile defines the policies and procedures for a device to link via Bluetooth to either a cellular phone or modem to send or receive a fax message. The GW provides modem emulation, and the DT provides a modem driver. Both sides have the appropriate modem dialing and control capability.

This profile has the same characteristics listed for the DUN profile. The Fax profile requires authentication and encryption. Audio feedback of the dialing process may be optionally provided via an SCO link initiated by the GW. HV3 packet support is mandatory when audio feedback is used. No support of voice calls is provided by this profile.

Headset (HS)

The HS profile is used for implementing the Ultimate HS (don't you love that name?) usage model. This profile applies to both the HS and to a computer, cellular phone, or an *audio gateway* (AG), which implements full-duplex audio capabilities for connection to the HS. The profile has the following characteristics:

- The Ultimate HS use case is the only one active between an HS and an AG.
- Monophonic CVSD modulation is used for the SCO link.
- One audio connection is enabled between an HS and an AG.

- The AG controls SCO link establishment and release.
- Multiple HS devices at the AG aren't supported by this profile.
- The HS user interface has at least one user-initiated action, such as a button press.
- Neither authentication nor encryption is mandated by this profile.

The AG may initiate a connection to the HS and then send a RING signal to alert the user. The RING can be either a bit pattern on an ACL link, in which case the HS itself will generate the RING tone, or it can occur as real-time digitized audio over an SCO link initiated by the AG.

Alternatively, the HS user can initiate the ACL connection with the AG by an action as simple as pressing a button on the HS. Of course, the HS must be preprogrammed to perform any inquiries and/or pages necessary to establish the link. The subsequent SCO link is still initiated by the AG, regardless of whether it is the master or slave.

The AG can also control the gain of the HS microphone and speaker over the ACL link. The actual volume levels set in response to the gain commands are implementation dependent. For a link that may be inactive for a long period, the park mode may be used to place both HS and AG into a low-power state.

Hands Free (HF)

The HF profile is one of the first profiles specifically directed at an automotive application. This profile defines the functions between a mobile phone and an HF device that controls the phone via a Bluetooth link. The HF device is assumed to be embedded in the vehicle as opposed to a portable HS covered by the HS profile. The usage model associated with the HF profile is "Operating a Phone via an In-Car Device." Wow.

The AG is typically the cellular phone that connects to the HF unit in the car. The profile has the following characteristics:

- Only one Operating a Phone via an In-Car Device use case applies to a link between HF and AG.
- Monophonic CVSD is required.
- One audio connection is enabled between an HF and an AG.
- Either HF or AG may initiate SCO link establishment and release.
- Multiple calls at the AG aren't supported in depth by this profile.

- Implementation guidelines are provided for the simultaneous presence of one Ultimate HS and one HF usage model between one HF and AG.
- Neither authentication nor encryption is mandated by this profile.

When the HF and AG meet, they form a *service level connection* over RFCOMM with which the HF uses modem AT commands to retrieve information such as call status indicators, hold services, and multiparty services from the AG. The devices can then either establish an audio link or perhaps enter the park state if no immediate user action occurs. The service level connection must exist before a SCO link can be established.

When an incoming call arrives, the AG sends a RING signal to the HF. The RING can be either a bit pattern on an ACL link, in which case the HF itself will generate the RING tone, or it can occur as real-time digitized audio over a SCO link initiated by the AG.

For outgoing calls, the phone number can be supplied by the HF either directly, through a memory dial process, or through a last-number redial. Call waiting and three-way calling are also possible. AT commands are available for the HF to activate voice recognition within the AG, if that feature is available. The AG can also remotely control the audio volume and microphone gain of the HF.

Video Conferencing

The Video Conferencing profile defines the protocols and procedures used by devices using Bluetooth for video conferencing based upon 3G-324M. The profile assumes that the two endpoints in the video conference are in two separate piconets, or perhaps in one 3G terminal and one piconet, and are separated by an intermediate non-Bluetooth network. Access by a DT is provided by a Bluetooth link to a GW, which in turn is connected to the non-Bluetooth network. This profile doesn't provide any remote control functions; however, the A/V Remote Control profile can be implemented for this purpose.

The DT can either originate a call or receive a call from the GW. The DT performs any required encoding and decoding of user data to or from the external network. The profile only supports point-to-point connections between a GW and DT, so there is no fixed master-slave role between the two devices. Use of authentication and encryption are optional, but their support is mandatory; that is, they must be available for use. Serial port emulation is used to transport modem AT control commands between the

GW and DT. User video is transported on a SCO link, which is configured to carry transparent data (refer to Chapter 11, "Transferring Data and Audio Information").

Profiles Based upon Object Exchange

The *Generic Object Exchange profile* (GOEP) works within the SPP and forms the framework for transferring *objects* between two Bluetooth-enabled devices. An object is defined as a small data file, which can be a business card or appointment information. Two devices that have GOEP capabilities begin by establishing an object exchange session between them, initiated by the client. The client then has the option of either 1) *pushing* one of its own files to the server, or 2) *pulling* a file from the server into its own memory. These operations fall under the name of OBEX.

The GOEP has the following characteristics:

- The profile supports point-to-point links only.
- Pairing and bonding must be supported when two devices meet for the first time.
- Although authentication and encryption must be supported, their use is implementation dependent and additional security may be required beyond the usual baseband security procedures.
- No fixed master-slave or power-down roles are specified in the profile.

Several other profiles are based upon the GOEP, as shown in Figure 11-1. The Bluetooth specification supports the non-Bluetooth *Infrared Data Association* (IrDA) object exchange, called IrOBEX, by adapting it to RF operation and naming it OBEX.

Object Push

The Object Push profile enables a Bluetooth-enabled device to *push* (send) an object into the inbox of another Bluetooth-enabled device. Contrary to implications within its name, the profile also enables a device to *pull* (retrieve) an object from another device if the device supports such a pull operation. The *exchange* of objects is also supported by the profile. The object content in a push can be a phone book or business card (vCard), calendar (vCalendar), messages (vMessage), or notes (vNote) as long as the devices can support these formats. Objects that are pulled or exchanged should be in the vCard format.

The server providing object exchange is called the *push server*, and its client is the *push client*. The client device always requires end-user interaction to initiate the push, pull, or exchange processes. In other words, these operations should never occur automatically. A push server must be able to receive multiple objects during a single OBEX session, but it is not mandatory for a push client to be able to send multiple objects.

File Transfer

The File Transfer profile extends the simple object push scenario into an exchange of files between two devices. The profile enables one device to connect to another and do the following, if authorized:

- Browse the file system, and open and view files and folders (directories).
- Transfer files from one device to another.
- Manipulate files and folders (such as creating, moving, or deleting) on the other device.

The *client* initiates the operation of pushing or pulling objects to or from a *server*. The server should have a *root folder* that is part of a *public directory* for the client to access. Different directories could be set aside for different clients. The profile requires the client to support file push and pull, but support for folder push and pull is optional. The server may have read-only files and folders to restrict the client's push options. The client is not required by the profile to support folder/file deletion or folder creation. Support for bonding is required, but its use is optional.

Synchronization

The Synchronization profile defines the capability of two devices to connect and update each other's files (phonebooks, calendars, e-mails, or notes) to the latest version that exists on either device. For example, a computer can connect to a mobile phone or *personal digital assistant* (PDA) to exchange *personal information management* (PIM) data to make the respective object stores identical, and the same process can be initiated by the phone or PDA. The synchronization process can be either automatic or initiated manually.

Bluetooth uses the *Infrared Mobile Communication* (IrMC) model for the Synchronization profile. The IrMC server is usually the mobile phone or PDA, and the IrMC client is usually the PC. However, the devices can exchange roles at times to meet protocol functionality requirements. Either device can initiate link and channel establishment procedures.

When an IrMC server and client meet for the first time, the client can perform either a limited or general inquiry, and the user selects the desired server from the resulting list. A PIN is entered into each device and, if implemented, an OBEX authentication password as well. Finally, the synchronization is processed. If the devices are bonded, then subsequent meetings for synchronization can occur either automatically or upon a user-selected synchronization command. Automatic synchronization can take place without user notification. Support of automatic synchronization is mandatory for a server but optional for a client.

Basic Printing

The Basic Printing profile enables mobile devices such as phones, pagers, and PDAs to print e-mail text and other short messages, as well as formatted documents. Optional support is given to structured data objects such as vCard and vCalendar. For traditional printing jobs originating in a laptop or desktop PC using specific printer drivers, the Hardcopy Cable Replacement profile should be used instead.

The *printer* is defined as (surprise) the server device, and the *sender* is the client that pushes an object to the printer. The information being printed can originate at the sender (direct printing) or can be referenced by the sender to another source, such as on the Internet (reference printing). Control over print formatting can be accomplished by either the sender or printer. Link-level authentication and encryption and device bonding are mandatory to support and optional to use. OBEX authentication is mandatory for a sender to support but optional to support or use for the printer.

The printer can be placed into the following modes:

- **Bluetooth Offline mode** The printer cannot receive print jobs over the Bluetooth link, and the printer cannot be discovered or paged. Support for this mode is optional.
- **Private Online mode** The printer can receive print jobs over the Bluetooth link only by senders that know the printer's BD_ADDR. Support for this mode is optional.
- **Public Online mode** The printer can receive print jobs over the Bluetooth link from senders that know the printer's BD_ADDR, and

the printer can be discovered through inquiry. Support for this mode is mandatory.

- **Bonding mode** The printer is ready for bonding with a sender. Support for this mode is mandatory.

The simplest way to print is for the sender to perform an OBEX PUT request (called FilePush in this profile) to send a basic job to the printer using the *simple push* transfer model. No status, error, or printer control is supported. The printer simply prints the information using its own formatting and settings. The printer can also retrieve a referenced object, such as an image, from the sender over a separate OBEX connection initiated by the printer.

A more complex print session can be initiated using the *job-based* transfer model, where the sender initiates a print session over which it can exercise more control over the printed output. Several print jobs can be transferred within a single print session. The printer can provide real-time job and printer status information to the sender while printing.

The optional *print-by-reference* (PBR) enables the sender to originate and control a print job where the information to be printed resides on a network instead of at the sender. The printer uses information passed to it by the sender to access the information to be printed.

Basic Imaging

The Basic Imaging profile sets up a series of constructs that enables Bluetooth-enabled devices to negotiate the size and encoding of images to be exchanged. Due to the large variety of image-encoding methods available, it is necessary for a negotiating process to be available so a received file can actually be used. Of course, all affected devices must support the type of image transferred for interoperability, so this profile requires that all Bluetooth-enabled imaging devices be able to receive JPEG thumbnail versions of an incoming image and be able to provide JPEG thumbnail versions of their stored images.

The *imaging initiator* begins a Basic Imaging session, and the responding device is called the *imaging responder*. Typical scenarios for this profile are

- Use of a PC to automatically download images from a digital still camera when the two devices come within range of each other
- Use of a printer to print images sent from a digital still camera
- Use of a laptop computer to send and control the projection of images

- Use of a remote device, such as a mobile phone, to control the shutter of a digital still camera
- Use of a mobile phone to send images to a wristwatch
- Use of a mobile phone to browse and retrieve images stored on a digital still camera

To accomplish these tasks, the Basic Imaging profile has the following six features, all of which should be controlled by a user and not performed automatically:

- **Image push** The imaging initiator pushes images to the imaging responder.
- **Image pull** The imaging initiator browses the images stored on the imaging responder, which downloads requested images to the initiator as requested by the user.
- **Advanced image printing** Used when the imaging responder is a printer, enabling the imaging initiator to specify print job options to control printer output.
- **Automatic archive** Triggers the imaging responder to download from the imaging initiator any of its stored images. The imaging responder decides which images to archive. This feature could be combined with a synchronization application to archive new images.
- **Remote camera** Enables the imaging initiator to view thumbnail monitoring images from the imaging responder, usually a digital still camera, and trigger its shutter.
- **Remote display** Enables the imaging initiator to push images to the imaging responder, usually a projector, and control the display sequence of those images.

Telephony Profiles

The Bluetooth 3-in-1 phone usage model envisions a single wireless phone that can be used as a cellular telephone, cordless telephone, and intercom. The latter two functions are enabled via Bluetooth, and a profile exists for each of these. The Cordless Telephony and Intercom profiles encompass the *Telephony Control protocol Specification Binary* (TCS-BIN), which is based upon a modified version of ITU-T Recommendation Q.931. This protocol doesn't discriminate between master and slave or even between the user and the network, but instead uses the term *outgoing side* to refer to the originator of the call and *incoming* side as the entity accepting the call.

The TCS has the following functionality:

- **Call control** Signaling for the establishment and release of voice and data calls. This includes direct communication with the LM for establishing and releasing a SCO voice channel.
- **Group management** Signaling for the management of groups of devices participating in calls.
- **Connectionless TCS** The capability to exchange signaling information without having to establish a call.

Both point-to-point and point-to-multipoint signaling methods are supported as part of the calling process. The former maps to a CO L2CAP channel and is used for signaling a single Bluetooth-enabled device, such as when a Bluetooth-enabled phone makes an outgoing call. The latter maps to a CL L2CAP channel using ACL broadcast packets and is used for multipoint call establishment or when a base station needs to alert all phones in a range that an incoming call has arrived. When a single phone answers the incoming call, then a point-to-point signaling channel is created.

Call control proceeds on a CO L2CAP channel and consists of a set of states that each side can enter. The set of states that are mandatory for Bluetooth TCS is a subset of those specified in Q.931, and this subset is called *Lean TCS*. The states are

- **General states** Used by either the outgoing or incoming sides
  - **Null** Both sides are idle.
  - **Active** Both sides are participating in a call.
  - **Disconnect request** Request to the other side that a call be disconnected.
  - **Disconnect indication** Indication from the other side that a call disconnect request has been issued.
  - **Release request** Sent in response to a disconnect indication.
- **Outgoing side states**
  - **Call initiated** Used to initiate a call.
- **Incoming side states**
  - **Call present** Indicates that a call has arrived from the outgoing side.
  - **Connect request** Answers the call.

Group management is performed by TCS through an entity called the *wireless user group* (WUG), which is simply a group of devices that support TCS and are within range of each other. The *WUG master* is a gateway (such as a phone base station) to an external network that can be accessed by the *WUG members*. Clearly, the WUG master must also be the piconet master. TCS provides a means for the WUG master to convey to the members the identity of the WUG master and all of the WUG members. Furthermore, once a WUG member has paired with the master, the member has access to all other members without further pairing action. Encryption is enabled on any CL L2CAP channel through use of a K_{master} (refer to Chapter 9). This key is also used for CO channels, and it should be changed periodically to prevent intrusion by phones that no longer have access authorization. Communication between individual WUG members may take place using unique link keys. The WUG can have more than seven members if the WUG master parks those members not actively participating in a call. Finally, TCS enables the establishment of a CL L2CAP channel for the exchange of control information not related to a specific call. Examples are control of speaker volume and microphone gain.

Various features in TCS are used in both the Cordless Telephony and Intercom profiles.

Cordless Telephony

The *Cordless Telephony profile* (CTP) is meant to standardize the way that a Bluetooth-enabled device can be used as a typical short-range cordless telephone. Unlike proprietary cordless telephones that are mostly incompatible across different manufacturers, phones using this profile will be able to communicate (if authorized) with any Bluetooth-enabled base station.

This profile defines the GW as a terminal endpoint that handles call setup requests to or from the external network. The home base station is an obvious example, but the GW can also attach to the cellular telephone network or even to a satellite link. The *terminal* (TL) is the wireless user terminal such as a cordless phone, 3-in-1 phone, or even a PDA or PC. The GW can be configured to support either a single TL or multiple TLs.

The user scenarios provided by the CTP center around the TL's capability to access the phone network through the GW, both to originate and receive calls. Both hook flash and *Dual Tone Multifrequency* (DTMF) signaling are provided so the TL can perform functions such as voice menu navigation. The profile also enables making direct calls between two TLs. This last scenario is an intercom function, so claiming support for CTP implies supporting the Intercom profile as well.

The GW is normally the master of the piconet when multiterminal use is enabled, so any TL wanting to initiate connection with a GW supporting multiple TLs will perform a master-slave switch after establishing contact with the GW (refer to Chapter 6, "Advanced Piconet Operation"). Operation may be degraded for other TLs already attached to the GW during the time a new TL is connecting. This profile can support a maximum of seven TLs.

The GW devotes as much time as possible to page scanning to facilitate timely TL connections. The TL periodically pages its GW until a connection is made, then the TL is parked while not engaged in a call. Because the TL is often battery powered, this scheme gives the TL extended battery life while providing a quick connection (required to be within 300 ms) to the GW when a call is initiated or received. Note that the GW can still broadcast messages to the parked TLs if those messages are placed into an appropriate beacon slot.

Security is maintained by requiring authentication between TL and GW, and all user data is encrypted. It is recommended that the 40-bit K_{master} used for TL-GW encryption be changed every week. Both TL and GW must support bonding.

As part of the Intercom profile, two TLs have the capability to connect to each other. They can remain connected to the GW only if they can support scatternet operations; otherwise, they must detach from the GW to perform intercom communication.

Intercom

The Intercom profile is used to implement the walkie-talkie part of the 3-in-1 phone usage model. This is a symmetrical use of the Bluetooth link so no specific roles are defined. Each participating device is called a TL. A TL may use the inquiry and page procedures to connect to another device that also implements the Intercom profile. Security is optional. The usual SCO link with CVSD audio (mandatory support) or another optional codec method is formed, and the intercom call takes place as a full-duplex voice link between users.

Audio/Video Profiles

Any device that distributes audio/video content over ACL channels must conform to the *Generic Audio / Video Distribution profile* (GAVDP). This profile specifies signaling procedures between two devices that want to set up, reconfigure, and terminate streaming channels. The *initiator* (INT) starts a signaling procedure with the *acceptor* (ACP). The roles are not necessarily fixed, but can be exchanged between the two devices should the need arise. A typical use would be a portable audio player (INT) sending streaming audio to a set of headphones (ACP). This profile uses the *Audio / Video Control Transport protocol* (AVCTP) for the transport mechanisms to control A/V devices, and the *Audio / Video Distribution Transport protocol* (AVDTP) for the transport mechanisms to distribute A/V information. There are no fixed master-slave roles in the GAVDP, and using link-level security is optional.

Summary of AVCTP Operation The AVCTP describes the transport mechanism used for command/response messages over point-to-point connections for discovering and controlling A/V devices. The messages themselves are defined in the applicable A/V control profile. The *controller* (CT) initiates an AVCTP transaction to the *target* (TG) device. Any applicable responses are returned by the TG to the CT.

AVCTP assumes that an ACL link already exists, upon which it can establish a CO L2CAP channel for A/V control. Because the CT and TG roles may reverse, the protocol must be able to support both of these roles on each device. AVCTP packets have their own packet formats, and each packet is placed into a single L2CAP payload. Long AVCTP messages may require more than one AVCTP packet and hence more than one L2CAP packet, especially if the L2CAP MTU is short.

Summary of AVDTP Operation The AVDTP applies point-to-point signaling over a CO L2CAP channel. The A/V streams are transported in an isochronous manner over L2CAP, and bandwidth can be shared between several A/V streams. Of course, bandwidth must also be shared with devices in the same piconet running other profiles, so A/V streaming may be degraded in these situations. A/V signaling is also transported over L2CAP for stream discovery, configuration, establishment, and transfer control.

The GAVDP provides the following capabilities for devices to distribute A/V information:

- Discovery of the A/V capabilities of a device and the means for negotiating A/V setup
- The means for establishing and terminating an A/V stream
- Mechanisms and formats for A/V streams, to include
  - Minimizing transmission latency
  - Optimizing available bandwidth

- Attaching timing information to the stream
- Reporting QoS and the status of A/V packets to the application

- Flexibility for use on devices of limited complexity
- Error recovery mechanisms
- Reducing overhead of transport protocol headers

The AVDTP defines a *stream* as a logical end-to-end connection of audio or video between two A/V devices. The stream is usually a single media channel, but exceptions (such as stereo audio) exist. Although the stream is unidirectional from *source* (SRC) to *sink* (SNK), bidirectional communication at the transport layer is possible for feedback information.

The AVDTP defines three states for users of the GAVDP:

- **Idle** An L2CAP signaling channel is open, but no streaming connection has been established. During this state, the INT retrieves the capabilities of the ACP unless already known and then selects and configures a capability. Next, L2CAP channels are established, and the devices move into the Open state.
- **Open** The streaming connection has been established between the two devices.
- **Streaming** The Start Streaming procedure is initiated by the user of INT, and both devices are ready for streaming.

The connection is released by the user of INT, and both devices return to the idle state. Release can be initiated from either the open or streaming states.

Advanced Audio Distribution

The *Advanced Audio Distribution profile* (A2DP) is used by an SRC to distribute high-quality mono or stereo audio content on an ACL channel to a SNK. This is distinguished from *Bluetooth audio*, which implements a toll-quality SCO channel for real-time, two-way voice. A typical use for the A2DP is streaming of music from a player to headphones or from a microphone to a recorder or public address system. Surround sound distribution is not part of this profile, nor are any remote control functions.

This profile enables the setup, control, and manipulation of streaming audio from SRC to SNK, but doesn't support synchronized point-to-multipoint distribution. Of course, for streaming to be successful, the required data rate must not exceed the aggregate Bluetooth data bandwidth available. Any content protection must be provided at the application level.

A2DP requires the support of the low-complexity *subband codec* (SBC) as defined in the profile documentation, and optional support can be provided for MPEG-1 and MPEG-2 audio, MPEG-2 and MPEG-4 AAC, and Sony's ATRAC family. A vendor-specific non-A2DP codec can also be defined and supported.

Video Distribution

The Video Distribution profile is intended to be used for streaming video between two devices. By combining this profile with A2DP, video with sound can be sent over the Bluetooth wireless link. As this is being written, the Video Distribution profile is still in its development phase.

Miscellaneous Profiles

The remaining profiles in this section are nested only within the GAP and cover a variety of functions such as remote control of audio/visual devices, mouse and keyboard connections with a computer, complex printing, and personal area networking.

Audio/Video Remote Control

The *Audio/Video Remote Control profile* (AVRCP) defines interoperability between Bluetooth devices that control audio and video within the A/V distribution scenarios. The profile specifies the scope of the *audio/video control* (AV/C) Digital Interface Command Set as defined by the 1394 Trade Association.[1] A user action is translated to the A/V control signal and sent by the CT to the TG over a Bluetooth link. The CT may be a PC, PDA, mobile phone, remote controller, or A/V device such as headphones or a player/recorder. TG devices can be an audio or video player/recorder, television, or home theater system. Conventional infrared remote controller functions can be realized using this profile. No A/V streaming is handled by this profile. Use of link-level security is optional, but support for authentication and encryption is mandatory.

The AVRCP can support remote control functions of a digital nature, such as changing a television channel, or of an analog nature, such as adjusting volume or picture brightness. Status information can be provided to the user either on the remote control itself or via an onscreen display at the device being controlled. The CT cannot, however, obtain status information directly from the TG using AVRCP on the Bluetooth link. Content streaming can take place between the TG and CT or from the TG to another device.

A point-to-point connection is assumed for control, but more than one CT may exist within a piconet. A single CT can support several TG devices, but the details of controlling such a network (such as target selection) is not defined in AVRCP. If the TG is a master, it should poll each CT at a recommended rate of 10 Hz to avoid excessive latency. A command can be sent from CT to TG either from a user input or automatically such as in response to a timer.

Although low-power modes are not mandated in AVRCP, the CT is apt to be battery powered. One of the advantages of IR-powered remote control devices is their negligible power consumption when not actively transmitting a control signal. To compensate for this, a Bluetooth-enabled CT should make use of low-power modes whenever possible.

The 1394 Trade Association defines several panel operations for A/V devices.[2] AVRCP requires support for at least one of these four A/V function categories:

- **Player/Recorder** This defines the basic operation of a player or recorder, independent of the type of media (tape, disc, solid state memory, and so on) or type of content (audio and video) it uses. At least the play and stop commands must be supported.
- **Monitor/Amplifier** Defines the basic operations of a video monitor or audio amplifier. At least the volume up and volume down commands must be supported.
- **Tuner** Defines the basic operation of a video or audio tuner. At least the channel up and channel down commands must be supported.
- **Menu** Defines the operation of a nested menu system for remote control. Display of the menu may be on the CT or TG. At least the root menu, up, down, left, right, and select operations must be supported.

Human Interface Device (HID)

The HID profile describes the protocols, procedures, and features used for the operation of Bluetooth-enabled HIDs that are used by humans to control computer systems. Early HIDs consisted of a mouse and keyboard, but now encompass any application that requires low-latency *input-output* (I/O) operations such as

- Keyboards, mouse applications (I refuse to say "mice"), trackballs, joysticks, and other pointing devices
- Front-panel controls such as knobs, switches, buttons, and sliders
- Complex gaming I/O devices such as special gloves and flight simulator controls
- Other HID-like uses that don't necessarily require human interaction such as bar code readers and telemetry applications

Profile Fundamentals The profile is based upon the USB definition to leverage a large existing class of HID. USB has several advantages when implemented in HID:

- Devices are grouped according to their *class*, and each class has a single software driver, eliminating the need for separate drivers for each device.
- Devices can *describe themselves* to their class driver, including information such as the controls they have and how they report data.
- Devices can be *hot swapped*, eliminating the requirement to power down the host and peripheral equipment.

The third characteristic is, of course, already part of a wireless link, but the first two provide significant advantages for connecting the large number of HIDs that are available in the marketplace. A separate HID class is provided in the USB specification. The HID profile defines how the *HID* connects to the *host*.

Profile Scenarios The HID profile defines the following scenarios as anticipated applications:

- **Desktop computing scenario** Use of wireless HID will free the desktop from multiple cables, enabling the HID to be placed in more convenient locations. A single HID can be used to control more than one computer. Unlike infrared applications, Bluetooth-enabled HID requires no *line-of-sight* (LOS) path to the host.
- **Living room scenario** Multiplayer gaming will be more convenient by removing tethers between players. Bluetooth can support audio I/O as well as control signals. Interactive television can be controlled with Bluetooth-enabled keyboards and other pointing devices without the

LOS path requirement typical of infrared control, and two-way capabilities are more easily supported.

- **Conference room scenario** Presentations can be more conveniently controlled using Bluetooth-enabled HIDs. The presenter can be located almost anywhere in the room rather than near the host computer or projector.
- **Remote monitoring scenario** Bluetooth-enabled battery powered remote monitoring has a large number of possible applications. Examples include temperature sensors, remote thermostats, security devices, and general telemetry. Long battery life is possible by using the low-power modes provided by Bluetooth, and the two-way link enables the host to control the remote monitoring devices and receive their data in a standardized way.

Profile Characteristics Although there are no required master-slave roles, the host should usually be the master of the piconet, especially when several HIDs are connected to a single host. In most cases, if an HID initiates contact with a host, then it will later request a master-slave switch. To ensure user satisfaction and acceptance, the Bluetooth-equipped HID should be easy and familiar to use, and the Bluetooth link shouldn't add more than about 10 ms of latency to that of a wired HID. Latency can be reduced by using L2CAP QoS, but even so, meeting HID latency requirements will be a challenge, especially when the Bluetooth link is supporting other traffic. The most common use for a particular HID is with a single host, so the HID can be put into the limited discoverable mode when the two first meet so the host can quickly obtain the HID's BD_ADDR. Subsequently, the HID can be paged for connection to the host.

Security can be maintained through the required support for authentication and encryption, which is initiated by the host. This is extremely important for devices, such as keyboards, which are often used to convey passwords and sensitive document input to the host. Therefore, support for authentication and encryption is mandatory for keyboards and keypads, and their use (with a combination key, not a unit key, refer to Chapter 9) is highly recommended. Upon first contact with the host, the HID establishes trust through pairing and bonding. For example, if the host has a display, then it can show a random PIN that the user can enter on the HID keyboard. If a single HID has a relationship with only one host, then it is said to communicate via *virtual cable*. Both HID and host have the capability to unbond themselves, without the presence of the other, to prevent an erroneous bonding situation. Security is further enhanced if the HID doesn't perform periodic inquiry scans. When no active connection is present, the HID may either stay in the page scan mode or shut down completely.

Replacing the cable between HID and host with a wireless link requires providing a power source for the HID. Bluetooth has a disadvantage in these applications because of the presence of both transmitter and receiver in the HID; most proprietary wireless links are one-way HID-to-host links only. If a battery provides power, then it must have a life of at least three months with three AAA or two AA alkaline cells. The HID should manage its own power. The host LM will always enable the HID to enter sniff and park modes, with hold optional. The profile defines several methods of balancing latency and power consumption.

One of the difficulties with implementing wireless mouse and keyboard links is that they are functional during the boot process in a typical PC before the Bluetooth protocol stack has been loaded. This functionality requires either updating the PC's *basic input/output system* (BIOS) or to develop a USB Bluetooth adaptor to emulate the operation of a wired peripheral during boot.

Hardcopy Cable Replacement

The *Hardcopy Cable Replacement profile* (HCRP) is directed mainly at *clients*, such as laptop and desktop computers, which communicate with *servers*, such as scanners and printers, although other uses are certainly possible. Data is rendered through a driver on the client device. The printing of pure images is not covered by this profile. Also, other profiles could be used for printing and scanning as well, but this profile has specialized characteristics making it more efficient for these tasks:

- It supports a IEEE 1284 ID string that identifies a host driver without regard to the transport layer employed, so the host can extend existing solutions to include Bluetooth.
- It implements flow control appropriate for the high data volume typical in printing and scanning.
- It provides a method for simple asynchronous notifications and simple control commands.
- It is connection oriented for more reliable fault recovery.
- Its implementation is low in the Bluetooth protocol stack to enhance throughput by avoiding the overhead of higher layers such as OBEX or RFCOMM.

Although one server, such as a printer, may support multiple clients over a wireless link, this operation is beyond the scope of the HCRP. Also, how the server handles multiple jobs is implementation specific, but this protocol requires a server that cannot support more than one client to make itself unavailable for service discovery or to refuse an additional connection while servicing a client.

Due to the sensitive nature of printed and scanned material, link-level security, including bonding, is mandatory to support and optional to use. No fixed master-slave role is defined, and no low-power modes are required. However, reliable data transfer is guaranteed so the L2CAP channel established for this profile will have an infinite flush timeout (refer to Chapter 8).

The server can be in one of four different modes, two of which must be supported in the profile. These modes are

- **Bluetooth offline mode** The server cannot receive data over the Bluetooth link; that is, it is nondiscoverable and nonconnectable. Support for this mode is optional.
- **Bonding mode** The server is ready to bond with a client. For this to occur, the server must be connectable, pairable, and either limited or general discoverable.
- **Private online mode** The server can be connected to and used only by clients that know the server's BD_ADDR. The server is connectable but nondiscoverable.
- **Public online mode** The server is either limited or general discoverable (or both) and is also connectable.

If a client has a print job, for example, it can use the inquiry process to obtain a list of print servers and select from the list or begin by paging a particular server if its BD_ADDR is already known. When a client connects to the server, it can discover what driver needs to be loaded through retrieval of the server's IEEE 1284 ID string or by another method. A *control* L2CAP channel is set up for these out-of-band control requests. If the requested driver is not available at the client, then the user should be informed. For a print job, data is rendered by the client using the driver and passed to the printer via a L2CAP *data* channel. (The HCRP doesn't define how data is sent over this channel.) Once the data is transferred, then the client disconnects. If the printer is busy or an error occurs, then this information should be sent to the client. The server uses an optional L2CAP *notification* channel for messages such as these.

A client can request *rich status* from the server by registering for these notifications. This enables the server to initiate a connection to the client and transfer status information as needed via an L2CAP notification channel. For example, a print job will probably complete its transfer to the server before the task of printing has ended. The client could disconnect from the server and await an end-of-job notification from the server at a later time.

As we mentioned in Chapter 8, L2CAP has no capability to control the flow on a single channel. Printing and scanning will probably require flow control on the data channel due to the high volume of data being transferred. HCRP uses a *credit-based flow control* mechanism for this purpose. Before a device is enabled to transmit data on the data channel, it must obtain *credit* from the other device. The client sends a *credit request message*, and the server responds with a *credit grant message* over the control channel. Credit is provided for a specific number of bytes (specified as a 4-byte value) and is granted to the client from the server based upon the server's buffer and processing capability. The client can also grant credit in response to a server request, such as if the server is a scanner.

Personal Area Networking (PAN)

Although Bluetooth-enabled devices communicate over a relatively short range, they may still need the capability to form wireless networks using OSI Layer 3 network protocols such as Ethernet. For this reason, the BNEP was developed to provide seamless communication over a Bluetooth wireless link. BNEP encapsulates packets from various network protocols, such as IPv4, IPv6, and IPX, for transport over a Bluetooth L2CAP channel. The protocol is implemented on a CO L2CAP channel and is at the same OSI MAC layer as Ethernet, Token Ring, ATM, and others. BNEP applies the same rules of network connectivity and topology defined in IEEE 802.3, and because the BD_ADDR is administered by IEEE 802, a Bluetooth network access point can be built as a bridge between Bluetooth and Ethernet. BNEP transports an Ethernet packet by removing its header and replacing it with a BNEP header and encapsulating the result within L2CAP and sending it over a Bluetooth link.

The client is called the *PAN user*, or PANU, in the profile. The servers provide either a *network access point* (NAP) or *group ad-hoc network* (GN) service. The NAP provides some of the features of an Ethernet bridge and can forward Ethernet packets between each of the connected PANU devices. A NAP device is connected to an additional network on which Ethernet packets are exchanged either via Layer 2 bridging or Layer 3 routing functions. The GN is also able to forward Ethernet packets to each PANU, but it does not provide access to any additional network.

The PAN profile describes how to use BNEP networking capabilities for Ethernet encapsulation, single piconet IP PAN, master forwarding of data, and providing NAP or GN service. Future versions of the PAN profile will address additional PAN issues.

The PAN profile has the following functional requirements:

- Defines and references dynamic ad-hoc IP-based personal networking
- Provides functions that are OS, language, and device independent
- Supports common networking protocols such as IPv4 and IPv6
- Supports NAPs for corporate LAN, GSM, or other types
- Accommodates memory, power, and user interface constraints typical of small, portable devices

To support these requirements, the PAN profile covers network discovery, forming a network manually, allocating addresses, resolving address and name issues, bridging and routing within the network, and security. The PAN profile does *not* cover automatic network formation, multiple piconet operation, and QoS issues.

Common ISDN Access

The Common Integrated Services Digital Network (ISDN) Access profile defines how applications access ISDN data and signaling over Bluetooth without modifying the legacy application itself. The *access point* (AP) is an ISDN terminal endpoint that communicates with up to seven *ISDN client* (IC) devices via Bluetooth. The AP handles the ISDN D-Channel protocol and routes requests and indications via *Common ISDN Application Programming Interface* (CAPI) messages to the IC. This enables the IC to use all of the standard ISDN features such as telephony, Internet, and fax.

The AP devotes as much of its free capacity as possible to page scanning. Following a page from an IC, a master-slave switch occurs to establish the AP as piconet master. Mutual authentication and encryption is required before data is transferred. A unit key is never used (refer to Chapter 9) for these purposes. The recommended encryption key size is the maximum 128 bits, but can be shorter if government regulations require. ISDN data is transferred across a CO L2CAP channel with infinite flush timeout (reliable channel). Multi-IC support at the AP is optional.

Summary

Bluetooth profiles are used to implement the various usage models by specifying how the appropriate protocols should function. This is done by taking a vertical slice through the protocol stack and examining the functionality of each layer. Functions can be mandatory, optional, conditional, excluded, or not applicable to the desired profile. In this way, the large set of capabilities set forth in the protocol layers is limited to only those needed for the particular function desired.

Unlike many other specifications, Bluetooth includes implementation details throughout the protocol stack, even up to the user interface level. In this way, it is hoped that the user of Bluetooth-enabled devices will be able to bypass some of the frustration experienced from operating devices using other protocols that required conformance to lower layers only. Bluetooth-enabled units conforming to the same profile are expected to be fully interoperable across manufacturers of any of the hardware and software within.

Several other profiles are still under development and will be released as they are completed. These include the Local Positioning, Still Imaging, and Extended Service Discovery profiles.

End Notes

1. 1394 Trade Association, AV/C Digital Interface Command Set—General Specification, Version 4.0, Document No. 1999026 (www.1394ta.org).
2. 1394 Trade Association, AV/C Panel Subunit, Version 1.1, Document No. 2001001 (www.1394ta.org).

CHAPTER 12

Module Fabrication, Integration, Testing, and Qualification

In response to the momentum that has built up behind Bluetooth, several manufacturers have developed *radio frequency* (RF) and *baseband* (BB) chipset solutions based upon a wide variety of design and fabrication processes. Some have created single-chip devices using *Complementary Metal-Oxide Semiconductor* (CMOS) alone, while others separate the RF and BB chips* and use different fabrication techniques for each. The chipsets can be combined with varying amounts of discrete components to provide Bluetooth capability in a wide variety of applications.

Bluetooth integration can be done in many, many different ways. The earliest systems consisted of a PC card that plugged into a PC and generally communicated with the host via HCI (see Figure 12-1). This technique proved to be the fastest to market, but lower costs can be realized by integrating Bluetooth into the host, perhaps by placing the entire protocol stack into the host and having it communicate only with the RF chip. On the other extreme, stand-alone Bluetooth-enabled systems, such as toys, may consist of only a microcontroller connected to a Bluetooth chipset and running all *input-output* (I/O) operations itself. Development and integration can be assisted by using a development kit that many chip and software manufacturers provide.

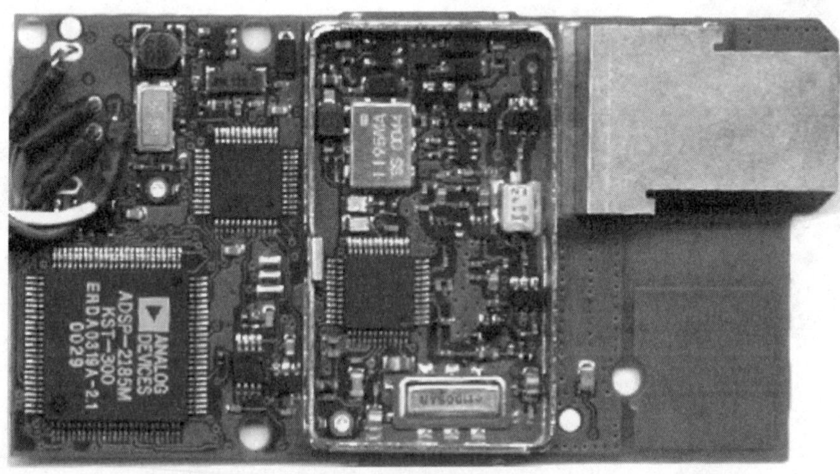

Figure 12-1 This PC card was one of the first Bluetooth products to hit the market. Note the high component count. (Source: National Semiconductor)

*Throughout this chapter, we will follow the manufacturer's convention and use the term *baseband chip* to refer to a single integrated circuit that contains the baseband and link controller (LC) functions as a minimum. In most cases, it also includes the LM, the host controller, and the HCI transport mechanism.

Because interoperability is extremely important for Bluetooth to be successful, the specification carefully outlines the performance criteria that must be met by any chipset if it is to achieve *Special Interest Group* (SIG) qualification. Government regulations must be met as well. Finally, the Bluetooth-enabled device itself must meet SIG protocol stack and profile conformance requirements before it can be marketed.

Testing the Bluetooth product can be difficult and requires considerable ingenuity due to the lack of standardized test point access on devices from different manufacturers. Developing a proper test program is important not only to obtain SIG qualification for the product, but also as a validation tool throughout the design process.

In this chapter, we will examine some of the hardware architecture that is employed in the design and construction of the Bluetooth chipset, followed by a look at various methods of integrating the result into the host. Next, we'll examine how SIG qualification is obtained, followed by how a module can be tested for the wide range of performance criteria listed in the specification. Following our usual convention, we won't discuss every test required by the specification, but instead, we will show how a module can be placed into the RF and BB test mode and how the device responds. Several of the transmitter and receiver performance criteria have been discussed in Chapter 3, "The Bluetooth Radio."

Hardware Architecture

The architecture behind the Bluetooth chipset generally encompasses circuit design, fabrication, and packaging. Design of the radio and baseband circuits for the Bluetooth chipset is, as you might imagine, extremely complex with tremendous pricing pressure. As a result, manufacturers have developed considerable intellectual property associated with these designs, especially at the radio level. Many of these manufacturers require a customer to sign a *nondisclosure agreement* (NDA) prior to being granted access to the datasheets, and several still withhold critical information from Bluetooth integration engineers even after the NDA is in place.

The Bluetooth Chipset

Most chipset solutions include the radio, BB, LM, and perhaps HCI along with its physical transport (see Figure 12-2). Some manufacturers

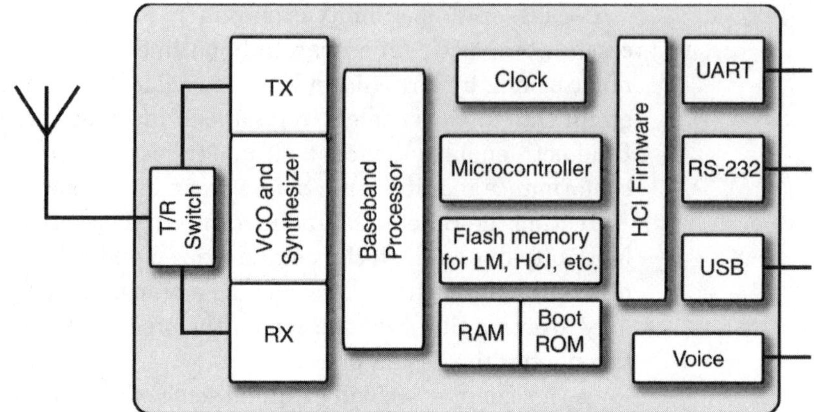

Figure 12-2 The Bluetooth chipset usually contains the radio, BB, LM, and HCI portions of the protocol stack, along with a physical means for communicating with the host.

implement all of these on a single chip, while others put the (analog) radio on one chip and the other (digital) functions on a separate chip. We'll discuss the advantages of each of these approaches in a later section.

A block diagram of a typical Bluetooth transceiver is shown in Figure 12-3. A *phase-locked loop* (PLL) synthesizer using a *voltage-controlled oscillator* (VCO) is at the heart of the radio, generating the RF carrier during the TX window and the *local oscillator* (LO) during the RX window. The receiver amplifies, filters, and detects a desired incoming signal and outputs the RX data stream. The transmitter modulates incoming data onto the correct 2.4 GHz hop frequency carrier, amplifies it, and sends it to the antenna.

To reduce cost and complexity, the VCO is often designed to operate at one-half the carrier frequency, and then its output is frequency-doubled prior to use. This technique also avoids unwanted interaction between the small-signal VCO and the large-signal transmitter *power amplifier* (PA) by having them operate on different frequencies. The synthesizer is under digital control by the baseband circuitry for hop channel selection.

During TX, the incoming data is Gaussian filtered and then applied to the VCO in such a way that its output changes to a higher or lower frequency corresponding to the voltage level of the baseband data signal. In other words, the VCO acts as a voltage-to-frequency converter during the modulation process. During RX, the frequency synthesizer and multiplier generate the LO frequency that is combined with the incoming modulated carrier in the receiver's mixer. The resulting *intermediate frequency* (IF) is amplified and detected.

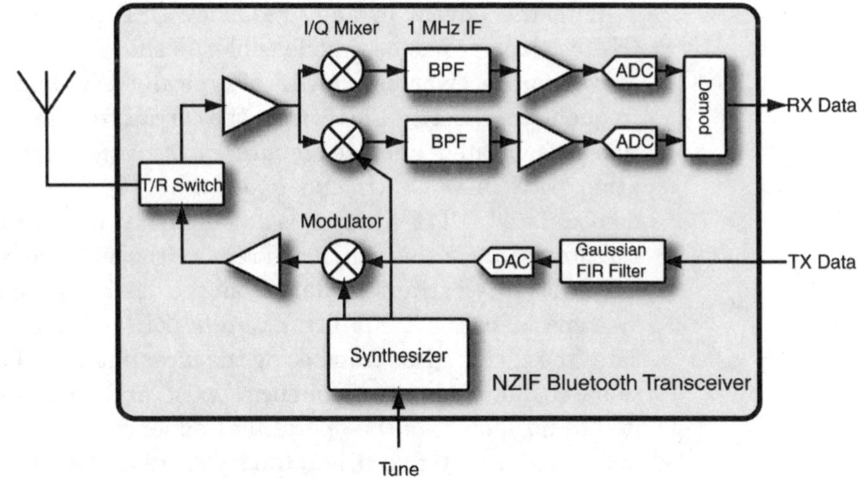

Figure 12-3 The Bluetooth radio transceiver consists of a frequency synthesizer, transmitter, and receiver.

Receiver Design Alternatives For several decades both analog and digital receivers used the classical *dual conversion* architecture, as shown in Figure 12-4. The first IF is high enough (for example, 100 MHz) so that images (refer to Chapter 3) can be effectively eliminated by the RF filter at the antenna, while the second IF is low enough (for example, a few hundred kHz) that inexpensive bandpass filters can be used to meet adjacent channel *carrier-to-interference* (C/I) requirements. Unfortunately, this design can have significant cost and size penalties for use in a Bluetooth radio. In particular, the first IF filter is typically an external *surface acoustic wave* (SAW) device, which is costly and physically large. Furthermore, two local oscillators must be included in the design, increasing cost still further.

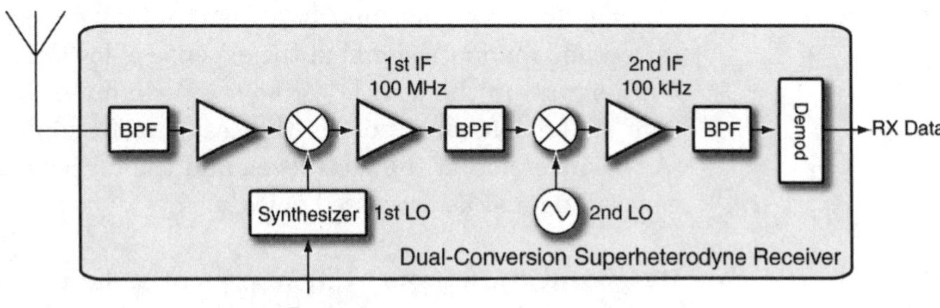

Figure 12-4 The dual-conversion receiver has been in use for many years. Its suitability for Bluetooth is limited because of its size and cost.

With the advent of high-frequency DSP technology the *near-zero IF* (NZIF) receiver has become feasible. As shown in Figure 12-3, this single-conversion receiver uses an IF of typically 1 MHz, so the local oscillator frequency is offset by 1 MHz from the carrier frequency. Noncoherent detection is facilitated, and the *signal-to-noise ratio* (SNR) is improved by creating both *in-phase* (I) and *quadrature* (Q) signals to be filtered and demodulated. All IF filtering is accomplished on chip, and subsequent digital processing of this signal can remove the effects of adjacent channel C/I and efficiently extract the data from the desired signal.

A special case of NZIF is the *zero IF* (ZIF) receiver, where the LO duplicates the carrier, thus translating the incoming signal in frequency directly to baseband. No image frequencies exist, and spurious responses are limited to harmonics of the operating frequency. The amount of RF circuitry and shielding is reduced, and the layout requirements are simplified. Most signal processing is done at baseband, not RF, so isolation, gain, filtering, and control are more easily accomplished.[1] ZIF architecture, however, has some significant challenges that must be overcome for it to be feasible. For example, if the LO is not carefully isolated or if the mixer has cross coupling between its terminals, then the receiver can jam itself. Unfortunately, this tendency can even depend upon external influences such as handling the receiver.[2] The resulting *direct current* (DC) errors from LO leakage or from nearby interferers can saturate the mixer unless its dynamic range is perhaps 20 dB higher than that required in an NZIF receiver.

Finally, many engineers are working toward what may be considered the ultimate receiver design, in which the RF signal at the antenna, perhaps after modest bandpass filtering, is directly digitized. All subsequent processing is performed via a *digital signal processor* (DSP) until the desired data stream emerges. This technique has the significant advantage of treating the entire radio beyond the front end as a digital device. Now the radio's performance is completely determined by its programming, which can change dynamically, depending upon its operational environment. For instance, if the receiver discovers interference in the next higher adjacent channel during a particular hop, then its filters can be quickly adjusted to reject that channel's signal at the expense of lower rejection from (nonexistent) signals in the next lower adjacent channel. Of course, the disadvantage of the software radio is the cost and feasibility of analog-to-digital (A/D) conversion at the GHz level, and the massive amounts of data thus generated must be processed quickly.

Baseband Functions Unlike present radio designs, which are either analog or mixed signal in nature, the Bluetooth baseband functions are completely digital. Many implementations are based upon a 32-bit *reduced instruction set computer* (RISC) core running at about 32 MHz to provide sufficient computation power to run the entire LC, LM, and HCI functions. Early designs required external flash memory, adding to the size and cost of the implementation. Later designs contain all processor functions within a single chip.

Device Fabrication

The fabrication and packaging of the Bluetooth radio and baseband circuits require a design that's easy and inexpensive to produce and is physically small. Many designers consider Bluetooth to be the most challenging wireless endeavor yet. A solution can consist of either a *system on a chip* (SoC) or a *system in a package* (SiP). As it implies, the SoC places as much functionality as possible within the IC itself, with (hopefully) a minimum of external components required. Design times, however, can be long because careful coordination must occur between the RF and BB sections of the chip. Updating the design and creating design variants are both major tasks with SoC. The SiP approach separates RF and BB functions into (typically) two different ICs, and a module is then built around these devices. We will discuss SoC and SiP in more detail later in this chapter.

Perhaps the ultimate SoC solution is found in the IR world, where a single transceiver chip contains everything except the power source (see Figure 12-5). For Bluetooth, however, things are not quite that simple, and most implementations require several external components in what is essentially a SiP realization.

There is wide consensus among the various manufacturers that CMOS is the best process for the implementation of digital circuits. On the other hand, disagreement prevails on how best to implement the Bluetooth radio. Some have built the transceiver out of CMOS as well, integrating both radio and baseband functions into a single chip. Others believe that combining bipolar transistors and CMOS onto a single chip (BiCMOS) is a more suitable design for the radio because it has better receive noise performance and sensitivity as well as higher transmit output power for a given size. BiCMOS has been widely used in cellular applications due to the higher performance required of handheld and base station transmitters and receivers, but the less stringent Bluetooth radio specification and small gate size has resulted in CMOS designs that work well.

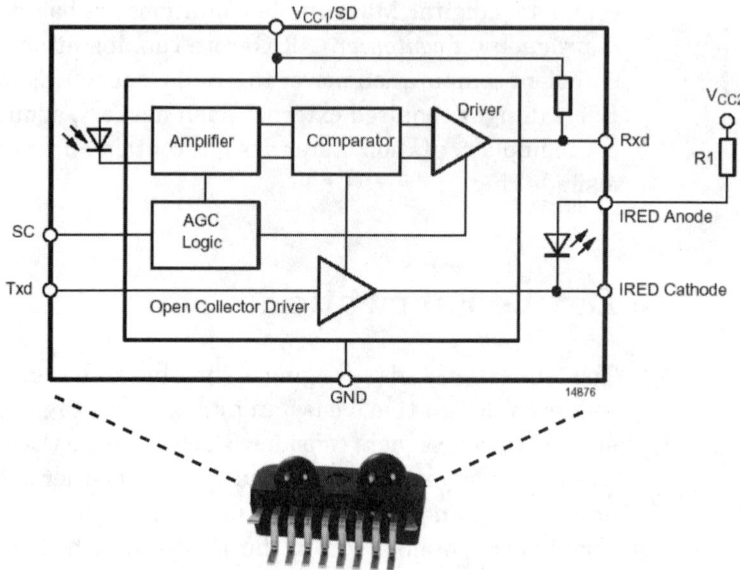

Figure 12-5 This IR transceiver contains the complete transmitter, receiver, and antennas within a single chip. Maybe RF will be this simple some day! (Source: Vishay Telefunken)

Technology Tradeoffs The design of a wireless communication device, such as Bluetooth, requires the engineer to consider several factors. Among these are the following:[3]

- Wireless usually contains both analog and digital circuitry.
- Technology choices are determined by the selected fabrication process.
- Cost-effective approaches to the resulting high system complexity may require modularizing designs for reuse in different systems.
- Market demands are high, and wireless devices may soon overtake the PC as the prime application area for chip fabrication.
- Pressure to keep costs low is intense.
- Varying international wireless standards must be considered.

In light of these tradeoffs, different manufacturers have selected different technologies for implementing Bluetooth solutions. These include several variations of CMOS, BiCMOS, *silicon-on-insulator* (SOI), *gallium-arsenide* (GaAs), and *silicon-germanium* (SiGe) placed in leaded or leadless plastic or ceramic packages that are in turn made into modules using *low temperature co-fired ceramic* (LTCC), thin film, or a standard *printed circuit board* (PCB). We'll now take a brief look at each of these.

CMOS For digital circuits, a new CMOS size standard generally appears about every six months, with gate sizes decreasing from 0.35 to 0.25 to 0.18 microns and so on. The physical size reductions can be substantial when the gate size decreases. For example, in 1999, a 0.25 micron CMOS IC with 1 million gates occupied an area of 40 mm,[2] but this is projected to reduce to only 3 mm^2 by 2003.[4]

For Bluetooth applications, straight CMOS designs fall into three different categories:[5] Bulk CMOS, RF-added CMOS, and SOI. Bulk CMOS is used for digital-only applications such as the Bluetooth baseband controller, RF-added CMOS enables the inclusion of the Bluetooth radio and its associated analog components on the same chip as the baseband processor, and SOI, the most exotic of the three, adds an insulator layer to traditional CMOS to increase isolation of the RF portion of the circuit.

Digital CMOS consists primarily of gate networks. However, when CMOS is used for analog applications such as RF, other components, such as resistors, capacitors, and inductors, must be integrated into the design. Combining digital and analog circuits on the same substrate presents several difficulties:[3]

- Yield rates may drop due to the wide variety of components needed for mixed signals.
- Analog circuit testing is more complex, and test points must be built into the design.
- Power supply voltages may be different for the digital and analog parts of the circuit.

The SOI process adds a *silicon dioxide* (SiO_2) insulating layer to a standard CMOS device, which provides the following benefits:[5]

- Drain capacitance is reduced for higher frequency response and lower parasitics.
- Threshold voltages are lower for reduced power consumption and battery count.
- Latchup potential and noise coupling from the substrate are reduced.
- Higher isolation enables more functions to be embedded on the same chip.

Wireless applications using SOI include the *low-noise amplifier* (LNA), transmit power amplifier, and synthesizer VCO.

SOI technology has two significant drawbacks. First, SOI *field-effect transistors* (FETs) have lower output impedance than their bulk CMOS brethren, and modeling this output impedance is difficult, so matching networks are more complex, and more uncertainty exists in their design. Furthermore, the cost of SOI can approach twice that of bulk CMOS for the typical core radio area of 10 mm^2.[6] The typical gain of a SOI FET at Bluetooth frequencies is about 15.7 dB, only slightly higher than the 15.2 dB provided by a 0.25 micron bulk CMOS FET.

One of the major hurdles that had to be overcome before all-CMOS RF devices became feasible for use in Bluetooth was the design of a suitable RF amplifier having a low noise figure in the receiver's front end. Fortunately, with the maturing of 0.25 micron CMOS fabrication, this amplifier could be built with reasonable gain and a noise figure of about 6 dB at 2.4 GHz. This situation is improving further with the use of 0.18 micron technology.

GaAs GaAs, pronounced "gas," was initially developed as a means for producing a low-noise, high-performance receiver front end for the military. GaAs devices are able to withstand a high *electromagnetic pulse* (EMP) without being permanently disabled. These devices are costly to produce because they require a fabrication platform that is incompatible with CMOS or bipolar, so they cannot be monolithically integrated with CMOS. Consequently, GaAs has lost favor among many Bluetooth chip designers.[7]

SiGe Early semiconductor devices were fabricated from doped germanium; these were fragile and inefficient, and entered thermal runaway at a relatively low temperature. The advent of the silicon substrate markedly improved semiconductor performance, and many engineers were quick to bid good riddance to germanium. Not so fast . . .

It was discovered in the late 1980s that silicon and germanium can be combined into a new semiconductor called SiGe (pronounced "sy-jee"). SiGe provides higher efficiency and more power in a radio transceiver that is smaller and has a lower noise figure than those made from conventional silicon. As a result, battery life can be extended, and greater integration can occur using BiCMOS at a modest cost increase over pure CMOS designs for a given die area.[7] In the late 1990s, semiconductor manufacturers began economic production of SiGe chips in quantity. Unlike GaAs, BiCMOS devices using SiGe can be built on the mature and reliable CMOS manufacturing platform. This enables fabrication of a BiCMOS radio that arguably performs better than a radio constructed from CMOS alone.

Summary of CMOS Fabrication Processes Table 12-1 gives a comparison of the number of masks and relative cost of several competing technologies for Bluetooth device fabrication. The base process is 0.25 micron

Table 12-1

Relative cost and complexity of silicon semiconductor technologies for Bluetooth[8]

Process	Number of Masks	Relative Cost
Digital CMOS	20	1.0
Mixed Signal CMOS	23	1.25
RF CMOS	25	1.45
BiCMOS	26	1.55
SiGe BiCMOS	29	1.65

digital CMOS. The cost of fabrication varies approximately linearly with process complexity and by the square of the die dimensions (linearly with die area). As an example, a 7 × 7 mm bulk CMOS chip costs about 19 percent more than a 5 × 5 mm SiGe BiCMOS device due to the increased die area. Furthermore, package size and pin count are often better indicators of actual cost than the IC technology itself.[8] Pin count can be reduced by using a serial instead of a parallel bus structure and integrating as many components as possible into the IC. The latter procedure reduces the number of pins and external components at the expense of increased die size for a given technology.

For Bluetooth applications that are primarily cost-driven, CMOS may be a better choice over BiCMOS if the die areas are nearly identical. Other applications that require low power consumption, better RX sensitivities, and/or higher TX power outputs will benefit from BiCMOS technology, perhaps using SiGe bipolar devices. For example, the SiGe process is well suited to the construction of a Bluetooth class 1 power amplifier, where efficiencies can exceed 60 percent using class AB operation.[7] Output power control can be easily integrated into the chip, reducing the need for additional external components.

Passive Components Design engineers sometimes make the mistake of concentrating on the size of the IC without devoting sufficient attention to the number, size, and cost of the passive components needed to make the IC function. Reducing the number of passive components can free valuable PCB space and reduce both nonrecurring costs of engineering the components into the circuit and recurring costs of purchasing and installing the components themselves.

Inductors are one of the main targets for integration into the chip itself. On-chip planar spiral inductors have typical tolerances of 2 percent, compared to 10 percent for externally wound components mounted on the PCB. Placing the inductors on-chip reduces occupied space, increases their self-resonant frequencies, and avoids unwanted front-end oscillation at Bluetooth frequencies.[6] A symmetrical layout for carefully controlled cross-coupling can be achieved with greater accuracy using on-chip inductors.

Using inductors with a high *quality factor* (Q, which is the ratio of reactance to resistance) is important for CMOS designs by compensating for the low transconductance of these devices compared to bipolar devices using the same bias current values. The Q for a typical bulk CMOS inductor has recently been improved from a poor 3 to a fair 7 (for a 2 nH inductor at 2.5 GHz), which is adequate for many wireless circuit applications.[6]

As CMOS technology continues to progress, more and more passive components will be fully integrated into the IC for the smallest possible footprint. For designs in which components external to the Bluetooth chip itself may be needed, such as receiver LNAs, class 1 transmit power amplifiers, T/R switches, and antenna matching networks (along with passive components that support these functions), an integrated module can be constructed on a traditional PCB or by using LTCC or thin-film techniques.

Packaging Options

Wireless applications, along with high-speed logic and bus communications, are areas where the choice of package for the IC is of critical importance for the success of the design. Many new packaging techniques were developed to handle frequencies above 1 GHz that are prevalent in modern wired and wireless communication systems.

The circuit package performs the following functions:[9]

- Protects the circuit from the external environment
- Protects the circuit from the PCB manufacturing process
- Provides an electrical interface between the circuit and PCB
- Provides a mechanical interface between the circuit and PCB
- Acts as a heat transfer mechanism between the circuit and heat sink
- Provides an interface for production testing
- Facilitates circuit integration into a small volume

Selection of the package material is a key design factor for meeting the product's performance requirements. Material selection is based upon *coefficient of thermal expansion* (CTE) relative to the PCB, heat conduction, environmental conditions, and cost.

In general, three types of material are used for package construction: plastic, ceramic, and metal. Molded plastic is commonly used for wireless carrier frequencies less than 6 GHz with power dissipations less than about 5 watts, so this package is prevalent in Bluetooth applications. Ceramic has somewhat better thermal conductivity than plastic, but metal conducts heat significantly better than ceramic. Both ceramic and metal can be hermetically sealed for superior environmental protection, but both are also more expensive than molded plastic.

Although a wide variety of package designs exist, they can be roughly divided into two categories: *leaded* and *leadless*. The most common leaded package is the *small outline integrated circuit* (SOIC), shown in Figure 12-6a. These are inexpensive to manufacture and are quite common but sometimes cause problems at RF due to the parasitic effects of the leads themselves. Fortunately, SOIC packages have adequate signal handling capability up to about 4 GHz, so they can be used for Bluetooth applications. However, relatively few standards exist for these packages beyond a common lead pitch and land size, so modeling their performance can be difficult.[9] A more complex chip design can be packaged with leads radiating from all four sides in a *quad flat package* (QFP) as shown in Figure 12-6b, or wrapped around the chip in a *plastic leaded chip carrier* (PLCC) design for a smaller footprint (see Figure 12-6c).

Leadless packages are also popular because of their small size, low cost, and performance advantage over leaded parts. Stray inductance and other parasitics are reduced by shortening the path between chip and board. One type of leadless package, called the *quad leadless pad* (QLP), is shown in Figure 12-7a. For a denser pin structure the *ball grid array* (BGA) can be used instead (see Figure 12-7b). The BGA style has the capability to include

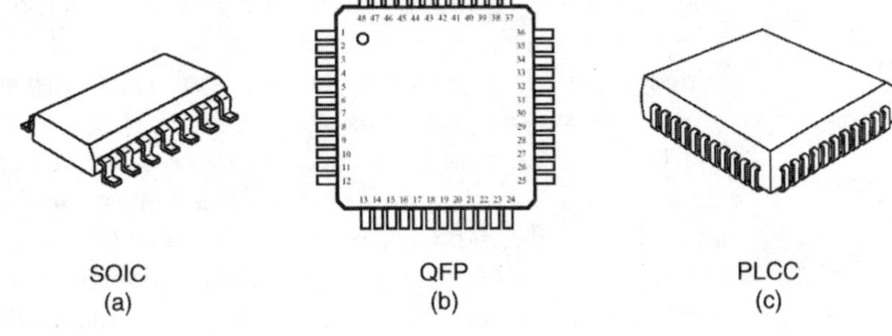

Figure 12-6
The SOIC package (a) is very common and inexpensive to manufacture and can be extended to the QFP for more complex chip designs (b), or the PLCC for smaller footprint (c). (Source: Allegro Microsystems and Atmel Temic)

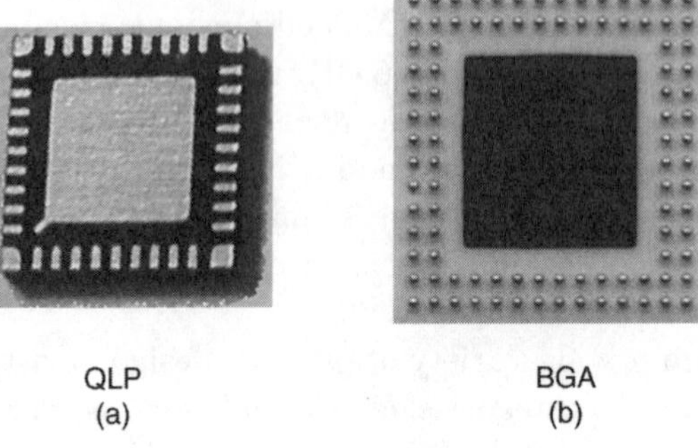

Figure 12-7 Leadless packages attach directly to the PCB and eliminate parasitics from leads that might be a significant fraction of a wavelength long. The QLP and BGA are simple, low-cost, reliable packaging techniques. Bottom views are depicted. (Source: Ericsson and Alcatel)

closely-spaced ground vias within the grid for decoupling and impedance matching.

Module Construction Using LTCC or Thin Film Both active and passive components can be built into a single module using either LTCC or thin film techniques. LTCC devices are constructed from layers of ceramic tape upon which a circuit is preprinted with the appropriate vias for interconnection (see Figure 12-8). The layers of ceramic are then sandwiched and fired at about 1000°C to create a module that looks much like a small, multilayer circuit board. Six to 10 layers are typically used, and cavities can be included for installation of active devices such as transistors or ICs and larger passive components. ICs are sometimes mounted upside down in a *flip chip* configuration within the ceramic cavity and connections made from the top of the chip directly to the module.

Inductors can be spiral wound on one layer with an underpass to the layer beneath. Built-in capacitors consist of parallel plates on different layers with a ceramic dielectric, and resistors can be printed on a layer. Ceramic is a Hi-Q material with a high dielectric constant, so good inductors and capacitors are relatively easy to construct. Metal patterns can be added, onto which either solder balls or lands are attached for module assembly onto the final PCB.

LTCC is a parallel construction process in that each layer can be constructed and tested on different fixtures before the composite structure is assembled and fired. Ceramics have a lower CTE than that of most PCB material, so if a large ceramic package is soldered to a PCB, then temperature extremes can place stress on the solder joints, resulting in an intermittent connection or, in extreme cases, fracture of the package itself. On the other hand, the low CTE of ceramic prevents large changes in circuit component values at temperature extremes.

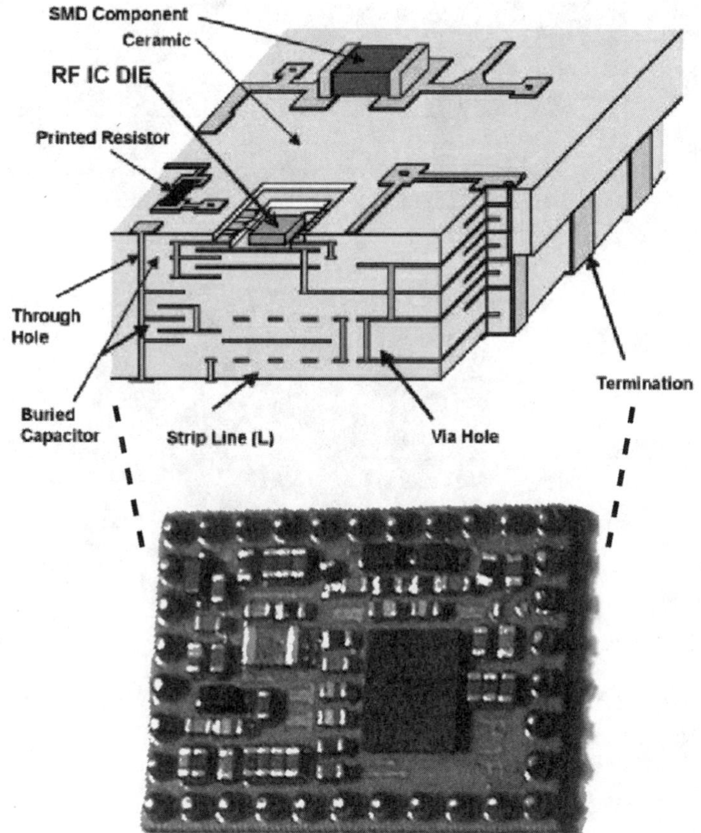

Figure 12-8 Several ceramic layers can be fused into a module using LTCC production techniques. (Source: Murata and Ericsson)

Thin-film components are created on a glass panel carrier, such as those used for *liquid crystal displays* (LCDs), although other substrates, such as alumina or silicon, can be used. As shown in Figure 12-9, a resistive material is deposited on the glass panel first and is then etched to form the resistive patterns. Next, metal is sputtered in two stages to form the lower and upper capacitor plates, and two different dielectrics can be added for both bypass and RF applications. A relatively thick copper layer is added for routing, and the device is capped by a dielectric passivation layer.[10]

Table 12-2 shows a comparison of LTCC and thin-film technology as of late 2000. LTCC provides several routing layers and presents a self-packaged module, which has a reputation for reliability due to the circuit's enclosure within layers of strong and chemically inert ceramic. Component tolerance in LTCC devices are typically less than that in thin film because of the shrinkage (10 to 15 percent) that occurs when the LTCC ceramic

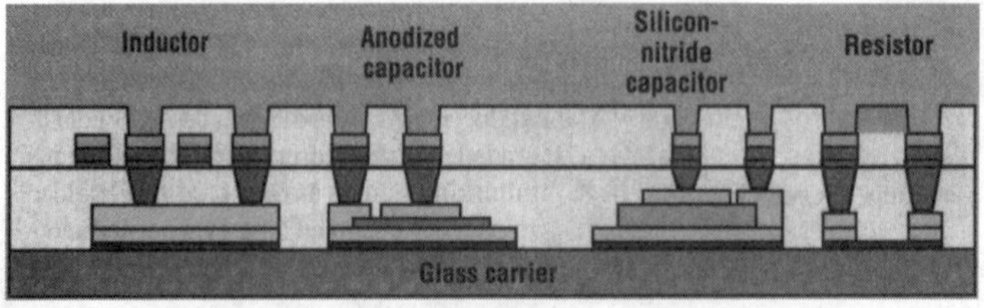

Figure 12-9 Thin film on glass can be used to create a number of passive components for RF applications. (Source: Intarsia)

Table 12-2

Comparison of LTCC and thin-film processes[11]

Parameter	LTCC	Thin Film
Line width	100 μm	15 μm
Via size	130 μm	30 μm
Capacitance density	1 pF/mm²/layer	100 to 500 pF/mm²
Inductor values	Up to 100 nH, 5% tolerance	Up to 150 nH, 2% tolerance
Resistor values	10 Ω to 100 kΩ, 30% tolerance	10 Ω to 500 kΩ, 10% tolerance
Number of routing layers	10 typical, 50 possible	2

layers are fired. There has been some progress on the development of *non-shrinkage LTCC* (NS-LTCC) to alleviate some of these tolerance irregularities. Thin film may have the potential for higher levels of integration due to the reduced component size, but is hindered by the smaller number of possible routing layers. Some engineers argue that thin film is easier to model in *computer-aided design* (CAD) circuit-design programs because it has essentially a two-dimensional architecture, compared to the multidimensional structure of LTCC.[10] The cost of the two processes is approximately equal for a given function.

Single-chip Versus Two-chip Solution

One of the most intense areas of disagreement between manufacturers is the single-chip versus two-chip Bluetooth solution. The single-chip solution places both the radio and baseband functions onto one monolithic CMOS IC, and the two-chip solution separates the radio and baseband functions into two ICs, with the radio usually constructed from BiCMOS and the baseband from bulk CMOS. It should be noted that earlier single-chip designs actually consisted of two chips, the radio/baseband, and an external flash memory for the protocol stack firmware. Flash has since been integrated into the radio/baseband chip, so we'll consider that configuration to be the true single-chip architecture.

Proponents of the single-chip solution point to the following advantages:

- Smaller physical size
- Reduced number of external components
- Potential lower cost from reduced component count
- Easier to integrate into a module
- Lower power consumption

Proponents of the two-chip architecture claim these advantages:

- Easier migration of lower protocols into the host
- Digital and analog circuitry can each be separately optimized
- Potential lower cost from increased flexibility
- Potential to select radio and baseband from different manufacturers
- Easier to test

Both single-chip and two-chip RF-BB solutions have been available for several years, so it's likely that the marketplace will find uses for both of these configurations.

BlueRF for Connecting Radio and Baseband Controller An advantage listed by the two-chip proponents is the capability to select a radio and baseband chip from different manufacturers. Furthermore, because process scaling has historically reduced the die size of digital circuits faster than that of analog circuits, separating the radio and baseband functions enables a new baseband chip to be designed and used apart from any redesign of the radio. The challenge, of course, is providing a means by which the baseband and radio chip can communicate with each other. (Yes, we're talking about yet another wired communication link as part of a Bluetooth implementation, but at least this one is over a very small distance—between two chips.)

Communication between the RF and BB chips can be done in one of three ways. The first is to design a custom link between the two. This is often done when a single manufacturer develops its own two-chip architecture, and the custom interface usually precludes using either chip with its partner from another manufacturer.

The second method of communicating between RF and BB is for the manufacturers to agree on a common interface. Once this is implemented then, in theory at least, any RF chip can talk to any BB chip or even to a host processor incorporating BB functionality, thus offering the greatest design flexibility. The *BlueRF* initiative was begun by Ericsson, Nokia, Mitel, Intersil, and Philips to create such an interface, and their efforts were coordinated by ARM. The goals behind BlueRF were to create a standard for physical interface signals, timing, electrical characteristics, and device behavior.[12]

BlueRF is a purely digital interface between RF and BB. Two physical connection methods were developed: an 8-line interface for a bidirectional port and a 14-line interface for a unidirectional port. When the BB and RF units first power up, BB interrogates RF for its unique manufacturer's code and part identifier, and then configures itself for operation with that device.

The RX interface in the RF chip can operate in one of three modes. In Mode 1, the radio provides only raw data to the BB chip; in Mode 2, the radio removes any DC offset before sending the signals to the BB; and in Mode 3, the radio extracts data and clock signals and also provides access code correlation. Obviously, higher modes require increased levels of digital signal processing within the RF chip.

Some manufacturers were unhappy with BlueRF for a number of reasons. Among these are[12]

- No simple method exists to measure the BER or jitter/glitch values in RX Modes 1 and 2.
- Access code correlation in Mode 3 is more efficiently performed in the BB processor.
- BlueRF lacks some useful control and timing signals between RF and BB.

- Some RF chips require analog data interfaces, but BlueRF is digital only.

Several manufacturers reacted to this situation either by retaining their custom interfaces between their own BB and RF solutions or by developing a customized, and hence incompatible, version of BlueRF. One fairly popular variant is sometimes called Mode 2.5, where the RF chip extracts some timing information from the incoming signal but leaves access code correlation to the BB chip. At any rate, BlueRF was not adopted by the SIG, and ARM no longer distributes the BlueRF specification.

A third method of communication between RF and BB is to place an *RF-to-BB bridge* between the two devices (see Figure 12-10). The bridge can be built from programmable logic that effectively translates commands and responses from one device to the other so they can operate together. Incompatibilities between RF and BB chips from different manufacturers usually involve control signals rather than data, so bridge firmware doesn't necessarily need to process every incoming signal.[13] On the other hand, a large number of RF-BB combinations are possible, each of which requires its own RF-to-BB bridge implementation.

Designing for Class 1 Transmit Power

For an increased range, the Bluetooth transmitter can be given the capability for output powers up to 20 dBm (100 mW). This power level is the

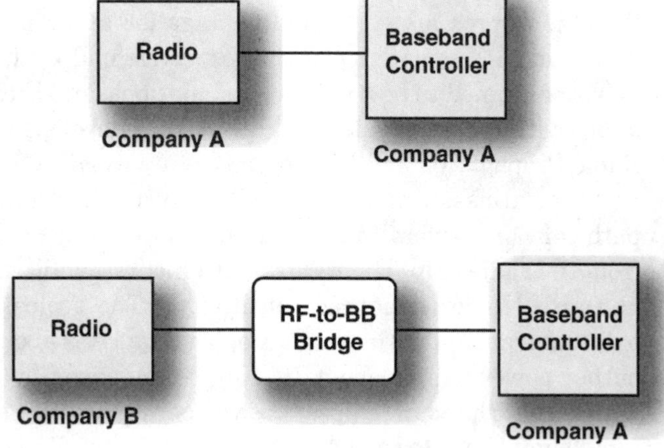

Figure 12-10 The RF-to-BB bridge can act as a translator between two units made by different manufacturers that cannot be directly connected together.

maximum allowed in Europe for spread spectrum in the 2.4 GHz band. The FCC allows power levels up to 30 dBm (1 W) for 2.4 GHz spread spectrum use in the United States, but Bluetooth devices will not receive SIG certification for power levels above 20 dBm.

Most Bluetooth-enabled devices communicate over distances less than 10 meters, so the class 2 or 3 transmitter is by far the most common implementation, and one of these is included in the one- or two-chip solutions. The higher power of the class 1 transmitter is most useful in a host that has one or more of the following characteristics:

- The host is likely to be in a fixed location, such as a printer or desktop computer.
- Power conservation isn't a primary concern.
- The host is a client that may occasionally need longer-range Bluetooth links.

The Bluetooth specification requires power control for transmit power levels in excess of 4 dBm, and power control in turn requires that the remote end of the link has RSSI capability and the capability to direct the local device's power control function through LMP packets. Also, note that higher-transmit power on one end of a Bluetooth link will have no effect on the range unless a) the corresponding receiver has an increased sensitivity level, and/or b) the transmitter at the other end of the link also has higher power capability. This is because the Bluetooth link is bidirectional and thus range is limited by the weaker of the two communication paths. Consequently, most class 1 transceiver designs also include a LNA in the receive path to provide an increased range when communicating with the more common class 2 or 3 transmitter with a standard receiver. Table 12-3 lists the approximate maximum range for different configurations of TX power and RX sensitivity at the respective endpoints. The path loss exponent is assumed to be 3.0 for all calculations, and these are made in accordance with the propagation characteristics developed in Chapter 2, "Indoor Radio Propagation and Bluetooth Useful Range."

A typical class 1 radio front end is depicted in Figure 12-11. The transmit path contains several matching networks and a PA to boost signal levels from the radio chip. The receive path includes a filter, matching networks, and an LNA for enhanced sensitivity.[14] The typical overall gain for the external transmit path using a single-stage PA is about 12 dB, so maximum output power will be about 16 dBm when driven by a class 2 transmitter within the Bluetooth chipset. Lower powers are obtained by reducing the input signal level to the PA, not by switching the PA out of the circuit, so the

Table 12-3

Approximate range limits for different receive sensitivities and transmit powers (n = 3.0)

		Device A		
		TX = 0 dBm RX = −70 dBm	TX = 20 dBm RX = −70 dBm	TX = 20 dBm RX = −80 dBm
Device B	TX = 0 dBm RX = −70 dBm	10 m	10 m	20 m
	TX = 20 dBm RX = −70 dBm	10 m	40 m	40 m
	TX = 20 dBm RX = −80 dBm	20 m	40 m	80 m

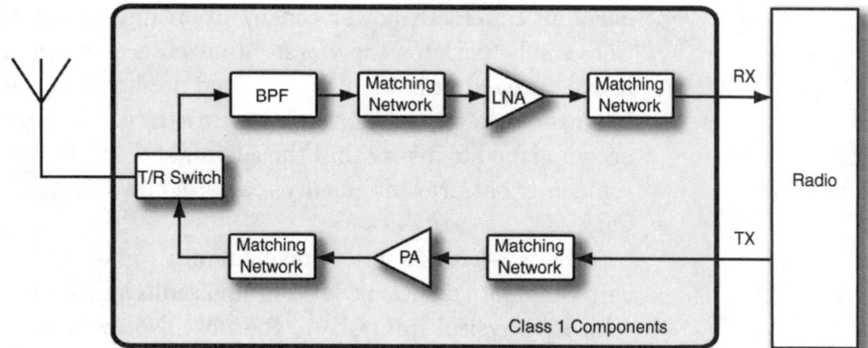

Figure 12-11
A typical class 1 front end consists of several external components in both transmit and receive signal paths, along with an external transmit/receive switch.

challenge is to design the PA to be efficient at widely varying input and output signal levels.

Power Consumption

When determining the power consumption of a particular Bluetooth solution, it's important to realize that this is a dynamic process, with high power needed for TX and RX, but much lower power is used when the device is idle or in one of the low-power modes. Total power consumption can be calculated by averaging the power values needed to perform a *cycle* of communication. For example, if a point-to-point piconet is exchanging single-slot packets, then power can be averaged over two slots; in an asymmetric five-slot situation, the power is averaged over six slots. If a device is in one of the low-power modes, then a cycle can last several seconds.

Average power consumption can be expressed in one of two ways: *average current* or *current per data bit*.[15] Although both of these are slight misnomers (current and power aren't the same), it's easy to convert current into power by multiplying by the supply voltage. Calculating accurate power consumption values is important for devices that are battery powered, less so for those using power from the mains. Battery-powered Bluetooth-enabled devices, of course, derive the power for their entire operation from the battery, and Bluetooth itself may consume only a small fraction of the total. Indeed, the best situation from a customer point of view occurs when the incorporation of Bluetooth has no discernable effect on battery life due to the presence of non-Bluetooth circuitry that requires much more power. If this is the case, then it makes little sense to spend lots of time deriving Bluetooth power consumption figures for all possible scenarios. Conversely, perhaps the worst situation occurs when Bluetooth is placed into a device that in the past required no separate power source, such as in a mouse or keyboard. In these cases, careful battery life analysis must be accomplished to insure that the customer won't be frustrated by constantly replacing batteries in devices where power considerations used to be ignored.

Table 12-4 lists some typical current consumption figures for a single-chip Bluetooth solution that contains radio, synthesizer, baseband, and HCI with its physical interface to the host. Notice that there are large differences in average consumption, depending upon the device's state, ranging from 15 μA when in the standby mode with no RX or TX to 60 mA when either sending or receiving an asymmetric DH5 channel. $V_{DD} = 1.8$ V for this device, so assuming that two 1,200 mA-h alkaline AA cells are used, battery life will be approximately 20 hours at a worst-case 60 mA current drain, but much longer if sleep is the primary activity. Of course, if the same battery is also powering other circuitry, then the additional current drain must be taken into account.

Table 12-4

Average current drain for the CSR BlueCore 2™ single-chip Bluetooth solution

Operation	Average Current Drain
SCO HV3 (slave)	27 mA
SCO HV3 (master)	28 mA
SCO HV1 (slave)	55 mA
SCO HV1 (master)	55 mA
ACL 115.2 kb/s UART (master)	18 mA
ACL 720 kb/s USB (slave or master)	60 mA
ACL Sniff 40 ms interval	4 mA
ACL Sniff 1.28 s interval	1 mA
Parked slave, 1.28 s beacon interval	1 mA
Sleep	15 μA
Peak current during TX	80 mA

Integrating Bluetooth into a Host

The first complete Bluetooth products to enter the market consisted of an add-on card for a PC application. The card contained the radio, baseband, LM, and HCI with its transport mechanism. The remaining Bluetooth protocol stack was placed onto the hard drive of the host, and together they could realize several Bluetooth profiles such as serial port, file transfer, dial-up networking, and synchronization. Although this approach proved to have the fastest time to market, the initial cost of the package was nearly $200. For Bluetooth to succeed as a ubiquitous cable replacement, technology costs must be reduced from that level significantly. To gain cost flexibility, other integration strategies should be considered. We will examine several of these in this section.

Selecting a Module

Because a large number of manufacturers are building Bluetooth modules, selecting the best one for a given application can be a daunting task. Depending upon how the module is integrated into the host, which we will examine shortly, the module could be as simple as a radio chip or as complex as the entire implementation from application to radio.

The target application will drive module selection to a high degree. These applications can be roughly divided into four categories:[16]

1. **Adapters** These include PC cards, USB or RS-232 dongles, a printer port attachment, and others. Adapters are characterized by fast time to market and relatively high cost.
2. **Appliances** These are largely self-contained, Bluetooth-enabled devices such as headsets, MP3 players, toys, and computer peripherals (HID). Low cost is extremely important in these applications.
3. **Embedded systems** These consist of computers or peripherals with Bluetooth capabilities built directly into them without using an abstraction such as HCI.
4. **Mobile phones** Bluetooth can be added to wireless phones by embedding it into the battery pack, into an adapter, or directly onto the main circuit board; hence, this category is a special case of the first or third category. This application has the potential for the highest sales volume, but cost pressure will be intense.

For each of these categories, different criteria must be emphasized when designing an integration strategy. The *module specification issues* can be supplied by the manufacturer of the chipset or software package and often need to be verified by the integration engineer. The *preintegration factors* are primarily determined and/or measured by the integration engineer. Table 12-5 lists these issues and assigns importance levels to each within the four application categories listed previously.[16] Of course, your situation may be different, so feel free to change the entries if you want.

Integration Strategies

The proper integration of Bluetooth capability into a host presents myriad possibilities, all of which requires careful study and perhaps many compromises. Add-on adapters generally communicate with the host through HCI, but a more cost-effective solution may be to fully integrate Bluetooth capability into the host itself. Other applications demand full integration as part of their initial design, such as the Bluetooth-enabled headset. Furthermore, placing Bluetooth into mobile phones presents its own set of challenges associated with the necessary small size, low cost, and the need for simultaneous interference-free operation of both phone and Bluetooth links.

Table 12-5

Module specification issues and preintegration factors

Issues and Factors	Category 1	Category 2	Category 3	Category 4
Module Specification Issues				
Size	**	***	**	***
Operating temperature range	***	**	***	***
Range of the wireless link	***	***	***	***
Profile type and data throughput	***	**	***	**
Radio performance	***	***	***	***
Physical and electrical characteristics	**	**	**	**
Test mode and access points	**	**	**	**
Regulatory preapproval	**	**	**	**
Preintegration Factors				
Battery life	*	***	*	***
Power consumption	**	**	**	**
Power supply noise	***	*	**	*
Over-the-air interference	***	***	***	***
Interference with host	*	**	*	***
Antenna radiation pattern	***	***	***	***

*** Very Important
** Somewhat Important
* Not Particularly Important or Not Applicable

From a macroscopic point of view, the integration strategy can be based upon the required time to market, size, interoperability, cost, and flexibility.[17] All of these are interrelated and can be summarized as follows:

- **Time to market** The fastest time to market can be realized by using an existing plug-in Bluetooth module because integration is simple and can be realized through HCI. Conversely, developing the Bluetooth solution by designing it as part of the host can be the most time consuming and involves increased risk.

- **Size** The smallest possible size for use in a PC is accomplished by placing the Bluetooth radio directly onto the host PCB and integrating everything from BB functions to the application into the host itself. For a stand-alone application, the smallest size will consist of a single *application-specific integrated circuit* (ASIC) containing the entire system, Bluetooth included. If size is less critical, then an external module operating through HCI can be used in a PC application or several chips working together for a stand-alone application.
- **Interoperability** Bluetooth-enabled devices from different manufacturers must be able to operate together in varying degrees for them to obtain SIG qualification, so it's important to insure that either the solution has already been tested or can be tested for interoperability from radio compatibility all the way to profile conformance. If a plug-in module solution is purchased, then looking for one that is already prequalified could prevent numerous headaches later. If the design includes embedded Bluetooth components, then a prequalified reference design can be equally valuable. Even if the entire Bluetooth protocol stack from BB to application is placed into the host, a prequalified RF module will increase the chance that the Bluetooth-enabled device will receive qualification quickly and easily.
- **Flexibility** For greatest flexibility, the approach using individual components is probably the best choice. Separating RF and BB functions enables each of these choices to be optimized, as long as the two can communicate with each other. Using a prexisting plug-in module has the least flexibility because the design features are already established for the RF, BB, LM, and HCI portions of the protocol stack.

Partitioning the Protocol Stack The basis for a typical Bluetooth integration is centered around where the protocol stack is to be partitioned between the module and the host. This partitioning can be structured in numerous ways, some of which are depicted in Figure 12-12. These cases are based upon those published by Walton and Kumar.[18] Depending upon where the protocol stack is partitioned, different methods are used for communication between the layers. An application can communicate with the upper layers of the Bluetooth protocol stack through an *application programming interface* (API). HCI is used if the protocol stack is partitioned between the L2CAP and the LM. Finally, as we discussed earlier, communication

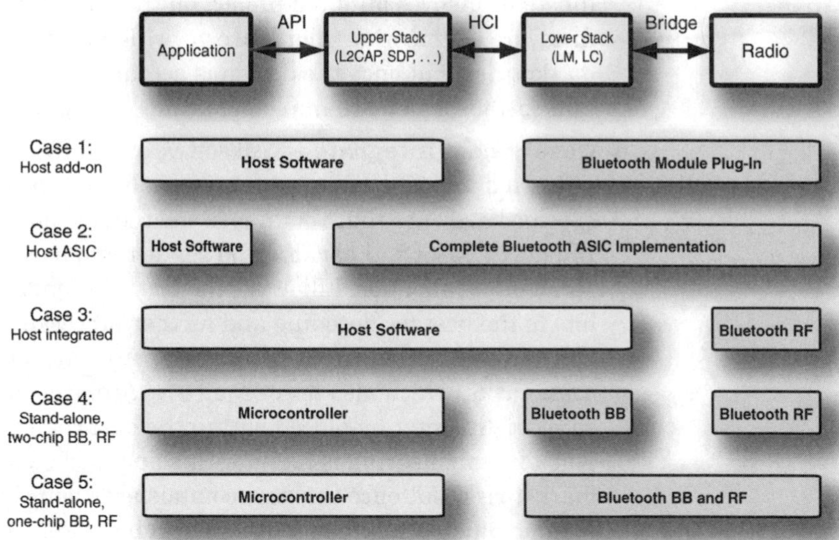

Figure 12-12 Several methods can be used to partition the Bluetooth protocol stack between host and module.

between BB and RF can take pace using BlueRF, a custom interface, or through an RF-to-BB bridge.

Although several other methods of partitioning the protocol stack exist, the five cases in the following list cover many of the most common implementation permutations. Three of these involve integrating Bluetooth capability into a host. The host can be a PC, major peripheral, or phone. The remaining two cases are for stand-alone integrations into devices such as HID, toys, and custom applications.

- **Case 1: host add-on** The host processor runs the application and communicates with the upper Bluetooth protocol stack through an API. The Bluetooth profiles and upper protocol stack are physically located on the host as part of its software package. The lower Bluetooth protocol stack from the LM to the radio is typically part of a plug-in module that connects to its host through HCI. This is an early, mature design that has the advantage of module portability between different hosts, but is also one of the most expensive implementations due to the cost of the module.
- **Case 2: host ASIC** The entire module and upper protocol stack in Case 1 is integrated into a single ASIC that communicates with the host through an API. This approach has the potential for an extremely low component count and can be placed on the host motherboard, with an antenna completing the design. The use of an ASIC for Bluetooth

functions insures minimal impact on the host CPU. Hardware and firmware engineering efforts are very high for this approach, increasing market risk, but once the design is accomplished, the additional per-unit cost for Bluetooth implementation will be relatively low.

- **Case 3: host integrated** All software and firmware in Case 1 is converted to host-based software kernels, and the only Bluetooth-specific hardware that remains is the radio itself. The host assembles Bluetooth baseband packets and sends them to the RF chip over a serial or parallel bus. Likewise, the RF chip sends raw incoming data bits to the host for decoding and further processing. Of course, the host CPU must have sufficient processing power available to accomplish these tasks, which also may require *real-time operating system* (RTOS) support. An antenna placed within the host completes the design. Software engineering is high for this approach, again increasing market risk, but once this is accomplished, then the additional per-unit cost for Bluetooth implementation can be extremely low.
- **Case 4: stand-alone, two-chip BB and RF** For stand-alone devices, such as headsets, HID, or toys, a microcontroller will probably be used to handle the Bluetooth upper protocol stack, profile functions, and I/O. For flexibility in the BB and RF solutions, a two-chip approach is taken, with added engineering required for these to communicate if they are from different manufacturers.
- **Case 5: stand-alone, one-chip BB and RF** As in Case 4, a microcontroller is used for implementation of the Bluetooth upper protocol stack, profile(s), and I/O. The single-chip BB/RF solution is used to reduce board space and perhaps cost to the lowest amount possible. The BB and RF portions of the circuit already communicate, so no additional engineering is required here.

Qualification and Testing

Bluetooth devices can be tested for proper operation at several stages during the manufacturing and integration processes. The first set of tests is performed by the chip manufacturer during the design, fabrication, and production of the IC. Next, if a module (SiP) is built around the chipset then it, too, undergoes extensive testing during design and production. Finally, for fully integrated implementations, the Bluetooth-enabled device itself must be tested for proper operation—indeed, this final set of tests is most important from a customer point of view. After all, a sensitive Bluetooth radio isn't any good if the integrated antenna is so poor that its range is frustratingly short or if it is susceptible to significant interference from the host processor.

The Bluetooth SIG requires a comprehensive set of tests as part of its qualification program. A Bluetooth-enabled device is required to be qualified by the SIG prior to being marketed. Therefore, it's important that the engineer remains oriented toward that goal while the project is in its design phase.

Test engineers enjoy speaking in abbreviations and acronyms, and this section is full of them. In some cases different abbreviations are used for essentially the same thing. For example, what should the tested device be called? Several names are used, such as *device under test* (DUT), *implementation under test* (IUT), *equipment under test* (EUT), and *system under test* (SUT). Sometimes the DUT refers to a chipset, while IUT, EUT, and SUT expand to include the entire Bluetooth-enabled system. Generally, the SIG qualification documentation uses the terms *EUT* or *IUT,* while the core specification covering the protocols uses the term *DUT*.

Keep in mind that Bluetooth qualification is a dynamic process that is undergoing changes as more products are qualified and as different needs arise. The testing of Bluetooth devices is also changing rapidly as the various test equipment manufacturers develop new solutions to some of the difficult, time-consuming test requirements. Although we cover the qualification and test processes in some detail here, changes will inevitably occur after this is written. Consequently, it's important to retrieve the latest documentation on these subjects before making decisions on how to proceed with your own test and qualification programs.

SIG Qualification

Every Bluetooth-enabled device must achieve *qualification* by the SIG before it can be marketed. The qualification program is the process by which a manufacturer demonstrates that a particular product meets the requirements of the Bluetooth specification. Although there are several overlapping areas between Bluetooth qualification and government regulatory approval, the latter process is outside the scope of Bluetooth SIG qualification.

Legal Aspects The Bluetooth specification and related documents contain a significant amount of intellectual property, and the SIG has been aggressive in obtaining patents and other legal protection for this property. Membership in the SIG is required before an entity is allowed to use the specification, not to mention developing and marketing a Bluetooth-enabled product. SIG membership requires signing an extensive set of legal documents regarding the promotion and use of Bluetooth. Membership as an adopter is without cost, but becoming an associate member requires the payment of an annual fee.

Once a product has been qualified, use of the patents and Bluetooth trademark are royalty free. On the other hand, no product using Bluetooth intellectual property can be developed and sold unless it has been qualified by the SIG. For example, a device cannot be built and marketed using a Bluetooth chipset but with a power amplifier that exceeds the 20 dBm class 1 transmit power limit or that hops or transmits data at a rate different than that listed in the specification because these products are ineligible for qualification.

Passing the qualification process demonstrates a certain measure of compliance and interoperability, but every product cannot possibly be tested for every aspect of operation in a reasonable amount of time. Therefore, passing the qualification process doesn't guarantee compliance, but only satisfies conditions of the Bluetooth license grant. Insuring that the product meets all compliance and interoperability requirements is still the responsibility of the manufacturer.

Qualification Process The mechanics of the qualification process are contained in the *Qualification Program Reference Document* (PRD) published by the SIG and available to members on its web site. Several companion documents are also published in which various details of qualification are given, which we will discuss shortly. Organization of the Bluetooth qualification process is shown in Figure 12-13. The following entities in the organization perform specific tasks:

- ***Bluetooth Qualification Review Board* (BQRB)** This group sets and maintains the qualification policy published in the PRD. Delegates have one vote per promoter company (3Com, Ericsson, IBM, Intel, Lucent, Microsoft, Motorola, Nokia, and Toshiba).
- ***Bluetooth Qualification Administrator* (BQA)** The person responsible for administering the qualification program for the BQRB.
- ***Bluetooth Qualification Body* (BQB)** These are individuals authorized by the BQRB to provide services to a particular SIG member who is seeking product qualification. The BQB checks

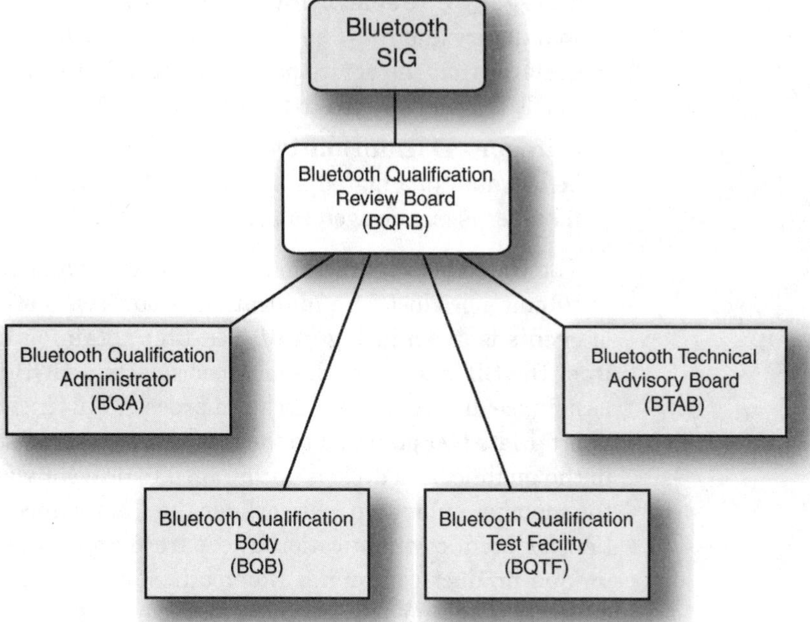

Figure 12-13 The organization of the Bluetooth qualification process

declarations and documents against requirements, reviews the product test reports, and lists qualified products in the official database.

- ***Bluetooth Qualification Test Facility* (BQTF)** A facility that's accredited by the BQRB to perform category "A" tests of Bluetooth products.
- ***Bluetooth Technical Advisory Board* (BTAB)** A forum of BQB and BQTF representatives that advises the BQRB on technical matters associated with Bluetooth testing and resolves technical questions regarding testing. In other words, the BTAB closes the loop in the qualification management process.

Qualification tests are divided into four different categories:

- **Category A (BQTF)** This test is mandatory and must be completed by an accredited BQTF.
- **Category B (declaration with evidence)** This test is mandatory and can be performed either by the SIG member or BQTF. Test results and setup conditions are reported to the BQB.

- **Category C (declaration without evidence)** This test is mandatory and must be performed by the SIG member. The member declares that the device passed the test, but no supporting evidence needs to be submitted to the BQB.
- **Category D (informative)** This test is not mandatory, but it could be elevated to a higher status at a later time by the BQRB. The member is encouraged to perform these tests.

The Bluetooth qualification process begins when a SIG member is ready to submit a product for qualification. An overview of the qualification chain of events is given in Figure 12-14. After obtaining the related documents from the SIG web site, the member selects a BQB to provide advice and assistance during the qualification process. This includes the creation of an appropriate test plan tied to the device being qualified. BQB selection early in the qualification cycle is encouraged to prevent wasted effort later. Next, the member submits paperwork for the Compliance folder and an application for product qualification. A separate application is required for each unique product containing Bluetooth. Finally, the member performs the required tests and submits a product to a BQTF for category A testing (see Figure 12-15). Test reports and other evidence of successful test completion are added to the Compliance folder. The BQB reviews the documentation and, if all is well, grants qualification and places the product into the *Qualified Products List* (QPL) database. A separate fee is charged by the SIG for the various aspects of the qualification program and for listing the product in the QPL.

Figure 12-14
The qualification sequence of events

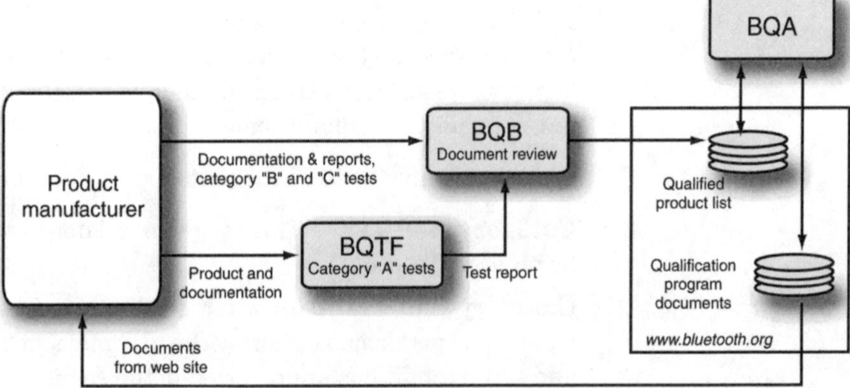

Figure 12-15 A rare look inside a BQTF laboratory. (Source: Bill Lae)

Test Documents The actual tests that need to be performed are listed within a document (actually, a spreadsheet) called the *Test Case Reference List* (TCRL). Don't you just *love* these abbreviations? The TCRL has separate sheets for each type of test, from RF to profiles, and the associated category (A through D) is given. Figure 12-16 shows the various tests listed in the TCRL and how they're organized.

The way in which each test is to be performed is spelled out in excruciating detail in a *test specification* (TS) for each layer in the protocol stack and for each profile. The TS includes, where applicable, *Implementation Extra Information for Testing* (IXIT), containing information related to the IUT and its testing environment.

The *Implementation Conformance Statement* (ICS) is a relatively simple document (finally!) with which the member claims conformance to a particular part of the specification. These come in two types: the *Protocol ICS* (PICS) and *Profile ICS* (ProfileICS). Each document consists largely of a set of "yes-no" check boxes for whether the IUT successfully meets each aspect of the protocol or profile. At the end of each TS document is a *test case mapping table* that shows the link between each test in the TS and its associated ICS entry. The ICS documents are to be included in the Compliance folder. Finally, a Declaration of Compliance is signed by the member and placed in the Compliance folder before it is reviewed by the BQB.

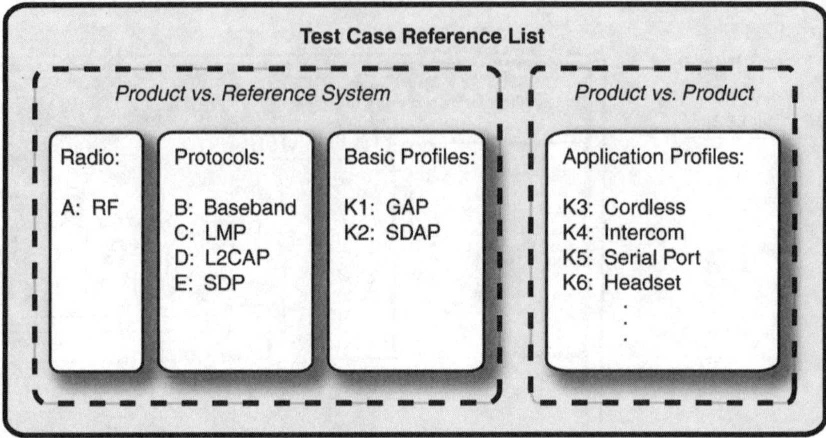

Figure 12-16 The TCRL contains the required tests that are needed for each protocol level and associated profile.

When the compliance process is completed ("all the yelling and screaming is over," as a colleague used to say), the Compliance folder becomes the official repository of all qualification documentation for a given product. This folder is maintained by the local BQB and contains

- A copy of applicable Bluetooth Member Agreement
- Description of the product, to include
  - Descriptive name
  - Model number
  - Hardware and software version numbers
  - Profiles supported
  - Preliminary user's manual or guide
  - Functional block diagram and product description
- Test plan
- Test report(s)
- ICS statements
- Declaration of Compliance

Test Categories Tests fall into two major groups: *conformance testing* and *interoperability testing*. Conformance testing is usually accomplished with test equipment that may or may not involve another Bluetooth module. Interoperability testing is done by connecting the IUT with a target product to ensure that they operate together as expected.

For interoperability testing at the radio, baseband, and basic LM levels, Blue Unit testing shall be used. A *Blue Unit* is a device that is supposed (but alas, not warranted) to be compliant to the specification, so most test irregularities are presumed to occur within the IUT. Unsuccessful testing with a Blue Unit doesn't necessarily equate to noncompliance, but it does require the member to document compliance by another means. A separate SIG document called "Test Specification: Blue Unit Test Cases" presents a list of tests against a Blue Unit that can quickly determine if the IUT has lower protocol interoperability problems. Blue Units will be phased out as other means for lower-level protocol interoperability testing become available.

Similarly, profile interoperability shall be tested against a listed product called a *Designated Profile Interoperability Tester* (DPIT). These are products that comply with the Bluetooth specification and implement one or more profiles. To avoid pandemonium, the number of listed DPITs will be kept small. If a SIG member applies to provide a DPIT, then the member must commit to make sufficient devices available for use in testing.

Conformance and interoperability tests can be segmented into the following categories:

- **RF testing** Performed using a combination of standard and special Bluetooth test equipment
- **Protocol conformance testing** Accomplished using a reference test system such as a protocol analyzer or another means to ensure conformance
- **Protocol interoperability testing** Performed using Blue Units at lower protocol levels
- **Profile conformance testing** Accomplished using an appropriate method to ensure conformance
- **Profile interoperability testing** Accomplished by connecting to another unit, such as a DPIT, supporting the same profile and verifying proper operation

A Bluetooth-enabled product is allowed to have functionality not specifically listed in any profile (such as wireless telemetry, toy, and so on). Functionality beyond an existing profile (such as including a proprietary audio codec with a headset) is permitted as long as interoperability is maintained with devices conforming to the profile being claimed. Development tools and demonstration devices not intended for sale to the public need not be qualified. If the product is not qualified, then that status must be clearly conveyed to the targeted audience or customer.

RF Test Setup

Testing the RF portion of the Bluetooth protocol stack is perhaps the most involved of the required test procedures for Bluetooth qualification. This is due to the need for rather complex, specialized equipment to perform a plethora of performance evaluations in both analog and digital domains.

Three phases of testing generally occur before a Bluetooth-enabled product enters the market:

- **Development testing** A manufacturer performs extensive tests on each component or set of components during the design of the Bluetooth device, whether it be a chipset, module, or fully integrated solution. The tests are customized to each stage in the design and are set by the manufacturer with qualification as a goal. Examples are measuring amplifier characteristics, power transfers, and correct operation of digital algorithms. Extensive access points are made available for connecting the test equipment and are limited only by the design architecture.
- **Qualification testing** Once a production-ready prototype has been created, it is tested for qualification. These tests only require access to the device's RF input/output, along with a means for placing the device into a special test mode. Probing the device's internals is not necessary for qualification, but the manufacturer can certainly access these additional test points for troubleshooting while preparing the device for qualification.
- **Production testing** After a device is qualified, it can enter production, with volumes reaching perhaps hundreds of units per day. Because it is impractical to run a full qualification test on each unit, the manufacturer will create a subset of tests based upon process engineering techniques that will insure proper performance. Of course, the manufacturer is still responsible for meeting all government and SIG requirements for each device sold.

Because the Bluetooth SIG is interested in qualification testing on a production-ready prototype, the test processes that are listed in the specification assume that access to the module's electronics is limited by what would normally be available during its operation, such as an RF connection for input/output and HCI (or similar) for control.

RF tests (and other tests, for that matter) can be efficiently performed with the DUT in a test setup such as that shown in Figure 12-17. The DUT is secured to a test jig, power and control (such as HCI) are attached, and its

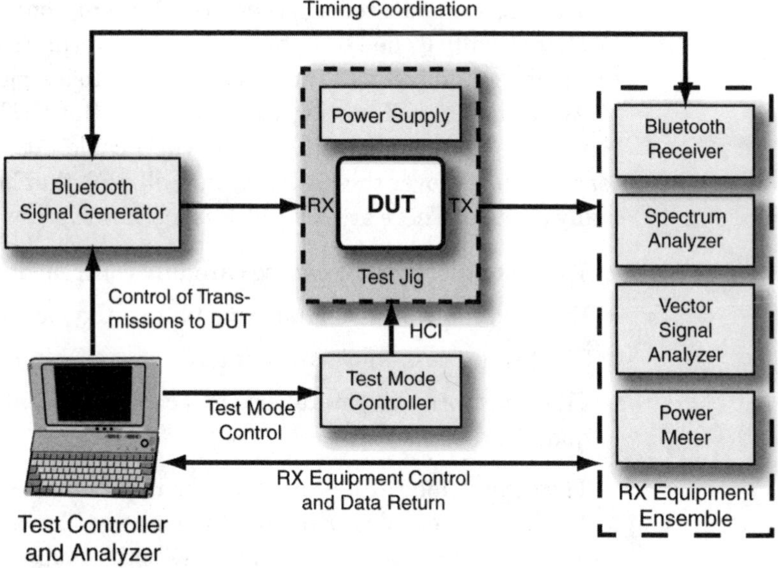

Figure 12-17 The DUT is placed under control of the test equipment for RF testing.

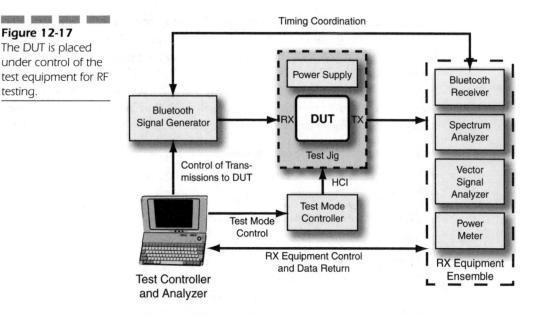

RF port(s) is connected to the test equipment. If necessary, the test jig can contain portions of the Bluetooth protocol stack not included in the DUT. The DUT transmitted signal can be analyzed by equipment such as a spectrum analyzer, vector signal analyzer, or power meter, any or all of which may be controlled by a PC. Connection from the DUT to this equipment can be with a cable or over an RF link for nominal testing, but for qualification, the Bluetooth RF TS requires connection via the antenna connector or a temporary 50 ohm connector if no antenna connector exists. RF signals to the DUT are provided by a Bluetooth signal generator, which is a specialized piece of equipment capable of creating both normal and certain out-of-specification signals and packets.

The Bluetooth receiver in the test equipment converts incoming signals to the baseband for analysis by the PC controller. The spectrum analyzer can display the power spectral density of the DUT transmitted signal over a variety of bands and bandwidths, so it can check for TX signal purity and spurious outputs. The vector signal analyzer captures both magnitude and phase of the DUT transmitted signal, so a more detailed analysis can be made in both time and frequency domains. Finally, the power meter can be used to determine transmitter power output values at the various hop frequencies. Some of this equipment may require special Bluetooth options to provide the proper functionality. For conceptual clarity, we show separate DUT transmit and receive paths, along with a separate test transmitter and receiver, but in reality, there is often only one RF connection to the DUT, especially if the DUT chipset has an internal T/R switch, and the test transmitter and receiver can be combined into a single chassis.

As we mentioned earlier, connections to the DUT transmitter and receiver can be made over the air or via a cable. At first glance, it may be tempting to use over-the-air links; after all, isn't that how the device will normally operate? There are several reasons, however, for using a cable instead:

- The test environment can be carefully controlled.
- External interfering signals won't affect the measurements.
- The test process won't cause interference to other devices.
- The Bluetooth TS dictates use of a cabled RF connection for qualification testing.

The second and third items could be important if several units are being tested simultaneously during production, as will most certainly be the case for high-volume products such as Bluetooth-equipped cell phones. Regardless of whether cabled or over-the-air signals are exchanged between DUT and the test equipment, the test engineer must compensate for various losses, mismatches, and the like within the test setup to insure accurate measurements are obtained. Certainly, at least some over-the-air testing should take place during production to ensure that the antenna operates correctly.

Test Mode To enable efficient testing, a Bluetooth device can be placed into a special *test mode* so it can operate in a nonstandard way to eliminate extraneous variables that could otherwise corrupt the results. For example, the frequency hop mechanism can be turned off when various stability and power measurements are being taken. The test mode is used mainly to test the transmitter, receiver, and baseband unit for compliance with the specification. Because the test mode supercedes normal operation, this mode must not be accessible by the end user.

For security reasons, the Bluetooth specification requires that the DUT test mode be *enabled* through some method other than an over-the-air command, but the method used can be implementation-dependent. For convenience, the HCI_Enable_Device_Under_Test_Mode command is provided for that purpose if an HCI (wired) link is available to the DUT. When this command is sent, the DUT responds with the Command Complete event, but does nothing else until the test mode is activated (the following paragraph explains how this is done). The DUT is then placed into the continuous page scan mode to await connection to the test equipment. The test mode is disabled through the HCI_Reset command.

After connecting as piconet master to the DUT, the tester *activates* the test mode via the LMP_Test_Activate command from the tester to the DUT. This command is provided by the Bluetooth signal generator in Figure 12-17. The DUT responds with LMP_accepted if testing has previously been enabled, and then it enters the test mode. At this point, the DUT will ignore all LMP commands not related to testing. (The DUT still responds to LMP power control commands and LMP_features_req.) The LMP_test_activate command has several parameters, each byte of which is XORed with 0x55 before it is transmitted and again when received to insure suitable whitening at the RF level. These parameters are

- Test scenario (8-bit unsigned integer)
  - 0: Pause test
  - 1: TX sequence 0000 . . .
  - 2: TX sequence 1111 . . .
  - 3: TX sequence 1010 . . .
  - 4: TX *pseudorandom bit sequence* (PRBS)
  - 5: Closed loopback—ACL packets
  - 6: Closed loopback—SCO packets
  - 7: ACL packets without whitening
  - 8: SCO packets without whitening
  - 9: TX sequence 11110000 . . .
  - 10–254: Reserved
  - 255: Exit test mode
- Hopping mode (8-bit unsigned integer)
  - 0: RX/TX on single frequency
  - 1: Hopping using Europe/USA (79-channel) pattern
  - 2: Reserved
  - 3: Hopping using France (23-channel) pattern
  - 4: Reserved
  - 5: Reduced hopping sequence (optional)
  - 6–255: Reserved
- TX frequency of DUT (8-bit unsigned integer)
  - $f = 2402 + k$ MHz

- RX frequency of DUT (8-bit unsigned integer)
  - f = 2402 + k MHz
- Power control mode (8-bit unsigned integer)
  - 0: Fixed TX power
  - 1: Adaptive power control
- Poll period (8-bit unsigned integer)
  - n × 1.25 ms
- Packet type (8-bit unsigned integer)
  - Same numbering system as in packet header
- Length of test sequence (16-bit unsigned integer)

The DUT exits the test mode either through LMP_Detach or by LMP_test_activate with the test scenario value set to 255. Alternatively, HCI_Reset can be used.

Testing the Transmitter

The DUT transmitter can be tested for output power, power control (if applicable), modulation characteristics, frequency accuracy and drift, power spectral density of the modulated signal, and spurious emissions. To test the DUT transmitter, it is placed into the test mode and sent a POLL packet by the tester. The DUT responds with a packet in the appropriate slave slot, as shown in Figure 12-18. The bit pattern in the packet payload is determined by the test scenario value. Frequency hopping is turned off, and the TX frequency is determined by the parameter in the LMP_Test_Activate command. Packet length is determined by the length of the test sequence value and may be longer than one time slot. Only packets without FEC should be used for TX testing, and whitening is automatically turned off.

Different test cases require different payload data structures. If a PRBS is selected for transmission, then a data string called *PRBS-9* is used. The maximum length of the sequence is 511 bits, and these appear to be random. Transmitted signals using this sequence will exhibit a power spectral density approximating that of real data. Other bit patterns available are 1111 . . . , 0000 . . . , 1010 . . . , and 11110000. . . . The last two are useful for checking the settling time of the Gaussian filter used in the transmitter. For example, the Bluetooth specification requires that the frequency deviation

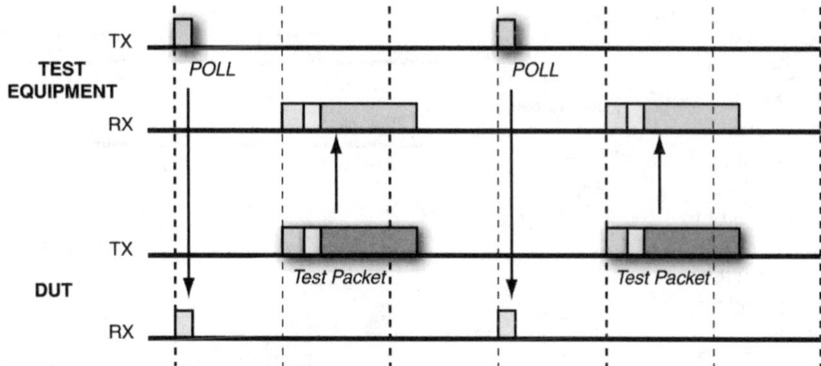

Figure 12-18 Transmitter test timing

for the 1010 sequence reach at least 80 percent of that when the 11110000 sequence is transmitted.

Faster testing of the radio over the entire Bluetooth frequency range can be done by selecting a reduced hop sequence. Five channels are selected (0, 23, 46, 69, and 93) according to the formula $f = 2402 + k$ MHz. Curiously, the highest frequency in this reduced sequence (2495 MHz) is outside the 2.4 GHz ISM band in many countries.

Testing the Receiver

The DUT receiver is tested for single- and multislot packet sensitivity, maximum input power level, C/I performance, blocking performance, and intermodulation. Testing the receiver using the setup in Figure 12-17 presents a special challenge because no direct access to the received data stream is provided. Instead, a process called *loopback* is used, where the signal generator sends baseband packets to the DUT, which in turn retransmits them back to the test equipment for baseband analysis. To eliminate test corruption from problems in the DUT transmitter, it should already have passed qualification and use maximum power for its retransmissions. The signal generator can change its power levels and packet structures for the various tests. Either ACL or SCO loopback can be activated via the test scenario field in LMP_Test_Activate command. For over-the-air production testing, a similar loopback mode between two Bluetooth-enabled devices can be activated using HCI_Write_Loopback_Mode.

An incoming test packet can be retransmitted by the DUT either in the next slave-to-master slot (*normal loopback*, see Figure 12-19a), or the DUT

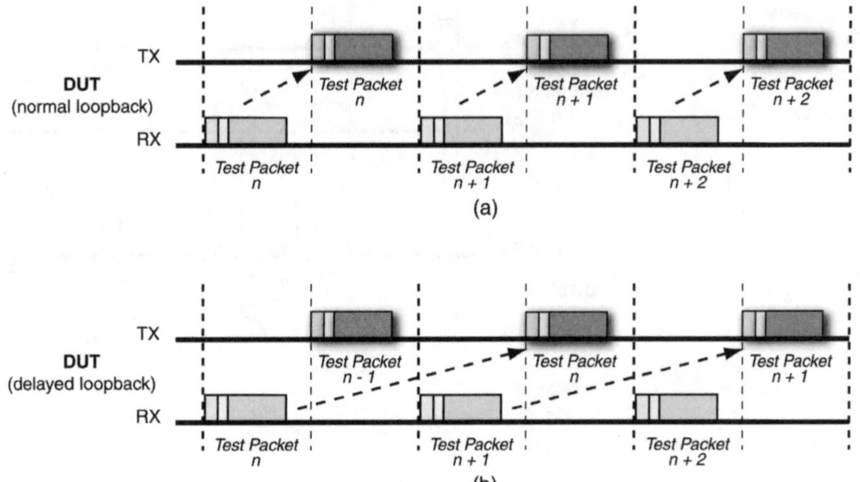

Figure 12-19 The loopback test timing. The DUT can loop back an incoming test packet in the next slave-to-master slot (a) or in the slave-to-master slot following the next incoming test packet (b).

can wait until the next test packet arrives and then loop back the previous packet in the following master-to-slave slot (*delayed loopback*, see Figure 12-19b). Delayed loopback can be implemented if the chipset needs extra time to process an incoming test packet before retransmission. For example, the tester may send the DUT a packet with FEC applied, in which case, the DUT must perform FEC payload decoding, perform a CRC, calculate a new CRC, and reencode the payload before retransmitting the loopback packet. Whether a Bluetooth chipset implements normal or delayed loopback is up to the manufacturer, so a test engineer needs to check this before initiating a loopback test to avoid frustration.

The pass/fail criteria for all receiver tests is whether the BER exceeds 10^{-3}. For example, the DUT receiver will pass the sensitivity test if loopback shows a BER of 10^{-3} or less when the signal generator sends it DH1 packets with a PRBS-9 payload at a -70 dBm input power level. For each of three operating frequencies, a total of at least 1,600,000 payload bits must be returned to the tester by the DUT and compared to what was sent to the DUT. The BER is the number of incorrect bits divided by the total number returned. Notice that Bluetooth qualification doesn't entail finding the actual RX sensitivity level—only that the receiver operates properly at an input signal level of -70 dBm. Of course, actual RX sensitivity can be determined by reducing the incoming signal to the DUT until the BER reaches 10^{-3}.

Other receiver tests are performed by sending the DUT *dirty packets* that have various faults listed in the TS. Once again, a BER of 10^{-3} or less indicates that the DUT passed the test. Because these tests are all validated through the DUT receiver's BER, the loopback mode can facilitate an efficient, automated approach to the tests.

Protocol Conformance Testing: The Protocol Analyzer

Testing conformance to the Bluetooth protocol stack, baseband and above, is often best accomplished with a device called a *protocol analyzer*. This typically consists of a PC with a Bluetooth plug-in card, along with an extensive software package. The protocol analyzer captures baseband packets and analyzes and displays the results. Figure 12-20 shows a simple exchange of baseband data between master and slave that was captured by a protocol analyzer. Higher-level protocol layers, such as LMP, L2CAP, RFCOMM, SDP, and so on can be derived from the baseband packets and tested for conformance to the specification.

The protocol analyzer can capture baseband packets in two ways. First, it can be an active participant in the piconet. For example, it can assume

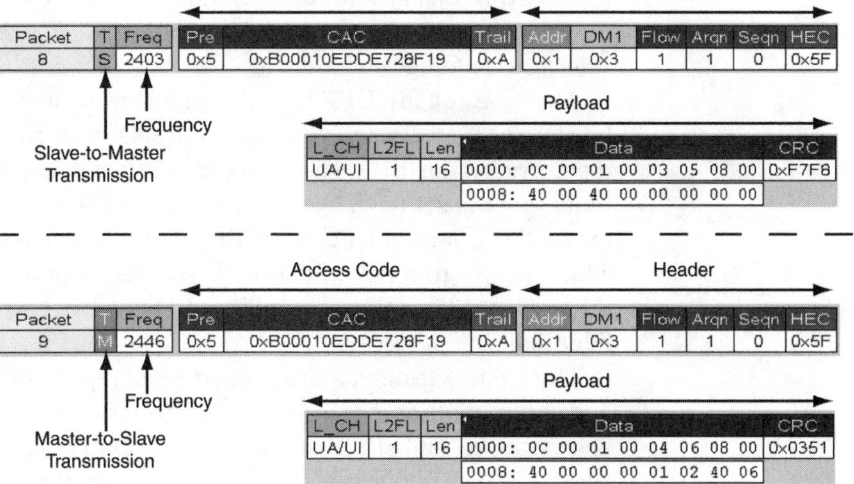

Figure 12-20 The protocol analyzer can capture, store, analyze, and display baseband packets. (Source: CATC)

the role of master, connect to the EUT, and test its compliance to the specification through a carefully choreographed set of packet exchanges, using packets that are known to be good and those with various predetermined flaws. Alternatively, the protocol analyzer can monitor piconet traffic between other devices without actively participating itself. Prior to doing this, though, the analyzer usually connects to each participant briefly to obtain address and timing information to facilitate efficient monitoring.

A well-designed protocol analyzer has almost unlimited versatility in testing protocol compliance. It can insert packets into the piconet that have various combinations of header and/or payload errors to gauge the response. It can send illegal commands or trigger errors only when certain conditions are met. Finally, it can filter the traffic and only display or analyze packets directly related to one protocol level or with one device.[19]

Interoperability Testing: The UnPlugFest

One of the most useful and efficient methods to test a Bluetooth-enabled device for interoperability is through a gathering of design and development engineers at an *UnPlugFest* (UPF). This SIG-organized event occurs several times per year at locations around the world and usually lasts for about four days. (The name was inspired by the "PlugFest" used for interoperability testing of different wired standards such as USB.) Participants from different manufacturers sit across from each other, assume determined expressions, place their products on the table, and try to get them to communicate. Basic operational testing (see Figure 12-21) needs to be done prior to attending the UPF.

Attendees at the UPF include engineers who are developing Bluetooth hardware and software, who can fix problems during the test, and who are members of the SIG. The implementation being tested must meet the requirements set forth by the UPF steering team, and these requirements may differ from one UPF to another. Participants must sign a document in which they agree not to use UPF results to promote their product, and they agree to keep the test results and prerelease product information confidential.

Third-party testers, marketing or sales representatives, students, press, and observers cannot attend. The UPF is focused on interoperability testing and is not intended to be a seminar or clinic, marketing event, or trade show. It is also not part of the qualification program, but attendance by eligible engineers is highly encouraged. Table 12-6 lists some of the differences between the UPF and the Bluetooth SIG qualification program.[20]

Figure 12-21 This test should already have been accomplished and is not usually part of the UnPlugFest agenda. (Source: Bill Lae)

Throwing caution to the wind, Phipps, Field-man, and Hanks give it the 'ole' smoke test.

Table 12-6 UnPlugFest versus Bluetooth qualification

Characteristic	UnPlugFest	Qualification Program
Reference implementation	None	Certified tester or test units
Test outcome	Confidential	Product listed or not listed
Test suites	Defined by organizers	Defined by BQRB
	Focus on interoperability	Focused on specification
	Short time available for testing	No time constraints on testing
Style	Informal	Formal
	Troubleshooting encouraged	No troubleshooting

The following are not considered part of interoperability and hence are outside the scope of the UPF:[20]

- Antenna testing
- Coexistence with other radiators such as 802.11
- Power management
- API or HCI testing
- Usability or user interface testing
- Mechanical and environmental testing

About six to eight weeks before an UPF is to be held, the SIG publishes a list of interoperability requirements for participants to examine. The testing procedures are described in the UPF test plan developed by the technical committee. Tests are based upon those required during qualification, but are usually shortened so they can be accomplished in a couple of hours. Participants can decide whether or not to use these procedures at the event. There are three testing categories, each with its own set of recommended tests:

- **Category 1** RF, BB, and LM
- **Category 2** L2CAP, SDP, RFCOMM, and TCS
- **Category 3** Profiles

A particular UPF may allow testing in all three categories, or it may limit testing to a subset of these categories, protocols, or profiles. A product must test successfully at each category level before it can proceed to the next.

Participants are expected to bring their EUT, laptop computers, and any specialized test equipment or jigs that they may need. Monitors, power supplies, basic test equipment, and miscellaneous lab tools are provided. Testing sessions are scheduled in blocks of one or two hours. The concentration level is intense because there never seems to be enough time to finish the test.

Summary

Many semiconductor manufacturers have designed Bluetooth chipsets that incorporate the radio, BB, LM, HCI, and host transport mechanism. Single-chip solutions employ CMOS designs, which have become practical with recent advancements in small gate size for high-frequency performance. Others have taken the two-chip route, with the radio built from BiCMOS and the separate baseband chip using digital bulk CMOS.

These chips can be used to provide Bluetooth capabilities by integrating them into a host or designing a stand-alone operation around them. For quick time-to-market, host integration can be as simple as a plug-in card or dongle, but unit cost is rather high for this approach. Unit cost can be reduced by creating an ASIC with the complete Bluetooth protocol stack onboard or by placing the entire protocol stack from baseband up onto the host itself and simply adding the radio IC. Engineering costs and time-to-market are both increased with these approaches.

The Bluetooth SIG requires that any device that is to be marketed using the Bluetooth name, trademark, and intellectual property must first be qualified through an extensive test program. This is to insure compliance with the specification, along with interoperability across the various manufacturers to insure a pleasant customer experience. Some tests must be performed by a BQTF, while the remaining tests can be performed at the manufacturer's facility and documented with or without evidence, depending upon SIG requirements.

Testing occurs during the design, qualification, and manufacturing phases of product development; the type of tests performed and the level of access to the unit for testing will differ during each of these phases. Testing for qualification at the RF and BB levels is performed by accessing only the RF and HCI parts of the DUT and placing them into a special test mode. The transmitter is tested by programming the DUT to emit test packets that can be used to analyze transmit performance. The receiver is tested by sending the DUT packets that it retransmits in a loopback fashion, after which the test equipment can compare both data sets and look for discrepancies and errors.

Conformance to the various Bluetooth protocols can be tested by use of a protocol analyzer. The analyzer can take either an active or passive role in the piconet, during which it collects and filters the packets exchanged among the piconet members, and then filters and assembles the results for analysis. For protocol and profile interoperability testing, the SIG has created the UnPlugFest, where design engineers can assemble in one location and test their products against those from other manufacturers in an attempt to analyze and fix any interoperability issues that may be discovered.

End Notes

1. Breed, G. "Receiver Architectures with No Intermediate Frequency," *Applied Microwave & Wireless,* April 1999.
2. Droinet, Y. "Advanced RF Technologies for the Wireless Market," *Microwave Journal*, September 2001.
3. Eynde, F. "Trade-offs of Silicon Technologies for Bluetooth and Other Wireless Apps," *Integrated Systems Design*, August 2001.
4. Svensson, C. and Gong, S. "System-in-a-Package Solution for Short-Range Wireless Communications," *Applied Microwave & Wireless*, November 2000.
5. McCullagh, M. "Cutting Bluetooth Costs with RF CMOS," *Communication Systems Design*, November 2000.
6. McCall, D. "What Application Developers Need to Know about Bluetooth Hardware," presented at the Bluetooth Developers Conference, San Francisco, CA, December 2001.
7. Derbyshire, J. "Silicon Germanium and Bluetooth," *Wireless Design & Development*, August 2000.
8. Nadler, A. "How to Select the Best Bluetooth Products," presented at the Bluetooth Developers Conference, San Francisco, CA, December 2001.
9. Mueth, C. "Packaging Options for Wireless IC Designs," *INSIGHT*, Vol. 6, Issue 3, 2001.
10. Faulkner, C. "RF Modules Enable an Integrated Approach to System Design," *Wireless Systems Design*, December 2000.
11. Arnold, R., et al, "Thin-Film Passive Integration Yields Tiny Bluetooth Module," *Wireless System Design*, August 2000.
12. Doron, M. "The Advantages of a Programmable RF-BB Interface in Bluetooth Devices," presented at the Bluetooth Developers Conference, San Francisco, CA, December 2001.
13. Agatep, A. "RF to BB Bridge," presented at the Bluetooth Developers Conference, San Francisco, CA, December 2001.
14. Wirth, M. "Driving First Class? Reaching 100 Meters in Four Microseconds," presented at the Bluetooth Developers Conference, San Francisco, CA, December 2001.

15. Linsky, J. "Beyond the Datasheet: Low-Power Design in Bluetooth Technology," presented at the Bluetooth Developers Conference, San Francisco, CA, December 2001.
16. "Investigating Bluetooth Modules: The First Step in Enabling Your Device with a Wireless Link," Agilent Application Note, 2001.
17. Walsh, K. "Strategies for Integrating Bluetooth Technology into Mobile Devices," *Wireless Design Online*, October 17, 2000.
18. Walton, S. and Kumar, R. "Reducing Time to Market for Bluetooth-enabled Products," *Communication Systems Design*, May 2001.
19. Boyett, K. "Protocol Tests Lay Groundwork for Bluetooth Success," *Evaluation Engineering*, August 2001.
20. Truntzer, F. "UnPlugFests," presented at the Bluetooth Developers Conference, San Jose CA, December 2000.

CHAPTER 13

Coexisting with Other Wireless Systems

With the development over the last few years of inexpensive RF fabrication technology that operates at high carrier frequencies, the 2.4 GHz ISM band has become the focus of attention for the deployment of a wide variety of communication systems. The low-cost aspect of these systems also means that a large number of them already exist, and these numbers will increase explosively as Bluetooth takes hold in the marketplace.

The 2.4 GHz ISM band has a lot going for it as a communications medium. It essentially has worldwide availability, the bandspread is large enough to be used for high data rates and/or interference avoidance, and transceivers are becoming extremely cheap. That's the good news. The bad news is that the band first came into existence as a residence for microwave oven emissions. This oven uses a magnetron transmitting several hundred watts of power at a nominal frequency of 2.45 GHz, so most governments decided long ago to set aside this band for general, unlicensed use. Certainly, no one will ever want to *communicate* using these ridiculously high carrier frequencies! How things change . . .

In this chapter, we will take a close look at the interference that Bluetooth must contend with, along with some of the approaches for solving the more critical situations. Of course, the most straightforward solution to the interference problem is to simply increase transmit power so that the receiver at the other end has a sufficient *carrier-to-interference* (C/I) ratio to extract the desired signal. If the solution were that easy, then this chapter would be extremely short. Squashing interference with higher transmit power is not an acceptable approach because of the increased interference caused to other users in the band, not to mention decreased battery life and higher component cost. Our goal isn't to put competing systems out of business, so we will treat interference from other users as a *coexistence* challenge. As such, our solutions will attempt to enhance reliable communications for all participants, not just those in our own Bluetooth piconet.

Coexistence solutions can be placed into two categories. The *collaborative* approach assumes that the interfering systems have a means of communicating between them to negotiate access to the medium. The most basic form of collaboration is to turn all devices off except for the one being used, but better methods are available to prevent excessive delays. If multiple wireless systems are collocated and controlled by a single PC, then collaboration is relatively simple. Matters become more complicated if a special wireless link is needed to coordinate operation between two or more wireless systems in different parts of a room or building.

The other approach to improving coexistence is through *noncollaborative* methods. This is more suited to wireless devices that aren't collocated. Each works independently in an attempt to improve its own throughput, hopefully without adversely affecting other links. For example, a device can monitor the communication channel and transmit only when it senses that no carrier is present. This is called *carrier sense multiple access* (CSMA) and is used in wired Ethernet (IEEE 802.3) and wireless Ethernet (IEEE 802.11). A FHSS system, such as Bluetooth, could check for interference and remove the offending channels from its hop frequency set, which is termed *adaptive frequency hopping* (AFH).

Both the IEEE and the SIG have recognized the need for modeling coexistence issues and recommending solutions, and they are working closely with each other in these endeavors. IEEE 802.15 Task Group 2 (802.15 TG2, sometimes called 802.15.2) has published several documents modeling coexistence and are working toward a set of recommended practices to coordinate solutions among the various manufacturers and other groups.

The Bluetooth SIG Coexistence Working Group is chartered to work within the Bluetooth community and with cross industry groups, such as the IEEE, to quantify the detrimental effect of interference on both Bluetooth and other systems. Aside from developing solutions, the information will also assist the manufacturers in providing realistic performance expectations to their customers and will provide valuable information to regulatory bodies. The group will publish "Best Practices" white papers and improvements to the specification and profiles as coexistence issues are solved.

Other Users in the 2.4 GHz ISM Band

Several different forms of transmission can take place in the 2.4 GHz ISM band. In the United States, FCC Part 15 rules enable narrowband, DSSS, and FHSS transmissions to be used as we discussed in Chapter 1, "Introduction." Bluetooth, 802.11, HomeRF, cordless phones, and other custom implementations all use this band. Amateur radio operators are licensed users that have access to these frequencies under Part 95 of the FCC rules. Part 18 of the FCC rules cover the operation of microwave ovens and other ISM equipment in this band, and even microwave lighting has been developed that could be an additional source of interference. In this section, we'll look at some of the users in this band that will likely have the most impact on the performance of the Bluetooth piconet.

IEEE 802.11

Of all the systems that communicate using the 2.4 GHz ISM band, IEEE 802.11 has perhaps generated the most interest as a potential interferer to Bluetooth. The IEEE adopted 802.11 as the first international *wireless LAN* (WLAN) standard in 1997. The standard defined a MAC layer and three PHY layers: DSSS and FHSS in the 2.4 GHz band and *diffuse infrared* (DFIR) as an optical alternative. For FHSS, the hop rates aren't specified, but most 802.11 equipment hops between 10 and 50 channels per second, depending upon the manufacturer. The hop channel set is identical to that used for Bluetooth. The DSSS implementation specifies 11 fixed channels, starting at 2,412 MHz and spaced every 5 MHz. All three implementations provide either 1 or 2 Mb/s raw data rates over distances up to about 100 meters for RF and room-wide for IR.

The goals of the standard were to describe a WLAN that delivers services equivalent to a wired network with high throughput, reliable data delivery, and continuous network connections. The 802.11 network is further enhanced by allowing for mobility and power savings that are transparent to the user. The general philosophy was to distribute decision making to the mobile stations to eliminate bottlenecks and provide fault tolerance. The architecture is flexible, supporting small or large networks that can be either semipermanent or permanent.[1]

After the original standard was developed and released, several task groups were formed to recommend improvements. These include

- **802.11a** Developed a 54 Mb/s enhancement for use in the U.S. *Unlicensed National Information Infrastructure* (U-NII) band at 5.1 to 5.8 GHz
- **802.11b** Developed an enhancement that increases data rates to 5.5 and 11 Mb/s while occupying the same bandwidth as 802.11 DSSS in the 2.4 GHz band
- **802.11e** Working to enhance time-critical data (such as two-way voice) over 802.11
- **802.11f** Standardizing roaming procedures among access points
- **802.11g** Developing an enhancement that increases data rates up to 54 Mb/s in the 2.4 GHz band
- **802.11h** Working on adding power control and dynamic frequency selection to 802.11a for use in Europe
- **802.11i** Addressing 802.11 security weaknesses

The 802.11b standard is extremely popular, having reached deployment levels of several million units. Almost all of the research in the WLAN-Bluetooth coexistence is directed toward the 802.11b implementation, so that is what we will focus on as well. For even faster data rates in the 2.4 GHz band, the 802.11g proposal has been developed, and this will shortly move into the final voting stage. An FCC rule change was required for the modulation technique used in 802.11g to be legal in the 2.4 GHz ISM band in the United States.

During early 802.11b deployments, it was soon discovered that equipment from different manufacturers wouldn't function together. To enhance 802.11b interoperability, the *Wireless Ethernet Compatibility Alliance* (WECA) was formed, and an interoperability test plan was released in April 2000. WECA created the commercial name *Wireless Fidelity* (*Wi-Fi*™) to indicate product interoperability. In a manner similar to Bluetooth, a manufacturer submits its 802.11b product to a third-party laboratory for testing, and if successful, the Wi-Fi logo can be affixed to the product. The majority of 802.11b products now carry the Wi-Fi label. For this reason, we will use the terms Wi-Fi and 802.11b interchangeably.

The 802.11a standard has recently become commercially viable with the advent of relatively inexpensive chipsets. The higher carrier frequencies dictate smaller antennas and shorter range for a given TX power compared to 802.11b, but the 54 Mb/s data rate is fast enough to convey *high-definition television* (HDTV) video. 802.11a products that pass their respective interoperability tests are given the name *Wi-Fi5* to emphasize their 5 GHz carrier frequencies and, hence, incompatibility with 802.11b Wi-Fi devices.

Other Bluetooth Piconets

Another source of interference to Bluetooth operation will soon become very common—that from other Bluetooth piconets in the same general area. Multiple piconets hop through the 79 FHSS channels using independent hop patterns, which means that every so often two transmitters will land on the same frequency and transmit at the same time. Jamming may or may not occur, depending upon whether an affected receiver's cochannel C/I requirements are violated. Likewise, multiple transmissions in adjacent channels could cause a jamming situation if a receiver's adjacent channel C/I is affected. We'll examine these situations in a later section.

Microwave Ovens and Other ISM Devices

The microwave oven is the reason that the 2.4 GHz band exists in the first place, but it can be a significant source of interference in this band. Of course, the oven itself is usually not affected by other devices transmitting nearby, so coexistence isn't an issue from the oven's point of view.

There are hundreds of millions of these ovens in use worldwide, and each generates many hundreds of watts of power at about 2.45 GHz, which is a resonant frequency of the water molecule. Unfortunately, these ovens tend to be prevalent in the very environments likely to contain Bluetooth and Wi-Fi communication systems. As we will discover later, the oven's magnetron RF source pulses on and off with the half-cycle of the AC mains, so bandwidths can be surprisingly high from startup and shutdown transients. Furthermore, as an oven ages, its leakage around the door will gradually increase.

Other ISM devices that operate using RF include industrial heaters, RF stabilized arc welders, medical diathermy (heat therapy) equipment, and ultrasonic machines of various types. The FCC places few restrictions on the operating frequency of these units, and Part 18 even allows unlimited radiating energy at several frequencies, including 2.45 GHz ± 50.0 MHz. Of course, other rules (such as health factors, see Chapter 1) limit the amount of radiation that can escape from the ISM equipment.

HomeRF

The HomeRF Working Group was formed to develop a 2.4 GHz WLAN that is similar to the 802.11 FHSS standard but at reduced complexity and lower cost. Furthermore, a DECT-like *time division multiple access* (TDMA) mechanism is added to support up to four toll-quality voice channels along with data. The hop rate is 50 channels per second, and TX power levels up to 250 mW are supported. Raw data rates are 1 Mb/s using 2FSK or 2 Mb/s using 4FSK. The HomeRF protocol is called *Shared Wireless Access Protocol* (SWAP).

As we pointed out in Chapter 1, in August 2000, the FCC allowed the use of *wideband frequency hopping* (WBFH) in the 2.4 GHz ISM band. This was in response to a petition by the HomeRF Working Group, who argued that the 11 Mb/s 802.11b modulation technique put FHSS systems at a disadvantage because there was no equivalent method to increase FHSS data rates without using a higher hop channel bandwidth. The FCC agreed, and as a result of the new rules, HomeRF developed SWAP version 2.0 to support data rates of about 10 Mb/s. This is done by using 15 hop channels, each 5 MHz wide, and 4FSK modulation.

HomeRF 2.0 could present significant interference to Bluetooth (and to Wi-Fi, for that matter) due to the relatively wide transmit bandwidth and high power. The greatest interference potential is probably between HomeRF and Wi-Fi, but it is unlikely that these two networks will exist in the same location because they both serve similar purposes.

Cordless Telephones

Cordless telephones in the United States are generally of proprietary design and meet the FCC Part 15 requirements. These phones use a wide variety of carrier frequencies from about 1.8 MHz up to and including the 2.4 GHz ISM band. Both digital and analog narrowband modulation are used, along with digital DSSS or FHSS if the phone operates in the 900 MHz or 2.4 GHz ISM bands. Sometimes one band is used for the uplink (handset to base), and the other is used for the downlink (base to handset). As a result, cordless telephone interference can take many different forms, depending upon the phone's make and model.

Bluetooth has the potential to be affected by any cordless telephone using the 2.4 GHz band for uplink, downlink, or both. Phones that use narrowband modulation, either analog or digital, have less interference potential because their power output is low (less than 1 mW), and their fixed frequency usually affects only one hop channel. On the other hand, phones using either DSSS or FHSS have much greater interference potential because of their wider bandwidths (several MHz) and higher transmit powers (10 to 100 mW).

Custom Devices

Aside from cordless telephones, other custom-built devices can create interference for Bluetooth. Part 15 rules enable TX power levels of 1 W when either DSSS or FHSS is used, and at these levels, Bluetooth could experience severe interference. Fortunately, the number of custom Part 15 devices (other than cordless phones) using the 2.4 GHz ISM band is low and will probably shrink further due to the high development cost compared to using Bluetooth itself for the same custom application.

Licensed Users

Amateur radio operators are licensed individuals that are allowed to operate a transmitter within the ham bands, one of which is coincident with the 2.4 GHz ISM band. Ham equipment can have transmit powers up to 1 kW, and a signal at this power level will wreak havoc with any nearby low-power communication system sharing the same band. Fortunately, the number of ham radio operators using this band are very few, and those that do tend to use low-power transmitters located outside in elevated areas and away from most environments in which Bluetooth is used.

Microwave Lighting

A new lighting system called the *sulfur lamp* has been developed that provides high efficiency and a spectrum closely matching that of sunlight. A magnetron operating at 2.45 GHz irradiates a mixture of sulfur and argon, which in turn emits a broad continuum of energy through a process called *molecular emission*. If the lamp isn't properly shielded, there is a potential for these devices to interfere with communications in the 2.4 GHz ISM band. The IEEE is working with the FCC to insure that microwave lighting and communication systems can coexist.[2]

Bluetooth and Wi-Fi: Can They Live Together?

Although we spend most of this chapter discussing technical issues related to the sharing of the 2.4 GHz ISM band between Bluetooth and Wi-Fi, another coexistence factor needs to be addressed: Is there a need for both in the marketplace? Strong opinions exist on either side of this argument. Some say that Bluetooth, with its relatively low data rate and slow market penetration, can't possibly compete with Wi-Fi. Others claim that Wi-Fi won't become widespread enough to perform the ubiquitous connectivity envisioned for Bluetooth.

It's important to realize that Wi-Fi and Bluetooth are indeed aimed at different markets. Bluetooth is meant to be a cable replacement technology that's cheap and robust. Wi-Fi is a more costly high-performance WLAN. In this light, we'll take a look at the characteristics of each, followed by reasons to employ both systems within a single host.

Wi-Fi Characteristics

When 802.11 was first approved as a specification, it had Bluetooth-like data rates (1 or 2 Mb/s) and a wide variety of PHY choices (DSSS, FHSS, and DFIR). Products were available using all three PHYs. The advent of DSSS-based 802.11b enabled data rates to increase to 11 Mb/s, and as a consequence, almost all manufacturers who wanted to stay in the 802.11 market developed 802.11b Wi-Fi products. The relatively high data rate of Wi-Fi, coupled with its significant market penetration, formed the foundation of arguments for those who support Wi-Fi over Bluetooth.

The traditional Wi-Fi network consists of a fixed *access point* (AP) that is often part of a backbone connection to a wired LAN, including a link to the Internet. A user can be in either a fixed (desktop) or semi-fixed (laptop) location and operate a Wi-Fi *station* (STA), usually a PC with an installed Wi-Fi *network interface card* (NIC). Stations communicate only with the AP in a *star topology*. This configuration is called the *infrastructure basic service set*, or simply BSS, as shown in Figure 13-1. Due to their channelized frequency structure, at most three independent Wi-Fi links can exist simultaneously in the same vicinity without mutual interference. Although some would argue that Bluetooth FHSS has better frequency management and

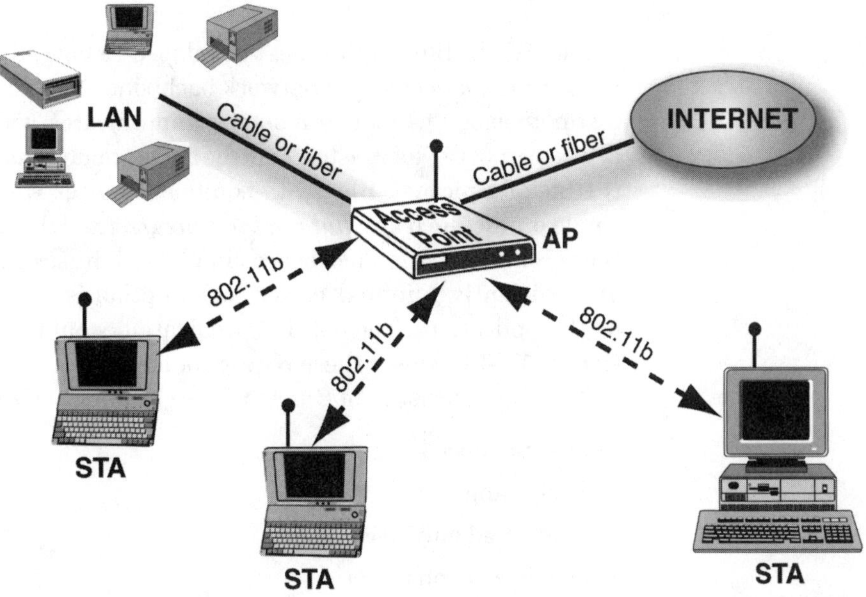

Figure 13-1 Most Wi-Fi networks are configured as a BSS, where STAs communicate with an AP, which in turn is connected to a backbone network.

interference rejection, its lower data rate makes it much less suitable than Wi-Fi for high performance WLAN applications.

Wi-Fi can also be connected in a *peer-to-peer* configuration, where one STA communicates directly with another without the presence of an AP. This type of link, called an *independent basic service set* (IBSS), seems at first glance to provide the same services as Bluetooth. However, many IBSS implementations are point-to-point only, at most three IBSS networks can exist within one area, and there is no simple equivalent to the Bluetooth page and inquiry processes. Furthermore, Wi-Fi packet retransmissions provide reliable data exchange, so using Wi-Fi for two-way voice can present latency problems in poor channel conditions.

The characteristics of Wi-Fi can be summarized as follows:

- High speed
- Long range
- Always connected
- Moderate power consumption
- Moderate cost
- Limited two-way voice support
- Cumbersome ad-hoc networking

Bluetooth Characteristics

Unlike Wi-Fi, Bluetooth wasn't developed to operate as a WLAN, where a fixed AP is connected to a network backbone. Although Bluetooth also uses a star topology when one master communicates with several slaves, any device can serve as master, and the master-slave roles can change during the life of a piconet. Bluetooth-equipped devices will often support only a single profile, such as a *human interface device* (HID) or remote control, and these communication needs are simple and must be accomplished cheaply and efficiently. Minimal power consumption is important in many Bluetooth applications. The 802.11 specification would require significant revision for Wi-Fi to meet these requirements.

The characteristics of Bluetooth can be summarized as follows:

- Medium speed
- Short range
- Connected only to complete a task
- Low power consumption

- Low cost
- Full real-time voice support
- Efficient ad-hoc networking

Collocated Nodes

In light of the preceding discussion, there is growing consensus in the industry that Bluetooth and Wi-Fi have a natural partition between them.[3] Wi-Fi is most suited for wireless access to Internet services requiring large amounts of data to be transferred quickly. Bluetooth is most suited for ad-hoc networks, two-way voice, and communicating between peripherals. As such, there may be good reasons to place both devices into a single PC. A user can wirelessly connect to the Internet through a Wi-Fi AP and enjoy fast web access and even full-motion video, as long as only a few other users are sharing the AP's 11 Mb/s bandwidth. The Bluetooth link would be used for such tasks as sending files to a printer, exchanging files between computers, and pulling scanned images or digital photographs from peripherals.

An obvious difficulty with collocated nodes is the mutual interference between Bluetooth and Wi-Fi that will inevitably occur when both are in simultaneous use. Although Wi-Fi uses DSSS and Bluetooth uses FHSS, they will occasionally overlap in frequency, and it's quite possible that either transmitter could block the other's receiver as well, reducing the throughput of both markedly. This is an area of intense research, and we'll discuss some of the results later in this chapter.

Characterizing Bluetooth Interference

Suppose a Bluetooth receiver is actively receiving and decoding an incoming baseband packet, called the *desired signal*, being sent from a *desired transmitter*. Suppose also that another transmission is present in the 2.4 GHz ISM band that has the potential to interfere with the desired signal. This second transmission, the *interfering signal*, comes from the *interfering transmitter*. The interfering signal may disrupt the desired receiver if it is strong enough to exceed the receiver's C/I specification. For example, the Bluetooth specification states that the receiver must meet a cochannel C/I value of 11 dB for qualification, so if the desired signal's power is less than 11 dB stronger than the same-channel interfering signal's power, then the BER at the desired receiver may exceed 10^{-3}.

The C/I performance for two different receivers is given in Table 13-1. The *standard receiver* C/I values represent the minimum required by the Bluetooth specification. By following the approach in Meihofer[4] we can define an *enhanced receiver* as having improved C/I as shown. Our analysis will take both types of receiver into account in an attempt to quantify the performance improvement offered by the enhanced receiver.

Interfering Transmitter Disruption Distance

The next step in characterizing the interference problem is to calculate how close an interfering transmitter needs to be before it has the potential to affect the desired receiver. Let's assume for simplicity that the interfering signal is also from a Bluetooth radio, and that both the desired and interfering transmitters have the same output power and antenna pattern. Suppose the interfering transmitter is close enough to just begin affecting the desired receiver. If the desired transmitter is d_d meters away from the desired receiver, then we can find the distance to the interfering transmitter, which we call d_i, by expressing the respective path losses given by Equation 2-4 in terms of C/I. The result is

$$C/I = PL_i - PL_d = 40 + 10n\log(d_i) - [40 + 10n\log(d_d)] \qquad (13\text{-}1)$$

which can be solved for d_i to yield

$$d_i = d_d 10^{\frac{C/I}{10n}} \qquad (13\text{-}2)$$

Table 13-1

Carrier-to-interference performance for a Bluetooth receiver

Requirement	Standard Receiver	Enhanced Receiver
Cochannel C/I	11 dB	8 dB
Adjacent (1 MHz) channel C/I	0 dB	−10 dB
Adjacent (2 MHz) channel C/I	−30 dB	−30 dB
Adjacent (≥3 MHz) channel C/I	−40 dB	−40 dB

Table 13-2

Maximum disruption distances for various C/I and path loss values, standard receiver, and Bluetooth interferer

C/I	n	Number of Hop Channels Affected	d_i/d_d
11 dB	2.0	1	3.6
0 dB	2.0	3	1.0
−30 dB	2.0	5	0.032
−40 dB	2.0	79	0.010
11 dB	3.0	1	2.3
0 dB	3.0	3	1.0
−30 dB	3.0	5	0.10
−40 dB	3.0	79	0.046

Table 13-2 gives the ratio of d_i to d_d for the standard receiver C/I values in Table 13-1 with a path loss exponent of 2.0 to represent free space and 3.0 to represent typical room clutter. As an example, suppose the desired transmitter is 10 meters from the desired receiver and free-space propagation applies. An interfering transmitter on the same hop channel just begins affecting the desired receiver from 36 meters away. The desired signal is affected when the interfering transmitter is 10 meters away, and it transmits either on the same channel or in either of the two adjacent channels. From 0.32 meters away, the interferer can affect two channels on either side of the desired hop channel, and from 0.1 meters away, the interferer has the potential to completely block the desired receiver. The latter situation is of critical importance when Bluetooth is collated with another wireless service such as Wi-Fi or a cell phone. (At 0.1-meter separation, the desired receiver's antenna is within the near-field of the interfering transmitter's antenna, so Equation 13-2 no longer strictly applies. Even so, our point remains—the interfering signal is *really* strong.)

It's also interesting to note that a higher path loss exponent means that the cochannel C/I situation requires that the interfering transmitter be closer to the desired receiver before it has the potential for disruption. The situation reverses when the C/I becomes negative; the interfering transmitter can be further away and still cause trouble.

The values in Table 13-2 are derived for a receiver that barely meets the requirements for qualification. Performance is better if the desired receiver has enhanced C/I values, as shown in Table 13-3. A lower C/I value means

Table 13-3

Maximum disruption distances for various C/I and path loss values, enhanced receiver, and Bluetooth interferer

C/I	n	Number of Hop Channels Affected	d_i/d_d
8 dB	2.0	1	2.5
−10 dB	2.0	3	0.31
−30 dB	2.0	5	0.032
−40 dB	2.0	79	0.010
8 dB	3.0	1	1.8
−10 dB	3.0	3	0.46
−30 dB	3.0	5	0.10
−40 dB	3.0	79	0.046

that the interfering transmitter must be closer to the desired receiver for disruption to be possible, sometimes significantly closer. Many engineers feel that improving receiver C/I performance is the single best thing that can be done to enhance packet throughput in a typical Bluetooth usage environment.

Types of Disruption

Now that we know how far away an interfering transmitter can be to cause disruption (at least for identical Bluetooth devices), we need to characterize the type of disruption that can occur. As we learned in Chapter 4, "Bluetooth Packets and Their Exchange," Bluetooth devices hop randomly throughout the 79 available channels, transmitting one packet per hop. Therefore, the probability of collision is partly determined by the fraction of hop channels that are affected by the interfering signal. For example, if one narrowband interfering signal is present and the C/I for that signal is 0 dB, and a standard desired receiver is being used, then disruption can occur each time the desired receiver hops into the interfering channel or into the channel on either side of the interference. This will occur on average about 3 times in every 79 hops.

In reality, though, the situation isn't quite that simple. As before, we will consider the interfering signal to be another Bluetooth device that has formed its own piconet. There are now several more variables that must be considered before a packet loss rate can be determined. Among these are

- The location of each desired piconet member
- The location of each interfering piconet member
- Different amounts of packet overlap from the lack of time synchronization between independent piconets
- Multislot versus single-slot packet transmissions
- Capability of the FEC to recover disrupted packets

The equations for modeling these conditions are extremely complex. Obtaining average Bluetooth throughput figures in an interference environment requires, among other things, plotting the distribution of possible locations for both the desired and interfering piconet members within a boundary such as a room, and then averaging the results. This often requires solving an integral equation, which can cause heart palpitations in both reader and author. Because we want to sell lots of books, we avoid calculus like the plaque (see Figure 13-2). Fortunately, the research community has already produced a quantity of useful results by modeling throughput reductions caused by Wi-Fi-on-Bluetooth, Bluetooth-on-Wi-Fi, and Bluetooth-on-Bluetooth situations. We examine these next.

Figure 13-2 We leave calculus to others. (Source: Bill Lae)

"Looks like a nasty Calculus build up here."

Approaches to Coexistence Analysis

The number of studies examining coexistence between Wi-Fi and Bluetooth and between separate Bluetooth piconets has grown considerably over the recent past. One of two approaches is generally used for modeling and analyzing coexistence. The first is the *analytical* approach. This can consist of a detailed *theoretical analysis*, where the coexistence problem is modeled mathematically, and a result is expressed as a function of input variables. For example, the input variables might be the size of the room and number of Bluetooth piconets, and the result is the average *packet error rate* (PER) or average throughput in such a situation. Another form of analytical approach is the *computer simulation*. This involves creating a network topology model (such as placement of users in a room) and running a PER simulation again and again, each time moving the users around to random locations to determine average performance. Sometimes the two analytical methods are combined, where a theoretical analysis is completed followed by a computer simulation to verify the results. The analytical approach lends itself well to modeling Bluetooth-on-Bluetooth interference because multiple piconets have the same signal structure and timing.

Coexistence can also be examined through the *empirical* approach, where wireless devices are actually deployed and measurements taken. The most straightforward empirical method is to equip some laptop computers with Wi-Fi and others with Bluetooth and move them about the room, checking data rates and packet error probabilities, in a *real-world* situation. A difficulty with this method is that the measurements can be influenced by additional variables such as unknown third-party interference. To avoid these problems, the empirical analysis can be done using a *laboratory simulation*, where signal generators and protocol analyzers are connected with cable in such a way that coexistence can be tested in a controlled environment. The empirical approach is often used for modeling coexistence between Bluetooth and Wi-Fi because it's usually easier to observe their interaction directly than perform a theoretical analysis on the vastly different signaling methods and timing used by each.

Coexisting with IEEE 802.11b (Wi-Fi)

As we mentioned earlier, there is increasing consensus within the wireless industry that a viable market exists for both Bluetooth and Wi-Fi devices, and their respective strengths are such that they will be required to coexist in many applications. It is therefore extremely important to discover the effect that each will have on the other when they operate together. Researchers have embraced this problem with enthusiasm, not only because it's important, but also because Bluetooth uses FHSS and Wi-Fi uses a modified DSSS, making the analysis more challenging.

Wi-Fi Signal Structure

The transmitted Wi-Fi signal consists of a carrier that is modulated using either BPSK or QPSK at a rate of 11 *million symbols per second* (Ms/s). Wi-Fi is backward compatible with DSSS 802.11 to achieve raw data rates of 1 Mb/s using BPSK or 2 Mb/s using QPSK. These lower data rates are obtained by superimposing upon each BPSK or QPSK symbol an 11-bit Barker sequence to spread the bandwidth of the transmitted signal by a factor of 11, thus meeting FCC requirements for a DSSS transmission (refer to Chapter 1). This enables 802.11 to legally use transmit powers above 1 mW for increased range.

The genius behind achieving higher data rates in 802.11b was the implementation of a process called *complimentary coded keying* (CCK) that replaces each 11-bit Barker code with a new spreading sequence selected from a set of sequences that represents additional data. Increased data rates can now be supported at 5.5 Mb/s or 11 Mb/s. The FCC approved CCK because it creates a transmitted signal with the same interference characteristics as an ordinary DSSS transmission.

Because of the reduced energy per bit, the effective range of 802.11b at the higher two data rates is also reduced. At 1 Mb/s and 20 dBm TX power, the typical indoor range is about 80 meters, but this decreases to about 15 meters at 11 Mb/s. An extremely useful feature of Wi-Fi is its capability to automatically select the highest of the four data rates that can be supported over a given path. Communication begins at a low rate, and then speed is increased if sufficient *signal-to-noise-and-interference ratio* (SNIR) is available. Speed is reduced if too many errors occur at the higher data rate. This can occur either from reduced signal strength over longer communication ranges or from lower C/I from increased interference. Either situation manifests itself by a gradual throughput degradation over the Wi-Fi link.

The power spectrum of a typical Wi-Fi signal is shown in Figure 13-3.[5] The bandwidth of the Wi-Fi transmitted signal remains the same regardless

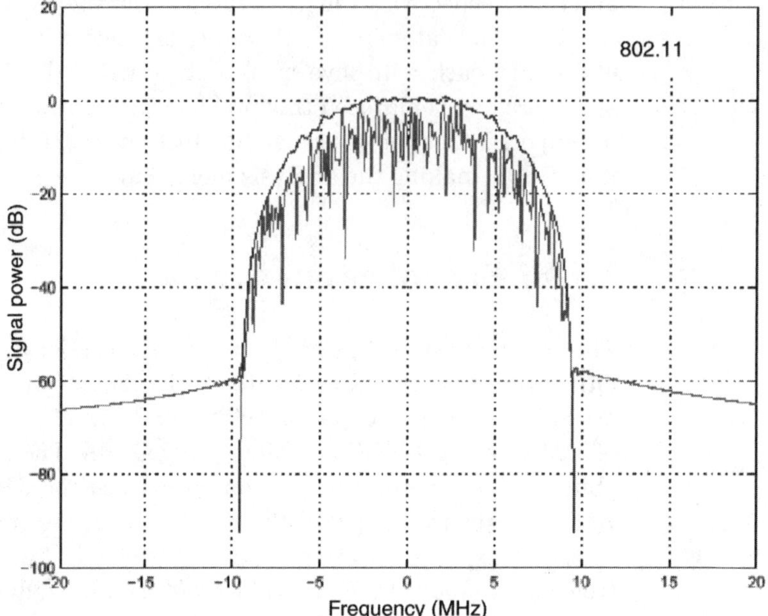

Figure 13-3 The power spectrum of the Wi-Fi transmitted signal. This spectrum is essentially identical for data rates of 1, 2, 5.5, or 11 Mb/s.[5]

of the data rate, which helps interference analysis immensely. The −20 dB bandwidth of this signal is about 16 MHz, which is the value used in many papers written on Wi-Fi coexistence. Another bandwidth commonly given to the Wi-Fi signal is 22 MHz, which is obtained using the rule of thumb that a digital signal's approximate bandwidth is twice the symbol rate for unfiltered baseband signaling. By assuming that the Wi-Fi signal's energy is uniformly distributed across 22 MHz, we can introduce some pessimism into the coexistence calculations, which is often preferred.

A total of 11 channels are designated for Wi-Fi use in the United States, and these are listed in Table 13-4. Although simple frequency agility is built into the 802.11 specification, very few implementations have this feature. The usual practice is to place a Wi-Fi AP on a single channel and leave it there. Separation between channels is 5 MHz, which means that there is significant bandwidth overlap between Wi-Fi networks using adjacent channels. Research has shown that if two Wi-Fi networks are deployed within earshot of each other and their bandwidths overlap, then their performance is severely affected. As such, Channels 1, 6, and 11 are commonly used for multiple Wi-Fi networks within a single building to avoid signal overlap as much as possible. Assuming a 22 MHz bandwidth, Figure 13-4 shows how these three channels are arranged within the 2.4 GHz ISM band.

Table 13-4
Wi-Fi channel set in the United States

Channel	Frequency (MHz)
1	2412
2	2417
3	2422
4	2427
5	2432
6	2437
7	2442
8	2447
9	2452
10	2457
11	2462

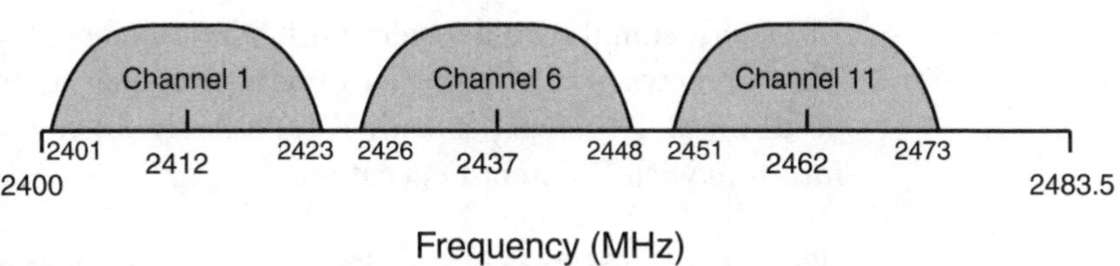

Figure 13-4 Three Wi-Fi channels can be occupied by separate networks without causing mutual interference.

Wi-Fi Effect on Bluetooth

A typical Bluetooth signal power spectrum is shown in Figure 13-5.[5] The −20 dB bandwidth is 1 MHz, as stated in the specification, and this value is almost universally used for coexistence analysis. If Bluetooth and Wi-Fi are both operating in the same area, then it's easy to see that they will occasionally interfere with each other.

When a Bluetooth piconet hops to a channel in which a Wi-Fi network is operating, some of the Wi-Fi signal energy will enter the passband of the Bluetooth receiver. If we assume that the Wi-Fi bandwidth is 22 MHz, then roughly 1/22 of the energy at the antenna will pass through the Bluetooth receiver's filters, so Bluetooth has a built-in 13 dB rejection of the Wi-Fi signal. We conclude, then, that for a 11 dB C/I to exist at a desired Bluetooth

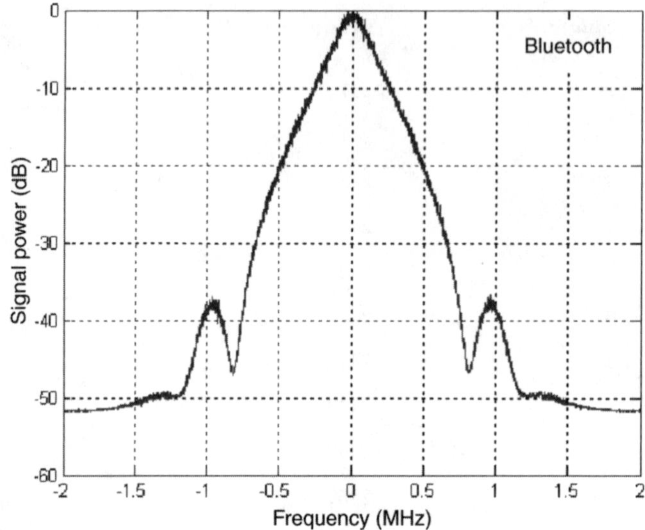

Figure 13-5 The power spectrum of the Bluetooth transmitted signal[5]

receiver, the total Wi-Fi interfering signal power can exceed the desired Bluetooth signal by 2 dB before it begins to disrupt the Bluetooth piconet. The bad news is that many Wi-Fi networks operate at 30 to 100 mW (15 to 20 dBm) of TX power, so they can be fairly far away and still potentially interfere with Bluetooth operation.

Worst-Case Analysis For a simple worst-case analysis, we can assume that a single Wi-Fi network is always transmitting and is within disruption range of the desired Bluetooth piconet. We'll also assume that if the piconet hops anywhere within the 22 MHz Wi-Fi signal bandwidth, then the affected packet is lost. What will the throughput degradation be for ACL packets? Be careful; this is a trick question. You may be tempted to say that the reduction will be 22/79, or 28 percent because that's the probability that the piconet hops into an occupied channel. Indeed, that figure often shows up in coexistence research papers. Remember, though, for a packet to be successful, both it and its acknowledgment must get through. Under our assumptions, then, we must calculate the probability that two successive hops are outside the Wi-Fi signal bandwidth, which is given by

$$\Pr(2\ \text{hopsOK}) = \left(\frac{79 - 22}{79}\right)^2 = 0.52 \qquad (13\text{-}3)$$

Thus, the throughput degradation is 48 percent compared to perfect operation. This assumes, of course, that interference is so strong that even a Bluetooth header protected by the rate 1/3 binary repetition code is lost. Under these worst-case assumptions, then, Bluetooth ACL performance is severely, but not catastrophically, affected by the presence of an interfering Wi-Fi network. The situation gets significantly worse, however, if interfering WLANs exist on Wi-Fi channels 1, 6, and 11. That situation leaves only 79 − 66 = 13 Bluetooth hop channels in the clear.

SCO Performance in Interference Bluetooth SCO performance is more difficult to analyze because audio quality is subjective to the listener and is most affected by how many bits are randomized within a burst error event. As such, packet loss criteria may not be particularly applicable for SCO performance analysis because a packet is considered lost over a wide range of bit error conditions. Even for loss events associated with catastrophic collisions, there is general disagreement on SCO packet loss rate versus unacceptable audio quality. Some claim loss rates of 2 percent are unacceptable, while others claim adequate audio quality still exists with packet loss rates as high as 20 percent. Audio quality degradation in the presence of an 802.11b network has been studied to some degree. One study has shown that as long as a class 3 SCO link is shorter than 2 meters, no audible degradation will occur from a WLAN operating in the vicinity.[6] Another study has shown that a SCO link will perform well if it is located at least 2 meters away from the nearest Wi-Fi AP or STA.[7]

The Bluetooth specification makes no provision for assessing SCO packet loss because no CRC is implemented. A modification has been proposed that provides both CRC and transmission frequency diversity for the SCO link.[8] In this algorithm, a 2-byte CRC is added to 25 bytes of voice data and the associated packet is sent twice on two different hop frequencies. If the first packet's CRC is good at the recipient's end, then the second transmission is ignored; otherwise, the redundant packet is received and its CRC checked. If both are bad, then an interpolation mechanism is used to replace the corrupted data. Channel utilization is high at 8 out of every 10 time slots, but SCO performance could improve markedly from a significant reduction in packet loss rates.

Empirical Studies We will now examine the results of two empirical studies where the effect of Wi-Fi on a Bluetooth piconet was analyzed and ACL performance figures were obtained. The first, performed by Texas

Instruments (the *TI study*) used the following equipment in a real-world test:[5]

- Wi-Fi equipment
  - **AP** Cisco Aironet Model 340, TX power 30 mW
  - **STA** Cisco Aironet Series 340 PC card in laptop computer, TX power 30 mW
- Bluetooth equipment
  - **Master** Digianswer PC card in laptop computer, class 1 TX power
  - **Slave** Digianswer PC card in laptop computer, class 1 TX power

To provide a maximum level of interference to the Bluetooth piconet, the Wi-Fi devices were placed about 0.3 meters apart, so any Bluetooth piconet member will receive interference from all communication across the WLAN link. The Bluetooth slave's position was fixed directly between the Wi-Fi AP and STA in a *collocated* topology (see Figure 13-6[a]) or 10 meters away from the Wi-Fi users in a *separated* topology (see Figure 13-6[b]). In both cases, the distance between master and slave was varied and throughput measurements taken. The results are given in Figure 13-7 for both topologies, along with a baseline throughput for the piconet without any Wi-Fi interference. All paths were LOS.

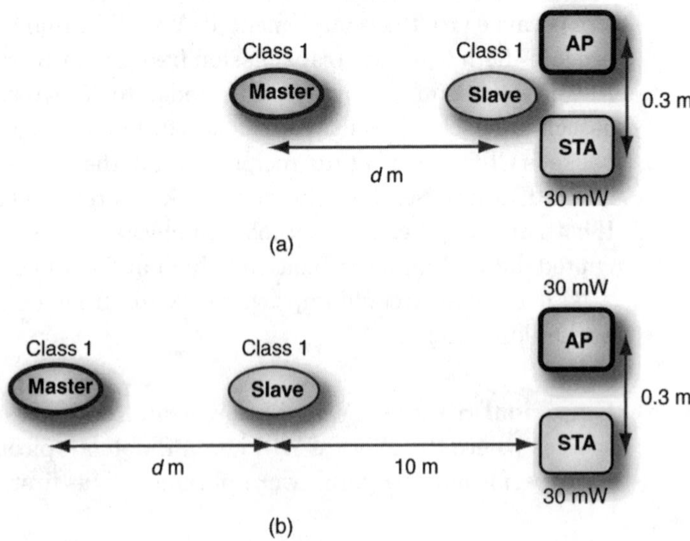

Figure 13-6 The Bluetooth slave can be either collocated with the Wi-Fi network (a) or separated from the network by 10 meters (b), TI study.[5]

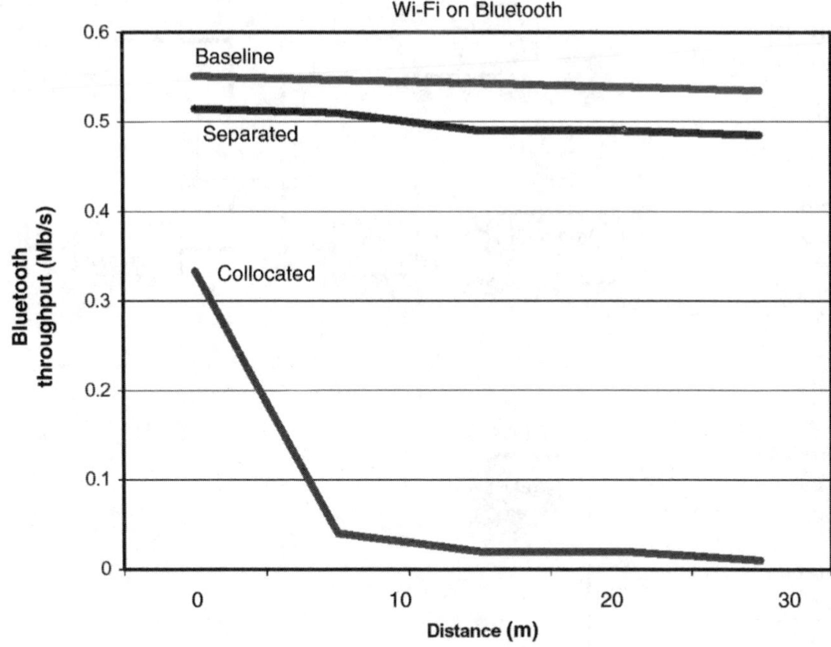

Figure 13-7 Bluetooth throughput as a function of master-slave distance for both collocated and separated topologies, TI study [5]

When one of the Bluetooth nodes is collocated with an operating Wi-Fi network, throughput is low even for small separations between master and slave. On the other hand, when the Bluetooth piconet is separated by at least 10 meters from any Wi-Fi node, then throughput is affected only slightly. Other studies have shown that separations of as little as 2 meters between Bluetooth and Wi-Fi still enable the Bluetooth piconet to perform acceptably well.[9] This point is also brought out in the following study.

The second empirical study was performed by the University of Wisconsin in conjunction with Eaton Corporation (the *UW study*).[10] The researchers connected cables between two Wi-Fi and two Bluetooth users, along with power adders, splitters, and variable attenuators to simulate different path loss values between the players (see Figure 13-8). A Bluetooth master was placed a simulated 7 meters east of the slave, and a single Wi-Fi node was located north of the slave and the distance between them varied (see Figure 13-9). Both Bluetooth devices used TX powers of 1 mW (class 3), and the Wi-Fi node was transmitting with a power of 25 mW. Packet loss probabilities are plotted in Figure 13-10 for one- and three-slot packets. Losses are above 0.5 in most cases when the Wi-Fi node is about 1 meter or less from the Bluetooth slave, but the loss probability drops rapidly until stabilizing at about 0.25 or so when the Wi-Fi node is 2 meters or further

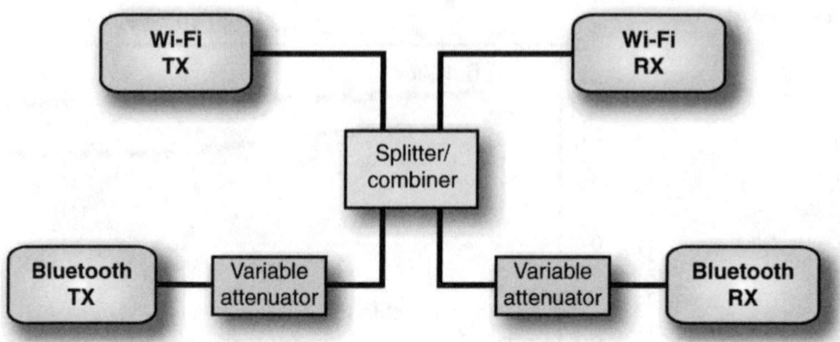

Figure 13-8 Bluetooth and Wi-Fi players are connected with cables, attenuators, splitters, and combiners to simulate different network topologies, UW study.[10]

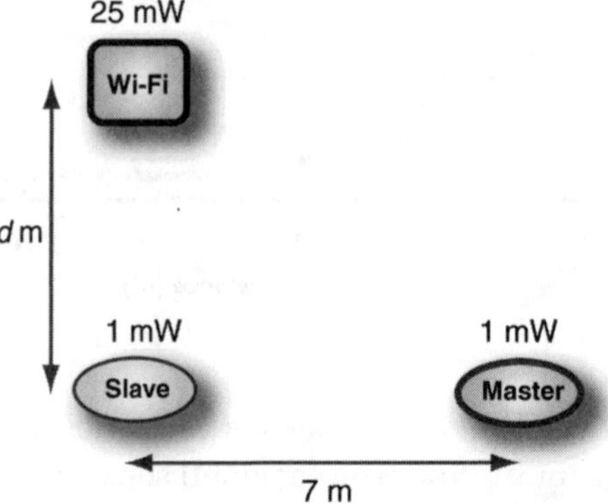

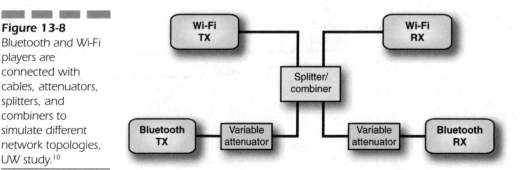

Figure 13-9 The coexistence topology used for Bluetooth packet loss measurements, UW study[10]

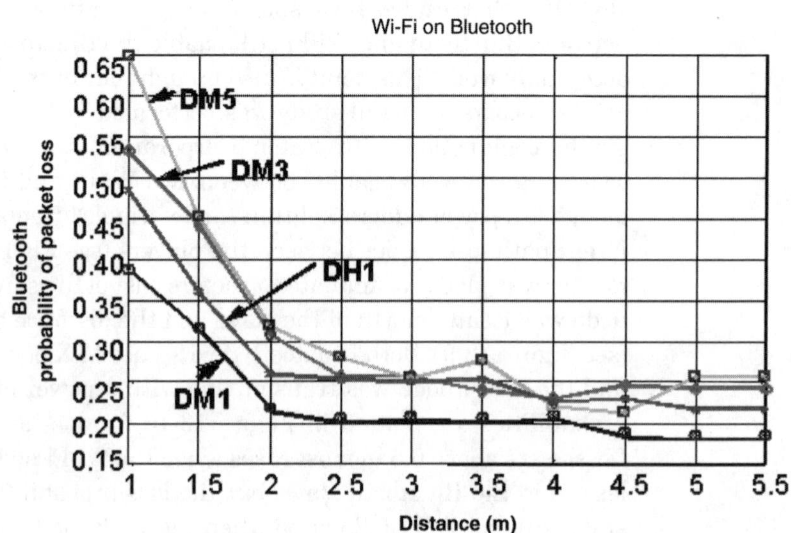

Figure 13-10 The probability of Bluetooth packet loss versus distance from a Wi-Fi node to the Bluetooth slave, UW study[10]

away. Once again, it's clear that a collocated Wi-Fi node will severely disrupt Bluetooth piconet operation. It's also interesting to note that, although FEC reduces the packet loss probability slightly for single-slot packets, the improvement isn't sufficient to offset the reduced throughput from FEC overhead.

Bluetooth Effect on Wi-Fi

Although it's already clear that Bluetooth and Wi-Fi cannot be collocated without taking some action to alleviate interference to the Bluetooth piconet, it would also be instructive to determine whether a Bluetooth piconet can significantly disrupt Wi-Fi throughput. If it does, will the disruption be greater or less than Wi-Fi's effect on Bluetooth?

To answer this question, we turn to the same two empirical studies presented in the preceding section. The TI study used the same equipment listed earlier, but the network topologies were changed to those shown in Figure 13-11.[5] In these scenarios, the Bluetooth master and slave are separated by 0.3 meters, so the Wi-Fi network is affected about equally by both master and slave traffic. In the *collocated* topology, the Wi-Fi STA was placed between the Bluetooth users, and the distance to the AP was varied (see Figure 13-11[a]). For the *separated* topology, the STA was moved to a position 10 meters from the piconet, and the distance between it and the AP

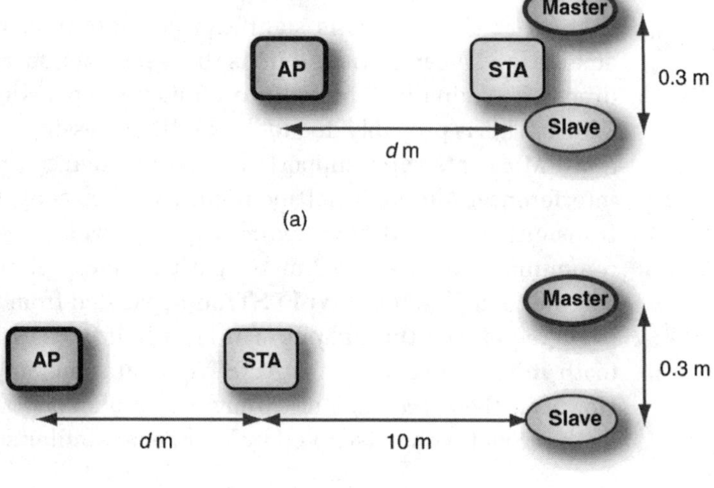

Figure 13-11 The Wi-Fi STA can be either collocated with the Bluetooth piconet (a) or separated from the piconet by 10 meters (b), TI study.[5]

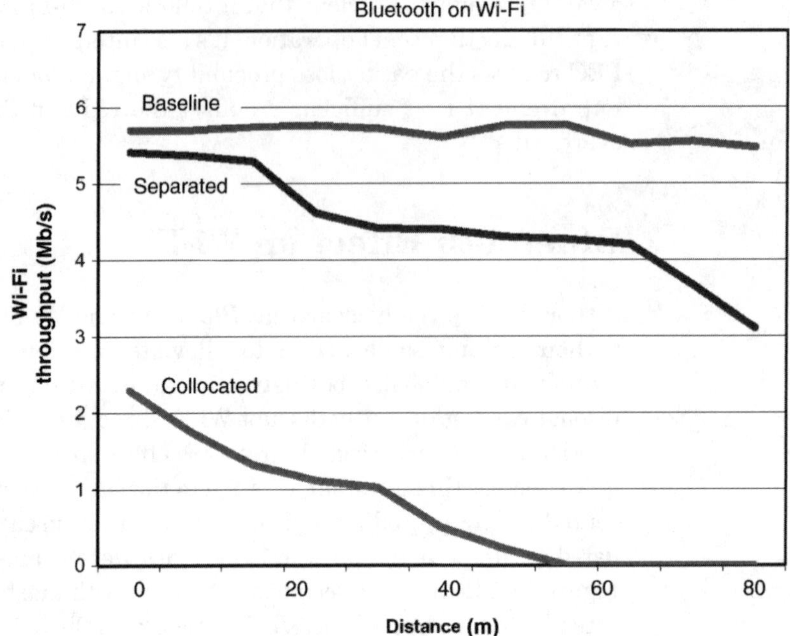

Figure 13-12 The Wi-Fi throughput as a function of AP-STA distance for both collocated and separated topologies, TI study [5]

was varied (see Figure 13-11[b]). Figure 13-12 plots the throughput figures for both of these situations, along with a baseline throughput that the Wi-Fi network can achieve without any Bluetooth interference. Once again, all paths were LOS.

When the Wi-Fi STA is collocated with a Bluetooth piconet, we notice that Wi-Fi throughput is significantly lower than the baseline, even for small AP-STA separations, but as the separation increases, the throughput drops off gradually. Even with a 30 meter separation, throughput is still 1 Mb/s. This is probably due to the 11 dB processing gain inherent in 2 Mb/s and 1 Mb/s data rates supported by Wi-Fi when confronted by narrowband interference. Although not mentioned in the study, the class 1 Bluetooth transmitters should have reduced their powers to 4 dBm or lower when communicating over a 0.3 meter path to be compliant with the Bluetooth specification. When the Wi-Fi STA is separated from the Bluetooth piconet by 10 meters, its throughput is at least half that achievable without Bluetooth interference, even for AP-STA separations approaching 80 meters. However, the *percentage* of throughput reduction exceeds that of a Bluetooth piconet being disrupted by Wi-Fi over similar separation distances.

The UW study once again employed the setup in Figure 13-8, but the simulated network topology was changed to examine the interference Bluetooth has on Wi-Fi.[10] A Bluetooth master was placed 1 meter east of the slave, and both establish a piconet and exchange HV1 packets at 1 mW TX power. This insures that the piconet is fully loaded; that is, a single-slot packet is transmitted in every time slot. A Wi-Fi AP with a TX power of 25 mW was placed 15 meters north of the Bluetooth slave. The Wi-Fi STA was located between the Bluetooth slave and Wi-Fi AP, and its distance north of the slave was varied. This scenario is shown in Figure 13-13.

The associated Wi-Fi packet loss probabilities are plotted in Figure 13-14. Losses approach 0.5 with the Wi-Fi node about 0.5 meters from the Bluetooth slave, but the loss probability drops rapidly as the separation increases. The loss rate is less than 0.2 for separations of 2 meters or more. We conclude that collocated Bluetooth and Wi-Fi nodes can severely disrupt each other's operation, but once they are separated by more than about 2 meters, performance becomes acceptable.

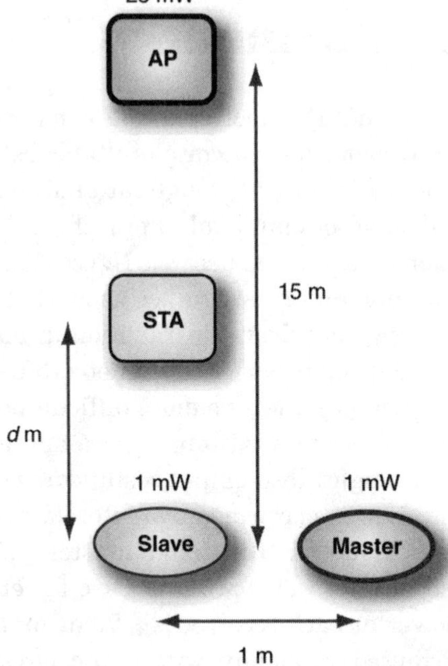

Figure 13-13 The coexistence topology used for Wi-Fi packet loss measurements, UW study [10]

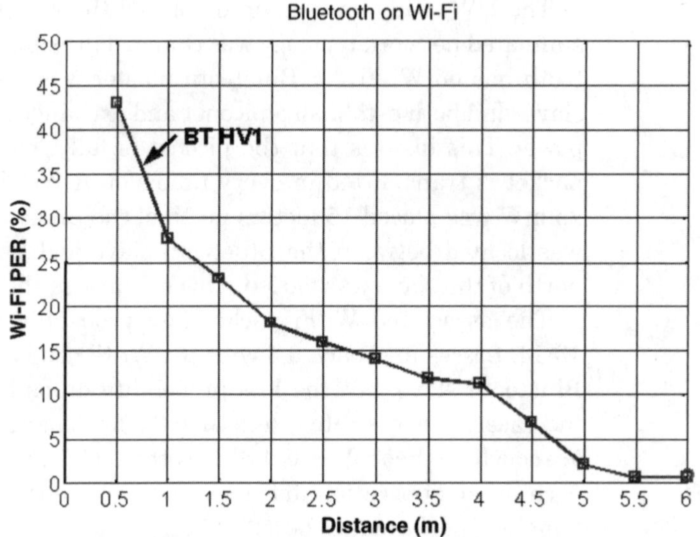

Figure 13-14 The probability of Wi-Fi packet loss versus the distance from a Wi-Fi node to the Bluetooth slave, UW study [10]

Coexisting with Other Bluetooth Piconets

We've already learned that Bluetooth piconets can interfere with each other when two or more transmissions occur simultaneously within a single hop channel. Bluetooth-on-Bluetooth throughput analysis is often easier to do via an analytical instead of empirical approach. A theoretical analysis or computer simulation is facilitated because the coexisting piconets all operate in a similar manner, and most studies show that several piconets can coexist without causing significant degradation in individual throughput. This feature, although good news for the Bluetooth user, makes the empirical approach to coexistence analysis more difficult because of the requirement to deploy several piconets within an area and move them around in such a way that meaningful throughput results are obtained.

An analytical study was performed by Motorola in which a circle with a 10 meter radius was assumed to contain a master and slave point-to-point link as the desired piconet.[4] The two units are far enough apart that the desired signal's power at each receiver is −70 dBm. Bluetooth interfering piconets are distributed randomly within the circle, as shown in Figure 13-15. All Bluetooth units use 1 mW TX power class 3, and the path loss exponent is $n = 3.5$, modeling an environment of moderate to severe clutter.

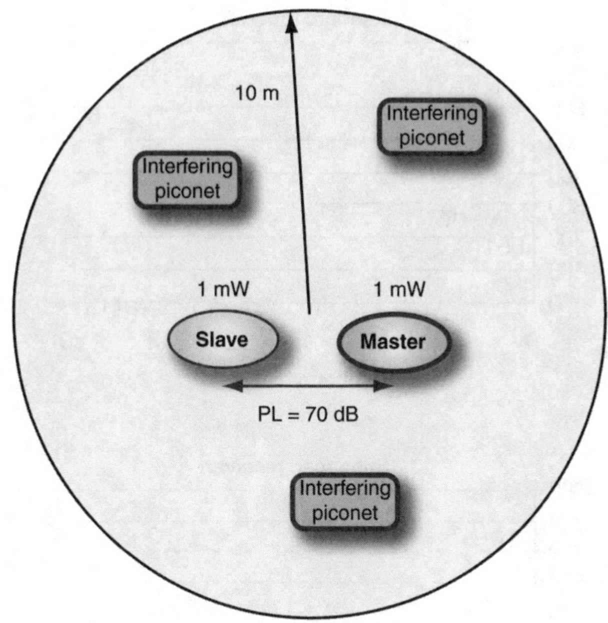

Figure 13-15 The Motorola study places a desired piconet at the center of a circle with a 10 meter radius and scatters interfering piconets randomly within the circle.[4]

Receivers in the desired piconet are assumed to have a sensitivity limit of −80 dBm, and the master and slave are exposed to the same amount of interference. To simulate a worst-case situation, all interfering piconets are considered to be fully loaded, transmitting in every available time slot. Both cochannel and adjacent channel interference are taken into account. An independent Rayleigh-faded path loss is assumed for each interferer and hop.

The results in Figure 13-16 show how throughput is affected as a function of the number of interfering piconets that exist within the 10 meter circle. Plots are made for one-, three-, and five-slot packets without FEC at a standard receiver (see Figure 13-16[a]) and at an enhanced receiver (see Figure 13-16[b]). The standard and enhanced receiver's C/I performance is listed in Table 13-1. Notice that the enhanced receiver, while having a C/I advantage of only a few dB, significantly outperforms the standard receiver when other piconets are present. Also, switching from DH5 to DH3 packets improves throughput once interference becomes sufficient to reduce DH5 throughput below about 400 kb/s. This is due to the higher retransmission penalty that must be paid when a DH5 packet is destroyed from a collision.

A comparison of DH1 and DM1 packet throughput values is shown in Figure 13-17(a), followed by the throughputs of DH3 and DM3 packets in

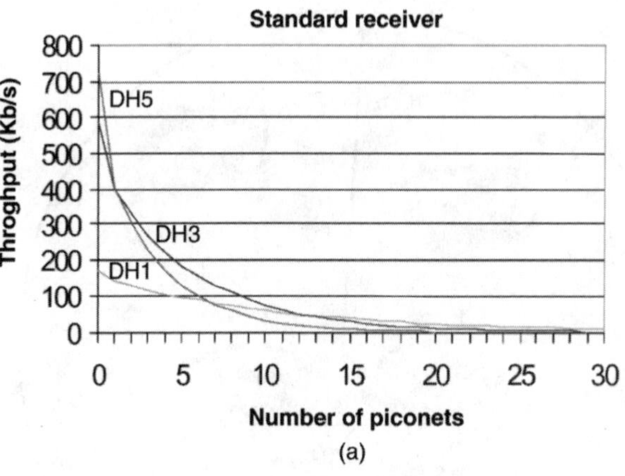

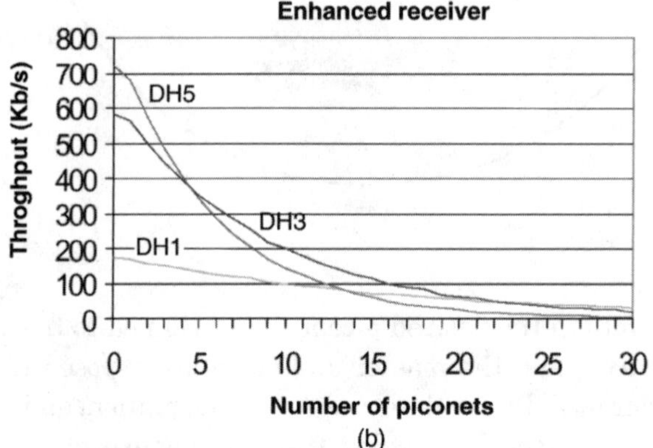

Figure 13-16 The throughput as a function of the number of interfering piconets within the 10 meter circle is plotted for a standard receiver (a) and enhanced receiver (b)[4]

Figure 13-17(b), again plotted as a function of the number of interfering piconets within the 10 meter circle. Do you remember the results of Figure 4-20 in Chapter 4 when we plotted throughput for the different ACL packets as a function of BER? No? Well, go ahead and look it up. I'll wait . . . Ah, welcome back. In Chapter 4, we discovered that the DM3 packet had the highest throughput when the BER was between about 10^{-4} and 10^{-2}, and bit errors were caused by random noise. However, when packets are lost from collisions with another Bluetooth piconet instead, then Figure 13-17 shows that error correction doesn't help very much. This is due to the fact that a collision tends to cause so many errors that the error correction code is overwhelmed, and the packet is lost anyway. Therefore, it makes little sense to use FEC when most packets are lost from collisions instead of from random noise.

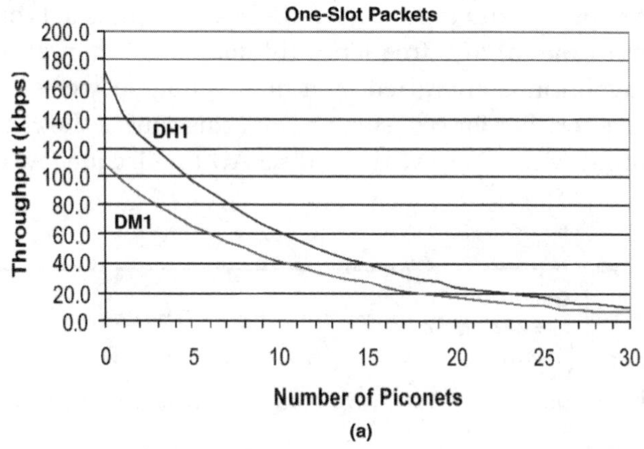

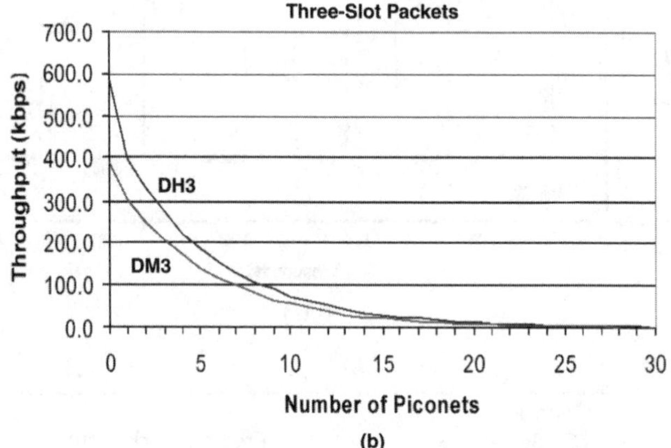

Figure 13-17 The throughput as a function of the number of interfering piconets within the 10 meter circle is plotted for single-slot packets (a) and three-slot packets (b). (Note the vertical axis scale change.) Packets without FEC have higher throughput when interference is the primary cause of packet loss.[4]

Coexisting with Microwave Ovens

Several million microwave ovens exist in the world, and many of them are physically located in the same areas in which Bluetooth devices will function. These ovens transmit several hundred watts at a frequency of 2.45 GHz, and this energy is ideally confined to the interior of the oven itself. In reality, some of the energy escapes, and thus the oven can be classified as an unintentional radiator of RF. Therefore, it's important to discover what effect these ovens will have on communications within the 2.4 GHz ISM band.

One of the most detailed studies on microwave oven emissions was performed by the *National Telecommunications and Information Administration* (NTIA) in 1994.[11] This study was mainly confined to consumer ovens of varying age and type. Figure 13-18 shows plots of typical oven emissions in both time (a) and frequency (b) domains. When an oven is operating, its magnetron is energized by a high voltage power supply that uses a half-wave rectifier for cost savings, and this causes the characteristic pulsed RF output with a period tied to the AC line frequency. In the United States,

Figure 13-18 Typical microwave oven emissions plotted in the time domain (a) and frequency domain (b)[11]

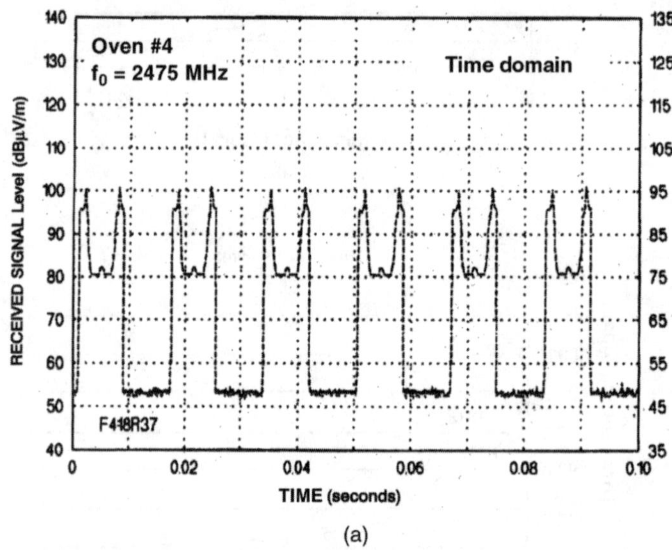

(a)

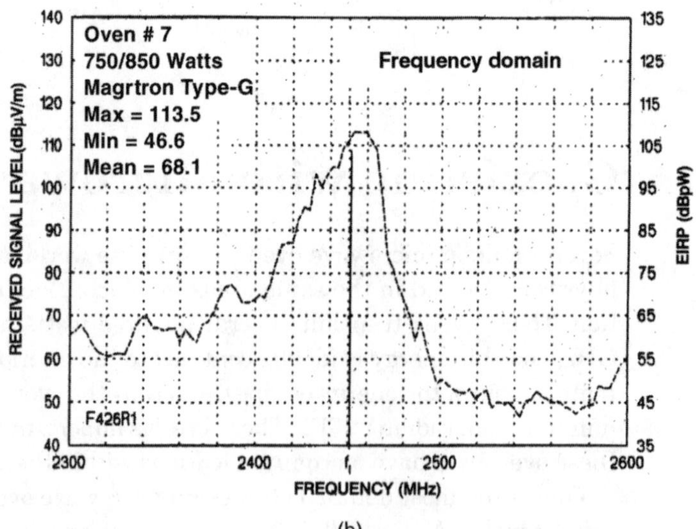

(b)

where the line frequency is 60 Hz, the magnetron is on for about 8 ms and off for about 9 ms while operating. The on-off transient power spikes shown in the figure are common in these ovens and will increase the bandwidth of the emitted interference.

Within the samples of ovens tested in the NTIA study,[12] EIRP emission averaged 5 dBm, but could be as high as 33 dBm (that's equivalent to 2 watts!). For the oven in Figure 13-18, the highest EIRP was 113.5 dBpW (dB relative to a picowatt) at a frequency of 2.45 GHz. To convert dBpW to dBm, just subtract 90 dB, giving an EIRP of 23.5 dBm for the sample oven. The −20 dB bandwidth for this oven is about 50 MHz, covering slightly more than half of the ISM band. Other ovens show composite bandwidths closer to 20 MHz.[12] Emissions are highest from a position directly in front of the door, but there's about a 10 dB drop in EIRP measured from behind the oven, so the typical oven acts as a mildly directional antenna in the azimuth plane.

There are reasons to believe that the interference to Bluetooth ACL links from consumer microwave ovens will be lower than might be expected from the measurements. First, the oven's magnetron is essentially a *continuous wave* (CW) device, so between on-off transients, the transmitted bandwidth may be as low as 2 MHz.[12] Second, the 9 ms off period during each power cycle is long enough that Bluetooth devices can perform several packet exchanges; for example, 6 pairs of single-slot packets or 2 five-slot packets coupled with a single-slot ACK. Even if the oven completely jams communications when its magnetron is energized, aggregate ACL throughput will drop only about 50 percent. Actual disruption will probably be less because Bluetooth will often use unaffected hop frequencies during magnetron operation. SCO performance degradation, however, may be noticeable due to the potential for burst error events, especially during magnetron on-off transients.

An empirical study of microwave oven interference on a Bluetooth point-to-point link showed that the oven 0.1 meter from one of the Bluetooth nodes caused throughput reductions of 45 percent to 65 percent, depending upon the distance between Bluetooth devices. When the oven was 5 meters or more from the Bluetooth nodes, impact was minimal.[13]

Unlike consumer ovens, commercial microwave ovens often contain two magnetrons to speed the heating process, and these are each operated from different AC line half-cycles. These ovens can cause greater interference to communications because no off time exists while the magnetrons are energized.[14] Commercial ovens are much more powerful than consumer ovens, and their RF emissions cover a wider bandwidth. Furthermore, they tend to be in operation for longer cooking times and over long hours.

Several proposals have been made to mitigate interference from microwave ovens to communication systems in the 2.4 GHz ISM bands. Among these are

- Deploying systems away from areas where ovens are located
- Using adaptive frequency hopping to avoid interference
- Using antenna polarization diversity in the communications equipment[15]
- Increasing oven shielding
- Using Bluetooth in the oven itself to control piconet operation while the oven is in use[16]

There is, of course, a real danger in placing too much intelligence into the microwave oven (see Figure 13-19).

Figure 13-19 Microwave oven intelligence has the potential to get out of hand. (Source: Bill Lae)

The revolution in "smart appliances" took an unexpected turn when Biff's microwave took Foo Foo hostage and threatened to give her 500 watts unless demands were met.

Noncollaborative Coexistence Solutions

When coexisting wireless systems are in the same general area, but not controlled by the same host, each must usually work independently to find some means for increasing throughput if needed. Such noncollaborative solutions are assumed to be implemented in such a way that the throughput of interfering devices isn't adversely affected. This is called the *good neighbor policy*. In other words, indiscriminate TX power increases are *verboten*.

Several noncollaborative options are available to Bluetooth systems for enhancing throughput in an interference-prone environment. The frequency hopping pattern can be adjusted to avoid areas in the band that are unusable. Packet lengths can be varied, FEC can be enabled or disabled, and flow control techniques can be used. To reduce interference to others, TX power can be set to the minimum level that still provides adequate QoS and latency. Finally, for different devices (such as Wi-Fi and Bluetooth) that are collocated within one host but don't collaborate, antenna isolation can be used in an attempt to prevent one transmitter from blocking the other's receiver. Most of our examples will be aimed at coexistence between Bluetooth and Wi-Fi, but many of these techniques can mitigate Bluetooth-on-Bluetooth interference as well.

Adaptive Frequency Hopping (AFH)

FHSS is considered to be an interference avoidance mechanism because of its capability to spread its transmissions out over as wide a bandwidth as needed so that at least some of the transmissions are in the clear. One of the advantages of FHSS over DSSS is that the total bandwidth over time can be made arbitrarily high simply by increasing the number of hop channels. Of the 83.5 MHz available in the 2.4 GHz ISM band, Bluetooth uses 79 MHz, so a good portion of the band can have significant interference without completely shutting down a piconet.

Bluetooth Specification 1.1 implements what might be called *classical FHSS*, where the devices hop in a pseudorandom pattern throughout the 79 channels using a pattern that doesn't repeat for about a day. Hence, the hop sequence can be accurately modeled as completely random, with about a

1 in 79 chance that the piconet will hop to any particular frequency and transmit a packet. If, say, *M* channels have significant interference, then the piconet has about an *M* in 79 chance that a particular transmission will be jammed. As we mentioned earlier, throughput can suffer even more because each ACL packet transfer requires two good hop frequencies to occur: one for the packet itself and one for the return ACK.

It should be obvious that performance can be improved significantly if the piconet members can avoid transmitting on unusable hop frequencies through AFH. Several different AFH implementations have been proposed, ranging from simply postponing a transmission when the hop generator is on a (known) jammed channel to completely redesigning the generator and reprogramming its hop channel set in response to changing interference conditions.

AFH can work well in combating interference that changes much more slowly than the Bluetooth hop rate, such as that produced by Wi-Fi networks, microwave ovens, and narrowband transmitters. It is much less useful for avoiding interference from other Bluetooth piconets without collaborating with them to discover their hop pattern, a procedure that is not allowed under current FCC rules.

FCC Rules on Adaptive Hopping FHSS rules in Europe and Japan allow the number of hop channels in the 2.4 GHz ISM band to be as few as 20, so removing up to 59 unusable channels from the Bluetooth hop sequence could be legal in those locations.[17] In the United States, though, implementing adaptive hopping has less flexibility. The Bluetooth hop generator was constructed to meet the FCC requirement to use no fewer than 75 hop channels in the 2.4 GHz ISM band, with an average occupancy time per channel not greater than 0.4 seconds within a 30 second period. Under these rules, no more than 4 frequencies can be removed from the Bluetooth hop channel set, severely limiting the effectiveness of AFH. It's important to realize that these rules apply to transmit powers that exceed about 1 mW (refer to Chapter 1), so Bluetooth class 3 transmitters can legally use any adaptive hop mechanism they choose, even to the point of selecting a clear frequency and remaining there for the life of the piconet. Of course, doing so would require a change in the Bluetooth specification.

For class 1 and 2 transmitters, however, the situation is more restricted. The FCC does allow the incorporation of intelligence to permit a FHSS system to recognize other users so that it can independently adapt its hopset to avoid occupied channels.[17] Any other coordination to avoid interference is prohibited, so adaptive hopping implementations must be essentially noncollaborative. Transmissions within a Bluetooth piconet are not continuous due to overhead from the hopping process, so violating the channel dwell-time requirements usually won't occur even with several hop frequencies eliminated from the channel set. The focus, then, is on how to retain the minimum 75 hop channels while avoiding transmissions on those that have interference. We will present one possible method to accomplish this in the following section.

As we mentioned in Chapter 1, the FCC has approved *wideband frequency hopping* (WBFH), allowing as few as 15 hop frequencies, as a way to increase FHSS data rates. The catch is that the total occupied bandwidth must still be at least 75 MHz, so the transmitted bandwidth must be 5 MHz in each of the 15 channels. Furthermore, maximum TX power is limited to 125 mW if fewer than 75 hop channels are used. The FCC has been asked to consider removing the total occupied bandwidth requirement and allow devices, such as Bluetooth, to hop on as few as 15 channels under the same limited TX power restriction.[7] This would provide AFH flexibility on par with that in Europe and Japan and wouldn't require any change in the 100 mW maximum TX power allowed in the specification.

AFH Implementation Process To properly implement AFH, the following steps are necessary[18,19]

- **Channel classification** A method for detecting interference on each of the hop channels
- **Link management** Distributing channel status information to piconet members
- **Hop sequence modification** Changing the hop sequence to avoid interference
- **AFH initiation** Begin using the AFH channel set
- **Channel maintenance** Periodically re-evaluating the channels to react to changing conditions

A Bluetooth device can perform *channel classification* in a number of ways. The simplest is to take RSSI readings on each of the 79 hop channels and make a *frequency usage table* (FUT) of good and bad channels based upon some threshold RSSI value. This method has the advantage of being both fast and passive; for example, a device not actively exchanging packets can evaluate up to two channels per time slot. To prevent transient interference (for example, another Bluetooth transmission) from triggering a bad verdict, a channel could be probed several times and the results averaged.

The RSSI method can be inaccurate because only the level of interference has been determined, not the channel's capability to successfully transport a Bluetooth packet. Making an evaluation based upon a percentage of lost ACL packets is more accurate, but is also slower and requires actually losing packets, thus reducing throughput, before a channel is deemed bad. This situation can be partially avoided if, after bringing a new slave into the piconet, the master initiates an exchange of test packets to assess each channel's performance.[20] For example, DH5 packets could be used to maximize packet loss rates and thus reduce the time required for packet loss statistics to be determined. Alternatively, classification of a channel as good or bad can be made based upon some BER threshold instead.

After each device in a piconet forms its FUT, this information can be sent from the slaves to the master as part of the *link management* process. A simple way to do this is through a new LMP command and response message, in which the master requests a slave's FUT and the slave responds.[21] The response contains a 10-byte parameter string in which 79 bits represent that particular slave's FUT. A bit corresponding to a channel is set to 1 if the channel is good and 0 otherwise.

Once the master has collected all slaves' responses, it can implement the *hop sequence modification* algorithm. The structure of the algorithm will depend upon government regulations and how the Bluetooth SIG decides to accommodate AFH. A somewhat complex approach is for the master to use LMP messages to convey a new hop sequence to each slave based upon some combination of the associated FUTs. Clearly, each slave must have the capability to modify its hop generator and acknowledge receipt of the new hop sequence message before AFH is initiated. For broadcast packet capability to be retained, the entire piconet must use the same hop channel set and hop sequence. The advantages of this technique include enhanced ACL throughput and increased SCO audio quality.

A simpler implementation method that requires no changes to the Bluetooth hop generator can be done in the following way:[21] When the master is about to send an ACL packet to a slave, it checks the next master-to-slave and slave-to-master frequencies in the hop sequence. The master sends the scheduled packet only if the master-to-slave hop channel is good in the corresponding slave's FUT *and* the slave-to-master hop channel is good in the master's FUT. If either frequency is bad, then the master waits until the next available master-to-slave slot and repeats the frequency check. In this way, AFH can be customized to each slave and the probability of a successful packet exchange is increased markedly. Of course, the method only works for ACL packets because latency issues dictate that a SCO packet be sent in the time slot in which it is scheduled.

Another argument against this simple implementation is that piconet throughput could be reduced because of the extra wait times involved. After all, why not just send the packet at every opportunity and use ARQ for retransmissions rather than skipping a transmission when a channel is listed as bad? The argument is a good one, but it applies only to single-slot packets. For multislot packets, it makes sense to wait for a clear frequency to be loaded into the hop generator before commencing transmission due to the increased penalty when one of these longer packets fails to get through. It can be shown that when this AFH algorithm is used in the presence of a nearby Wi-Fi network, average delay for successfully transmitting a DM1 packet increases by about 60 percent, but for DM3 the average delay drops by about 15 percent, and for DM5 the average delay is about 30 percent lower.[21] Therefore, this particular AFH transmission algorithm should be used only for multislot packets.

Finally, *AFH initiation* takes place via an LMP packet exchange. A similar exchange is used to terminate AFH. Once AFH is implemented and initiated, *channel maintenance* begins. This is accomplished by having each member of the piconet continue to update its FUT, with slaves periodically sending results to the master so that channel usage closely matches actual conditions. In this manner, AFH can serve as a long-term performance enhancement to multislot Bluetooth packets as well as potentially increasing the throughput of other users in the 2.4 GHz band that are no longer competing with Bluetooth in certain parts of the ISM band.

Unfortunately, AFH could possibly force Bluetooth piconets into a small part of the band if several Wi-Fi networks, or even a malicious user, occupy much of the available frequency set. For instance, if Wi-Fi networks exist on 802.11 Channels 1, 6, and 11, then as few as 13 hop channels could remain for Bluetooth devices that employ AFH and can sense the presence of the Wi-Fi signals. In other words, AFH could enable other users to preempt bandwidth. Pandemonium could result if several Bluetooth piconets attempt to use this small channel set simultaneously. To avoid this situation, a limit should be placed on the number of bad channels that can be identified during the channel classification process, and only those having the highest levels of interference should be removed.

Is FUT-based AFH legal under existing FCC rules? It could be argued that the rules are violated if a FUT contains more than 4 channels listed as bad, which drops the hop channel count below 75. However, if these bad channels are probed for reassessment at least once every 30 seconds, then the full 79 hop channels are still being used even if a FUT lists only a few of these as good.[17]

Adaptive SCO Frequency Selection A variation of adaptive hopping could improve the performance of a Bluetooth SCO link.[22] When a HV3 SCO link is established, a pair of these packets is exchanged every six time slots. Four of these slots, then, don't contain traffic associated with this particular SCO channel and are therefore available for ACL or additional SCO links.

Many situations exist where two Bluetooth-equipped devices will connect with just one SCO point-to-point link (such as a headset), so four out of every six time slots will be unused. For improved SCO performance, these slots could be made available for SCO packet exchange if the scheduled HV3 slots contain interference. The master could use its own FUT, along with that from the slave, to determine the best two consecutive frequency slots to use within each six-slot time period. The associated slave simply listens on each master-to-slave slot until the master's SCO packet appears, and then it responds with its own return SCO packet in the following slave-to-master slot. In this manner, SCO performance is improved at the expense of a slight increase in the slave's power usage.

Adaptive Packet Selection

As we already know, Bluetooth units have the capability of selecting three different ACL packet lengths and whether or not to use a rate 2/3 FEC. Packet selection can affect throughput because the price to pay for retransmitting a long packet is greater than sending a short packet again. A Bluetooth device could attempt to mitigate interference by matching ACL packet type against some performance criteria, such as average delay, throughput, PER, or BER, and adjust the packet type accordingly. Some of these criteria, such as average delay, are easier to monitor at the higher protocol layers than others (such as BER).

When establishing a SCO link, the master has the choice of HV1, HV2, or HV3 packets, depending upon the level of FEC desired. When communicating in an interference-prone environment, many packet losses occur from catastrophic collisions, and this significantly reduces the effectiveness of FEC as a recovery technique. As such, it is usually preferable to make use of HV3 packets for SCO communication to reduce overall channel traffic and associated interference to other users.

TX Power Control

The Bluetooth class 1 transmitter must operate with a feedback mechanism from the device at the remote end of the link for power control implementation. The LMP_incr_power_req and LMP_decr_power_req are used for this purpose, commanding the local transmitter to increase or decrease power, respectively, to keep the remote receiver's RSSI within the Golden Receive Power Range (refer to Chapter 3, "The Bluetooth Radio"). The Bluetooth specification requires that RSSI alone be used for power control in accordance with the good neighbor policy. The downside to this, however, is that a strong interfering signal may generate a LMP_decr_power_req, further encumbering communication over the desired Bluetooth link. Wouldn't it make more sense to use C/I instead of RSSI to determine TX power levels?

Suppose our desired piconet implements transmit power control in response to C/I levels instead of RSSI, and interference is caused by another piconet also equipped with C/I-based power control. If the remote receiver in the desired piconet senses decreased average C/I over several hop channels, then it could request an increase in the local TX power, causing C/I to improve. Of course, this will hinder the performance of the other Bluetooth piconet, but that's not our concern, right? Ah, but the other piconet also has its power control tied to C/I. What will its response be to our TX power increase? That's right, it will raise its TX power level. Now our piconet senses decreased C/I, calling for another power increase, and . . . a transmit power war has begun. It will likely end when both piconets are transmitting at maximum power, and C/I is still poor for everyone involved.

Antenna Isolation

If two wireless systems are collocated in the same host, then the potential for interference is severe. As we will shortly discover, interference can be greatly reduced by using collaborative techniques between the systems, but this type of implementation can be complex. For the time being, let's assume that both collocated systems operate independently. This situation will occur if, for example, a user installs Wi-Fi from one manufacturer and Bluetooth from another on the same laptop. How can their transmissions be isolated for reduced mutual interference? One possibility is to place their respective antennas as far apart as possible. This option is available if the wireless modules have external antenna connectors or if the laptop manufacturer has already installed the antennas away from each other.

Two antennas can be installed into a computer with vertical separation, horizontal separation, or in an echelon configuration, as shown in Figure 13-20. For each of these configurations, we can find the approximate isolation in dB between the two.[23] Vertically separated antennas have isolation given by

$$L_v = 28 + 40\log\left(\frac{h}{\lambda}\right) \tag{13-4}$$

while antennas that are separated horizontally have isolation given by

$$L_h = 22 + 20\log\left(\frac{d}{\lambda}\right) - (G_t + G_r) \tag{13-5}$$

and the echelon configuration has isolation given by

$$L_e = L_h + \frac{2(L_v - L_h)}{\pi}\tan^{-1}\left(\frac{h}{d}\right) \tag{13-6}$$

Suppose two 2.4 GHz quarter-wave antennas are put behind the display in a laptop lid that measures 26 cm high by 30 cm wide. For a wavelength of 12 cm, each antenna is about 3 cm long. Allowing for antenna length, they can be separated vertically by about h = 20 cm and horizontally by about d = 30 cm. Assuming that each antenna has a gain of 0 dB, we find vertical isolation L_v = 37 dB, horizontal isolation L_h = 28 dB, and echelon isolation L_e = 31 dB.

Suppose Bluetooth with a TX power of 0 dBm and Wi-Fi with a TX power of 15 dBm are both installed in this laptop computer, with each connected

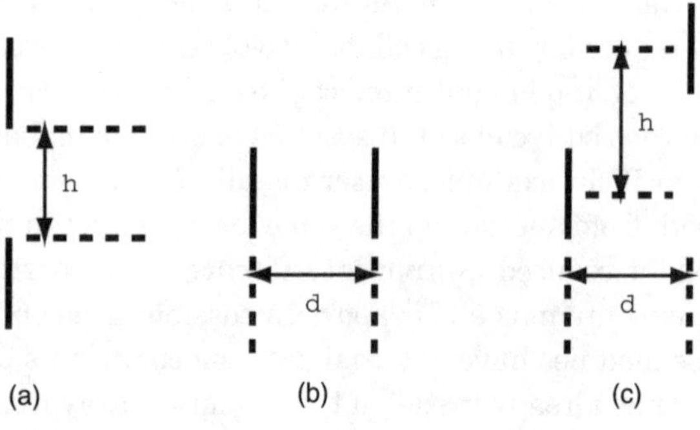

Figure 13-20 Antenna configurations can be vertical (a), horizontal (b), or echelon (c).

to its own antenna. Let's assume that the two antennas are separated vertically for greatest isolation, and that no additional coupling between the two devices exists. Is the antenna isolation sufficient to enable coexistence with minimal disruption? The Wi-Fi signal power at the Bluetooth antenna will be 15 − 37 = −22 dBm, and with an additional 13 dB rejection in the Bluetooth narrowband RX filter, interference power is −35 dBm. An incoming Bluetooth signal must be 11 dB above that level to meet cochannel C/I criteria. The required −24 dBm desired signal received power level is too high to enable a reasonable range to the remote Bluetooth device when the hop channel is within the Wi-Fi transmitted bandwidth.

The cochannel C/I for Wi-Fi is 10 dB, but the entire Bluetooth transmitted signal will enter the Wi-Fi receiver's passband if their frequencies overlap. Therefore, the 37 dB antenna isolation with a 0 dBm Bluetooth TX power requires the incoming Wi-Fi signal to be −27 dBm, still too high to enable Wi-Fi to have a reasonable range. We conclude, then, that antenna isolation alone isn't sufficient to enable one receiver to operate while the other transmitter is on. Furthermore, actual isolation will probably be less than 37 dB due to signal reflections from nearby conductive surfaces, including those in the computer itself.

Coexistence Solutions for Wi-Fi

When Wi-Fi is placed in an interference-prone environment, it can take a number of measures to improve its performance. Among these are

- **Dynamic channel selection** The 802.11 specification allows automatic selection of the best frequency using some performance metric such as PER, SNR, or RSSI. Most implementations, however, require manual channel selection. As the level of interference in the 2.4 GHz ISM band increases, it's possible that dynamic channel selection will find its way into Wi-Fi implementations. This method isn't suitable for avoiding Bluetooth transmissions because channels cannot be changed quickly.
- **Rate scaling** Wi-Fi can support data rates of 11, 5.5, 2, and 1 Mb/s, and the selected rate can also be based upon such criteria as PER, SNR, or RSSI. Using the lower bit rates in an interference environment increases the energy per data bit, which in turn lowers the BER.
- **Adaptive fragmentation** The 802.11 specification provides a means for fragmenting a large data packet into several smaller pieces, each of which is transmitted and acknowledged separately. When interference

is present, then the shorter fragments have increased probability of success, and overall throughput may be enhanced. Fragmentation can be used to transmit data between magnetron cycles in a microwave oven.

- **Frequency excision** When an incoming Wi-Fi transmission is corrupted by narrowband interference, often the interference can be rejected by a notch filter through frequency excision. With the availability of powerful DSP devices, it's possible that excision may soon be common in Wi-Fi receivers, and may even have sufficient processing power to track and excise multiple narrowband FHSS interferers. The Wi-Fi node could improve performance by determining the hop pattern for each Bluetooth piconet in advance.

Collaborative Coexistence Solutions

We've already discovered that antenna isolation isn't sufficient to ensure seamless coexistence for Bluetooth and Wi-Fi that are operating very close to each other. An implementation that is almost sure to become common is when both are physically on the same host. Same-host operation does have a significant advantage, though, because the two wireless systems can be controlled together in such a way that each encumbers the other as little as possible. Furthermore, the control is accomplished through a nonwireless (that is to say, wired) link, so interference won't affect control signal reliability.

Wireless devices that aren't collocated on the same host can also use collaborative techniques to limit their interference to each other, but such collaboration will probably be controlled over the wireless link itself, decreasing reliability. Fortunately, at least in the case of Wi-Fi and Bluetooth, we discovered earlier that even relatively short separations (such as 2 meters) is sufficient to reduce interference to a tolerable level without adding the complexity of collaborative coexistence.

The goal of collaboration is to implement some type of reservation scheme for transmission of data that is fair to both systems by enhancing the throughput of each without requiring excessively long delays. We will examine several collaborative approaches, starting from the simple, low-performance techniques and working up to complex, high-performance solutions.[20]

Manual Switching

The simplest collaboration method between Wi-Fi and Bluetooth is for the user to manually switch between them, depending upon which is needed at any particular time. One common and cumbersome implementation of manual switching is through the use of separate PC cards, one with Bluetooth hardware and the other with Wi-Fi hardware. The desired card is installed, the matching software launched, and communication can commence.

Many wireless software packages include the ability to turn off the communication hardware, so perhaps it would be more convenient to insert both PC cards into the host and activate only one at a time. Although many computers have two PC card slots, it's often difficult to install both together because of the extra height needed for the antenna, and even if they could both be installed, the proximity of the two antennas within a few millimeters of each other would severely affect both antenna radiation patterns.

Driver-Level Switching

If a host can physically accommodate both Wi-Fi and Bluetooth hardware, then the two wireless systems can be switched on and off electronically at the driver level. Switching thus becomes a type of TDMA that is controlled by the host *operating system* (OS), as shown in Figure 13-21.[20] The depicted implementation has two antennas, but with the proper *transmit/receive* (T/R) switching, a single antenna is also feasible.

It's likely under most conditions that a user will require access to only one of these wireless systems at any one time; for example, most web

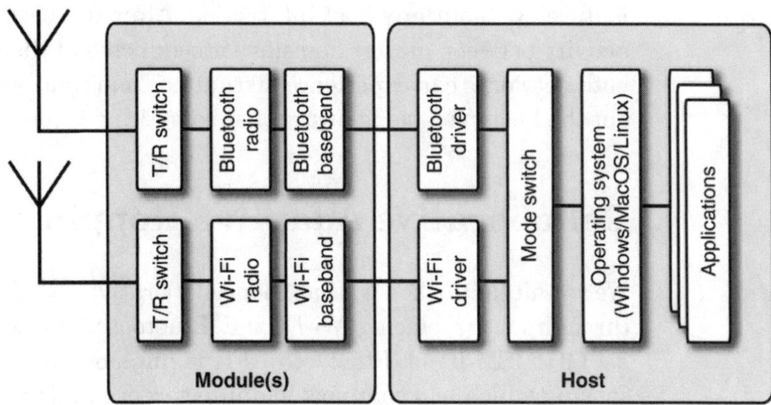

Figure 13-21 Driver-level switching for multiplexing operation between Bluetooth and Wi-Fi[20]

browsing (Wi-Fi) occurs without simultaneously sending a file to a printer (Bluetooth). However, during those times when both systems are operational, the OS attempts to prevent one device from transmitting while the other is actively receiving data. Suspending the operation of one radio, of course, can affect its throughput, and the resulting decrease may be apparent to the user. An added difficulty exists when a Bluetooth SCO link is present because it's likely that noticeable audio disruptions will occur during Wi-Fi operation.

A link can be suspended either with or without notifying others in the network. Suspension without notification can be done quickly: When the OS determines that a packet is arriving on network A, it simply locks out network B's transmitter. This form of suspension could have an adverse effect on the other participants in network B, though, because any incoming data won't be acknowledged, retransmissions will occur, and the other users will experience reduced throughput. On the other hand, it may be difficult to notify other users in network B of the suspension because, once a network A packet begins to arrive, it's too late to energize network B's transmitter.

Driver-level switching reaction time is limited by the time between OS notification of an incoming packet and its response in disabling the other radio's transmitter. This time is highly variable and can routinely exceed the duration of the incoming packet itself,[20] which invalidates this entire process. Long switching delays equate to slow switching times, further reducing potential throughput. The situation is alleviated somewhat when the local host initiates communication and can disable the unused transmitter first.

Aside from the normal response-time delays associated with the busy host processor, many situations exist where wireless transmissions occur without host notification. For example, Bluetooth devices respond to inquiries, pages, and can even convey their Bluetooth name and supported features without any host interaction. After the piconet is formed, polling activity between master and slave is conducted at the LM level. As a result, both networks can experience disruption from collocated interference without the host even being aware that something is amiss.

Collaborative Adaptive Hopping

We've already shown that noncollaborative AFH can often increase throughput by placing Wi-Fi and Bluetooth into separate parts of the 2.4 GHz ISM band. Bluetooth AFH is implemented noncollaboratively by sensing which hop channels had interference and avoiding data transmissions on those channels. When Wi-Fi and Bluetooth are collocated, collaborative AFH is much easier to implement because the host knows the channel on which its Wi-Fi node is operating, and the affected frequencies can be removed from the Bluetooth hop channel set. Simple, yes?

Not so fast. There are a number of difficulties with this approach. First, FCC rule changes are needed for this type of AFH to be legal for Bluetooth TX powers above about 1 mW (actually, above −1.3 dBm EIRP). Second, the AFH channel set must be conveyed to other piconet users, and they all must have AFH capability and also agree to use that particular set. What if another Bluetooth piconet member has a collocated Wi-Fi node operating on a different channel? We conclude that collaborative AFH has great potential to reduce interference between collocated Wi-Fi and Bluetooth nodes, but several hurdles must be crossed before its implementation can be successful.

MAC-Level Switching

The MAC-level switching concept is very similar to driver-level switching, but because the process takes place much lower in the respective protocol stacks, response time is much faster and switching can be matched more closely with dynamic changes in the two networks.[20] At this level, it's possible to always prevent one node from transmitting while the other is receiving, and other techniques, such as collaborative AFH, can be implemented efficiently.

MAC-level switching presents some significant challenges. Perhaps the greatest challenge is the requirement to fully integrate both Wi-Fi and Bluetooth at these lower protocol levels and ensure that proper cross-coordination takes place. This is no small feat, given the extensive differences between the two systems at just about every protocol level. Another challenge to MAC-level switching is that the lower protocol layers cannot optimize communication scheduling to satisfy any system requirements that the host may have, such as prioritizing one link over the other, without additional interaction between the host and lower protocol layers.

If the Bluetooth device is the piconet master, then it can implement a form of TDMA between Wi-Fi and Bluetooth operation at the MAC level.[22] As part of its normal operation, the Wi-Fi AP transmits a short beacon packet to mark the beginning of each frame. The time between successive beacons can be divided between Wi-Fi and Bluetooth if both are operating together. Although collisions are eliminated, Bluetooth SCO links can't be supported with this technique.

System-Level Integration

Finally, the highest-performing and most complex method of collocating Bluetooth and Wi-Fi is by including coordination at all protocol layers through system-level integration. This collaborative approach is the most complex because it requires integrating both Bluetooth and Wi-Fi so closely that they occupy the same silicon. On the other hand, performance of both Bluetooth and Wi-Fi in a collocated environment can be increased significantly without requiring any changes to either specification.

One such system-level implementation is shown in Figure 13-22.[24] MAC-layer switching is accomplished through the SIMOP-D block that can coordinate timing and (if enabled) AFH to prevent cross interference. The SIMOP-A block provides additional coordination at the PHY layer and has a core functionality to support Wi-Fi operation when a Bluetooth SCO link is active, and to assist in operating low data-rate Bluetooth activities, such as polling peripherals (for example, HID), to reduce their impact on Wi-Fi communications.

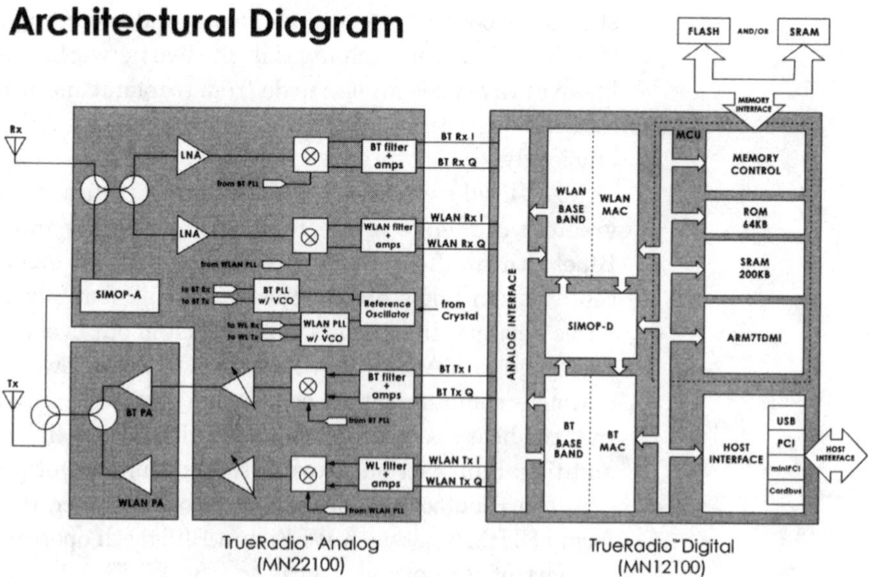

Figure 13-22 The block diagram of Mobilian's True Ratio™[24]

Bluetooth Operation Aboard Commercial Aircraft

Most government entities that regulate commercial aviation, including the *Federal Aviation Administration* (FAA) in the United States, prohibit the use of RF transmission equipment by passengers during flight. These prohibitions can extend to receivers as well. These rules exist in part to prevent the possibility of interference between these devices and critical aircraft communication and navigation equipment.

An empirical study on the feasibility of using 2.4 GHz Bluetooth class devices in aircraft was performed by Intel on a Boeing 747, Boeing 727, and Gulfstream V to represent wide-body and narrow-body commercial aircraft as well as a common business-class jet.[25] The first part of the study examined whether 2.4 GHz was a viable band for communication within the confines of the passenger compartment. In this environment, multipath components provided adequate signal strength even when the LOS path was severely cluttered. Deep single-frequency RSSI nulls were present throughout the compartment, but typically more than 80 percent of the band was usable in any location. As such, Bluetooth FHSS would enable adequate throughput to occur regardless of the locations of the piconet members within the cabin. Average path loss at 10 meters of separation in a Boeing 727 varied between 60 and 70 dB, depending upon location. This is consistent with a path loss exponent of about 3.0. An additional path loss of 15 dB typically occurred between the cabin and baggage compartment.

The second part of the study concentrated on whether these devices could coexist with the numerous communication and navigation radio systems used during aircraft operation. Measured emissions from a Bluetooth transmitter were matched against worst-case propagation losses to the aircraft systems antenna in question, and an interference analysis accomplished against the lowest level of desired signals received from the associated ground-based communication or navigation transmitter. From these figures, a safety margin can be computed. The study showed that these margins were adequate for the operation of all tested avionics, ranging from 10 dB to over 30 dB, depending upon the system. The study had insufficient data to assess the coexistence of Bluetooth with the *Global Positioning System* (GPS), but the test engineers estimated the safety margin to be greater than 10 dB.

In one test, six Bluetooth piconets were operated within the passenger cabin, five with 0 dBm TX power, and one with 30 dBm TX power (10 dB higher than the maximum allowed by the specification). All piconets had a duty cycle of 75 percent or greater. Aircraft communication and navigation systems were operated in various modes, with input signals generated by test equipment or distant ground-based stations from which the signal level was just above the minimum usable. The equipment operated normally, and no change was observed whether or not the Bluetooth piconets were in operation.

Because Bluetooth is meant to be included within a host device that may itself be either an intentional or unintentional radiator, tests were made to determine if the host's emission characteristics changed after Bluetooth was added. Only unintentional radiators (such as a laptop computer and PDA) were tested, and no increase in unintentional emissions was discovered.

In summary, operating Bluetooth piconets within the passenger cabin of the tested aircraft had no effect on operation of either communication or navigation avionic equipment. Therefore, the study recommends that FAA rules be amended to allow the operation of Bluetooth-equipped devices during noncritical phases of flight.

Summary

The 2.4 GHz ISM band is rapidly filling with communication systems, and Bluetooth devices will need to coexist with these and maintain reasonable throughput without adversely affecting the other users. Bluetooth's FHSS transmission is well suited to avoiding interference provided by Wi-Fi, microwave ovens, HomeRF, and other Bluetooth piconets, but mechanisms, such as AFH, can use intelligence, rather than statistical methods, to further increase throughput and reduce its interference on others. Coexisting with Wi-Fi is important because interference between the two can become severe if they are within about 2 meters of each other.

Coexistence solutions fall into two broad categories. Noncollaborative methods are used by one system to enhance its own throughput without adversely affecting other independent users. These include using AFH, adaptive packet selection, and antenna isolation. Collaborative methods include manual switching, driver-level switching, AFH, MAC-level switching, and integrating all of these into a systems-wide approach. Because Bluetooth and Wi-Fi are aimed at different markets, it's likely that both will find their way into the same host. Severe disruption to each is possible if both operate indiscriminately within such close proximity, but collaborative methods can be used to enhance their capability to coexist.

For operation of Bluetooth onboard aircraft, a comprehensive study has shown that aircraft avionics are unaffected by the presence of several Bluetooth piconets within the passenger compartment.

End Notes

1. O'Hara, B. and Petrick, A. *IEEE 802.11 Handbook: A Designer's Companion*, IEEE Press, 1999.
2. Zyren, J., et al. "Letter to Secretary of FCC on Alternative Emission Limit for Microwave Lighting Devices," IEEE Document IEEE 802.1199/083, March 10, 1999.
3. Bray, J. "Bluetooth vs 802.11b: And the Winner Is?" *Portable Design*, January 2002.
4. Meihofer, E. "The Performance of Bluetooth in a Densely Packed Environment," presented at the Bluetooth Developers Conference, San Jose, CA, December 2000.
5. Shoemake, M. "Wi-Fi (IEEE 802.11b) and Bluetooth: Coexistence Issues and Solutions for the 2.4 GHz ISM Band," Texas Instruments white paper, Version 1.1, February 2001.
6. Haarsten, J. and Zürbes, S. "Bluetooth Voice and Data Performance in 802.11 DS WLAN Environment," Ericsson white paper, May 1999.
7. Lansford, J. "Adaptive Hopping," presented at the Bluetooth Developers Conference, San Francisco, CA, December 2001.
8. Shellhammer, S. "SCORT—An Alternative to the Bluetooth SCO Link for Voice Operation in an Interference Environment," IEEE 802.15 TG2 Document IEEE 802.1501/145 r1, March 2001.
9. Lansford, J. "Wi-Fi (802.11b) and Bluetooth Simultaneous Operation: Characterizing the Problem," Mobilian Corporation White Paper, 2000.
10. Mitter, V., et al. "Empirical Study for IEEE 802.11b WLAN & Bluetooth Coexistence in UL Band," IEEE 802.15 TG2 Document IEEE 802.1501/148 TG2, January 2002.
11. Gawthrop, P., et al. "Radio Spectrum Measurements of Individual Microwave Ovens," NTIA Report 94-303-1, March 1994.

12. Buffler, C. and Risman, P. "Compatibility Issues Between Bluetooth and High Power Systems in the ISM Bands," *Microwave Journal*, July 2000.

13. Sega, K. "Measuring Interference Among Bluetooth, Microwaves and IEEE 802.11b," *Nikkei Electronics Asia*, October 2001.

14. Kamerman, A. and Erkoçevic, N. "Microwave Oven Interference on Wireless LANs Operating in the 2.4 GHz ISM Band," presented at the 8th annual IEEE Symposium on Personal, Indoor, and Mobile Radio Communications, Helsinki, September 1997.

15. Neelakanta, P., et al. "Bluetooth-Enabled Microwave Ovens for EMI Compatibility," *Microwave Journal*, January 2001.

16. Neelakanta, P. and Sivaraks, J. "A Novel Method to Mitigate Microwave Oven Dictated EMI on Bluetooth Communications," *Microwave Journal*, July 2001.

17. Golmie, N. "Dialog with FCC," IEEE 802.15 TG2 Document IEEE 802.1501/00144 r0, March 2001.

18. Meihofer, E. "Enhancing ISM Band Performance Using Adaptive Frequency Hopping," Motorola white paper, December 2001.

19. Treister, B., et al. "Adaptive Frequency Hopping: A Non-Collaborative Coexistence Mechanism," IEEE 802.15 TG2 Document IEEE 802.15-01_252r0, May 2001.

20. "Wi-Fi (802.11b) and Bluetooth: An Examination of Coexistence Approaches," Mobilian white paper, 2001.

21. Golmie, N. and Liang, J. "Non-Collaborative MAC Mechanisms," IEEE 802.15 TG2 Document IEEE 802.1501/316 r0, June 2001.

22. Liang, J. "Proposal for Non-Collaborative BT and 802.11b MAC Mechanisms for Enhanced Coexistence," IEEE 802.15 TG2 Document IEEE 802.1501/026 r1, March 2001.

23. Timiri, S., "RF Interference Analysis for Collocated Systems," *Microwave Journal*, January 1997.

24. "Sim-Op—Unleashing the Full Potential of Wi-Fi and Bluetooth Coexistence," Mobilian white paper, 2001.

25. Schiffer, J. and Waltho, A. "Safety Evaluation of Bluetooth Class ISM Band Transmitters on Board Commercial Aircraft," Intel white paper, December 28, 2000.

CHAPTER 14

The Future of Bluetooth

People working within the Bluetooth industry are almost universally convinced that the future of Bluetooth is bright. A large amount of market momentum has built behind the concept, and support comes from both well-established firms and companies created specifically to develop new Bluetooth hardware and software. There has been some discord from perceived competition between Bluetooth and Wi-Fi, but many recognize a market need for both. Fortunately, there doesn't yet appear to be any real challenge to Bluetooth from another cable-replacement technology that has the potential to divide manufacturers and force customers to select between two incompatible alternatives.

In this chapter, we will take a brief look at some of the opportunities provided by Bluetooth wireless in several areas. We'll resist trying to become too specific, though, because the road is littered with past technology predictions that look ludicrous from today's point of view. After all, weren't each of us supposed to have a personal flying machine in our driveway by now?

Gaining Customer Confidence

Clearly, the most important criterion to enhance the future of Bluetooth is customer confidence in the technology. The SIG has been focused on this aspect of Bluetooth ever since the organization was formed, and that was one of the reasons that the Bluetooth specification covers all aspects of operation from the radio to the user interface.

The success or failure of Bluetooth in the marketplace depends upon whether the customer finds the technology useful, cheap, reliable, secure, and easy to operate. For Bluetooth to be *useful*, it must enable a task to be performed better than would otherwise be the case. The simple elimination of a cable is certainly useful in many applications, such as file synchronization and digital camera data transfers. However, if a set of batteries is needed to substitute for the cable (such as mouse and keyboard), then careful power management is critically important to a successful design.

Achieving the market penetration that many manufacturers envision for Bluetooth requires it to be extremely *cheap*. Bluetooth itself is merely a communication medium, so it usually manifests itself as additional circuitry and, hence, additional cost, within a product. Many of these products are already inexpensive in their traditional manifestations, such as the remote control or headset, and customers may balk at paying significantly higher prices for these items with Bluetooth capabilities.

To be *reliable*, the Bluetooth-equipped device must have an adequate range and not be temperamental when asked to connect. Wireless devices can occasionally be troublesome (witness the cellular phone), but only rarely should this occur with Bluetooth, and the fix should be easy, such as changing location slightly. Devices from different manufacturers should operate together seamlessly.

The operation of Bluetooth-equipped devices must be *secure* and accepted as such by the customer. Link-level security built into Bluetooth should help this situation by addressing security issues and providing a consistent platform for authentication and encryption across a wide range of applications. As consumers become more technically literate, they begin to understand the limitations of technology as well as its benefits. Most wireless phone users are aware of the potential for eavesdropping, and they also realize that security is important for other wireless implementations as well, including Bluetooth.

Finally, a customer needs to find Bluetooth so *easy to operate* that its use is obvious right out of the box. The actual operation of Bluetooth is quite complex, as we've pointed out on several occasions throughout this book. A well-designed set of controls and a properly written user manual insulates the consumer from that complexity. This is a difficult challenge because new technology often places high demands on the consumer when it first arrives on the market. Is there anyone who hasn't been frustrated by the infernal VCR clock? I finally put a piece of black electrical tape over mine to hide the flashing 12:00.

In a nutshell, if the customer operates a product without being aware of the Bluetooth link, then its integration can be considered a success.

Future Applications

Because Bluetooth is so new, just about any application could be called futuristic, but we will take a look at a few of those that could be considered an extension or addition to the Bluetooth profiles that have been published as of early 2002. Some of these are already being developed by the SIG. These applications take advantage in varying degrees of Bluetooth's capability to be always on, always connected, mobile, and easy to use. Throughout all of this, of course, it's important not to lose sight of the requirement to get basic applications to operate properly.[1]

- **Local positioning** Because the *global positioning system* (GPS) doesn't operate reliably inside buildings, Bluetooth-equipped devices could communicate with another unit that is in a location for favorable reception of GPS positioning information. By using a combination of

RSSI, several GPS/Bluetooth nodes, and perhaps even signal arrival timing, a roaming Bluetooth-equipped device can pinpoint its location with high accuracy. This application could be extremely useful in tracking people and animals as well as tracking and controlling robots and products on an assembly line.

- **Universal remote control** By replacing traditional IR links with Bluetooth RF, the remote control no longer needs to be pointed at the appliance being controlled. Home theater systems, for example, are often deployed in several locations within the room, or even in adjacent rooms, making Bluetooth the ideal controlling medium.
- **Proximity detection and access control** By using the paging process, two Bluetooth-equipped devices can determine whether they're in close proximity to each other. RSSI and timing enable the two devices to determine the distance between them with reasonable accuracy. Authentication verifies the identity of the two units to prevent unauthorized access. Applications include controlled entry to an automobile, hotel room, airliner, or secure area, and as automatic access authorization to a computer. Home security systems could be automatically enabled and disabled using this method.
- **Automotive applications** Aside from keyless entry, Bluetooth can be used for diagnostics, entertainment, access to other information, and navigation. Wireless communication between subsystems within the car reduces manufacturing cost by reducing the number of cable runs, connectors, fasteners, and access holes. Areas that are difficult to monitor, such as tire pressure, could be done via Bluetooth.
- **Interactive games** Connecting games via a Bluetooth link gives players the ability to compete against each other rather than against the computer.
- **Wireless pen** A Bluetooth-equipped writing instrument that automatically digitizes its motions can be employed to enable the user to easily e-mail a drawing or handwritten note to any recipient through an accompanying cellular phone.
- **Electronic payment** A Bluetooth-equipped *personal digital assistant* (PDA) could enable its user to conveniently pay for items without queuing at the checkout counter. Instead, the purchaser scans items into the PDA using perhaps a barcode reader; then the PDA communicates with the cell phone and debits the purchaser's account. This process can be extended to general e-commerce such as buying airline tickets, paying for a hotel room, vending machine purchases, and entry to a theme park or sports event.

- **Medical applications** The world of medicine can use Bluetooth for such tasks as real-time data collection and biometrics. Nursing homes could use the technology to track their residents, and hospitals can monitor the location and condition of their ambulatory patients and use a Bluetooth link for automatic drug dispensing. Such a link enables drug doses to be changed remotely in response to a doctor's or nurse's direction or patient's condition.
- **Bio implant** *Radio frequency identification* (RFID) technology is already being used for pet identification, and a smaller, injectable RFID unit has been developed that is compatible with human tissue for use in pacemakers, defibrillators, and artificial joints.[2] By replacing the RFID tag with a Bluetooth transceiver, both range and versatility are enhanced. For example, the implant could monitor a person's health and provide an alert via a PDA or wristband if medication is needed, or even automatically place a call through an external cell phone to medical authorities in an emergency, using a position location capability to direct personnel to the victim.

Operational Enhancements

What is in the future for enhancements to the basic operation of Bluetooth? Certainly, new profiles will be introduced as they are conceived, created, and approved. These will be released over time in the form of updates to the existing specification. Furthermore, sets of critical errata will be published as problems surface. However, most developers have found the Specification 1.1 to be stable, so future Bluetooth-enabled products for the general consumer will conform to this version of the specification.

As Bluetooth matures and its capabilities and limitations are experienced in actual applications, attention will invariably turn to how its performance can be improved. Here are some of the possibilities for future enhancements to Bluetooth's operation, with emphasis on the lower protocol levels:

- **Faster data rates** Some would argue that Bluetooth needs to increase its data rate by at least a factor of 10 to be competitive. Others argue that the present data rate is fast enough for its intended purpose. This is an area of intense debate, but it's important to realize that faster data transmission carries with it penalties in the form of shorter range, higher power consumption, and/or higher cost. At typical realized data rates (400 kb/s), 100 pages of printed text can be transferred in about 10 seconds and a 600 kB high-resolution digital photo in about 12 seconds. Are higher data speeds worth the penalties? To answer this question, the Radio2 working group has been formed within the Bluetooth SIG to investigate higher data rates.

- ***Adaptive frequency hopping*** **(AFH)** This topic was covered extensively in the last chapter, and we concluded that AFH can reduce packet loss if properly implemented. Many feel that AFH will eventually become part of the Bluetooth specification.
- **Store-and-forward capability** In its present form, Bluetooth has no formal means for increasing its range by relaying messages through third parties. Of course, the feature can be implemented in higher protocols, but this can be cumbersome. Placing a store-and-forward capability into the specification will provide built-in commands that can be used to specify the destination and, if necessary, the route that a packet should take through intermediate Bluetooth nodes. Advantages include an effective range limited only by device availability along the route and lower transmit power for reaching closer nodes. Disadvantages include lower reliability, increased interference from multiple transmissions, reduced throughput, and higher node complexity.
- **Smart antennas** If sufficient signal processing capability is developed to control an electronically steerable array antenna, then the potential exists to greatly increase performance by directing the antenna's main lobe toward the desired node. Furthermore, the antenna's nulls can be placed in the direction of arriving interference. As a result, C/I will improve markedly, possibly enabling reductions in transmit power. Also, interference to other users will be further reduced from the nulls aimed in their direction. Low-power directed-beam transmission could potentially decrease both the range of vulnerability to eavesdropping and the range of susceptibility to jamming. Disadvantages, of course, are the cost, complexity, and power requirements of such an implementation.

Summary

Go wireless!

Aftermath.

End Notes

1. Miller, B. "Emerging and Future Applications for Bluetooth Wireless Technology," presented at the Bluetooth Developers Conference, San Francisco, CA, December 2001.
2. Murray, C. "IC Opens Door to Human Bar Code," *EE Times*, January 7, 2000.

ACRONYMS AND ABBREVIATIONS

ACK	Acknowledgment
ACL	Asynchronous connectionless
ACO	Authenticated ciphering offset
A/D	Analog-to-digital
AF	Attenuation factor
AFH	Adaptive frequency hopping
AM	Amplitude modulation
AM_ADDR	Active member address
AMPS	Advanced mobile phone system
AP	Access point (802.11)
API	Application programming interface
AR_ADDR	Access request address
ARQ	Automatic repeat request
ASCII	American Standard Code for Information Interchange
ASIC	Application-specific integrated circuit
ASK	Amplitude shift keying
ATM	Automated teller machine
AU_RAND	Authentication random number
AWGN	Additive white Gaussian noise
BB	Baseband
BC	Broadcast
BCH	Bose-Chadhuri-Hocquenghem
BD_ADDR	Bluetooth device address
BER	Bit error rate (same as PBE)
BFSK	Binary frequency-shift keying
BGA	Ball grid array
BiCMOS	Bipolar transistor and complementary metal oxide semiconductor
BPF	Bandpass filter

BPSK	Binary phase shift keying
BSS	Basic service set (802.11)
BTAB	Bluetooth Technical Advisory Board
BQA	Bluetooth Qualification Administrator
BQB	Bluetooth Qualification Body
BQRB	Bluetooth Qualification Review Board
BQTF	Bluetooth Qualification Test Facility
b/s	bits per second (also bps)
BW	Bandwidth
CAC	Channel access code
CAD	Computer-aided design
CCITT	International Telegraph and Telephone Consultative Committee
CCK	Complimentary coded keying
CDMA	Code division multiple access
C/I	Carrier-to-interference ratio (also CIR)
CID	Channel identifier
CL	Connectionless
CLK	Bluetooth clock
CLKE	Bluetooth estimated clock
CLKN	Bluetooth native clock
CMOS	Complementary metal oxide semiconductor
CO	Connection-oriented
CPU	Central processing unit
CRC	Cyclic redundancy check
CSMA	Carrier sense multiple access
CTE	Coefficient of thermal expansion
CTS	Clear to send
CVSD	Continuously variable slope delta modulation
CW	Continuous wave
D/A	Digital-to-analog
DAC	Digital-to-analog converter
DAC	Device access code

dBd	Decibels relative to a dipole antenna
dBi	Decibels relative to an isotropic source
dBm	Decibels relative to one milliwatt
dBp	Decibels relative to one picowatt
DCE	Data circuit-termination equipment or data circuit endpoint
DCI	Default check initialization
DCID	Destination channel identifier
DFIR	Diffuse infrared
DH	Data high speed
DIAC	Dedicated inquiry access code
DM	Data medium speed
DoS	Denial-of-Service
DPIT	Designated profile interoperability tester
DPSK	Differential phase-shift keying
DS/CDMA	Direct sequence code division multiple access
DSP	Digital signal processor
DSR	Data set ready
DSSS	Direct sequence spread spectrum
DTE	Data terminal equipment or data terminal endpoint
DTR	Data terminal ready
DUT	Device under test
DV	Data-voice
ECC	Error correction code
EIA	Electronic Industries Association
EIRP	Effective isotropic radiated power
EMP	Electromagnetic pulse
EN_RAND	Encryption random number
ERTX	Extended response timeout expired
ESCE	External security control entity
EUT	Equipment under test
FAA	Federal Aviation Administration
FAF	Floor attenuation factor
FCC	Federal Communications Commission

FCS	Frame check sequence
FEC	Forward error correction
FET	Field-effect transistor
FDD	Frequency division duplexing
FH/CDMA	Frequency hop code division multiple access
FHS	Frequency hop synchronization
FHSS	Frequency hop spread spectrum
FIR	Finite impulse response
FM	Frequency modulation
FSK	Frequency-shift keying
FUT	Frequency usage table
GaAs	Gallium-arsenide
GAP	Generic access profile
GFSK	Gaussian-filtered frequency-shift keying
GHz	Gigahertz
GIAC	General inquiry access code
GOEP	Generic object exchange profile
GPS	Global positioning system
HC	Host controller
HCI	Host controller interface
HDTV	High-definition television
HEC	Header error check
HID	Human interface device
HV	High-quality voice
HVAC	Heating, ventilation, and air conditioning
IAC	Inquiry access code
IBSS	Independent basic service set (802.11)
IC	Integrated circuit
ICS	Implementation Conformance Statement
IEE	Institution of Electrical Engineers (UK)
IEEE	Institute of Electrical and Electronics Engineers
IF	Intermediate frequency
Im	Imaginary

IMD	Intermodulation distortion
IN_RAND	Initialization random number
I/O	Input/output
IP	Intellectual property
IP	Internet protocol
IR	Infrared
IRQ	Interrupt request
ISI	Intersymbol interference
ISM	Industrial, scientific, and medical
ISO	International Standards Organization
ISP	Internet service provider
ITU	International Telecommunications Union
IUT	Implementation under test
IXIT	Implementation Extra Information for Testing
kb/s	Kilobits per second (also kbps)
kHz	Kilohertz
L2CA	Logical link control and adaptation
L2CAP	Logical link control and adaptation protocol
LAN	Local area network
LAP	Lower address part
LC	Link controller
L_CH	Logical channel
LCID	Local channel identifier
LFSR	Linear feedback shift register
LIAC	Limited inquiry access code
LK_RAND	Link key random number
LM	Link manager
LMP	Link manager protocol
LNA	Low-noise amplifier
LO	Local oscillator
LOS	Line-of-sight
LSB	Least significant bit
LTCC	Low-temperature cofired ceramic

MAC	Medium access control layer in the OSI model
Mb/s	Megabits per second (also Mbps)
MHz	Megahertz
MMI	Man-machine interface
MPE	Maximum permissible exposure
MS	Master-slave
MSB	Most significant bit
MSK	Minimum shift keying
Ms/s	Megasymbols per second
MTU	Maximum transmission unit
mW	Milliwatt
NAK	Negative acknowledgment
NAP	Nonsignificant address part
NDA	Nondisclosure agreement
NIC	Network interface card
NS-LTCC	Nonshrinkage low-temperature cofired ceramic
NTIA	National Telecommunications and Information Administration
NZIF	Near-zero intermediate frequency
OBEX	Object exchange
OCF	OpCode command field
OET	Office of Engineering and Technology
OGF	OpCode group field
OLOS	Obstructed line-of-sight
OOK	On-off keying
OS	Operating system
OSI	Open systems interconnection
OUI	Organizationally Unique Identifier
PA	Power amplifier
PAN	Personal area network
PB	Packet boundary
PBE	Probability of bit error (same as BER)
PCB	Printed circuit board

PCM	Pulse code modulation
PDA	Personal digital assistant
PDU	Protocol data unit
PER	Packet error rate
PHY	Physical layer in the OSI model
PICS	Protocol Implementation Conformance Statement
PIFA	Planar inverted "F" antenna
PIN	Personal identification number
PL	Path loss
PLCC	Plastic leaded chip carrier
PLL	Phase-locked loop
PM	Phase modulation
PM_ADDR	Parked member address
POTS	Plain old telephone service
ppm	Parts per million
PRBS	Pseudorandom binary sequence
PRD	Program reference document
PSK	Phase-shift keying
PSM	Protocol service multiplexer
QAM	Quadrature amplitude modulation
QFP	Quad flat package
QLP	Quad leadless pad
QoS	Quality of service
QPSK	Quadrature phase shift keying
Re	Real
RF	Radio frequency
RFCOMM	Radio frequency communication (Bluetooth serial port)
RFID	Radio frequency identification
RISC	Reduced instruction set computer
RSSI	Received signal strength indication
RTOS	Real-time operating system
RTS	Request to send
RTX	Response timeout expired

RX	Receive
RXD	Receive data
SAR	Specific absorption rate
SAW	Surface acoustic wave
SCID	Source channel identifier
SCO	Synchronous connection-oriented
SDAP	Service discovery application profile
SDP	Service discovery protocol
SIG	Special Interest Group (Bluetooth)
SiGe	Silicon-germanium
SiO_2	Silicon dioxide
SiP	System in a package
SNIR	Signal-to-noise-and-interference ratio
SNR	Signal-to-noise ratio
SoC	System on a chip
SoI	Silicon-on-insulator
SOIC	Small outline integrated circuit
SP	Scan period
SPI	Serial port interface
SR	Scan repetition
STA	Station (802.11)
SUT	System under test
SWAP	Shared wireless access protocol
TCB	Telecommunications Certification Body
TCS	Telephony control protocol specification
TCP	Transport control protocol
TCRL	Test Case Reference List
TDD	Time division duplexing
TDMA	Time division multiple access
TID	Transaction identifier
TO	Timeout
T/R	Transmit/receive
TS	Test specification

TX	Transmit
TXD	Transmit data
UA	User asynchronous
UAP	Upper address part
UI	User isochronous
U-NII	Unlicensed National Information Infrastructure
US	User synchronous
UART	Universal asynchronous receiver/transmitter
UPF	UnPlugFest
USB	Universal serial bus
UTF-8	Unicode Standard Transformation Format Eight
VCO	Voltage-controlled oscillator
W	Watt
WAP	Wireless application protocol
WBFH	Wideband frequency hop spread spectrum
WECA	Wireless Ethernet Compatibility Alliance
WEP	Wired equivalent privacy
WHO	World Health Organization
Wi-Fi	Wireless Fidelity (802.11b interoperability)
Wi-Fi5	Wireless Fidelity Five (802.11a interoperability)
WLAN	Wireless local area network
WPAN	Wireless personal area network
WRC	World Radiocommunications Conference
XOR	Exclusive-OR
ZIF	Zero intermediate frequency

REFERENCES

"2.4 GHz and 5 GHz WLAN: Competing or Complementary?" Mobilian White Paper, 2001.

"A Place for Everything," *Bluetooth World*, December 2001.

"A Wireless Connectivity Technologies Comparison," Infrared Data Association White Paper, September 10, 1998.

Adraw, R. "Perform Bluetooth RF Measurements," *Supplement to Penton's Electronics Group*, Winter 1999–2000.

Aggarwal, A. "The Continued Evolution of Bluetooth Wireless Technology," presented at the Bluetooth Developers Conference, San Francisco CA, December 2001.

Ajluni, C. "Can Bluetooth and 802.11b Co-Exist?" *Wireless Systems Design*, February 2001.

Andersen, P. "Bluetooth Development, Test and Debug," presented at the Bluetooth Developers Conference, San Francisco CA, December 2001.

Anderton, D. and Stanton, S. "Developing Measurement Solutions for Bluetooth," *Supplement to Penton's Electronics Group*, Winter 1999–2000.

Beckers, S. and Zoltek, T. "Designing Bluetooth Packages for Easy Testing," *Wireless Systems Design*, September 2001.

Bibaud, S. "Bluetooth Host Based Solutions," presented at the Bluetooth Developers Conference, San Jose CA, December 2000.

Bradley, R. "Antenna Design," presentation by GigaAnt, Inc., October 2000.

Bray, J. and Sturman, C. *Bluetooth: Connect Without Wires,* Upper Saddle River, NJ: Prentice Hall, 2001.

Breed, G. "Antenna Basics for Wireless Communications," *RF Design,* October 1995.

Breed, G. "A Primer on Antenna/Human Body Interaction," *Applied Microwave & Wireless,* November/December 1998.

Cain, P. and Alistair, M. "Getting a Grip on Bluetooth Testing," *Communications Systems Design*, December 2000.

Carey, T. "Fading and Multipath Testing in Communications Systems," *Microwave Journal*, November 1996.

Chen, W. "Motorola's Bluetooth Solution to Interference Rejection and Coexistence with 802.11," Motorola White Paper, December 2001.

Cheung, K., et al. "A New Empirical Model for Indoor Propagation Prediction," *IEEE Transactions on Vehicular Technology*, August 1998.

Dempsey, M. "The Physiological Effects of 2.4 GHz Frequency Hopping Radios," WLI Forum, www.wlif.com, September 1998.

Dykes, P. "In the Neighbourhood," *Bluetooth World*, March 2000.

Elder, B. and Hay, B. "Formulating Bluetooth Manufacturing Test Strategies," *Evaluation Engineering*, September 2001.

Frenzel, L. "Single-Chip Bluetooth Transceiver Speeds Design," *Electronic Design*, March 19, 2001.

Glisic, S. and Pajkovic, M. "Rejection of an FH Signal in a DS Spread-Spectrum System Using Complex Adaptive Filters," *IEEE Transactions on Communications*, May 1995.

Golmie, N. "Using a Combined MAC and PHY Simulation Model to Measure WLAN Interference on Bluetooth (Part I)," IEEE 802.15 TG2 Document IEEE 802.1500/388r0, November 2000.

Heftman, G. "Bluetooth Meets WLANs—Can They Live Together?" *Microwaves & RF*, August 2000.

Heider, G. "A Medium-Term Vision of Bluetooth," presented at the Bluetooth Developers Conference, San Francisco CA, December 2001.

Heraeus Inc. "A Low Loss LTCC System for Wireless Applications," *Microwave Journal*, November 2000.

Hill, R. "A Practical Guide to the Design of Microstrip Antenna Arrays," *Microwave Journal*, February 2001.

Horne, J. and Vasudevan, S. "Modeling and Mitigation of Interference in the 2.4 GHz ISM Band," *Applied Microwave & Wireless*, March/April 1997.

Howitt, I. "WLAN and WPAN Coexistence in UL Band," *IEEE Transactions on Vehicular Technology*, July 2001.

Hui, P. "Design of Integrated Inverted F Antennas Made of Asymmetrical Coplanar Striplines," *Applied Microwave & Wireless,* January 2002.

Kamerman, A. "Coexistence between Bluetooth and IEEE 802.11 CCK Solutions to Avoid Mutual Interference," IEEE 802.11 document 802.1100/162, July 2000.

Kawanda, H. "Bluetooth Modules: Application-Specific Components Speed Product Integration," *Microwave Product Digest*, April 2001.

Keese, W. "Developing a Flexible High Performance Bluetooth Radio," presented at the Bluetooth Developers Conference, San Jose CA, December 2000.

Langlois, P. "Bluetooth in the Fast Evolving Wireless World: Challenges and Solutions," presented at the Bluetooth Developers Conference, San Jose CA, December 2000.

Lansford, J. and Stephens, A. "TG2 Mobilian Draft Text," IEEE 802.15 TG2 Document IEEE 802.15-01300r1, July 2001.

Lee, D. J. Y. and Lee, W. C. Y. "Propagation Prediction in and Through Buildings," *IEEE Transactions on Vehicular Technology*, September 2000.

Lombardi, M. "Uncovering Bluetooth Packet Errors," *Electronics Products*, October 2001.

Maxwell, C. "Integrating Bluetooth into a Handheld Computer Platform," presented at the Wireless Symposium, San Jose CA, February 2001.

McCartney, D. "Embedded Antennas—The Engine for Differentiation," *Microwave Product Digest*, September 2000.

McCartney, D. "Embedded Bluetooth Antennas," *Wireless Design & Development*, September 2000.

Meivert, J. "Application Migration from Host-Based to Embedded Environments," presented at the Bluetooth Developers Conference, San Francisco CA, December 2001.

Milios, J. "Baseband Methods for Power Saving," presented at the Bluetooth Developers Conference, San Jose CA, December 2000.

Morrow, R. "Site-Specific Engineering for Indoor Wireless Communications," *Applied Microwave & Wireless*, Vol. 11 No. 3, March 1999.

Nord, L. "Making the Optimum Radio Choice for your Bluetooth Functionality Requirements," presented at the Bluetooth Developers Conference, San Francisco CA, December 2001.

Paillard, C. "Bluetooth Technology Integration: Which Path Should You Choose?" *Wireless Systems Design*, January 2001.

"Performing Bluetooth RF Measurements Today," Agilent Application Note 1333, November 2000.

Reeves, J. "A Dream Test Solution for Bluetooth," *Applied Microwave & Wireless*, March 2001.

Rios, C. "A Proposal for 802.11b and Bluetooth Coexistence in Enterprise Environments," IEEE 802.15 TG2 Document IEEE 802.15-TG2-363r0, June 2001.

Robinson, A. "On Your Marks for Testing Bluetooth," *Test & Measurement World*, September 2000.

Rosener, D. "Bluetooth Integration Poses Challenges for Developers," *Microwaves & RF*, May 2000.

Savage, W. "Bluetooth SoC Integration," presented at the Bluetooth Developers Conference, San Francisco CA, December 2001.

Schneiderman, R. "Interoperability Tops Bluetooth Vendor Issues," *Supplement to Penton's Electronics Group*, Winter 1999–2000.

Schweber, B. "RF-Channel Simulators," *EDN*, September 11, 1998.

Sherman, K. "SOC - Hardware Issues," presented at the Bluetooth Developers Conference, San Francisco CA, December 2001.

Shellhammer, S. "IEEE 802.15.2 Clause 5.1—Description of the Interference Problem," IEEE 802.15 TG2 Document IEEE 802.15-091r0, September 1999.

Shoemake, M. and Lowry, P. "IEEE 802.11b Coexistence Testing Data," IEEE 802.15 TG2 Document IEEE 802.15-<084>, January 2001.

Sizer, T. "Bluetooth SIG Coexistence Working Group," IEEE 802.15 TG2 Document IEEE 802.1501/158r0, March 2001.

Soltanian, A. and Van Dyck, R. "802.11b Deterministic Frequency Nulling to Mitigate Bluetooth Interference," IEEE 802.15 TG2 Document IEEE 802.1501/079r1, March 2001.

"Some Hints and Tips for Simplifying the Production of Bluetooth Enabled Products," CSR White Paper, September 2001.

Stein, J. "Indoor Radio WLAN Performance. Part II: Range Performance in a Dense Office Environment," Harris Semiconductor White Paper, 1997.

Stevenson, C. "Radio2 Working Group," presented at the Bluetooth Developers Conference, San Jose CA, December 2000.

Tang, Y. and Sobol, H. "Measurements of PCS Microwave Propagation in Buildings," *Applied Microwave & Wireless*, Winter 1995.

Treister, B., et al. "Overview of Coexistence Mechanisms," IEEE 802.15 TG2 Document IEEE 802.15-TG2-363r0, June 2001.

Van Dyck, R. and Soltanian, A. "IEEE 802.15.2 Clause 14.1—Collaborative Co-Located Coexistence Mechanism," IEEE 802.15 TG2 Document IEEE 802.15-, July 2001.

Woo, H. "Bluetooth-IC Testing Meets Chip Design," *EDN*, May 3, 2001.

Xavier, B. and Malone, J. "Relative Advantages of Current Bluetooth Wireless Technologies, Chip Sets & Architectures," presented at the Bluetooth Developers Conference, San Francisco CA, December 2001.

Yestrebsky, T. "MICRF001 Antenna Design Tutorial," MICREL Application Note 23, July 1999.

Zavosh, F. "Design of High Gain Microstrip Antennas," *Microwave Journal*, September 1999.

Zyren, J. "Reliability of IEEE 802.11 WLANs in Presence of Bluetooth Radios," IEEE 802.15 TG2 Document IEEE 802.15-073r0, September 1999.

Zyren, J. and Gandolfo, P. "Effects of WBFH Interference on Bluetooth Receiver Reliability," Intersil White Paper, September 1999.

INDEX

Symbols

A

C

D

G

H

I

M

N

O

P

S

T

U

V

W

ABOUT THE AUTHOR

Robert Morrow, Ph.D., is the president of Morrow Technical Services and the creator of two of the leading Bluetooth courses in the wireless industry. He has given tutorials on Bluetooth, 802.11, and RFID at numerous conferences, contributes frequently to trade journals and periodicals, and has received a patent on a spread spectrum communication system. Dr. Morrow is a retired Air Force pilot and professor living in Centerville, Indiana, where he enjoys amateur astronomy.